KU-484-998

Molecular Mechanisms to Regulate the Activities of Insulin-like Growth Factors

Molecular Mechanisms to Regulate the Activities of Insulin-like Growth Factors

Proceedings of the 4th International Symposium on Insulin-like Growth Factors, at Tokyo International Forum, Tokyo, Japan, 21–24 October 1997

Editors:

Kazue Takano
Department of Internal Medicine
Institute of Clinical Endocrinology
Tokyo Women's Medical College
Tokyo, Japan

Naomi Hizuka
Department of Internal Medicine
Institute of Clinical Endocrinology
Tokyo Women's Medical College
Tokyo, Japan

Shin-Ichiro Takahashi
Department of Applied Animal Sciences
Graduate School of Agriculture and Life Sciences
The University of Tokyo
Tokyo, Japan

1998

ELSEVIER

Amsterdam – Lausanne – New York – Oxford – Shannon – Singapore – Tokyo

International Congress Series No. 1151
ISBN 0 444 82524 X

This book is printed on acid-free paper.

Published by:
Elsevier Science B.V.
P.O. Box 211
1000 AE Amsterdam
The Netherlands

Printed in The Netherlands

Preface

On the 40th anniversary of the discovery of sulfation factor by Salmon and Daughaday, we were very honored to hold the 4th International Symposium on Insulin-like Growth Factors (IGFs) in Tokyo, Japan. During 40 years, the name of sulfation factor was renamed somatomedin in 1972 and insulin-like growth factors in 1987 according to the function and structure. The existence of binding proteins was described in the 1970s and since then ten various kinds have been reported.

The purpose of this symposium was to bring researchers throughout the world together again to facilitate exchange of their knowledge and most recent findings. At this symposium there were thirteen symposia of invited presentations on selected topics, two special lectures and three workshops addressing the methodologies critical to research progress. One hundred and sixty-six free communications were presented at the poster session. Selected twenty-six papers were presented in their respective symposia. All young investigators could receive travel grants. We arranged the Banquet in cruising Tokyo Bay by the ship "Symphony". All the participants enjoyed a part of Japanese culture and built up a friendship in academic and social aspects during the symposium.

There had been three international symposia on IGFs and were held in Nairobi (1982), San Francisco (1991) and Sydney (1994). All these symposia served as a forum for exchange of information on a worldwide basis and fulfilled the demands of experts keenly interested in this field. Since then, there has been considerably great progress in the research on IGFs especially on gene knockout models of IGF, insulin and their receptors, IGFBP proteolysis, IGF-independent IGFBP action, and IGF specific signaling pathway distinguished from insulin signaling pathway, using molecular and cellular biological techniques. Of course, advances in other research fields also shed a light on molecular mechanisms to regulate the activities of IGF at various steps. All of these interesting presentations by symposists were included in this book. In understanding all of these complicated mechanisms we can point out the way how to apply IGFs to various diseases in the very near future. We also discussed the nomenclature of IGFBPs during this symposium, which will be published later.

We hope this volume will continue to contribute elucidation of physiological roles of IGF and their related proteins until the Fifth International Symposium in Brighton, 1999.

Kazue Takano, MD
Naomi Hizuka, MD
Shin-Ichiro Takahashi, PhD
Editors, The Foundation for Growth Science
Organizers, The 4th International Symposium on IGFs

Acknowledgements

The committee Members of "The 4th International Symposium on Insulin-like Growth Factors" take this opportunity to express our gratitude to Professor Kazuo Shizume, Chairman of the Foundation for Growth Science, in organizing this Symposium. We would like to thank Sumitomo Pharmaceutical Co., Ltd., Eli Lilly Japan K.K., Novo Nordisk Pharma, Japan Chemical Research (JCR) Pharmaceuticals Co., Ltd., Pharmacia & Upjohn, Serono Japan Co., Ltd., Fujisawa Pharmaceutical Co., Ltd., Daiichi Radioisotope Laboratories, Ltd., Cosmic Corporation, Eiken Chemical Co., Ltd., and Nikken Chemicals Co., Ltd. for their generous sponsorship of this Symposium.

Convex Incorporation contributed as the Secretariat and Japan Amenity Travel contributed as the travel agent, for which we are deeply indebted.

Special thanks go to all the participants for their cooperation and also to the people who were concerned with the symposium for their assistance of various kinds.

Committee Members

Organizer:

The Foundation for Growth Scinece
(Chairman of Directors: K. Shizume, MD)

Chairperson: K. Takano (Japan)

Secretary General: N. Hizuka (Japan)

Deputy Secretary Generals: K. Chihara, H. Kuzuya, T. Tanaka (Japan)

Scientific Committee:

R.	Baxter	(Australia)	M.	Binoux	(France)
D.	Clemmons	(U.S.A.)	C.	Conover	(U.S.A.)
K.	Fujieda	(Japan)	L.	Giudice	(U.S.A.)
P.	Gluckman	(New Zealand)	D.	Hill	(Canada)
N.	Hizuka	(Japan)	J.	Holly	(U.K.)
P.	Holthuizen	(Holland)	T.	Kadowaki	(Japan)
W.	Kiess	(Germany)	D.	LeRoith	(U.S.A.)
J.	Martin	(Australia)	F.	Minuto	(Italy)
F.	de Pablo	(Spain)	D.	Powell	(U.S.A.)
L.	Read	(Australia)	R.	Rosenfeld	(U.S.A.)
P.	Rotwein	(U.S.A.)	S.	Shimasaki	(U.S.A.)
J.	Sussenbach	(Holland)	S.	Takahashi	(Japan)
J.	Zapf	(Switzerland)			

International Advisory Committee:

W.	Blum	(Germany)	W.	Daughaday	(U.S.A.)
R.	Froesch	(Switzerland)	L.	Fryklund	(Sweden)
K.	Hall	(Sweden)	R.	Hintz	(U.S.A.)
F.	Matsuzaki	(Japan)	V.	Sara	(Australia)
E.M.	Spencer	(U.S.A.)	J.L.	Van den Brande	(Holland)
J.	Van Wyk	(U.S.A.)			

The 4th International Symposium on IGFs (Oct. 21st—24th, 1997, Tokyo, Japan).

Contents

Preface v

Acknowledgements vi

List of committee members vii

A. From sulphation factor to IGF-1, 40 years of research on the regulation of cartilage growth
W.H. Daughaday 1

B. Diabetes and the heart: physiological and therapeutic aspects of IGF-I
E.R. Froesch, M.A. Hussain, M.Y. Donath and J.L. Zapf 11

C. Regulation of gene expression and synthesis of IGF and IGFBP

Methods to analyze tissue-specific promoters for IGF and IGFBP gene expression
C.T. Roberts 23

Involvement of AP-1-like motifs in the IGF-I gene regulation
Y. Kajimoto, Y. Umayahara, Y. Fujitani, Y. Yamasaki and M. Hori 31

Insulin-like growth factor binding protein-1 gene activation in human endometrium
L. Tseng, J. Gao and H.H. Zhu 39

Regulation of ALS gene expression by growth hormone
Y. Boisclair, S. Bassal, M.M. Rechler and G.T. Ooi 49

Insulin-like growth factor gene targeting
W. Won and L. Powell-Braxton 57

Genetic disruption of IGF binding proteins
J. Pintar, A. Schuller, S. Bradshaw, J. Cerro and A. Grewal 65

Targeted mutations of insulin and IGF-1 receptors in mice
D. Accili, H. Kanno, Y. Kido, D. Lauro and K.I. Rother 71

D. Regulation of IGFBP action

Methods to detect and analyze modified insulin-like growth factor binding proteins
S.M. Firth 79

IGFBP proteases — Physiology/pathophysiology
J.M.P. Holly, L.A. Maile, S.C. Cwyfan Hughes, J.K. Fernihough and S. Xu 89

Proteolytic fragments of IGF binding protein-3: Physiological significance
M. Binoux, C. Lalou, S. Mohseni-Zadeh, P. Angelloz-Nicoud, C. Daubas and S. Babajko 99

IGFBP regulation by proteases
C.A. Conover 107

Regulation of insulin-like growth factor I actions by insulin-like growth factor binding protein-5
D.R. Clemmons, Y. Imai, B. Zheng, J. Clarke and W.H. Busby Jr 115

IGF-independent actions of IGFBPs
Y. Oh, Y. Yamanaka, H.-S. Kim, P. Vorwerk, E. Wilson, V. Hwa, D.-H. Yang, A. Spagnoli, D. Wanek and R.G. Rosenfeld 125

The regulation and actions of ALS
P.J.D. Delhanty 135

E. Tissue-specific regulation of IGF activity

Insulin-like growth factors in the development of the pancreas
D.J. Hill, J. Petrik, E. Arany, W. Reik and J.M. Pell 145

The early embryonic neuroretina: A CNS site of production and action of (pro)insulin and IGF-I
F. de Pablo, B. Díaz, M. García-de Lacoba, E. Vega and E.J. de la Rosa 155

IGF-I and uterine growth
O.O. Adesanya, J. Zhou, C. Samathanam, L. Powell-Braxton and C.A. Bondy 163

Bone morphogenetic proteins and IGF system
S. Mohan and D.J. Baylink 169

Regulation of IGF and IGFBP gene expression in bone
R. Okazaki 179

The IGF system in the ovary
G.F. Erickson, T. Kubo, D. Li, H. Kim and S. Shimasaki 185

The roles of IGFs and IGFBP-1 in non-pregnant human endometrium and at the maternal: placental interface during human pregnancy
L.C. Giudice 195

The prostatic IGF system: New levels of complexity
A. Grimberg, R. Rajah, H. Zhao and P. Cohen 205

Evolutionary aspects of the IGF system
C. Collet, J. Candy and V. Sara 215

The IGF system in the brain — response to injury and therapeutic potential
P.D. Gluckman, J. Guan, A. Scheepens and C.E. Williams 225

The IGF-1 paradox in cerebellar granule cells: prevention of death via apoptosis opens the route to death via glutamate-triggered necrosis
P. Calissano, M.T. Ciotti, C. Galli, D. Mercanti, N. Canu, L. Dus, C. Barbato, O.V. Vitolo and C. Zona 231

IGFs and IGFBPs in organogenesis: Development of normal and abnormal kidneys
V.K.M. Han and D.G. Matsell 243

Expression of IGF-I and IGFBPs in rat kidney and the effect of GH and nutrition
S. Kobayashi and H. Nogami 251

Regulatory role of interleukin1β and nitric oxide on vascular smooth muscle cell proliferation in primary culture
T. Bourcier and A. Hassid 261

F. IGF signal transduction

The yeast two-hybrid system to investigate IGF-I receptor signal transduction
R.W. Furlanetto, K. Frick, B.R. Dey, W. Lopaczynski, C. Terry and S.P. Nissley 269

Signal transduction mechanism of insulin and growth hormone
T. Kadowaki, T. Yamauchi, K. Tobe, K. Ueki, H. Tamemoto, Y. Kaburagi, R. Yamamoto-Honda, T. Tsushima and Y. Yazaki 279

The insulin-like growth factor-I receptor and cellular signaling: Implications for cellular proliferation and tumorigenesis
D. LeRoith, A.P. Koval, A.A. Butler, S. Yakar, M. Karas, B.S. Stannard and V.A. Blakesley 285

Differences between insulin and IGF-I signaling
P. Nissley, B.R. Dey, K. Frick, W. Lopaczynski, C. Terry and R.W. Furlanetto 291

G. IGF in diseases and a possible clinical application of IGF

The multiple roles of the IGF-I receptor in cell growth
R. Baserga, M. Prisco and M. Resnicoff 301

Role of IGF-II in Wilms tumourigenesis and overgrowth disorders
A.E. Reeve 309

The IGF system in breast cancer
D. Yee, J.G. Jackson, C.-N. Weng, J.L. Gooch and A.V. Lee 319

rhIGF-I/IGFBP-3 (SomatoKine) therapy for the treatment of osteoporosis
S. Adams, D. Rosen and A. Sommer 327

Role of insulin-like growth factor-I in gastrointestinal growth and repair
C.-B. Steeb, C.A. Shoubridge, J. Lamb, G.S. Howarth and L.C. Read 331

Metabolic actions of insulin-like growth factors
J. Zapf, C. Schmid and E.R. Froesch 341

IGF-1 treatment during early HIV infection: Immunological and hormonal effects
F. Sattler, J. LoPresti, M. Dube, A.B. Montgomery, P. Jardieu, C. Spencer, M. Saad, J.T. Nicoloff and R.G. Clark 351

IGF-I treatment of growth hormone insensitivity
R.G. Rosenfeld 359

IGF-I therapy for patients with extreme insulin resistance syndromes in Japan
M. Kasuga and Extreme Insulin Resistance Syndromes Research Group 365

Insulin-like growth factor-I (IGF-I): Therapeutic potential in neuromuscular and other neurological diseases
J.M. Farah Jr 371

H. Directions for research into the insulin-like growth factor system as the millennium approaches; closing remarks to the IVth IGF Symposium
E.M. Spencer and R. Sapolsky 385

Index of authors 393

Molecular Mechanisms to Regulate the
Activities of Insulin-like Growth Factors
K. Takano, N. Hizuka and S-I. Takahashi (Editors)

FROM SULFATION FACTOR TO IGF-1, 40 YEARS OF RESEARCH ON THE REGULATION OF CARTILAGE GROWTH

W. H. Daughaday

Division of Endocrinology and Metabolism, University of California, Irvine, P.O. Box 157, Newport Beach, CA 92662, U.S.A.

1. BACKGROUND

Modern GH research in America can be dated to 1915 when Herbert M. Evans accepted an invitation from the University of California at Berkeley to Chair its Department of Anatomy. He made the central focus of his department the isolation of the hormones of the anterior pituitary gland and the characterization of their function (1). Growth hormone attracted early attention and by 1922 Evans with Long (2) had produced gigantism in rats with crude pituitary extracts. Evans was fortunate that one of the two holdovers from the old department was Philip E. Smith, a skilled experimental surgeon. Smith made a major contribution to pituitary research by developing simple hypophysectomy of rats using the parapharyngeal approach (3). The availability of hypophysectomized rats made possible sensitive and specific assays for anterior pituitary hormones. Pituitary extracts induced a dose-dependent widening of the proximal tibial growth plate cartilage which provided the basis for a sensitive assay for GH (4,5). The assay was sensitive to a total dose of 20 ug of purified bGH divided into four equal daily injections.

Clinical investigators attempted to use this assay to measure GH in human serum. Kinsell et al. (6) reported that serum from a patient with acromegaly caused a greater widening of the growth plate than did normal serum. Gemzel et al. (7) found that plasma fractions from a patient with acromegaly and another with gigantism led to growth of the growth plate, but similar fractions from plasma from normal subjects were inactive. Because we now know that the amounts of GH which could have been present in these sera and plasma fractions were insufficient to produce the observed growth. It seems likely that the widening resulted from IGFs in these sera and plasma extracts and not from GH.

In the early 1950s ^{35}S-sulfate became available to biomedical investigators. It was found that the incorporation of this isotope into cartilage chondroitin sulfate provided a useful index of one the most important synthetic activities of this tissue. Ellis et al. (8) reported in 1953 that there was a fall in the uptake of ^{35}S-sulfate into costal cartilage *in vivo* after hypophysectomy which could b restored partially by three daily bGH injections. These findings were confirmed and extended in 1955 by Denko and Bergenstal (9). They injected hypox rats with ^{35}S-sulfate for 8 days with and without bGH. An increase in isotope uptake after bGH occurred in all cartilage sites tested. These studies suggested to us that the uptake of ^{35}S-sulfate might be a useful assay parameter of GH action. In initial studies with Murphy and Hartnett (10) rats were injected with ^{35}S-sulfate for various periods after hypophysectomy at intervals the rats were killed and the uptake of isotope in the entire proximal tibial epiphysis was measure. A progressive fall in uptake occurred in the first two weeks after the operation. In the next study rats two weeks after hypophysectomy were given two doses of bGH 24 hours apart. ^{35}S-sulfate was also injected with the second dose of bGH. Bovine GH treatment resulted in a dose dependent increase in ^{35}S-sulfate uptake in the epiphysis (Fig.1). A significant increase in ^{35}S-sulfate uptake required a total of 10 μg of bGH which represented only a minor improvement in sensitivity as compared to the standard tibial plate width assay.

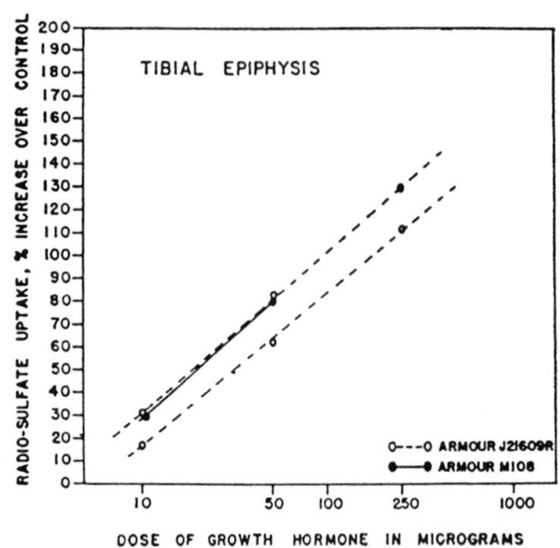

Fig. 1. Hypox rats were injected with the indicated doses of bGH given as two injections 24 hours apart. With the second injection the rats also received ^{35}sulfate. After 48 hours the uptake of isotope in the proximal tibial epiphysis was determined. The results of three separate experiments with two different sources of bGH are shown. Reprinted with permission from Murphy WR, Daughaday WH, Hartnett C, J Lab Clin Med 47:715-722,1956. Mosby-Year Book, Inc., publishers.

At this time I was joined in the laboratory by William D. Salmon who undertook the development of an *in vitro* cartilage assay for GH with the hope that it might be sufficiently sensitive for clinical specimens. First it was demonstrated that hypophysectomy reduced the uptake of 35S-sulfate by costal, nasal and xiphoid cartilage, and normal uptake could be restored by bGH injections (11). Costal cartilage was selected for subsequent studies because multiple segments could be obtained from a single rat. We were disappointed to find that the addition of bGH in concentrations up to 50 μg per ml produced less than 20% stimulation of ^{35}S-sulfate uptake which was much less than could be achieved with injections of bGH. Negative results were obtained when 140 ng/ml of bGH was added to a medium containing a high concentration of hypox rat (11).

2. BIRTH OF THE SOMATOMEDIN HYPOTHESIS

To explain the high potency of bGH when injected into hypox rats to increase ^{35}S-sulfate uptake into cartilage and its ineffectiveness when added *in vitro* we next considered that bGH was acting indirectly *in vivo* by inducing the appearance in serum of a secondary growth factor.

At the time there was much talk of new growth factors at Washington University. Rita-Levi-Montalcini and Victor Hamburger(12) had made their pioneer observations which led to recognition of nerve growth and Stanley Cohen, who also played a major role in the isolation nerve growth factor, observed that some of his crude preparations of nerve growth factor accelerated the opening of eyelids of neonatal rats, an observation which led to the discovery of epidermal growth factor (13). Our first experiments were successful in showing that normal rat serum was highly potent in stimulating the uptake of ^{35}S-sulfate by hypox rat costal cartilage segments *in vitro,* but incubations with hypox rat plasma were without effect and sometimes actually inhibitory (11). The response to normal rat plasma was dose dependent at serum concentrations from 1.4 to 71% (Fig. 2).

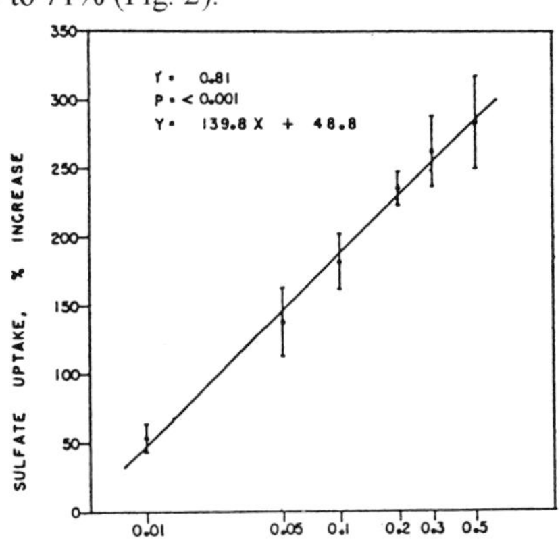

ML OF SERUM IN T.V. OF 0.7 ML

Fig. 2. The effect of different concentrations of normal rat serum on the uptake of ^{35}S-sulfate by costal cartilage segments of hypox rats. The incubations were for 24 hours. Reprinted with permission from Salmon Jr, WD and Daughaday WH, J Lab Clin Med 49:825-826,1957, Mosby-Year Book, Inc., publishers.

The time course of appearance of stimulatory activity in serum of hypox rats after bGH injections was studied by incubating this serum with costal cartilage segments from untreated hypox rats (Fig. 3). Stimulatory activity appeared in serum of hypox rats 6 hours after bGH injection. Peak stimulation occurred at 12 hours with a slight decline at 24 hours. The costal cartilage of the GH treated hypox rats exhibited a stimulation of ^{35}S-sulfate uptake within 6 hours and continued to rise steeply at 12 and 24 hours. This experiment provided a good correlation between the appearance of stimulatory activity in the serum and the increase in uptake in cartilage.

On the basis of these experiments we concluded that "Serum from normal rats contains a factor which promotes sulfate uptake, termed sulfation factor, which promotes sulfate uptake *in vitro* by cartilage from hypophysectomized rats. The sulfation factor appears in the plasma of hypophysectomized rats after the administration of growth hormone."

3. FROM SULFATION FACTOR TO SOMATOMEDIN TO INSULIN-LIKE GROWTH FACTOR

In 1960 I reported with C. Reeder (14) that the daily injection of hypox rats with bGH led to a remarkable stimulation of uptake of ^{3}H-thymidine measured *in vitro*. The effect was evident

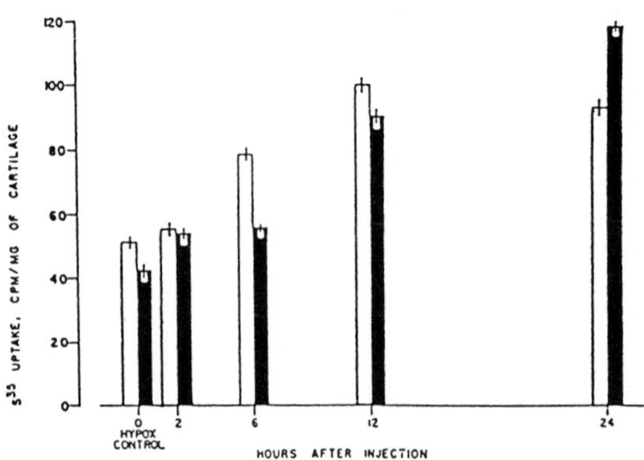

Fig. 3. Hypox rats received 500 μg of bGH intraperitoneally. At the indicated times the rats were killed and serum obtained. The costal cartilages from these rats were incubated with ^{35}S-sulfate and the uptake measured (solid columns). Sera obtained from bGH treated rats at the designated times were incubated with costal cartilage from untreated hypox rats. The resulting uptake of labeled sulfate is shown (open columns). Reprinted with permission from Salmon Jr., WD, Daughaday WH, J Lab Clin Med 49:825-836,1957. Mosby-Year Book publishers.

after 24 hours and reached as much as 26 fold stimulation at 48 hours. Thereafter, incorporation decreased to a lower level which remained constant for six days of the study. Incubation with normal rat serum but not hypox rat serum induced a concentration dependent increase in the uptake of ^{3}H-thymidine (Fig. 4). Under the conditions of study extremely high concentrations of bGH added to the incubation medium failed to stimulate ^{3}H-thymidine uptake.

The isolation of sulfation factor from plasma of acromegalic patients was undertaken by Kirstin Hall of Stockholm who was joined by Judson Van Wyk of Chapel Hill during a sabbatical year. They assayed their purified extracts with both the hypox rat costal cartilage and an embryonic chick cartilage assay which Hall had developed (15). They observed that their purified fractions stimulated both sulfate uptake and thymidine uptake (16, 17). It was clear that the name "sulfation factor" did not reflect the range of actions of this GH dependent serum factor. For this reason the name "somatomedin" was proposed (18).

Concurrent with the early study of sulfation factor, Rudolph Froesch of Zurich and various collaborators began a series of studies of the insulin-like activity of serum which could not be neutralized with insulin antibody which they called "non-suppressible insulin-like activity (NSILA)". They undertook the purification of the acid alcohol form of NSILA using a rat epididymal fat pad test system (19). When it was determined that their purified extracts stimulated both 35S-sulfate uptake by hypox rat cartilage and glucose uptake by epididymal fat the identity of somatomedin-C and NSILAs was suspected (20). The purification of NSILAs culminated in the isolation of two peptides and the elucidation of their primary amino acid sequences by Rinderknecht and Humbel (21, 22). It was recognized that the sequence of the two peptides had close homology with that of proinsulin and for that reason they were named "insulin-like growth factor I (IGF-I)" and "insulin-like growth factor II (IGF-II)". Because IGF-I proved to be most GH dependent the somatomedin hypothesis could be restated as follows: GH acts on a major target organ, namely the liver, to stimulate the synthesis and secretion of IGF-I which reaches its skeletal targets as a true endocrine agent.

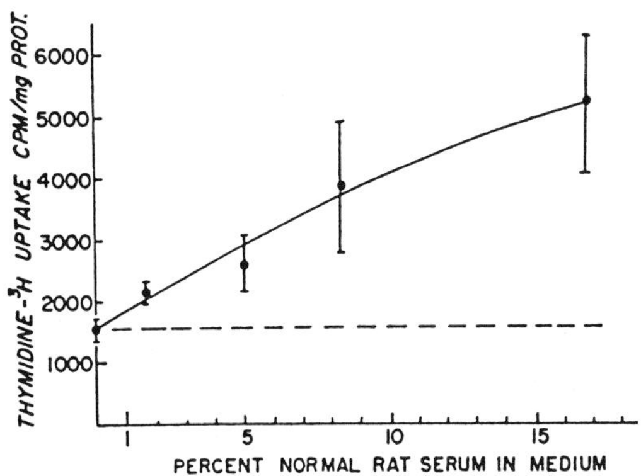

Fig. 4. Hypox rat cartilage was incubated with different concentrations of normal rat serum for 24 hours before the addition of ^{3}H-thymidine. Incubation was continued for an additional 24 hours. Reprinted with permission from Daughaday WH, Reeder C, J Lab Clin Med 68:357-368,1966, Mosby-Year Book, Inc, publishers.

4. AUTOCRINE/PARACRINE ACTION IGF-1 AS AN ALTERNATIVE TO ENDOCRINE ACTION OF IGF-1

D'Ercole et al. (23) first reported that multiple fetal mouse tissues produced IGF-I as detected by RIA. Later this same group reported that GH treatment of hypox rats increased the IGF-I content of many organs as measured by RIA (24). Isaksson and his coworkers in Goteburg Sweden have provided much information about the autocrine/paracrine action of IGF-I in cartilage and have challenged the contribution of IGF-I as a hormonal agent (25). They first reported that injections of hGH into the proximal tibial epiphysis of hypox rats stimulated tibial growth (26) . These results were confirmed by Russell and Spencer (27) who found that local infusion both of hGH and IGF-1 could produce widening of the cartilage growth plate of hypox rats. Later Schlechter et al.(28) were able to block the local effect of hGH on the epiphysis by the simultaneous infusion of IGF-1 antiserum . Thus, providing evidence that IGF-I was an essential intermediate of GH's action on growth cartilage. In further support of this conclusion, Nilsson et al. (29) found that both systemic and local injections of GH increased the expression of IGF-1in growth plate chondrocytes as detected by an immuno histochemical method. Isgaard et al. (30) found that systemic GH treatment increased the IGF-1 mRNA content of growth plate cartilage.

The Isacksson group has conducted a number of experiments with dissociated growth plate chondrocytes which suggest that GH has an additional mitogenic action on chondrocytes precursor cells which is not mediated by IGF-1(25). Such an action might explain the observation that GH replacement appears to be more effective than IGF-1 replacement in restoring normal growth in GH deficient rodents and human patients. IGF-1 must have some mitogenic action on precursor cells because IGF-1 is able to restore long term growth in GH resistant patients, Laron syndrome (31). Isaksson and his collaborators (25) concluded that, "GH stimulates longitudinal bone growth by stimulating the differentiation of epiphyseal growth plate precursor cells and indirectly by increasing the responsiveness to IGF-1 and enhancing the local production of IGF-1 that stimulates the clonal expansion of differentiating chondrocytes".

The apparent ability of GH to stimulate linear growth of hypophysectomized rats (32) and spontaneously GH deficient mice (33) with little or no rise in serum IGF-1 has provided more support for the conclusion that IGF-1 acts primarily as an autocrine/paracrine agent. In Fig. 3 which showed a prompt rise in sulfation factor after GH injection of hypox rats the cartilage assay is sensitive to free and loosely protein bound IGF-1 but most IGF-1 RIAs measure total IGF-1 in serum. Free IGF-I which is the critical component for most of the biologic action is less than 1% of total IGF-1. Over 90% of the serum IGF-1 is normally present in a large ternary complex including IGF, IGFBP-3 and an acid-labile subunit. The ternary complex is markedly reduced in hypox rat serum and it takes to more than a week to be reconstituted after GH treatment. For this reason total IGF-1 as measured by RIA rises sluggishly after GH treatment.

5. DETERMINING THE CONTRIBUTION OF THE ENDOCRINE ACTION OF IGF-1 *IN VIVO*

The experimental evidence, which has been briefly reviewed, clearly establishes that IGF-1 can act both as an endocrine factor and as an autocrine/paracrine factor in stimulating cartilage growth. The characteristics of the hypox rat costal cartilage systems permit evaluation of each pathway in regulation of growth in the animal. These costal cartilage segments are composed of an essentially uniform cell population devoid of vascular, lymphatic and neural elements. This avoids the necessity of dissociation and selective cell culture which distorts vital regulatory cell-cell interaction. Both in the animal and in culture cartilage is dependent on diffusion to sustain its largely anaerobic metabolism. This permits a good correlation of hormonal regulation *in vivo* with that observed *in vitro.*

A difficulty in such comparison has been that under the original conditions GH had a small and inconstant action *in vitro*. Salmon and Burkhalter (34) have recently reexamined the direct effects of GH on the uptake of ^{35}S-sulfate by hypox rat costal cartilage segments with minor but important changes in the original 1957 conditions. HEPES buffer enriched with amino acids and albumin were substituted for a simple KPS buffer. A significant change was also made in the period of isotopic labeling. After variable periods of preincubation ^{35}S-sulfate was added for a two hour pulse. In the original study had been added for the entire period of incubation. These changes improved recognition of the somewhat delayed stimulation of uptake when GH was added *in vitro*. Under these new conditions both bGH and IGF-1 stimulated ^{35}S-sulfate uptake in a dose dependent manner but the slope of the response with IGF-1 was much steeper than occurred with GH (Fig. 5).

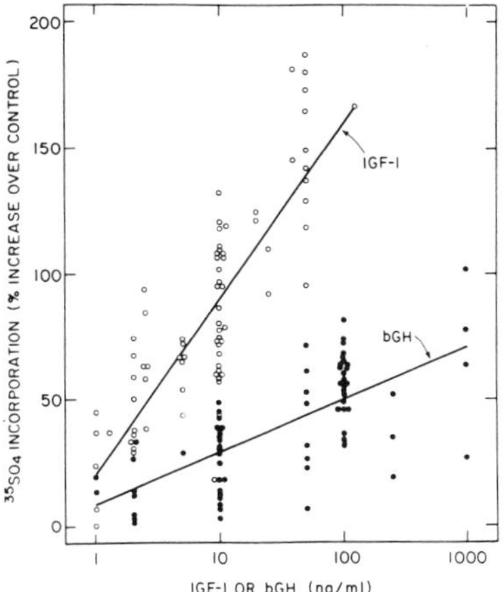

Fig. 5. Pooled results of experiments in which bGH and IGF-1 were added to incubations of hypox rat costal cartilage and the uptake of ^{35}S-sulfate determined. Reprinted with permission from Salmon, Jr. WD, Burkhalter VJ, J Lab Clin Med 129:430-438, 1997. Mosby-Year Book, Inc., publishers.

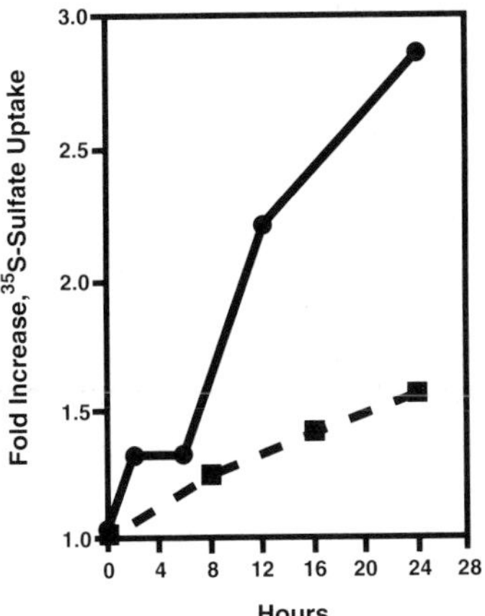

Fig. 6. The ability of GH injections of hypox rats to increase the uptake of ^{35}S-sulfate by costal cartilage segments (solid line) derived from results shown in Fig. 3. The new results of incubation of hypox rat costal cartilage with bGH (dashed line) were derived from Fig. 4 of Salmon and Burkhalter (34).

It is instructive to compare these results with those obtained with cartilage from hypox rats given bGH *in vivo*. The response of cartilage from rats given bGH intraperitoneally was earlier and greater in magnitude than when bGH was added directly to cartilage incubation (Fig. 6). The difference in these two curves represents the contribution of hormonal IGF-1 to the cartilage response. The rise in uptake of sulfate when bGH was added *in vitro* is largely due to the autocrine/paracrine action of cartilage IGF-1.

Salmon and Burkhalter (34) also reported that bGH added to incubations of hypox rat cartilage caused significant stimulation of ^{3}H-thymidine uptake, a response that had been missed in the original report of Daughaday and Reeder (14), but again the response was much less than followed bGH injection of hypox rats.

These studies do not provide any insight on the possible effects of bGH on chondrocytes progenitor cells that are unmediated by IGF-1 because costal cartilage segments have few such cells.

6. SUMMARY AND CONCLUSIONS

The existence of sulfation factor (IGF-1) was first suspected because of the inability of bGH added to incubations of hypox rat costal cartilage segments to reproduce the marked stimulation of 35S-sulfate uptake of such cartilage from rats receiving bGH injections, and the finding of a highly potent intermediary of GH action in serum. We now know that serum IGF-1 is largely of hepatic origin. It was later found that GH dependent IGF-1 was produced in the tibial cartilage growth plate which has local mitogenic and anabolic effects. In addition a possible unmediated effect of GH on chondrocytes precursor cells has also been proposed. Recently modification of the hypox rat costal cartilage assay system has allowed comparison of the relative contribution of IGF-1 reaching cartilage as an endocrine growth factor with that produced locally and acting as an autocrine/paracrine growth factor. The results suggest that both routes of IGF-1 action contribute to GH response but that the endocrine pathway makes the greatest contribution.

REFERENCES

1. M.M. Grumbach, J. Clin. Endocrinol. Metab. 55 (1982) 1240.
2. H.M. Evans and J.A. Long, Proc. Natl. Acad. Sci. U.S.A. 8 (1922) 38.
3. P.E. Smith, Am. J. Anat. 45 (205) 1930.
4. H.M. Evans, M.E. Simpson, W. Marx and E. Kibrick, Endocrinology 32 (1943) 13.
5. F. S. Greenspan, C.H. Li, M.E. and H.M. Evans, Endocrinology 45 (1949) 455.
6. L.W. Kinsell, G.D. Michaels, C.H. Li and W.E. Larsen, J. Clin. Endocrinol.8 (1948) 1013.
7. C.A. Gemzell, F. Heijkenskjold and L. D. Strom, J. Clin. Endocrinol. Metab. 15 (1955) 537.
8. S. Ellis, J. Huble and M.E. Simpson, Proc. Soc. Exp. Biol. Med. 84 (1953) 603.
9. C.W. Denko and D.M Bergenstal, Endocrinology 57 (1955) 76.
10. W.R. Murphy, W.H. Daughaday and C. Hartnett, J. Lab. Clin. Med. 47 (1956) 715.
11. W.D. Salmon, Jr and W.H. Daughaday, J. Lab. Clin. Med. 49 (1957).
12. R.Levi-Montalcini and V. Hamburger, J. Exp. Zool. 116 (1951) 321.
13. S. Cohen, Proc. Natl. Acad. Sci. U.S.A. 46 (1960) 302.
14. W.H. Daughaday and C. Reeder, J. Lab. Clin. Med. 68 (1966) 357.
15. K. Hall, Acta Endocrinol. 63 (1970) 338.
16. J.J. Van Wyk, K. Hall, J.L. Van den Brande and R.P. Weaver, J. Clin. Endocrinol. Metab.

32 (1971) 389.
17. K. Hall and K. Uthne, Acta. Med. Scand. 190 (1971) 137.
18. W.H. Daughaday, K. Hall, M.S. Raben et al., Nature 235 (1972) 107.
19. E.R Froesch, H. Burgi, E.B. Ramseier et al., J. Clin. Invest. 42 (1963) 1816.
20. A.E. Zingg and E.R. Froesch, Diabetologia 9 (1973) 472.
21, E. Rinderknecht and R.E. Humbel, J. Biol. Chem.253 (1978) 2769.
22. E. Rinderknecht and R.E. Humbel, FEBS Lett. 89 (1978) 283.
23. A.J. D'Ercole, G.T. Applewhite and L.E. Underwood, Dev. Biol. 75 (1980) 315.
24. A.J. D'Ercole, A.D. Stiles and LE. Underwood, Proc. Natl. Acad. Sci. U.S.A. 81 (1984) 935.
25. O.G.P. Isaksson, A. Lindahl, A. Nilsson and J. Isgaard, Endocr. Rev. 8, (1987) 426.
26. O.G.P. Isaksson, J.-O. Jansson and I.A.M. Gause, Science 2i6 (1982) 1237.
27. S.M. Russell and E.M. Spencer, Endocrinology 116 (1985) 2563.
28. N.L. Schlechter, S.M. Russell, E.M. Spencer and C.S. Nicoll, Proc. Natl. Acad. Sci. U.S.A. 83 (1986)793
29. A. Nilsson, J. Isgaard, A. Lindahl et al., Science 233 (1988) 1515.
30. J. Isgaard, C. Moller, O.G.P. Isaksson et al. Endocrinology 122 (1988) 1515.
31. J. Guevara-Aguirre, A.L. Rosenbloom, O. Vasconez et al., J. Clin. Endocrinol. Metab.82 (1997) 629.
32. C.C. Orlowski and S.D. Chernausek, Endocrinology 122 (1988) 44.
33. S.E. Gargosky, P. Tapanainen and R.G. Rosenfeld, Endocrinology 134 (1984) 2267.
34. W.D. Salmon, Jr. and V.J. Burkhalter, J. Lab. Clin. Med. 129 (1997) 430.

Molecular Mechanisms to Regulate the
Activities of Insulin-like Growth Factors
K. Takano, N. Hizuka and S-I. Takahashi (Editors)

Diabetes and the Heart: Physiological and Therapeutic Aspects of IGF I

E.R. Froesch, M.A. Hussain, M.Y. Donath and J.L. Zapf

Division of Endocrinology and Diabetes, Department of Medicine, University Hospital, CH-8091 Zürich, Switzerland

1. INTRODUCTION

The physiology of IGF I is not yet understood in every detail because some major questions have not yet been answered among which I shall choose just four:

- In which form is IGF I biologically active on cells?
- Do IGFBPs decrease or enhance the cellular effects of IGF I?
- To what extent is the IGF I serum level representative for its bioactivity?
- How does IGF I influence insulin sensitivity and is this an important issue in the pathogenesis of NIDDM?

To question No 1 there appears to be a simple answer. When one adds increasing amounts of IGFBP-3 to the incubation medium containing a constant amount of IGF I, the effects of the latter on adipose tissue continuously decrease down to baseline. These early observations may be criticized because IGF I affects adipose tissue via the insulin receptor. However, other cells responding to IGF I through the type 1 receptor behave similarly. Therefore, most of the effects of IGF I in vitro are due to the interaction of free IGF I with the type 1 IGF receptor or in some instances with the insulin receptor (for review see 1). Compared to these in vitro findings, IGF I bioactivity is usually thought to parallel total IGF I concentration in serum where between 80 and 90% of total IGF I is, under normal circumstances, bound to the 150 kD IGF I complex (2). We tend to relate total IGF I levels in serum to IGF I bioactivity assuming that the organism has some means to bring this 150 kD complex in contact with type 1 IGF receptors by mechanisms responsible for the dissociation of the complex into free IGF I, IGFBP-3 and the acid-labile subunit. However, we still ignore how this process, if it exists, comes about. It is much easier to conceive a mechanism by which a cell produces and releases IGF I and stimulates itself or a neighboring cell with IGF I in the free form. The latter autocrine/paracrine mechanism undoubtedly plays a role in cell-cell interaction but little is known about its importance at the level of the whole organism. The other question regarding possible enhancer effects of IGFBPs on IGF I activity is timely and investigators try to answer by studying in many different in vitro model systems. There are some hints that IGFBP-5 may actually be an enhancer of IGF I bioactivity in bone but this still has not been proven in a definitive manner. In general, IGFBPs are thought to have the following major effects: they serve as a reservoir of IGF I prolonging the half-life of serum IGF I from minutes (free IGF I) to many hours. Moreover, they compete for the biologically active sites of IGF I on the type 1 and on the insulin receptor. At this point it may be argued that none of the other 5 binding proteins (BP-1, -2, -4, -5 and -6) form this specific complex containing IGF-I, BP-3 and the acid-labile subunit and that IGF-I bound to these other IGFBPs does actually cross the capillary barrier and, therefore, may come into contact with cell receptors. This is certainly true but appears to be important mostly in situations in which IGF I is administered exogenously or else in which IGF I or IGF II is produced under the influence of stimulators other than growth hormone. In acromegaly or during growth hormone therapy, IGF I levels rise in parallel with the levels of BP-3 and ALS. Therefore, the absolute rise of free IGF I in serum of acromegalics is small compared to the marked increase of the 150 kD IGF complex. This is, however, the classical situation of growth stimulation, where chondrocytes as well as osteoblasts proliferate and produce matrix. At the same time, however, IGF I and insulin actions on insulin-sensitive tissues are not enhanced but rather diminished. In fact, acromegalics are insulin-resistant rather than insulin-sensitive and often exhibit diminished glucose tolerance or even frank diabetes mellitus. The reverse is true when IGF I is administered exogenously. Then, free IGF I levels increase and a relatively large portion of IGF I is bound to IGFBPs that do not associate with the acid-labile subunit and display a molecular weight around 30'000 to 50'000 in which they cross the capillary barrier and are accessible to cells. In this situation, as has been shown in hypophysectomized rats (3,4), growth hormone-deficient mice (5), diabetic rats (6), normal human (7) as well as diabetic subjects (8), IGF I exerts both growth-promoting as well as insulin-like effects. In these situations, IGF I does not only mimic the effects of insulin but does also enhance the effects of insulin (Table 1).

The last question regarding the enhancement of insulin sensitivity by IGF I touches on an issue that may be important for the pathogenesis of NIDDM as shall be discussed in the next two paragraphs.

1.1. The food and famine theory: role of IGF I

In 1963, Rabinowitz and Ziehler (9) formulated the so-called "food and famine theory" on the respective roles of insulin and GH in the diurnal metabolic rhythms in humans as follows: "We propose that, during a day, metabolism is dominated alternately by the action of insulin, or of human GH, or of the combined effects of the two, in a three-cycle phase determined by the intake of food. Exposure to insulin in the immediate postprandial period encourages storage of carbohydrate and fat, exposure to human GH plus insulin in the delayed postprandial period encourages protein synthesis and exposure to human GH in the remote postprandial phase encourages mobilization and peripheral oxidation of fat and the translocation of glucose into muscle and adipose tissue."

In the same year of the publication of this hypothesis, nonsuppressible insulin-like activity (NSILA) was described. NSILA was later purified from plasma and characterized as insulin-like growth factor I and II (10).Serum levels of IGF I depend on nutrition, GH and insulin. Some of the anabolic effects of GH, insulin and IGF I are enhanced when these hormones act in concert (see Table 1). In the three-cycle phase of human metabolism determined by food intake, insulin is the major player in the immediate postprandial period. Insulin enhances glucose and amino acid uptake and storage by cells as well as lipid synthesis and storage. IGF I induces sensitivity towards the action of insulin and suppresses proteolysis. In the early postprandial period, all three hormones promote anabolism with IGF I as the main modulator in mediating and/or supporting the effects of both insulin and GH. In the remote postprandial period, IGF I attenuates the induction of insulin resistance induced by GH and simultaneously promotes lipolysis by reducing insulin secretion, thus allowing for increased combustion of lipids. In addition, IGF I inhibits proteolysis and together with GH decreases protein oxidation. For the discussion of the functions of IGF I in the food and famine theory, it is important to remember that two extremely insulin-sensitive tissues, namely the liver and adipose tissue, do not express type 1 IGF receptors. Since IGF I decreases insulin secretion and does not by itself act on the liver and on adipose tissue, it is the anabolic hormone which will switch the organism from the use of glucose to the use of free fatty acids because lipolysis is no longer inhibited by insulin (11).

Table 1
Endocrine and metabolic effects of administration of growth hormone and insulin-like growth factor I, respectively

	Endocrine effects of		
	GH	IGF I	
Insulin secretion	↑	↓	opposite
Insulin sensitivity	↓	↑	
	Metabolic effects		
Glycemia	↑	↓	opposite
Glucose tolerance	↓	↑	
FFA	↑	↑	additive
β-OH-butyrate	↑	↑	
Lipid oxidation	↑	↑	
Energy expenditure	↑	↑	
Protein oxidation	↓	↓	
Proteolysis	↓	↓	
Amino acids levels	↓	↓	
Glucose utilization	↓	→	

FFA, free fatty acids; GH, growth hormone; IGF I, insulin-like growth factor I

1.2. Insulin resistance, NIDDM, IGF I, and the carnivore connection (12) (Figure 1)

It has lately been hypothesized that the NIDDM explosion in certain ethnic groups like American Indians, Polynesians and other populations who have lived mostly on protein because they had not developed agriculture, became diabetic when suddenly confronted with an "omnivorous dietary regime", i.e. junk food of Western civilization. They did not have the time to adapt to the sudden surplus of carbohydrates. The theory says that during the many centuries when most of the food consisted of protein and fat, blood glucose had to be synthesized by the liver via gluconeogenesis from amino acids, insulin requirement was low,

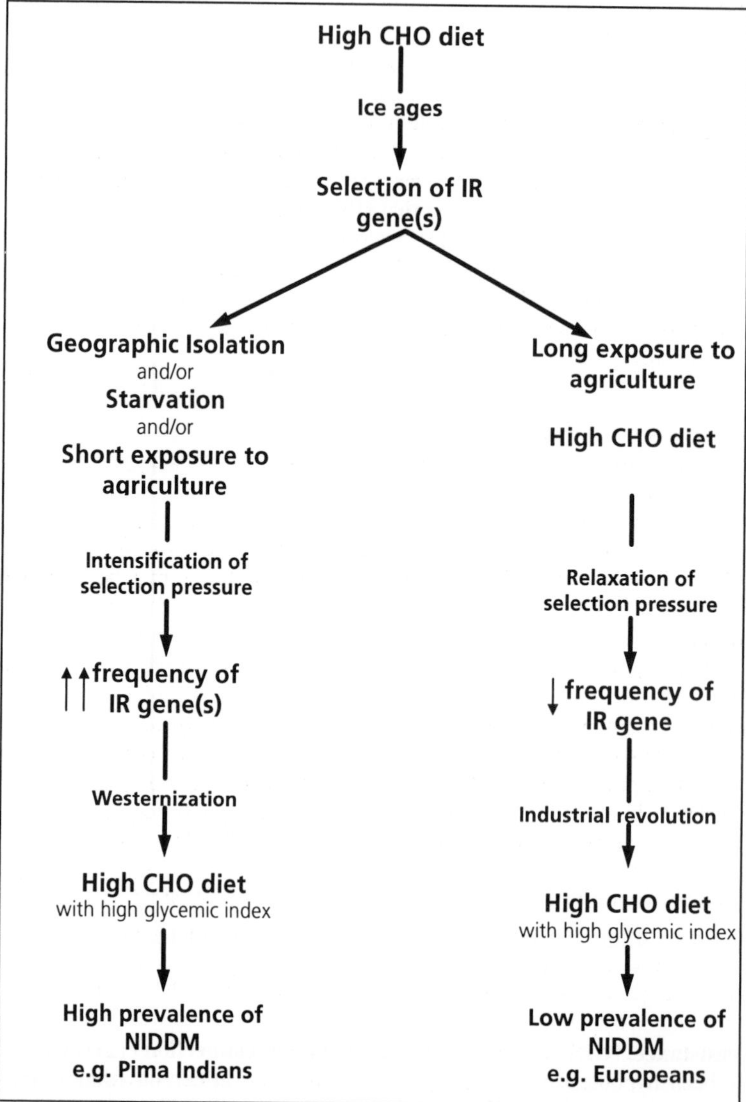

Fig. 1. The carnivore connection. IR, insulin resistance; NIDDM, non-insulin-dependent diabetes mellitus. From (12).

and genes related to insulin resistance had a positive effect on survival. When these humans were all of a sudden exposed to Western food in which protein is mixed with a lot of fat and carbohydrates, they were incapable of mastering the glucose load, became hyperglycemic, hyperinsulinemic, developed obesity and later on NIDDM. This theory is compatible with the thrifty genotype of NIDDM, postulated by Neel (13). Many isolated ethnic populations lived on hunting and fishing and, therefore, ate mostly protein and fat. Food supply depended on the success of hunting and fishing and times of food surplus alternated with long periods of famine. During fasting, insulin levels were low and insulin resistance developed. GH in this situation was important for the mobilization of fat from adipose tissue, and IGF I helped to maintain muscle protein by inhibiting muscle proteolysis. Therefore, a certain degree of insulin resistance together with fat accumulation during times of food surplus was a necessity for survival during a famine. These favorable survival factors combined with a certain degree of insulin resistance became hazardous in our Western society where there is a constant surplus of food which increases the probability to develop NIDDM.

It is conceivable that a special setting of the GH-insulin-IGF I axis is responsible for relative insulin resistance leading to NIDDM. As we will see later, the administration of IGF I increases insulin sensitivity to a considerable extent, so that patients with NIDDM respond very well to IGF I therapy.

1.3. IGFs and extrapancreatic tumor hypoglycemia (14,15)

IGFBP-1 concentrations in serum depend on insulin secretion. When insulin secretion is stimulated by glucose, BP-1 levels fall acutely. It has been hypothesized that through this fall of BP-1 levels, IGF I may become more easily accessible to tissues so that IGF I may actually be a glucohomeostatic peptide. This hypothesis is still under debate. BP-1 is a small fraction of the total pool of BPs in serum so that it is difficult to believe that it may be of importance for glucose homeostasis. Nevertheless, IGFs can become hypoglycemic substances when they are produced in large amounts, such as is the case in extrapancreatic tumor hypoglycemia. Zapf et al (15) and Daughaday et al (14) have shown that most of the large tumors causing hypoglycemia are characterized by the following set of metabolic and endocrine abnormalities: hypoglycemia, very low or unmeasurable insulin levels, markedly decreased or unmeasurable GH levels, inhibited lipolysis and suppressed glucose production from the liver. It was found that these tumors produce 'big' IGF II in large amounts. 'Big' IGF II is an incompletely processed IGF II which binds to binding proteins and to the type 1 IGF receptor as well as to the insulin receptor. When produced in large amounts, 'big' IGF II inhibits insulin and GH secretion directly at the level of the β-cell and the somatotrophic cell, respectively. Due to the fact that GH secretion is inhibited, the liver ceases to synthesize the acid-labile subunit so that 'big' IGF II now is no longer bound in the 150 kD complex but rather present as the free peptide or bound to the various non-GH-dependent IGF binding proteins in serum. Those can cross the capillary barrier so that big IGF II may react with the type 1 IGF and the insulin receptor and thereby causes hypoglycemia.

In this context, it is important to recall that IGF I synthesized and released by the liver under the control of GH does not lead to hypoglycemia or increased insulin sensitivity but rather to insulin resistance and impaired glucose tolerance as is typical for acromegaly. These effects on glucose metabolism are mostly due to increased secretion of GH leading to a stimulation of lipolysis and of gluconeogenesis while the sensitivity of muscle towards insulin is decreased.

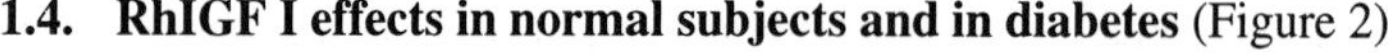

1.4. RhIGF I effects in normal subjects and in diabetes (Figure 2)

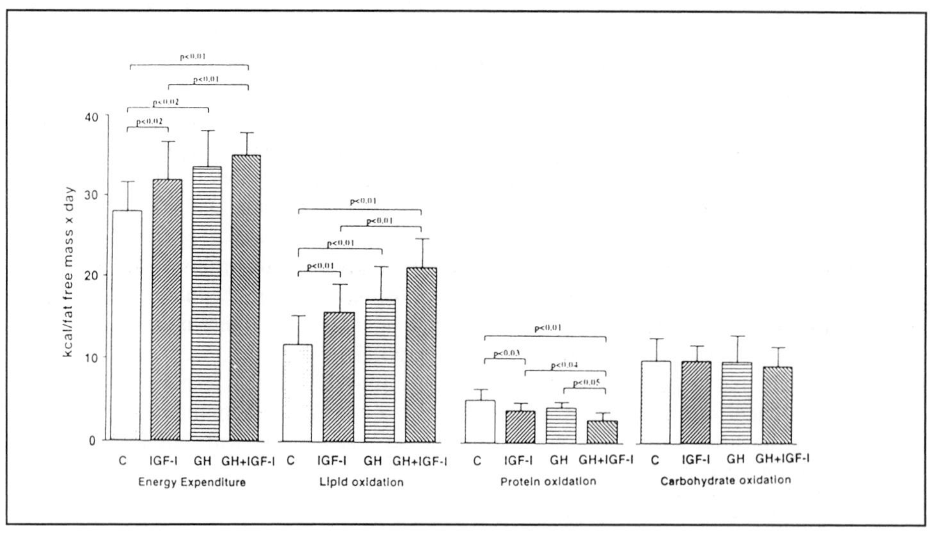

Figure 2. Resting energy expenditure and lipid, protein, and carbohydrate oxidation rates in eight growth hormone-(GH) deficient subjects on day 7 of indicated treatment (mean ± SD). IGF, insulin-like growth factor; C, control. From (22).

In large doses, rhIGF I is a hypoglycemic substance. IGF I induces a short-lived hypoglycemia in normal and in hypophysectomized rats. In man, the hypoglycemic potency of rhIGF I is approximately 1/12 of that of insulin. When 100 μg of IGF I are administered per kilogram body weight as an intravenous bolus, the blood sugar falls very rapidly and to the same extent as in a classical insulin tolerance test with 0.15 U/kg body weight (7)). The recovery from hypoglycemia occurs at about the same rate after either insulin or rhIGF I. When large doses of rhIGF I are infused subcutaneously or intravenously, total glucose consumption by the organism is increased. This increase of glucose oxidation and nonoxidative glucose disposal was quantitated during euglycemic clamping with both hormones and again a potency ratio of 12:1 in favor of insulin was found (16). The question arises whether rhIGF I acts mostly through the type 1 IGF or through the insulin receptor. This question can be tentatively answered. Free fatty acid levels fall rapidly proving that IGF I in large concentrations inhibits lipolysis in vivo as it does in vitro. Hepatic glucose release is also inhibited by large doses of rhIGF I (17). Since neither tissue contains type 1 IGF receptors, this action of rhIGF I is bound to be mediated via the insulin receptor. Generally speaking, large concentrations of free IGF I in serum have acute insulin-like effects, mostly via cross-reaction with the insulin receptor. The controversial data in the literature regarding the effects of IGF I on hepatic glucose output and lipolysis can be explained by differences in dosing of rhIGF I. Very large doses always have an insulin-like effect also on the liver and on adipose tissue whereas small doses of IGF I do not. This statement is also supported by findings in patients with nonfunctional insulin receptors. Schoenle et al (18).administered bolus injections of rhIGF I to three patients with type A insulin resistance. In all three patients, no acute hypoglycemia occurred while there was a very slow fall of the blood sugar. The most likely explanation for the lack of an immediate hypoglycemic response to large doses of rhIGF I in patients with nonfunctional insulin receptors is that rhIGF I cannot crossreact with the defective insulin receptor and that the slow effects on glucose metabolism observed in these patients are mediated via the type 1 IGF receptor. The results of these experiments also support the notion that the signal transduction through the type 1 IGF receptor is different from that through the insulin receptor. In any case, the Glut-4 response which is so closely connected to the binding of insulin to its receptor does not appear to be shared by the interaction between IGF I and the type 1 receptor. In this context one may recall the results of experiments in which high doses of rhIGF I were administered to streptozotocin-diabetic rats. These rats responded to rhIGF I with a growth response in spite of the fact that hyperglycemia, glucosuria and overall glucose metabolism were not improved by rhIGF I treatment (6). These data support the notion that even though rhIGF I does crossreact with the insulin receptor, this is not a major component of the organism's response to rhIGF I.

1.5. Observations and reasons in favor of rhIGF I treatment in diabetes (Figure 3)

Why should rhIGF I ever become a player in the therapy of diabetes mellitus if it cannot be administered in doses in which it directly acts on glucose uptake, thereby lowering blood glucose? In the following we would like to discuss several observations which make us believe that rhIGF I may eventually play a role in the therapy of different forms of diabetes mellitus.

1. As a consequence of the metabolic disturbances of diabetes, i.e. of the ineffectiveness of insulin, GH-hypersecretion and poor utilization of nutrients, the synthesis and secretion of IGF I from the liver is decreased. This has only little to do with the insulin dose that is administered to diabetics but rather with the metabolic disturbance resulting from hypoglycemia alternating with hyperglycemia and the poor response of the "diabetic" liver to insulin. In very well controlled type 1 diabetics, serum IGF I levels are normal and so is insulin sensitivity of tissues. However, with the present schemes of insulin therapy, it is practically impossible to achieve ideal metabolic control so that most type 1 diabetics do have a certain degree of IGF I deficiency. The same holds true for type 2 diabetics who could theoretically be in optimal metabolic control if they lost weight on a reducing regimen. However, this appears to be particularly difficult for many type 2 diabetics who may have inherited a set of "insulin resistance" genes with a tendency to hyperphagia.

2. Uncontrolled type 1 diabetics and obese type 2 diabetics are relatively insulin-resistant. Insulin resistance is aggravated in NIDDM by obesity. Most diabetics characteristically have elevated levels of insulin in the peripheral blood. Thus, type 1 diabetics treated with one or two daily injections of insulin subcutaneously have for the most part of the day higher peripheral insulin levels than normal subjects. Insulin peaks are lower but last longer. The same is true for patients with NIDDM who have increased endogenous insulin levels while fasting, during the night and in the morning but respond slowly to an increase of glycemia after meals and characteristically exhibit delayed insulin peaks. Despite elevated insulin levels, most of these patients are hyperglycemic during most of the day, and hyperglycemia aggravates the sluggishness of the response of the β-cells to hyperglycemia and other stimuli.

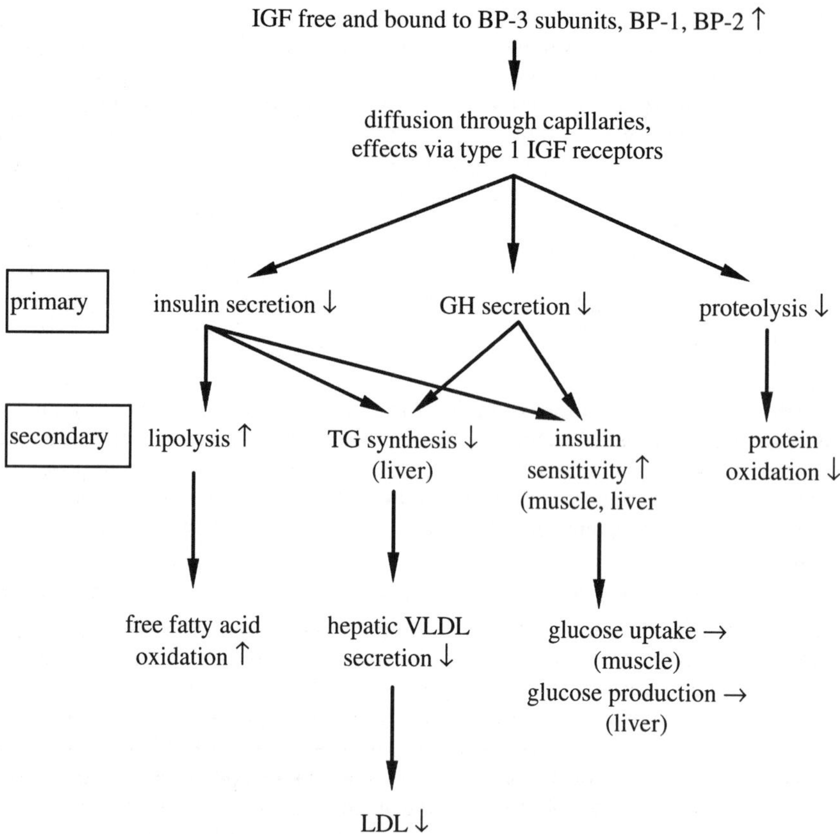

Figure 3. Metabolic and endocrine sequelae of insulin-like growth factor (IGF I) therapy. BP, binding protein; VLDL, very low-density lipoprotein; LDL, low-density lipoprotein. From (19).

3. rhIGF I administered intravenously or subcutaneously has two major endocrine effects: a) it diminishes insulin secretion (20,21,22), i.e. fasting as well as secretory burst mass but not frequency (23), and b) it decreases the number and the height of the GH peaks during the night and decreases GH secretion after secretory stimuli (24). One might expect that the partial inhibition of insulin secretion might result in hyperglycemia. However, the reverse is the case. In normal subjects as well as in early type 2 diabetics, the administration of rhIGF I leads to an improvement of glucose tolerance at comparatively lower insulin secre-

tion rates (25). Only when insulin secretion is further inhibited by much larger doses of rhIGF I, fasting hypoglycemia and subsequent glucose intolerance may result.

4. In order to determine whether rhIGF I therapy acts mostly via suppression of GH secretion which, when excessive, is known to induce hyperinsulinemia and glucose intolerance, rhIGF I was administered to GH-deficient adults. In GH-deficient patients, rhIGF I therapy also improved glucose tolerance at decreased insulin levels indicating that inhibition of GH secretion is not the sole and major reason why rhIGF I therapy improves glucose tolerance and increases the organism's sensitivity to insulin (24).
5. In both, normal and GH-deficient subjects, the administration of recombinant IGF I in low doses leads to a rise of fasting and postprandial free fatty acid concentrations because of the decreased insulin levels. By loosening the insulin brakes on lipolysis, free fatty acid levels and ketone body levels are increased by rhIGF I therapy. According to the classical glucose fatty acid cycle of Randle, an increase of free fatty acids and ketone bodies should lead to decreased glucose utilization and loss of insulin sensitivity. However, this is not the case during the administration of rhIGF I. In normal subjects as well as in GH-deficient subjects, the following observations were made: rhIGF I increases total energy expenditure of the organism on the expense of increased lipid oxidation while glucose oxidation and nonoxidative glucose utilization remained constant and protein oxidation was reduced at decreased amino acid levels (22,24). Therefore, the administration of rhIGF I switches the organism to lipid oxidation at unchanged glucose utilization rates and increased insulin sensitivity. Increased insulin sensitivity was actually demonstrated by euglycemic, hyperinsulinemic clamping during constant rhIGF I administration. At comparable insulin concentrations, glucose utilization was clearly increased during rhIGF I administration.
6. Carroll et al (26) have used rhIGF I as a substitution therapy giving one dose of rhIGF I at night to adolescent type 1 diabetics. IGF I levels were raised from below normal to normal levels, glycemic control was improved at insulin doses which were decreased by 45%. Glycosylated hemoglobin levels were significantly improved after one month of IGF I therapy and no side-effects were noted at these low doses of rhIGF I. These preliminary results of rhIGF I substitution therapy in adolescent type 1 diabetics make it likely that rhIGF I therapy as an adjunct to insulin therapy may become a very useful tool in rendering glycemic control easier for type 1 diabetics. rhIGF I therapy has been experimentally tried in a small number of NIDDM patients who had previously been on dietary control plus or minus sulfonyl ureas. These patients received relatively larger doses of rhIGF I, responded very well with respect to a significant improvement of glycemic control at decreased endogenous insulin levels and exhibited an improvement of the lipid profile. Within 5 days, fasting VLDL-triglyceride and LDL-cholesterol levels were significantly decreased (25). Recently, it has been reported that the postprandial metabolism of a fat load is not altered during rhIGF I therapy (T. Bianda, personal communication).

1.6. Cardiovascular effects of infused rhIGF I

Evidence is accumulating that IGF I exerts specific effects on the heart. In vitro, it increases β-myosin heavy chain and skeletal α-actin expression (27,28), enhances myofibrillar development and decreases smooth muscle α-actin (29) and atrionatriuretic factor expression (30) in cultured cardiomyocytes. In the isolated perfused rat heart it has positive inotropic effects (31). It improves myocardial function in normal adult rats (32) and acts in a cardioprotective manner on damaged rat myocardium (33,34,35). Similar effects have been reported for

GH (for review see 36). In humans, infused IGF I increases forearm blood flow (37,38,39) as well as cardiac output and stroke volume (40)). Donath et al. (40) have injected rhIGF I sc (60 μg/kg) in healthy male volunteers and investigated cardiac function and performance by Doppler echocardiography. They found that after 2-4 h, IGF I significantly increased stroke volume, cardiac output and the ejection fraction without changes in blood pressure suggesting that IGF I has a positive inotropic effect. In a recent study, similar effects were observed in patients with chronic heart failure (41). During a 4 h iv infusion of 60 μg/kg of rhIGF I, cardiac output and stroke volume in these patients increased and systemic vascular resistance, right atrial pressure and pulmonary artery wedged pressure decreased. Thus, short-term IGF I administration, which was well tolerated, improved cardiac performance in these patients most likely by decreasing peripheral vascular resistance, as also observed with GH (for review see 42). Extrapolating from the stimulatory effects of IGF I on myofibrillogenesis shown in rat cardiomyocytes in vitro (29), it does not appear unlikely that IGF I may stimulate impaired myofibrillogenesis in hearts of patients with chronic cardiac failure if treatment were continued over several weeks. Stimulation of the formation of new myofibrils may then further enhance the positive inotropic effect of IGF I and contribute to recompensation of overload heart hypertrophy.

2. Conclusions

The term NSILAs was coined in 1960 and the two peptides responsible for non-suppressible insulin-like activity were isolated, characterized and sequenced by Rinderknecht and Humbel in 1978 (43). From then on IGF research exploded and it became possible to solve many questions regarding the synthesis, secretion, regulation, biological effects and physiological importance of the IGFs. IGF I has been found to be the most important mediator of the action of pituitary GH. The knowledge about this function of rhIGF I has resulted in the successful treatment of GH-insensitive Laron dwarfs with rhIGF I. Whereas GH-deficient dwarfs grow well when treated with pituitary GH, GH-insensitive Laron dwarfs grow well with rhIGF I.

Whereas glucose homeostasis is fully dependent on the second-to-second regulation of insulin secretion by the β-cell, IGF I may be involved in the overall sensitivity of the organism to insulin. When nutrition is inadequate, endogenous IGF I levels fall and the organism develops a certain degree of insulin resistance. IGF I plays a major role in increasing insulin sensitivity again during refeeding and in the fed state. The balance between insulin sensitivity and insulin resistance is influenced by IGF I. This can be shown convincingly when relatively insulin-resistant type 1, type 2 or type A diabetics are treated with rhIGF I. During treatment with rhIGF I, the organism becomes more sensitive to insulin and glucose utilization occurs at considerably lower insulin levels. A relative IGF I deficiency may be related to the so-called insulin resistance genes which are held responsible for the NIDDM explosion in certain ethnic groups which had lived mostly on protein, not having developed agriculture, and then were suddenly transferred into the Western world with a nutritive surplus of carbohydrate and fat.

The physiology and pathophysiology of IGF I has become particularly complex when it was found that IGF I is not only an endocrine hormone but also a very important autocrine/paracrine cytokine functioning for instance as an estramedin in the process of remodeling of bone (44). Even more difficult is the interpretation of the physiological and pathophysiological role of the six different IGFBPs. These latter aspects of the physiology of IGF are now being investigated in a large number of laboratories all over the world.

Although many questions remain to be answered, we believe that rhIGF I will find a place as an important player in diabetes therapy. It increases insulin sensitivity, helps burning fat, lowers LDL-cholesterol levels, Lp(a) and fasting VLDL-levels while the removal of ingested fat (chylomicrons and chylomicron remnants) from blood remains unchanged. Since rhIGF I diminishes peripheral arteriolar resistance, thereby enhancing blood flow through several organs, it may prove useful in diabetic ischemic disorders.

Finally, rhIGF I has acute inotropic effects on the healthy as well as on the diseased heart. Furthermore, rhIGF I increases the expression of contractile elements not only in cultured heart cells but also in the impaired left ventricle of animals. Speculations about the possible usefulness of rhIGF I in the therapy of cardiac disorders are premature.

References

1. E.R. Froesch and J. Zapf. Diabetologia, 28 (1985) 485.
2. R.C. Baxter and J.L. Martin. Proc. Natl. Acad. Sci. USA, 86 (1989) 6898
3. E. Schoenle, J. Zapf and E.R. Froesch. Diabetologia, 16 (1979) 41.
4. H.-P. Guler, J. Zapf, E. Scheiwiller and E.R. Froesch. Proc. Natl. Acad. Sci. USA, 85: (1988) 4889.
5. S. Van Buul-Offers and J.L. Van den Brande. Acta Endocrinol., 92 (1979) 242.
6. E. Scheiwiller, H.-P. Guler, J. Merryweather, C. Scandella, W. Maerki, J. Zapf and E.R. Froesch. Nature, 323 (1986) 169.
7. H.P. Guler, J. Zapf and E.R. Froesch. N. Engl. J. Med., 317 (1987) 137.
8. P.D. Zenobi, S.E. Jaeggi-Groisman, W. Riesen, M. Roder and E.R. Froesch. J. Clin. Invest., 90 (1992) 2234.
9. D.K. Rabinowitz and L. Zierler. Nature, 199 (1963) 913.
10. R.E. Humbel. Eur. J. Biochem., 190 (1990) 445.
11. M.A. Hussain, O. Schmitz and E.R.Froesch. NIPS, 10 (1995) 81.
12. J.C. Brand Miller and S. Colagiuri. Diabetologia, 37 (1994) 1280.
13. J.V. Neel. Am. J. Hum. Genet., 14 (1962) 353.
14. W.H. Daughaday and M. Kapadia. Proc. Natl. Acad. Sci. USA, 86 (1989) 6778.
15. J. Zapf, E. Futo, M. Peter and E.R. Froesch. J. Clin. Invest., 90 (1992) 2574.
16 I. Turkalj, U. Keller, R. Ninnis, S. Vosmeer and W. Stauffacher. J. Clin. Endocrinol. Metab. 75 (1992) 1186.
17. R. Jacob, E. Barrett, G. Plewe, K.D. Fagin and R.S. Sherwin. J. Clin. Invest., 83 (1989) 1717.
18. E.J. Schoenle, P.D. Zenobi, T. Torresani, E.A. Werder, M. Zachmann and E.R. Froesch. Diabetologia, 34 (1991) 675.
19. E.R. Froesch and M. Hussain. J. Intern. Med., 234 (1993) 561
20. H.P. Guler, C. Schmid, J. Zapf and E.R. Froesch. Proc. Natl. Acad. Sci. USA, 86 (1989) 2868.
21. P.D. Zenobi, S. Graf, H. Ursprung and E.R. Froesch. J. Clin. Invest., 89 (1992) 1908.
22. M.A. Hussain, O. Schmitz, A. Mengel, A. Keller, J.S. Christiansen, J. Zapf and E.R. Froesch. J. Clin. Invest., 92 (1993) 2249.
23 N. Tørksen, M.A. Hussain, T.L. Bianda, B. Nyholm, J.S. Christiansen, P.C. Butler, J.D. Veldhuis, E.R. Froesch and O. Schmitz. Am. J. Physiol. (Endocrinol. Metab. 35), 272 (1997) E352.

24. M.A. Hussain, O. Schmitz, A. Mengel, Y. Glatz, J.S. Christiansen, J. Zapf and E.R. Froesch. J. Clin. Invest., 94 (1994) 1126.
25. P.D. Zenobi, P. Holzmann, Y. Glatz, W.F. Riesen and E.R. Froesch. Diabetologia, 36 (1993) 465.
26. Carroll PV, M. Umpleby, G.S. Ward, S. Imuere, E. Alexander, D. Dunger, P.H. Sönksen and D.L. Russell-Jones. Diabetes, 46 (1997) 1453.
27. J.R. Florini and D.Z. Ewton. Growth Regul., 2 (1992) 23.
28. H. Ito, M. Hiroe, Y. Hirata, M. Tsujino, S. Adachi, M. Shichiri, A. Koike, A. Nogami and F. Marumo. Circulation, 87 (1993) 1715.
29. M.Y. Donath, J. Zapf, M. Eppenberger-Eberhardt, E.R. Froesch and H.M. Eppenberger. Proc. Natl. Acad. Sci. USA, 91 (1994) 1686.
30. B.A. Harder, M.C. Schaub, H.M. Eppenberger and M. Eppenberger-Eberhardt.. J. Mol. Cell. Cardiol., 28 (1996) 19.
31. D.M. Johnston, P.D. Gluckman, C.M. Bratt, G.R. Ambler and W.F. Lubbe. Growth Regul., 4 (1994) 44 (abstract).
32. A. Cittadini, H. Stömer, S.E. Katz, R. Clark, A.C. Moses, J.P. Morgan and P.S. Douglas. Circulation, 93 (1996) 800.
33. G.R. Ambler, B.M. Johnston, L. Maxwell, J.B. Gavin and P.D. Gluckman. Cardiovasc. Res., 27 (1993) 1368.
34. R.L. Duerr, S. Huang, H.R. Miraliakbar, R. Clark, K.R. Chien and J. Ross. J. Clin. Invest., 95 (1995) 619.
35. M. Buerke, T. Murohara, C. Skurk, C. Nuss, K. Tomaselli and A.M. Lefer. Proc. Natl. Acad. Sci. USA, 92 (1995) 8031.
36. A. Cittadini, S. Fazio, B. Ferravante, L. Saccà and P.S. Douglas. Endocrinol. Metab., 4 (1997) 27.
37. Fryburg DA. Am. J. Physiol. (Endocrinol. Metab. 30), 267 (1994) E331.
38. A.C. Copeland and K.S. Nair. J. Clin. Endocrinol. Metab., 79 (1994) 230.
39. W. Kiowski, K. Stoschitsky, J.H. Kim, M.A. Hussain and E.R. Froesch. Circulation, 90 (1994) I-242 (abstract).
40. M.Y. Donath, R. Jenni, H.-P. Brunner, M. Anrif, S. Kohli, Y. Glatz and E.R. Froesch. J. Clin. Endocrinol. Metab., 81 (1996) 4089.
41. M.Y. Donath, G. Sütsch, XW. Yan, B. Piva, H.P. Brunner, Y. Glatz, J. Zapf, F. Follath, E.R. Froesch and W. Kiowski. The Endocrine Society 79th Annual Meeting, June 11-14, Minneapolis, USA, (1997) (abstract)
42. S. Fazio, A. Cittadini, B. Biondi, G. Riccio, D. Sabatini and L. Saccà. Endocrinol. Metab., 4 (1997) 33.
43. E. Rinderknecht and R.E. Humbel. J. Biol. Chem., 253 (1978) 2769.
44. M. Ernst, Ch. Schmid and E.R. Froesch and J. Zapf. Diabetologia, 28 (1985) 485.

Molecular Mechanisms to Regulate the
Activities of Insulin-like Growth Factors
K. Takano, N. Hizuka and S-I. Takahashi (Editors)

Methods to analyze tissue-specific promoters for IGF and IGFBP gene expression.

C. T. Roberts, Jr.

Department of Pediatrics (NRC-5), School of Medicine, Oregon Health Sciences University, Portland, OR 97201

1. ABSTRACT

Activation of the IGF signaling system depends upon the action of the IGF ligands functioning in autocrine, paracrine, and endocrine modes. The bioactivity of the IGFs is modulated by the IGFBPs, which are synthesized by a variety of tissues. Thus, the ability of the IGFs to exert their biological effects in a particular setting will depend upon the local synthesis of IGFs and IGFBPs, as well as the contribution of IGFs and those IGFBPs derived from the circulation. The regulated expression of the IGF and IGFBP genes, occurring in a tissue-specific fashion, is, therefore, a major factor influencing IGF system activity. These are multiple approaches for analyzing levels of IGF/IGFBP gene expression that vary in their sensitivity, resolution, quantitativeness, and suitability in a particular experimental system. General techniques for analyzing gene expression with respect to mRNA levels per se include Northern analysis, RNase protection (RPA) or S1 nuclease assays, reverse-transcription polymerase chain reaction (RT-PCR), and *in situ* hybridization. These methods are applicable to both tissues and cells in culture. Techniques that distinguish transcriptional from post-transcriptional control include analyses of pre-mRNA by RPA or RT-PCR, nuclear run-on assays and mRNA stability determinations. These can be done in tissues but are more feasible in cultured cells. The definition of specific genetic regulatory regions can be achieved through transient transfection approaches using putative promoter sequences fused to reporter genes such as chloramphenicol acetyl transferase (CAT), or luciferase, in conjunction with gel-shift and DNase I footprinting assays employing shorter DNA sequences and purified proteins or nuclear extracts. Judicious application of an appropriate combination of such approaches can delineate the specific DNA elements and regulatory proteins whose interactions govern the tissue-specific synthesis of IGFs and their binding proteins.

2. LEVELS OF CONTROL OF GENE EXPRESSION

The expression of genes in eukaryotic cells involves a complex sequence of events beginning with DNA itself and culminating in the appearance of an active gene product (generally protein). The entire process can be envisioned as occurring at a series of levels, each of which is theoretically subject to regulation that may vary in a tissue-specific fashion. These include processes that can be clarified as pre-transcriptional (chromatin remodeling and imprinting), transcriptional (initiation and elongation by RNA polymerase II and associated factors), post-transcriptional (RNA splicing, polyadenlation and 3' end formation, nucleocytoplasmic transport and mRNA degradation), translational (initiation and elongation), and post-translational (precursor processing and secretion, protein degradation and covalent modification such as phosphorylation, glycosylation and acetylation). While all of these mechanisms contribute to the expression of a given gene, the function of the promoter is to direct the assembly of pre-initiation complexes at specific sites of transcription initiation so that primary transcripts can be generated in a regulated fashion. This short review will therefore focus on those techniques that are useful to determine mRNA levels, methods to distinguish transcriptional versus post-transcriptional control (either of which may influence mRNA levels), and, finally, approaches to analyze promoter activity and regulation by tissue-specific factors.

3. METHODS TO DETERMINE mRNA LEVELS

3.1 Northern blot hybridization

There are a number of techniques that can be employed to ascertain mRNA levels and their potential variation from tissue to tissue. The classic technique is Northern blot hybridization, in which RNA (total or poly A(+)-enriched) is resolved by size on agarose gels using glyoxal or formaldehyde as a denaturant to ensure that RNA secondary structure does not result in anomalous migration. The size-separated species are then transferred to a nitrocellulose or nylon membrane by capillary, pressure or electrophoretic transfer. Specific immobilized RNA species can then be detected by hybridization with oligonucleotides, cDNA or antisense (cRNA) probes that have been radioactively (generally with ^{32}P) or chemically (i.e., biotin or DIG) labeled. Hybridization is then evaluated by autoradiography or enhanced chemiluminescence. There are both advantages and disadvantages to this standard technique. Northern analyses are relatively sensitive, yet do require a reasonable amount of RNA for each sample to be analyzed. Additionally, the integrity of the RNA (as assessed by a 2:1 ratio of the apparent fluorescence of EtBr-stained 28 and 18S rRNA bands and the presence of the 45S rRNA precursor) is critical for accurate mRNA quantitation by Northern blotting. This is because any nicks introduced into an mRNA molecule prevent it from migrating at a single position consistent with its length. Thus, any degradation of mRNA samples rapidly reduces their suitability for Northern analysis. Finally, detection of larger mRNA species (which are also obviously more susceptible to degradation) is influenced by the efficiency of the transfer from gel to membrane.

3.2 RNase protection assays

A second approach for determining mRNA levels is the solution hybridization/RNase protection assay (RPA). In this procedure, a cDNA fragment is subcloned into a vector that contains prokaryotic RNA polymerase promoter sequences (i.e., T3, T7 or SP6) flanking the multiple cloning site. Linearization of the plasmid at the 5' end of the cDNA insert and *in vitro* transcription from the promoter at the 3' end of the insert produces a labeled antisense RNA probe that can be hybridized to target RNA samples in solution. After digestion of the hybridization mixtures with a combination of RNass A and T1, antisense RNA probe that was protected from RNase digestion by virtue of its hybridization to an exactly complimentary target mRNA sequence can be quantitated by electrophoresis on a polyacrylamide gel followed by autoradiography. This technique is significantly more sensitive than Northern analyses and several probes can be used simultaneously as long as the sizes of the protected probe bands are resolved by the gel system employed. The RPA is significantly more sensitive than Northern hybridization and does not require that the target RNA samples be undegraded, since the antisense RNA probe is usually complimentary to only a portion of the entire mRNA sequence. This is an advantage when analyzing large transcripts or when attempting to detect specific mRNAs in degraded samples. Some disadvantages are that RPAs are technically more demanding than Northern analyses and can produce data from samples that are seriously degraded. For the latter reason, it is prudent to independently assess the integrity of the RNA samples to be assayed in RPAs by running analytical amounts on a Northern gel. Additionally, optimal use of RPA requires exactly complementary probes and target sequences, so that the use of heterologous probes is impractical. An obvious difference between Northerns and RPAs is that the former provide a direct demonstration of the size of the hybridizing mRNA species while the latter generally do not.

3.3 Polymerase chain reaction

The most recently adopted technique for determining mRNA levels is reverse transcription-coupled polymerase chain reaction (RT-PCR). This is the most sensitive technique currently available, but the most difficult to perform in a routinely quantitative fashion. Briefly, RNA samples are hybridized with a oligonucleotide primer complementary to a selected region of the target RNA some distance from the 5' end of the transcript. Extension with reverse transcriptase then generates first-strand complementary DNA (cDNA). This product is then incubated with two additional primers; one that corresponds to a sequence upstream of the sequence targeted by the original primer and one complimentary to a sequence just upstream of the initially targeted site. Subsequent cycles of standard PCR generate a DNA fragment whose yield can predict the quantity of target mRNA in the original sample. The specificity of the process is facilitated by choosing primers that bracket one or more introns to preclude the production of a PCR product from contaminating genomes' DNA. Additionally, the use of a second internal, or nested, primer for the PCR step increases the specificity of the product obtained. There are two issues that must be addressed to ensure that this method is quantitative. The first is the reproducibility of the reverse transcription step itself. The most appropriate way to control for variability is to add to each sample a synthetic RNA corresponding to at least the region of the authentic mRNA sequence that is being amplified, but that differs in length or by a single nucleotide change that removes or introduces a restriction enzyme site in the resulting PCR product. This internal control (essentially a competitor) will presumably be reverse transcribed and amplified in parallel with the authentic target RNA sequence, and the yield of RT-PCR product can be used to normalize the yield of the product generated from the authentic mRNA target. The second consideration in RT-PCR analyses is to ensure that the number of PCR cycles is still in the exponential phase of amplification, so that yield is proportional to the amount of input RNA.

3.4 *In situ* hybridization

The techniques described above are appropriate for the determination of specific mRNA levels in RNA extracted from different cell lines or tissues, but are less applicable when one wishes to discern expression patterns within a given tissue or organ that may have a complicated architecture. This sort of information is best obtained from *in situ* hybridization studies, in which fixed tissue sections are hybridized with ^{35}S-labeled antisense RNA probes to accurately localize gene expression to specific cell types within a given organ or tissue or from tissue to tissue (i.e., in embryos).

4. TRANSCRIPTIONAL VERSUS POST-TRANSCRIPTIONAL CONTROL

4.1 Nuclear run-on assays

If one has obtained presumptive evidence for tissue-specific gene expression using one of the techniques described above, the next question is whether the observed differences arise as a result of differences in promoter activity or partially through post-transcriptional mechanisms. Described below are three techniques to distinguish promoter activity from other levels of gene regulation. The standard technique for determining the transcriptional activity of intact genes is the nuclear run-on assay. In this procedure, nuclei are isolated and incubated in the presence of high concentration of ^{32}P-UTP and α-amanitin, which inhibits further transcription initiation. Elongation by pre-existing transcription complexes is allowed to proceed for a short period to label nascent transcripts, and this labeled RNA is then quantified by hybridization to target DNA corresponding to the transcribed gene of interest that has been immobilized on a membrane (typically in a slot blot format).

4.2 Pre-mRNA analysis

A second, more recently developed, approach is a variation of the RPA in which the antisense RNA probe is complementary to an exonic sequence as well as part of an adjacent intron. This probe will hybridize fully only to pre- (i.e., unspliced) mRNA (or to contaminating genomic DNA), whose levels would more directly reflect transcriptional rate than changes in the levels of fully processed, cytoplasmic mRNA. When directly compared in the context of the same gene, both approaches produce comparable data, with the RPA being more sensitive.

It must be emphasized, however, that neither of the approaches measures the rate of transcription initiation directly, since the data obtained with either technique can be influenced by changes in the rate of transcription elongation or attenuation that may occur upstream of the region corresponding to the target DNA sequences employed in the nuclear run-on assay or to the exon-intron junction incorporated in the cRNA probe in the RPA. Given this concern, it is advisable to use sequences as close as possible to the site of transcription initiation to minimize possible contributions from variations in elongation rate.

4.3 mRNA stability

While these two techniques give some indication as to potential changes in apparent promoter activity, a major level of post transcriptional control that can influence the overall levels of specific mRNA spewn is mRNA stability. Regulated degradation of mRNA is observed much more frequently than control at the level of transcription elongation, and can be rather readily discerned, but only practically in cells in culture. Changes in the rate of mRNA degradation will effect the half-life of the mRNA species in question, but the latter can only be determined in the absence of *de novo* transcription. In practice, this is accomplished by incubating cells in the presence of a transcriptional inhibitor such as actinomycin D or, more recently, DRB, and RNA extracted at various times subsequently and specific mRNA levels determined at each time point using one of the techniques described in section 3 above. An analysis of this sort will provide information on the half-life of a given mRNA species and any variations in these parameters from cell line to cell line or as a result of a specific treatment of a single cell line. Although this procedure is not applicable to tissues *per se* it does reveal if a particular transcript is subject to differential degradation, which could contribute to tissue-specific regulation *in vivo*. It must be emphasized that transcriptional and post-transcriptional mechanisms of gene regulation are not mutually exclusive, but may both contribute to the overall level of expression of a given gene. Indeed, genes whose products require complex regulation, such as those encoding IGFs and IGFBPs, would be expected to be subject to multiple levels of control.

5. BASIC APPROACHES TO DEFINING PROMOTER ACTIVITY

The preceding sections describe approaches to detect apparent changes in mRNA levels and to ascertain to what extent such changes can be ascribed to changes in promoter activity. This final section will outline a series of technical approaches that can be employed to evaluate the activity of defined promoter regions.

5.1 Transient transfection assays

The activity of defined promoter regions can be conveniently assessed by subcloning fragments of genomic DNA corresponding to regions containing sequences that flank the transcription start site or sites of the gene of interest into vectors that contain a suitable reporter gene adjacent to the multiple cloning sites. Earlier versions of reporter vectors employed the bacterial CAT gene, since this enzymatic activity is normally absent from mammalian cells. Most current studies employ vectors in which the inserted promoter sequence drives the expression of the firefly luciferase gene, which is also absent from normal mammalian cells but which confers significantly more sensitivity on the assay. When active promoter sequences are cloned upstream of one of these reporter genes and the resulting construct introduced into a cell line by calcium phosphate or lipid-mediated transfection or by electroporation, the introduced DNA is maintained in the transfected cells over several days, during which time the product of the reporter gene accumulates. The level of the reporter gene product assayed in cell extracts is considered to reflect the intrinsic activity of the DNA sequence cloned into t he reporter vector. In general, a second plasmid encoding another reporter gene (such as β-galactosidan, growth hormone, or thymidin kinon) driven by a constitution promoter (i.e., CMV, RSV, TK, or SV40) is co-transfected in order to monitor transfection efficiency from plate to plate. Promoter activity is then repressed as the level of reporter gene activity normalizes for the level of the activity encoded by the transfection control vector. Using this approach, a given promoter region can be evaluated in cell lines derived from different tissues, or particular regions of the promoter that are necessary for activity in a given cell line can be identified by comparing the activity of a series of deletion constructs of a specific promoter.

While transient transfection studies provide useful information on the intrinsic activity of putative promoter sequences and may allow one to infer the location of particular regions that appear to be required for optimal activity in a particular cell line or differentiate activity in lines derived from, for example, different tissues, it is necessary to corroborate such data with that obtained from studies focusing on a particular promoter sequence and the proteins with which they interact. The latter information is typically obtained from gel-shift and footprinting assays of the type described in the following section.

5.2 Gel-shift assays

Gel-shift assays (also termed gel retardation or electrophoretic mobility shift - EMSA - assays) rely upon the ability of transcriptional regulatory proteins to reduce the apparent mobility (in active polyacrylamide gels) of DNA fragments to which they are bound. In practice, one can incubate labeled DNA fragments corresponding to specific regions of the promoter with either purified proteins or nuclear lysates from various tissues or cell lines. Subsequent electrophoresis of the protein-DNA complexes on native polyacrylamide gels will detect the presence of protein bound to the DNA sequence being analyzed by virtue of the reduced mobility of a portion (or all) of the probe, which itself would migrate toward the bottom of the gel. These gel-shifted bands correspond to probe molecules bound to specific DNA-binding proteins, and differences in the patterns observed with extracts from different sources may provide clues to the actions of proteins that may be responsible for differential expression of the gene that is controlled by the promoter sequence being analyzed.

5.3 DNA footprinting

The gel-shift assays described above reveal the existence of DNA-protein interactions but, at least when using larger DNA fragments, provide little information on the specific sites involved in interaction with putative transcriptional regulatory proteins. While potential regulatory motifs can be identified based upon this similarity to defined consensus recognition sites for particular transcription factors, a more functional assessment can be made by the use of DNA footprinting assays. In this technique, DNA fragments up to a few hundred base pairs in length are labeled at one end and incubated with nuclear extracts containing transcriptional regulatory factors. Subsequent treatment of these reactions with limiting amounts of DNase I will produce a set of cleavage products representing the distance between the labeled end of the DNA fragment and all those bases accessible to DNase. Bound regulatory proteins will preclude cleavage at those sites by DNase I, so that when the pattern of cleavage products is analyzed by polyacrylamide gel electrophoresis, gaps in the "ladder" of bands generated by DNase I correspond to sites that are interacting with proteins in the nuclear extract. Through the use of this techniques, particular DNA sequences can be identified that may be targets for transcription factors. Comparison of the data generated with nuclear extracts from different tissues can provide an indication of specific motifs that may be necessary for tissue-specific gene expression.

None of the techniques described above gives a complete picture of promoter function.

Rather, an integrated approach that employs multiple procedures to generate internal consistent data is required to elucidate the tissue specific regulation of a promoter and the particular sequences and regulatory proteins that effect this control.

The techniques discussed in this short review are each described in a myriad of references in the recent research literature, but the general references listed below provide useful, practical descriptions of these various approaches for the investigator who is not already experienced in the molecular biology of gene regulation.

REFERENCES

1. Adamo, M.L., Stannard, B., LeRoith, D., and Roberts, C.T., Jr. Approaches for the purification, quantitation, and analysis of hormone and receptor mRNAs. In: Handbook of endocrine research techniques. DePablo, F., Weintraub, B., and Scanes, C.(eds.), Academic Press, San Diego, 1993.
2. F.M. Ausubel, R. Brent, et al (eds.), Short Protocols in Molecular Biology, John Wiley & Sons, New York, 1992.
3. L.G. Davis, W.M. Kuehl and J.F. Battey, Basic Methods in Molecular Biology, Appleton & Lange, Norwalk, Connecticut, 1994.
4. J. Sambrook, E.F. Fritsch and T. Maniatis, Molecular Cloning: A Laboratory Manual, Cold Spring Harbor Laboratory Press, Plainview, N.Y., 1989.

Molecular Mechanisms to Regulate the
Activities of Insulin-like Growth Factors
K. Takano, N. Hizuka and S-I. Takahashi (Editors)

Involvement of AP-1-like motifs in the IGF-I gene regulation

Y. Kajimoto, Y. Umayahara, Y. Fujitani, Y. Yamasaki, and M. Hori

First Department of Medicine, Osaka University School of Medicine, 2-2 Yamadaoka, Suita 565, Japan

The expression of insulin-like growth factor-I (IGF-I) is regulated by various hormones, oncogenes, and other growth factors. In early 1990s, the entire structures of the IGF-I genes were characterized and two gene promoters (P1 and P2), located 5' to exon 1 and exon 2, respectively, were identified. Among those two promoters, the P1 promoter is known to be more active than the P2 promoter in most tissues including adult liver and thus is called "the major promoter". As a step toward understanding the molecular basis of the IGF-I gene regulation, we showed that PKC is a potential activator of the P1 promoter. Phorbol ester (TPA) causes an increase in the DNA-binding activity for the AP-1-like sequences both in the human and chicken IGF-I gene P1 promoters. In terms of the chicken AP-1 motif, c-Fos and c-Jun were identified as the binding factors. Those two proteins were also shown to be involved in the estrogen responsiveness of the promoter through the post-translational modification. In contrast to these observations in chicken, however, it was not the c-Fos/c-Jun heterodimer that binds to the AP-1-like motif in the human IGF-I gene. Instead, the TPA-inducible gel-shift complex was recognized by an anti-C/EBPβ antibody, thus suggesting that TPA-responsive activation of C/EBPβ is implicated in the promoter activation. Unlike the AP-1 motif in the chicken IGF-I gene, the C/EBPβ-binding motif in the human gene did not mediate the estrogen responsiveness. Interestingly, the C/EBPβ-binding motif was shown to be essential to generate a strong transcriptional activity in a tumor cell-line, SK-N-MC.

1. Introduction

Insulin-like growth factor I (IGF-I), a 70-residue single-chain growth-promoting polypeptide, plays an essential role in embryonic and postnatal growth in mammals (1,2). Its expression is regulated by hormonal, nutritional, tissue-specific, and developmental factors in a coplicated manner. In rats and humans, the single-copy six-exon genes are transcribed by two promoters into nascent RNAs with different 5' leader sequences that undergo both alternative RNA splicing and differential polyadenylation to yield multiple mature transcripts (2). In 1991, the chicken IGF-I gene promoter became the first regulatory sequences of IGF-I gene ready to be studied (3). Since then, we have analyzed the molecular mechanisms for the IGF-I gene regulation using the chicken promoter as a model, and have found several clues to understanding the entire machinery. In this article, we would like to review those observations especially focusing on the regulating mechanisms by PKC and estrogen.

Activation of PKC may be involved in IGF-I gene regulation. The results of nuclear run-off assay have shown that TPA-treatment of human macrophage-like cells increased transcription rate of the IGF-I gene (4). A rapid increase in diacylglycerol concentration was observed after GH treatment in proximal tubular cells from canine kidney (5) and in preadipocyte Ob1771 cell (6), thus suggesting that PKC may also play a role in signaling of GH action.

It has been well known that estrogen is a regulator of IGF-I expression; it enhances the IGF-I expression in certain estrogen-sensitive cells and tissues such as osteoblast cells, uterus, ovary or breast cancer cells. As the estrogen regulation of IGF-I expression may be relevant to cancer growth or post-menopausal osteoporosis, the mechanisms underlying the estrogen regulation of IGF-I expression would be of clinical interest.

2. Identification of the TPA responsiveness in the chicken IGF-I gene promoter

In 1991, the entire structure of the chicken IGF-I gene was determined (3). Multiple transcription start sites were dispersed within about 70 bp region. Analysis of the nucleotide sequence near the transcription initiate region revealed that the chicken IGF-I gene lacks apparent "TATAA" box or "CAAT" box (3). Instead, the nucleotide sequences flanking this start site bore striking similarity to the "initiator" sequence described by Smale and Baltimore (7) as an essential motif for the accurate transcription of the mouse terminal deoxynucleotidyltransferase gene. This motif was later identified in some other "non-housekeeping" genes which lack a TATA box as well.

In the 5'-flanking region of the chicken IGF-I gene where the promoter activity was identified, several putative cis-active elements as can be assumed from literature were identified (3). Among those, we focused on an AP-1 motif located in -420 - -427. First we examined whether the chicken IGF-I gene promoter can be activated in response to TPA. The results of reporter gene analyses showed that promoter activities of both 2100-bp and 600-bp 5' flanking region were increased about 4-fold after TPA (10^{-7} M) treatment (8). When point mutation was introduced into the AP-1 site by PCR-mediated site-directed mutagenesis, TPA-induced promoter activation was diminished significantly (4.9-fold to 1.7-fold) and basal promoter activity was lost by 74%. These results indicate that the AP-1 site plays a major role in generating both the TPA-responsiveness in promoter activity and the basal promoter activity. Moreover, gel mobility shift assay showed the TPA inducible specific protein binding to the AP-1 site (8). Taken together these results indicate that the AP-1 site in chicken IGF-I gene functions as an authentic AP-1 binding site and possibly mediates PKC action.

3. Estrogen regulation of the insulin-like growth factor-I gene transcription involves an AP-1 enhancer

3.1. Backgrounds

In 1988, the basic mechanism for the estrogen signal transduction was revealed (9). The estrogen receptor (ER) mediates estrogen actions basically by simply binding as a homodimer to specific target DNA sequences known as the estrogen-responsive element (ERE), and thus stimulates transcription of a target gene (classical pathway of estrogen signaling). The results of RNA analyses using cycloheximide have suggested that the estrogen regulation of IGF-I gene expression does not require *de novo* protein synthesis in rat uterus (10) or in osteoblasts in culture (11), leading to the suggestion that an ERE exists in the regulatory regions of IGF-I genes and plays a central role in the phenomenon (12). Contrary to expectation, however, no consensus ERE has been identified to date in the characterized portion of IGF-I gene promoters (3,13,14). This suggests that some novel mechanisms may underlie the estrogen regulation of IGF-I gene expression.

3.2. The chicken IGF-I gene promoter is estrogen-responsive

First we performed a series of gene transfer studies and examined whether and how the IGF-I gene promoter is a target of estrogen regulation (15). The IGF-I-luciferase reporter plasmids containing segments of the promoter of the chicken IGF-I gene were transfected into HepG2 cells which exogenously expressed ER. HepG2 cells, which do not express endogenous ER and are very easy to be transfected with exogenous genes, were used as host cells for the study. We transfected human ER-expressing plasmid into the cells and found in hormone binding assay that the human ER-expression plasmid (HE0)-transfected HepG2 cells had 2.2 fmol/mg protein of specific estradiol binding sites, while the untransfected HepG2 cells had no binding site (undetectable). In such exogenous ER-expressing HepG2 cells, a series of reporter gene analyses was performed in order to examine whether the IGF-I gene promoter is a target of regulation by estrogen.

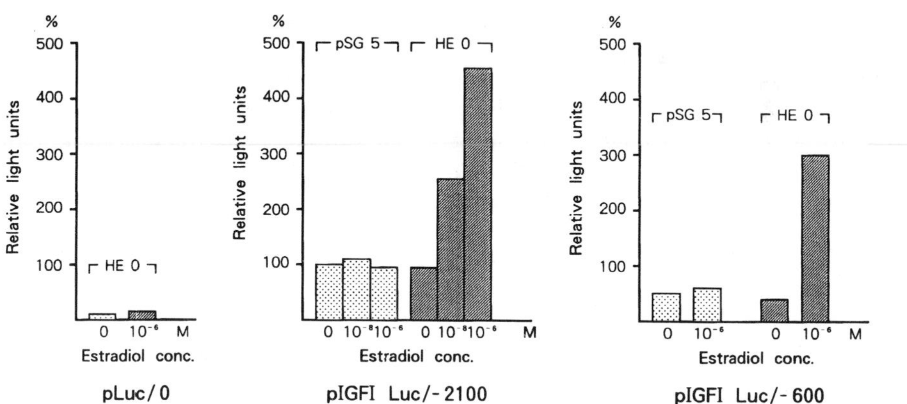

Figure 1. Estrogen responsive activation of IGF-I gene promoter in ER-expressing HepG2 cells.

The reporter plasmids were individually cotransfected with HE0 (ER-expressing plasmid) or pSG5 (control) into HepG2 cells, and 24 h later, 10^{-6} M or 10^{-8} M (pIGFI Luc/-2100 only) 17β-estradiol was added. Luciferase and CAT assays were performed 24 h after addition of estradiol.

As shown in Fig. 1, when 10^{-6} M 17β-estradiol was added, promoter activities of 2100-bp and 600-bp DNA fragments increased 4.6-fold and 8.6-fold, respectively, in the HE0-transfected HepG2 cells. A smaller increase in the promoter activity of the 2100-bp DNA fragment was observed when 10^{-8} M 17β-estradiol was added, indicating that a physiological concentration of estrogen readily causes this phenomenon. On the other hand, neither the 2100-bp nor the 600-bp fragment displayed a response to estrogen in terms of its promoter activity in control vector (pSG5)-transfected HepG2 cells (Fig. 1). Thus it was demonstrated that the 600 bp 5'-flanking DNA of chicken IGF-I gene contains enough cis-active elements to mediate estrogen-induced promoter activation, and that, like any other estrogen action, the phenomenon is mediated by ER.

Although there were no consensus ERE existed in the 600 bp 5'-flanking DNA of chicken IGF-I gene, one region which was 70% homologous to the ERE consensus sequence was identified between nucleotide positions +5 and +17. However, the mutation analysis revealed this region could not function as an authentic ERE (data not shown). Thus, no conventional ERE existed within the 600 bp 5'-flanking DNA of chicken IGF-I gene despite the estrogen responsiveness of its promoter activity.

3.3. Involvement of the AP-1 motif in the estrogen responsiveness

The AP-1 enhancer, which could be identified in the chicken IGF-I gene, may also mediate estrogen effects: the association of AP-1-binding transcription factors and nuclear receptors had been suggested (16). Accordingly, we investigated whether the AP-1 enhancer plays a role in the estrogen-responsive IGF-I gene activation. Another mutation analysis was performed using a reporter plasmid whose AP-1 motif had been destroyed. Then, the destruction of the AP-1 motif caused complete loss of estrogen responsiveness of the IGF-I promoter (0.81-fold induction), suggesting an essential role for the motif in mediating estrogen effects on the IGF-I gene transcription.

To examine the changes on protein binding to the AP-1 motif by estrogen, a gel mobility-shift analysis was performed. The results showed that specific protein-bindings to the AP-1 probe increased from as early as 0.5 h after estradiol was added, and the increase continued at least until 6 h (data not shown). These results suggested that estrogen enhances specific protein binding to the AP-1 motif and thus activates the chicken IGF-I gene promoter.

To examine whether the increased c-Fos, c-Jun binding to the AP-1 motif depends on the transcriptional or translational regulation of those proteins by estrogen, changes in c-fos, c-jun mRNA and in c-Fos, c-Jun protein amounts were examined. The results revealed that estradiol treatment induces neither gene expression nor protein synthesis of those AP-1 binding factors (data not shown). These results suggested that post-translational modulation of Fos-Jun activity is involved in the estrogen-responsive IGF-I gene activation.

3.4. Requirement of DNA-binding domain of ER for mediating estrogen-responsive IGF-I activation

Generally, a DNA-binding domain of ER facilitates ER binding to ERE located in target genes. We determined whether the DNA-binding domain of ER is required for the estrogen responsiveness of chicken IGF-I gene promoter. A DNA-binding domain-deficient ER was allowed to be expressed in the HepG2 cells and the estrogen responsiveness was examined. The results revealed that the DNA-binding domain of ER is of great importance for the signal transduction, in spite of estrogen responsive chicken IGF-I gene promoter lacks ERE.

3.5. No de novo protein synthesis necessary for estrogen-induced IGF-I gene activation

The necessity of the DNA-binding domain naturally suggested that the estrogen-ER complex binds to an ERE located in a regulatory region of a gene encoding an unknown mediator of estrogen action, and that the mediator would be involved in post-translational modulation of the Fos-Jun activity. To test this hypothesis, we examined whether *de novo* protein synthesis is required in estrogen induced protein binding to the AP-1 site. A series of gel-mobility shift assay, which was performed using the protein synthesis inhibitor cycloheximide, revealed that the protein-DNA complex induced by estrogen was not affected by the addition of cycloheximide (data not shown). This observation indicates that *de novo* protein synthesis is not necessary for the estrogen activation of chicken IGF-I gene promoter, and thus denies the involvement of an unknown mediator with its gene transcription being regulated in an estrogen-responsive manner. Therefore, binding of the estrogen-ER complex to an ERE does not seem to be necessary and the DNA-binding domain of ER was considered to have some important function other than facilitating the binding to ERE. It should be noted that this observation was consistent with the previous observation that IGF-I gene activation by estrogen did not need *de novo* protein synthesis.

3.6. Anti-ER antibody inhibits gel-shift complex formation

Specific antibodies against c-Fos, c-Jun and ERT were used in order to identify the proteins constituting the gel-shift complex induced by estrogen. As shown in Fig. 2, both gel-shift complexes induced by estradiol and by TPA revealed the same mobility in the gel (lanes 2,7), and both were eliminated by antibodies against c-Fos and c-Jun (lanes 3,4,8,9). Thus it was shown that the DNA-binding proteins constituting the gel-shift complexes induced by estrogen and by TPA are identical: both complexes contain c-Fos and c-Jun and lack the estrogen-ER complex. In spite of this, the gel-shift complex induced by estrogen, but not that induced by TPA, was eliminated when a specific antibody against ER had been added to the binding reaction (Fig. 2; lanes 5 and 10). This indicates that an intact estrogen-ER complex must be present in the binding reactions when the Fos-Jun and the AP-1 motif form the complex in an estrogen-dependent manner.

Taken together, these results suggested that a direct or indirect post-translational modulation of c-Fos, c-Jun by estrogen-ER complex, which seems to require the DNA-binding domain of ER, plays a major role in the chicken IGF-I promoter activation (15).

3.7. Generalized aspects of estrogen responsive pathways through AP-1 site

It has been known for years that the anti-estrogen drugs such as Tamoxifen has paradoxical estrogen-like effects (17). The estrogen-like effects of tamoxifen include an increase in cervical hyperplasia, an increased risk of

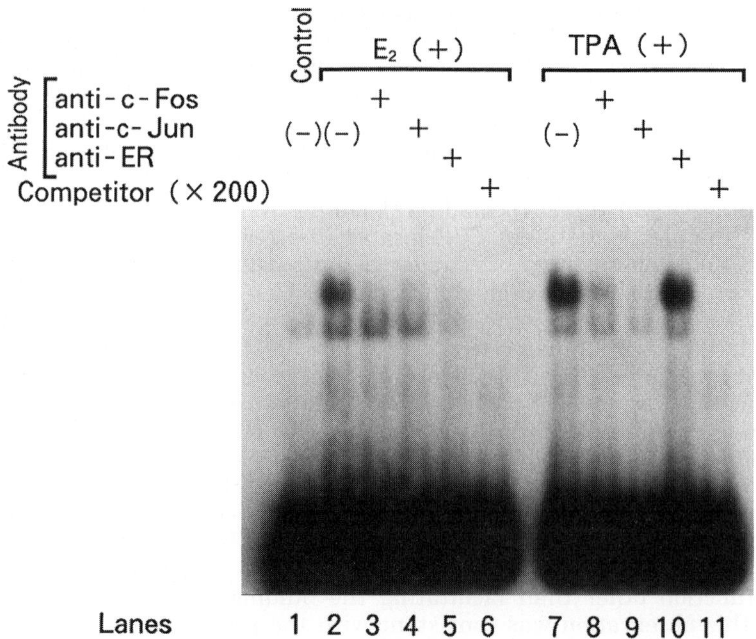

Figure 2. The estrogen-induced gel-shift complex at AP-1 motif contains c-Fos and c-Jun but not ER.

A gel-shift assay was performed using whole cell extracts isolated from ER-expressing HepG2 cells which were untreated (lanes 1) or pretreated with 10^{-6} M 17β-estradiol for 2 h (lanes 2-6) or 10^{-7} M TPA for 2 h (lanes 7-11). A specific antibody against c-Fos (c-Fos(Ab-1)) (lanes 3,8), c-Jun (c-Jun/AP-1(Ab-2)) (lanes 4,9), or ER (lanes 5,10) was added, when required, to the binding mixtures.

uterine cancer, and unpredictable stimulation of breast tumor growth during tumor progression (18). The reason Tamoxifen has both estrogen-like and anti-estrogen effects is suggested to be relevant to the fact that there are two transcription activation domains in ER; AF-1 and AF-2. AF-1 located in the receptor amino-terminal domain is hormone dependent, whereas AF-2 in the C-terminal is hormone independent (19). It was considered that tamoxifen exerts its anti-estrogen effect by repressing the AF-2 activity. Tamoxifen allows ER binding to the ERE as does estrogen, but inhibits gene expression because it does not allow AF-2 to function. It was therefore assumed that tamoxifen may have agonistic activity in those cells in which the AF-1 becomes active, and this expectation seemed to be fulfilled when tested with classical EREs. However, in various cells such as endometrial cells or cervical cells for which the estrogen-like effect of tamoxifen often causes troubles, the agonistic effect of tamoxifen sometimes exceeds that of estrogen (20). These observations are hard to reconcile with the tamoxifen action exerted through classical EREs.

Recently, Webb et al. (16) reported that estrogen-like effect of tamoxifen is mediated by AP-1 site of target gene and the mechanism is the same with estrogen effect via AP-1 site. They transfected various mutated ERs and reporter genes containing AP-1 site or ERE and evaluated the estrogen and tamoxifen responsiveness of the promoter. Then they found that AP-1 site plays a major role in the gene activation by tamoxifen. They also found that there are two different pathways in tamoxifen activation of the gene via AP-1: a pathway which requires the DNA-binding domain of ER and another which does not. In HeLa cells or Ishikawa cells, DNA-binding domain dependent pathway seems to be dominant, while in many of breast cancer cells DNA-binding domain independent pathway is dominant.

Despite the fact that ER does not bind DNA directly, DNA-binding domain is required in the this pathway as we showed in the chicken IGF-I gene regulation by estrogen. Naturally, it is expected that DNA-binding domain has some important function other than facilitation of the ER binding to DNA. In nuclear receptor family members including glucocorticoid receptor (GR), DNA-binding domain is suggested as being involved in interaction with other proteins including Fos and Jun (21). Considering the structural similarities in the DNA-binding domains of these steroid receptors including the zinc finger motif, it may be reasonable that the DNA-binding domain of ER also preserves some function in facilitating some protein-protein interaction. Those interactions may activate Fos, Jun activity directly, or activate the cascades which finally activates those protein activity .

Webb et al. reported that protein-protein interaction is also involved in estrogen-induced transcriptional activation of target genes via the DNA-binding domain independent pathway (16). In contrast to the DNA-binding domain dependent pathway in which transactivating activity of Fos and Jun activates AP-1 enhancer, it may be ER itself that activates the AP-1 enhancer in this DNA-binding domain independent pathway. It has been assumed that transactivation function of ER and that of Jun (and Fos) act on AP-1 site synergistically in this pathway (16).

4. Conclusion

As an initial approach toward understanding the molecular basis of the IGF-I gene regulation, the PKC regulation was chosen as a target of the initial study. Our present data revealed the existence of a unique AP-1 motif in the chicken IGF-I gene promoter, which seems to be involved also in the estrogen regulation. Despite the fact that ER does not need to directly bind to the IGF-I gene regulatory sequences, the pathway requires the intact DNA-binding domain of the ER, suggesting that the domain has some important roles in addition to that originally named for. Although the human IGF-I gene is also known to be a target for the estrogen regulation and the impariment of the regulating machinery may be relevant to some human diseases such as osteoporosis, the mechanism underlying the estrogen-responsiveness is yet to be determined. Because the human gene also lacks an ERE as does the chicken counterpart, we may able to find an AP-1 motif somewhere in the human IGF-I gene.

REFERENCES

1. W.H. Daughaday, and P. Rotwein, Endocr. Rev., 10 (1989) 68-90.
2. P. Rotwein P, Growth Factors, 5 (1991) 3-18.
3. Y. Kajimoto and P. Rotwein, J. Biol. Chem. 266 (1991) 9724-9731.
4. I. Nagaoka, B.C. Trapnell and R.G. Crystal, J. Clin. Invest. 85 (1990) 448-455.
5. S.A. Rogers, M.R. Hammerman, Proc. Natl. Acad. Sci. USA 85 (1989) 8998-9002.
6. A. Doglio, C. Dani, P. Grimaldi and G. Ailhaud, Proc. Natl. Acad. Sci. USA 86 (1989) 1148-1152.
7. S.T. Smale and D. Baltimore, Cell 57 (1989) 103-113.
8. Y. Kajimoto, R. Kawamori, Y. Umayahara, N. Iwama, E. Imano, T. Morishima, Y. Yamasaki and T. Kamada, Biochem Biophys Res Commun 190 (1993) 767-773.
9. R.M. Evans, Science 240 (1988) 889-895.
10. L.J. Murphy and J. Luo, Mol. Cell. Endocrinol. 64 (1989) 81-86.
11. M. Ernst and G.A. Rodan, Mol. Endocrinol. 5 (1991) 1081-1089.
12. L.J. Murphy and A. Ghahary, Endocr. Rev. 11 (1990) 443-453.
13. S.W. Kim, R. Lajara and P. Rotwein, Mol. Endocrinol. 5 (1991) 1964-1972.
14. L.J. Hall, Y. Kajimoto, D. Bichell, S.W. Kim, P.L. James, D. Counts, L.J. Nixon, G. Tobin and P. Rotwein, DNA Cell Biol. 11 (1992) 301-313.
15. Y. Umayahara, R. Kawamori, H. Watada, H. Imano, N. Iwama, T. Morishima, Y. Yamasaki, Y. Kajimoto and T. Kamada, J. Biol. Chem. 269 (1994) 16433-16442.
16. P. Webb, G.N. Lopez, R.M. Uht, P.J. Kushner, Mol. Endocrinol. 9 (1995) 443-456.
17. R.P. Kedar, T.H. Bourne, T.J. Powles, W.P. Collins, S.E. Ashley, D.O. Cosgrove and S. Campbell, Lancet 343 (1994) 1318-1321.
18. M. Morrow and V.C. Jordan, Arch. Surg. 128 (1993) 1187-1191.
19. N.J. Webster, S. Green, J.R. Jin and P. Chambon, Cell 54 (1988) 199-207.
20. J.L. Albert, S.A. Sundstrom and C.R. Lyttle, Cancer Res. 50 (1990) 3306-3310.
21. M. Phal, Endocr. Rev. 14 (1993) 651-658.

Molecular Mechanisms to Regulate the
Activities of Insulin-like Growth Factors
K. Takano, N. Hizuka and S-I. Takahashi (Editors)

INSULIN-LIKE GROWTH FACTOR BINDING PROTEIN-1 GENE ACTIVATION IN HUMAN ENDOMETRIUM

L Tseng*, J Gao and HH Zhu

Department of Obstetrics/Gynecology and Reproductive Medicine, State University of New York at Stony Brook, Stony Brook, New York, 11794-8091

ABSTRACT

Synthesis and secretion of human insulin-like growth factor binding protein-1 (IGFBP-1) are upregulated during decidualization of human endometrium. We have studied the tissue specific regulation of the IGFBP-1 gene in human endometrial stromal cells treated with progestin and relaxin (RLX). During the progressive decidualiztion, numerous genes are either transiently or constitutively activated. Among them, induction of the IGFBP-1 gene is the most active one, and it is constitutively expressed in decidual cells in the hormonal milieu of pregnancy. The secretion rate and steady-state levels of IGFBP-1 mRNA increase by 10^3- to 10^4-fold in *in vitro* decidualized stromal cells. Functional analysis of the IGFBP-1 gene promoter showed that the activity increased exponentially to 10^4-folds in cells treated with medroxy progesterone acetate (MPA) and RLX over a 13-day culture period. Two regions in the IGFBP-1 gene promoter are responsible for this extraordinary induction of the IGFBP-1 gene. The proximal promoter region between -1 to -300 bp, contains multiple functional elements, CCAAT, progesterone/glucocorticoid response element (PRE/GRE)s and cAMP binding protein response element (CRE). The distal promoter region, between -2.6 to -3.4 kb, mediates 95% of the total promoter activity in decidualized stromal cells. Functional and binding analysis in the distal promoter region showed that multiple Sp1 elements interacting with a novel Sp3 transcription factor activate the IGFBP-1 gene promoter in human endometrium.

INTRODUCTION

IGFBP-1 in the fetal-maternal compartment

Human IGFBP-1 is primarily synthesized in the liver and gestational endometrium. After conception, decidual cells secrete IGFBP-1 which causes a rise of amniotic fluid IGFBP-1 from <1 to 150 μg/ml in early through mid gestation (1). The function of this extraordinarily high level of IGFBP-1 in the fetal-maternal interface is currently under vigorous investigation. It has been implicated with convincing evidence that IGFBP-1 regulates the attachment of trophoblastic cells to the maternal endometrium and that it prevents excessive invasions of the placenta (2-4). We have found that IGFBP-1 also inhibits the IGF-I or progestin-stimulated DNA synthesis in endometrial stromal cells dose-dependently. 50% inhibition occurs at equi-molar concentration of IGF-I and IGFBP-1. Phosphorylated IGFBP-1 is more potent than the non-phosphorylated form (5). These observations indicate that IGFBP-1 regulates IGF-I

---

*Corresponding author. This work was supported by grant from NIH HD 19247

function via autocrine and paracrine mechanisms in the maternal-fetal unit. In addition, IGFBP-1 alone promotes cell migration through the interaction of its RGD sequence with the cell surface receptor, $\alpha5\beta1$ integrin (6), strongly suggesting that IGFBP-1 plays an important role in embryo implantation and placentation.

Regulation of amniotic fluid IGFBP-1 levels

In pregnant maternal serum, significant variation of IGFBP-1 levels has been observed. In general, IGFBP-1 levels are elevated up to 200 ng/ml during the early pregnancy and maintained at a high level throughout the rest of gestation (1,7). Amniotic fluid levels, which peak at mid gestation, are 100- to 1000- fold higher than that of maternal serum (1). Since the endometrial stromal/decidual cells are the major source of IGFBP-1 in amniotic fluid, the production rate of this protein in endometrium is the major controlling factor of its level in the amniotic fluid. It is generally accepted that progesterone induces the endometrial cell production of IGFBP-1 during early pregnancy. Indeed, in non-pregnant women, it has been shown that progestin-containing intrauterine devices induce IGFBP-1 secretion in human endometrium (8). In addition, locally produced factors also regulate IGFBP-1 gene activation. After conception, progesterone initiates the induction of numerous cellular and secretory factors, which cause the transformation of endometrial stromal cells into decidual cells. Thus, IGFBP-1 gene activation is mediated by multiple endometrial factors, and is differentiation-dependent. The antiprogestins, mifepristone and onapristone prevent the differentiation of endometrium and disrupt implantation and pregnancy. They also drastically alter the IGFBP-1 gene activation (9-14).

Progressive decidualization of human endometrial stromal cells

To evaluate endometrial stromal/decidual IGFBP-1 gene activation, it is essential to assess the process of decidualization. Decidualization occurs after conception and extends through gestation. We have developed a culture system to decidualize the stromal cells in the absence of embryo and endometrial glandular epithelial cells, as have other investigators. The long-term exposure of endometrial stromal cells to progestin and RLX induces the progressive decidualiztion characterized by the induction of multiple genes, cell growth and differentiation. We have observed that progestin exerts a strong stimulation on DNA synthesis and cell growth in 2-5 days' culture and a stimulation that is less active in 5-10 days' culture. Progestin inhibits DNA synthesis after 15 days' incubation (4). Thus, we view this culture system as a growth phase (0 to 2 to 10 days' progestin-exposure) and a differentiation phase (beyond 10 days' progestin-exposure). Stromal cells are morphologically transformed into predecidual cells and then into decidual cells (13). Transmission and scanning electron microscopy showed that *in vitro* decidualized stromal cells exhibit a morphology almost identical to that of decidual cells at 10 weeks' gestation (15). Thus, the stromal/decidual cell culture provides a system to evaluate the hormonal requirements, autocrine regulation, biochemical changes and gene activation during progressive decidualization. We have previously shown that numerous genes, aromatase, fibronectin and IGF-I are transiently activated, and that prolactin (PRL), PRL-receptor, progesterone receptor A and B, IGF-II, and IGFBP-1 are persistently activated during decidualization (16-26). Among them, the most active genes are PRL and IGFBP-1 which have often been used as markers of decidual cells. In fact, inductions of these two genes resemble each other though the mechanisms are fundamentally different (14, 16). We believe that the endometrial factors and embryonic factors have great impact on the IGFBP-1 gene regulation in endometrium.

INDUCTION OF IGFBP-1 GENE IN ENDOMETRIAL CELLS

Endometrial stromal/decidual cell IGFBP-1 secretion and mRNA level

In vivo, the amniotic fluid IGFBP-1 level increases from <1 to 150 μg/ml from conception through mid-gestation (1). Such a dramatic increase reflects a rapid growth and differentiation of stromal/decidual cells. It also suggests the induction of IGFBP-1 in decidual cells. Induction of IGFBP-1 has been clearly demonstrated in endometrial stromal cell culture (14,18,21-25). The production pattern of IGFBP-1 shows a two-stage induction (Table 1). In the growth phase, the secretion rate of IGFBP-1 is moderate. After sequential administration of MPA and RLX for a month, daily production of IGFBP-1 increased to >100 μg /10^6 cells (18). The phosphorylated forms are the predominant IGFBP-1 isoforms (26) similar to that found in HepG2 cell culture (27).

Table 1. Daily production of IGFBP-1 of cultured endometrial stromal/decidual cells.

Culture condition (differentiation state)	Days in culture	IGFBP-1 production rate (μg/10^6 cells/day)
control (endometrial stroma cells)	1-10	n.d. to 0.01
MPA or MPA+RLX (stromal/predecidual cells)	1-5	0.01-0.1
	5-10	0.1-1.0
MPA+RLX or MPA/RLX (decidual cells)	>10	10-150

n.d.: not detectable, MPA: 0.1 μM, RLX: 20 ng/ml. Data were adapted from Bell et al., 1992, ref. (18).

Analysis of the steady-state IGFBP-1 mRNA levels showed that the increase of mRNA levels correlated with the IGFBP-1 secretion rate (14,18). Stromal cells treated with progestin for 20 to 30 days caused an exponential increase of the IGFBP-1 mRNA (~1000-fold). Relaxin, which is secreted from the corpus luteum in gestation and is locally produced in decidual and placental cells, enhanced the induction of IGFBP-1 mRNA levels. Maximal mRNA levels reached ~8 pg IGFBP-1 mRNA/ug total RNA, ~ 20-25% of the total mRNA in cells treated with MPA and RLX (Fig.1). The antiprogestin, RU 486, caused a transient superinduction of the mRNA level and the transcription rate (14). The constitutive expression, however, depends on the presence of progesterone in the culture medium since prolonged incubation with RU 486 eventually diminished the mRNA to the basal level (Fig. 2). Insulin and IGF-I, which are structurally related to RLX, reduced the IGFBP-1 secretion and mRNA level in progestin-treated cells (29).

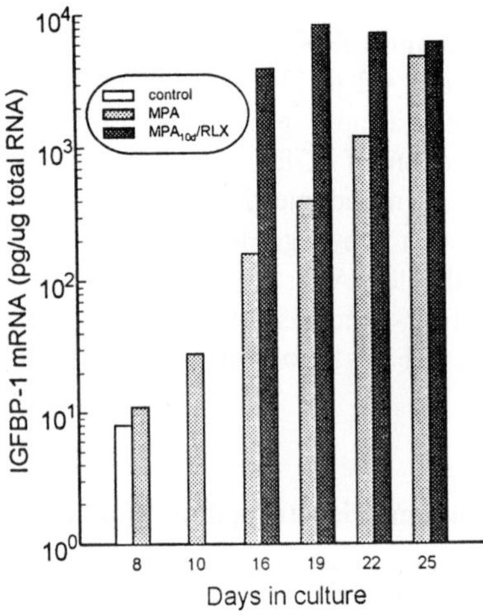

Fig. 1. *IGFBP-1 mRNA levels in human endometrial stromal cells.* Stromal cells were incubated with or without the hormone. RNA was subjected to Northern blot analysis. The amount of mRNA was estimated by comparing the radioactivities in each hybrids to a standard curve made of truncated sense IGFBP-1 mRNA hybridized in parallel. Data represent mean values of two separate hybridizations. Adapted from Tseng et al., Biol Reprod 1992, ref. (14).

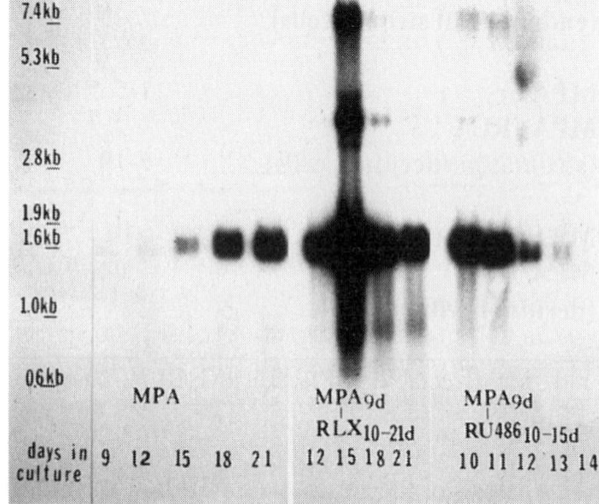

Fig.2. *Northern blot analysis of IGFBP-1 mRNA in human endometrial stromal cells* Stromal cells were incubated with MPA, MPA/RLX or MPA/RU 486 in sequence. Total RNA was isolated from each individual sample at various incubation times and subjected to Northern blot analysis. IGFBP-1 mRNA locates at 1.6 kb based on the standard RNA ladders, 1 kb to 7.4 kb, which were run in parallel. Reprinted from Tseng et al., Biol Reprod 1992, ref. (14).

IGFBP-1 gene promoter activity in endometrial stromal cells

Analysis of the function and binding characteristics of the IGFBP-1 gene promoter provides information on decidual cell-specific gene activation. Recently, we have analyzed the activity of the IGFBP-1 gene promoter in an endometrial/decidual cell system (21-25). A time study over a 13-day culture period showed that the promoter activity increased exponentially to $>10^4$ folds in cells treated with MPA and RLX (Fig. 3).

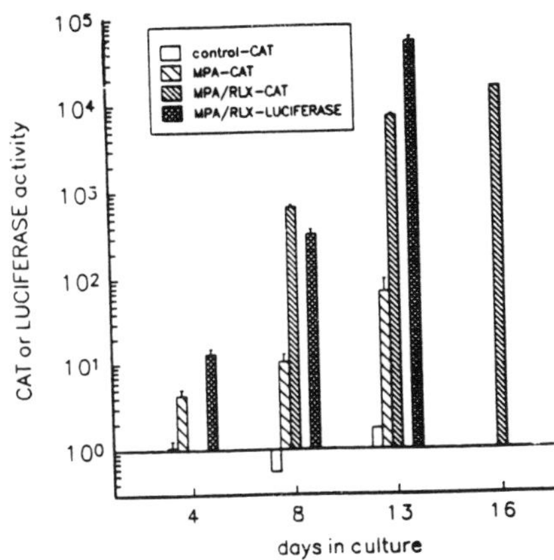

Fig. 3. *Time study on CAT and luciferase activity in cells transfected with IGFBP-1 promoter p3.6CAT and p3.6Luc.* Endometrial stromal cells were transfected with p3.6CAT or p3.6Luc and continuously cultured for 13 or 16 days with no hormone (control), 0.1 μM MPA or MPA and 20 ng/ml RLX. CAT activities (% of conversion) were normalized by luciferase activity from pSRV-Luc (mean ±s.d., n=3). Reprinted from Gao et al., Mol Cell Endocrinol 1994, ref. (21).

Deletion analysis showed that two regions (-1 to 1.2 kb and 2.6 to 3.4 kb) in the IGFBP-1 promoter are responsible for the activation of the IGFBP-1 gene (Fig. 4). The major difference between IGFBP-1 gene activation in the gestational endometrium and the hepatic system lies in the regulatory sequence in the distal promoter region, between -2.6 to -3.4 kb, which mediates >95% of the total promoter activity derived from -3.3 kb to +68bp (24). When the distal promoter region was activated in fully differentiated decidual cells, the driving force derived from the proximal promoter region becomes insignificant.

IGFBP-1 gene activation in hepatic and endometrial cells shares the same transcription initiation sites (14). The promoter of the IGFBP-1 gene has been extensively studied in the hepatic system by Powell et al., and other investigators (29-35). To compare the induction pattern of the IGFBP-1 gene in the hepatic and decidual cell systems, a schematic representation of the cis-elements and trans-regulatory proteins identified in the HepG2 and stromal/decidual cells is presented in Fig. 5. The proximal promoter region between -1 to -300 bp (Fig. 5) contains multiple sections of regulatory sequence, including cis-elements CCAAT (-72 bp), multiple sites of progesterone/glucocorticoid response element (PRE/GRE) homologous, insulin response element (IRE) and CRE. In HepG2 cells, HINF-1 is the transactivator that interacts with CCAAT (30). However, our preliminary observation showed that HINF-1 is not present in the stromal/decidual cell system. Instead, multiple factors in stromal/decidual nuclear extracts bound to CCAAT. Thus, different sets of transactivators mediate the CCAAT activities in these two types of tissues.

Fig. 4. *Activity of promoter fragments and mutant of the IGFBP-1 gene.* CAT activities of 14 promoter/CAT constructs (2 pmol/transfection) were transfected into endometrial stromal cells. Cells were cultured with MPA and RLX for 12 days. Normalized CAT activity was measured at the end of culture (mean ± s.d., n=3). The upper panel shows the relative activities derived from the longer fragments of the promoter (-1.2 to 3.4 kb), mutant of p1.2 (CCAAT mutated to TTAAT) and a construct of 3.4 kb deleted the fragment between (-2600-2732 bp). The lower panel shows the activities derived from the proximal promoter fragments (+68 bp to -1.2 kb)

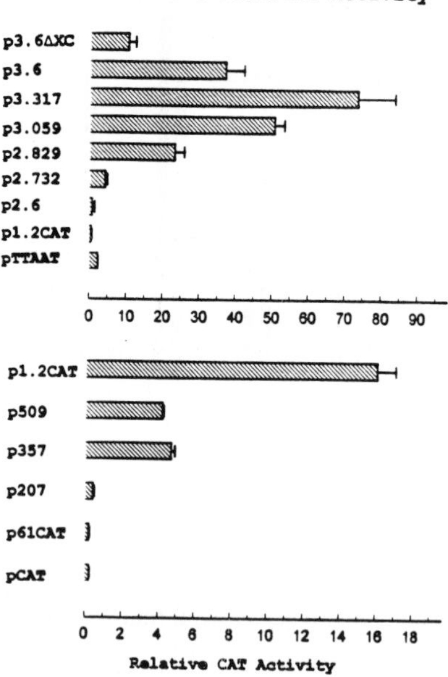

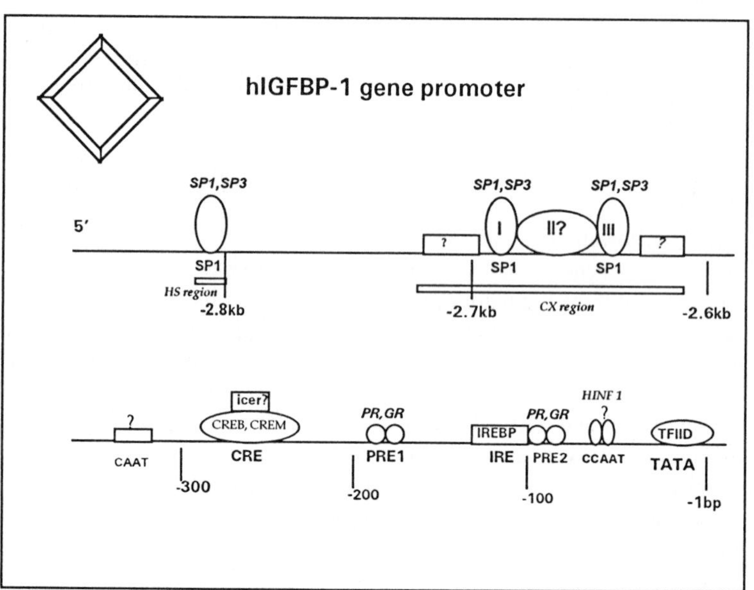

Fig. 5. *Organization of the IGFBP-1 gene promoter.* Schematic representation of elements identified in the first 300 bp of the IGFBP-1 proximal promoter and distal promoter between -3.4 to -2.6 kb and transcription factors which they bind in endometrial stromal cells or HepG2 cells. PRE: progesterone response element. IRE and IREBP: insulin response element and its binding protein. CRE, CREB, CREM, ICER: cAMP response element and its binding proteins. Sp1: Sp1 family protein response element.

The PRE/GRE 1 and 2 sites were active in dexamethasone treated-HepG2 cells co-transfected with a glucocorticoid receptor construct (33). The positive effect of progestin on IGFBP-1 gene transcription in endometrial cells indicates that the endogenous progesterone receptor (PR) (20) exerts a direct stimulation on the IGFBP-1 gene promoter. We found that PRE 1 and 2 sites mediate the progestin-activated promoter activity (Gao et al., unpublished observation). However, as shown in the time study (Fig. 4), progestin induces a moderate increase 4- to 40-fold. It should be noted that PR also exerts a suppressive effect derived from the 59 bp (-2630 to -2688 bp) mediated through protein-protein interaction (25). IRE is located in the proximal promoter region (33,34). Insulin inhibits the IGFBP-1 gene expression in both hepatic and endometrial systems (28,32). CRE mediates a weak activation in HepG2 cells (29). In stromal/decidual cells, the CRE site appears to be one of the critical regions that mediate the IGFBP-1 transcription by RLX activated factors whereas RLX is inactive in hepatic system.

The dramatic increase of the IGFBP-1 transcription mediated by the distal promoter region appears to be actviated through an induction of progestin-associated decidual nuclear factor(s) which activates the SP1 sites in the distal promoter region (24). Preliminary results showed that the activation of the distal promoter region depends on the proximal region, -300 to +68 bp. It appears that decidual nuclear factors play important role in the synergistic effect between the distal and proximal promoters.

Within the distal promoter region, the CX region (Fig.5), 132 bp located between -2.732 to -2.6 kb, controls the activity derived from -3.3 to -2.7 kb since deletion of CX diminished 80% of the total activity (Fig.4). In this region, three *cis*-elements, I, II and III (Fig.5) were active in HEC-1B and HepG2 cells (24). However, only *cis*-element II (-2660 to -2638 bp) was active in endometrial stromal cells. Promoter activity derived from pCXCAT including the *cis*-elements I, II and III was repressed in HEC-1B, HepG2 cells and unstimulated stromal cells. During decidualization, the repression was gradually diminished which was mediated through *cis*-elements I and III (24).

To search for the decidual cell-specific nuclear factor capable of derepressing the IGFBP-1 gene, the electrophoretic mobility shift assay (EMSA) was carried out. Two specific binding complexes formed by the nuclear extracts of stromal cells to *cis*-element I and III. An additional complex, C1, was formed only by the decidual nuclear extracts and not by the extracts of unstimulated stromal cells, HEC-1B or HepG2 cells (Fig. 6).The C1 complex was also formed by 30 bp (-2830 to -2800 bp) in the HS fragment (Fig.5) which contains Sp1 binding motif (24). HS fragment is also transcriptionally active. Sp3 antibody abolished the C1 complex (24) suggesting that Sp3 family protein activates IGFBP-1 gene. Western blot showed a novel decidual cell protein, 21-23 kDa Sp-like protein, which may be the major transregulatory protein that activates the IGFBP-1 gene. These observations showed that the interaction between multi-Sp1 sites and the Sp3 family protein mediates the major transcriptional activation of IGFBP-1 gene in decidual cells.

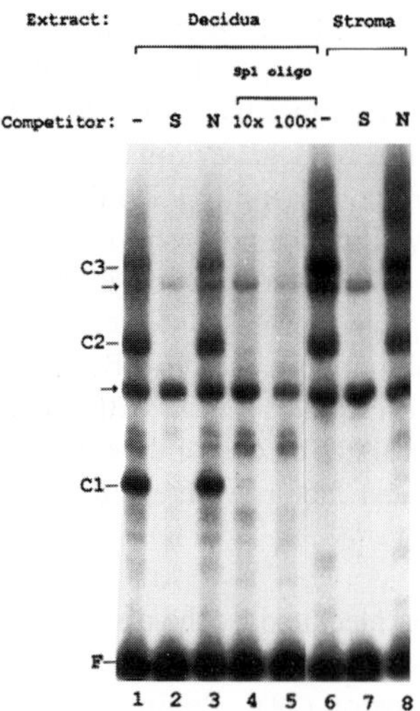

Fig. 6. *EMSA with nuclear extracts of unstimulated endometrial stromal cells and Decidual Tissue.* EMSA was carried out as described in reference (25). The binding reaction mixtures contained 6 μg nuclear extracts of decidual tissue (lanes 1-5) or unstimulated stromal cells (lanes 6-8) and the $1x10^4$ cpm ^{32}P-labeled BR probe. C1, C2 and C3: major specific DNA-protein complexes. Arrows indicate nonspecific DNA-protein complexes. F: free probes. Specific (S) and nonspecific (N) competing DNAs were the unlabeled probe and the 220-bp *Nde*I/*Hpa*I fragment (100x). Ten and 100- folds of Sp1 oligonucleotide were also used as competing DNA (lanes 4 and 5) respectively.

In summary, the nuclear proteins in hepatic and endometrial cells specifically determine the magnitude and pattern of human IGFBP-1 gene transcription. Further characterization of endometrial stromal/decidual nuclear proteins is essential to better understand tissue-specific IGFBP-1 gene expression.

ACKNOWLEDGMENTS

We thank the clinical staff of the Department of Obstetrics/Gynecology and Reproductive Medicine and the Department of Pathology at the State University of New York at Stony Brook, and physicians and pathologists at St. Charles Hospital (Port Jefferson, NY) for providing us with viable endometrial specimens and histological diagnosis of the human endometrial specimens.

REFERENCES

1 Bell SC, Hales MW, Patel SR, Kirwan PH, Drize JO, Milford-Ward A. BR J Obstet.. Gynecol 93;909-15,1986
2. Giudice L. IGFBP-1 and maternal defense in implantation. 3rd International conference of the uterus, endometrium and myometrium. NY, NY.1996.
3. Ritvos O, Tapio R, Jalkanen J, Suikkari A-M, Voutilainen R, Bohn H, Rutanen EM. Endocrinology 122:2150-2157, 1988.
4. Frost RA, Mazella J, Tseng L. Biol Reprod 49: 104-111 1993.
5. Irving JA, Lala PK Exp Cell Res 217: 419-427,1995.
6. Jones JI. Gockerman A. Busby WH Jr. Wright G. Clemmons DR Proc Nat Acad Sci USA 90:10553-10557, 1995
7. Bohn H. Kraus W.Archives of Gynecology. 229:279-91, 1980.

8. Pekonen F. Nyman T. Lahteenmaki P. Haukkamaa M. Rutanen EM. J Clin Endocrinol Meta 75:660-664, 1992.
9. Cameron ST. Critchley HO. Thong KJ. Buckley CH. Williams AR. Baird DT. Human Reproduction. 11(11):2518-26, 1996
10. Cameron ST. Critchley HO. Buckley CH. Chard T. Kelly RW. Baird DT. Human Reproduction. 11(1):40-9, 1996
11. Dao B. Vanage G. Marshall A. Bardin CW. Koide SS. Contraception. 54(4):253-8, 1996
12. Bygdeman M. Swahn ML. Gemzell-Danielsson K. Gottlieb C. 1:121-5, 1994
13. Lane B, Oxberry W, Mazella J, Tseng L Human Reproduction. 9: 259-266, 1994
14. Tseng L, Gao JG, Chen R, Zhu HH, Mazella J, Powell DR Biol Reprod 47:441-450, 1992.
15. Wynn R, Am J Obstet. Gynecol 118:652-660, 1974.
16. Huang JR, Tsang L, Bischof P. Janne OA Endocrinology, 121:2011-2077, 1987.
17. Tseng, L Endocrinology, 115: 833-83, 1984.
18. Bell SC, Jackson JA, Ashmore J, Zhu HH, Tseng L, J Clin Endocrinol Metab. 72:1014-1024,1991.
19. Zhu HH, Huang JR, Mazella J, and Tseng L. Human Reproduction 7:141-146, 1992.
20. Tseng L, Zhu HH, Biology Reprod (in press).
21. Gao J, Mazella J, Tseng L Mol Cell Endocrinol 104:39-46,1994.
22. Gao J, Mazella J, Powell D, Tseng L, DNA Cell Biol 13: 829-837,1994.
23. Gao J, Mazella J, Tseng L Mol Endocrinol 9: 1405-1412 1995.
24. Gao J, Tseng L, Mol Endocrinology 10:613-621,1996
25. Gao J, Tseng L, Mol Endocrinology 11:973-979,1997
26. Frost RA, Tseng L. 72nd Annual Meeting of Endocrine Society Meeting, Abstract # 19. Atlanta Georgia, 1990 and J Bio Chem 1991; 266:18082-18088
27. Jones I, D'Ercole AS, Camacho Hubner C, Clemmons DR. Proc Natl Acad Sci ; 88:7481-7485,1991.
28. Tseng L, Frost R, Chen R, Zhu HH, Mazella J, Lane B, and Bischof, Role of insulin family growth factors, relaxin, IGFs and insulin on the development of endometrial cells in Current "Concept in Fertility Regulation and Reproduction" eds: Cantor, Van Look, Wiley-Eastern pp379-394, 1994
29. Powell DR. Lee PD. Suwanichkul A, Advances in Experimental Medicine & Biology. 343:205-14, 1993.
30. Suwanichkul A, Cubbage ML, Powell DR, J Biol Chem 265:21185-21193,1990.
31. Suwanichkul A, DePaolis LA, Lee PD, Powell DR, J Biol Chem 268:9730-9736,1993
32. Suwanichkul A. Allander SV. Morris SL. Powell DR, J Biol Chem. 269(49):30835-41, 1994.
33. Powell DR. Allander SV. Scheinmann AO. Wasserman RM. Durham SK. Suwanichkul A. Progress in Growth Factor Research.6:93-101, 1995.
34. Goswami R, Lacson R, Yang E, Sam R, Unterman T, Endocrinology 134:736-743,1994.
35. Babajko S, Tronche F, Groyer A, Proc Natl Acad Sci USA 90:272-276,1993.

Molecular Mechanisms to Regulate the
Activities of Insulin-like Growth Factors
K. Takano, N. Hizuka and S-I. Takahashi (Editors)
1998 Elsevier Science B.V.

Regulation of ALS gene expression by growth hormone

Y. Boisclair[a], S. Bassal[b], M.M. Rechler[b], and G.T. Ooi[b*]

[a]Department of Animal Science, Cornell University, Ithaca, NY 14853, USA

[b]Growth and Development Section, Molecular and Cellular Endocrinology Branch, National Institute of Diabetes and Digestive and Kidney Diseases, National Institutes of Health, Bethesda, MD 20892, USA

1. INTRODUCTION

In plasma, the majority of the insulin-like growth factors (IGFs) is present as a ~150 kDa ternary complex. This complex is comprised of IGF-binding protein-3 (IGFBP-3), the predominant IGFBP in plasma, and an acid-labile subunit (ALS) that does not directly bind IGFs (1). Plasma also contains lower molecular mass complexes of ~50 kDa which are made up of several IGFBP species that are incompletely saturated with IGFs leaving virtually no free IGF in the circulation (2).

Unlike free IGFs and IGFs bound in the ~50 kDa complexes which can cross the capillary endothelium, the 150 kDa complex is confined to the circulation. The ability of ALS to recruit IGFs to the 150 kDa complex therefore helps to prolong the half-lives of the IGFs in the circulation, and allows IGFs to be stored in plasma at high concentration to facilitate their endocrine actions and minimize local effects such as activation of the insulin receptor with possible resulting hypoglycemia (3,4). ALS is thus an important determinant of the endocrine actions of IGFs on target cells. As IGFBP-3 is widely distributed, whereas ALS is limited to the plasma, ALS appears to be the crucial component that determines the formation of the 150 kDa complex.

ALS is synthesized exclusively in the parenchymal cells of liver (5,6), and like IGF-I and IGFBP-3, is stimulated by growth hormone (GH) (7-10). GH stimulates ALS mRNA in liver and ALS protein levels in the circulation (5-11). This regulation occurs at the level of gene transcription in rat liver. The transcriptional effects of GH are direct as GH increases the abundance of ALS mRNA and the secretion of ALS protein in isolated primary rat hepatocytes (11,12). This article describes recent studies in our laboratories on the molecular mechanism underlying the transcriptional regulation of the ALS gene by GH. Using the H4-II-E rat hepatoma cell model system, we show that GH induction of ALS gene transcription is mediated by the binding of members of the Signal Transducers and Activators of Transcription (STAT) family to a single cis-regulatory element resembling a γ-interferon activated sequence (GAS) in the ALS promoter.

* To whom correspondence should be addressed: Bldg. 10 Rm 8D14, 10 Center Drive MSC 1758, Bethesda, MD 20892-1758, USA. email: guckooi@helix.nih.gov

2. A SINGLE GAS ELEMENT IN THE ALS PROMOTER MEDIATES GROWTH HORMONE STIMULATION OF THE ALS GENE IN H4-II-E RAT HEPATOMA CELLS

2.1. 5'-deletion analysis and site-mutagenesis

We previously reported that ligating a 1953 bp fragment of the mouse ALS 5'-flanking region (nt -2001 to -49 with respect to A_{+1}TG) to a promoterless firefly luciferase reporter gene enables it to be regulated by GH in a dose-dependent manner when the construct was transiently transfected into H4-II-E rat hepatoma cells (12,13). We have now mapped the region of the promoter that confers GH-responsiveness to the ALS gene using a series of 5'-deletion mutants. This is schematically shown in Fig. 1. The luciferase activity of the construct ending at nt -2001 was stimulated 2.7 ± 0.5 fold (mean ± se) by GH. Deletion of the region from nt -2001 to nt -653 was without effect. Stimulation by GH was lost, however, in deletion constructs terminating at nt -483, -323 or -245, suggesting that the 170 bp region between nt -653 and nt -483 contains the regulatory sequences which are essential for GH-stimulation of the ALS promoter. This 170 bp region contains two sites that resemble the GAS consensus sequence, TTNCNNNAA (14), which has been shown to mediate the effects of various cytokines including GH on the transcription of other genes (15). The first site, TTCCTAGAA (ALS-GAS1), is located between nt -633 and nt -625; the second site, TTAGACAAA (ALS-GAS2) is located between nt -553 and nt -545. In order to ascertain the functional importance of each of these two ALS-GAS sites, they were mutated independently in the context of the luciferase constructs containing the nt -703 to nt -49 promoter fragment which retain full responsiveness to GH (2.4 ± 0.2 fold stimulation) (Fig. 2). Block mutation of ALS-GAS2, in which the 9 bp native sequence between nt -553 and -545 was replaced by an EcoR I linker (5'-CGAATTCGC-3'), did not affect GH-responsiveness of the ALS promoter. However, similar block mutation of the ALS-GAS1 site abolished the ability of the promoter to respond to GH, indicating that ALS-GAS1 is essential for GH-stimulation.

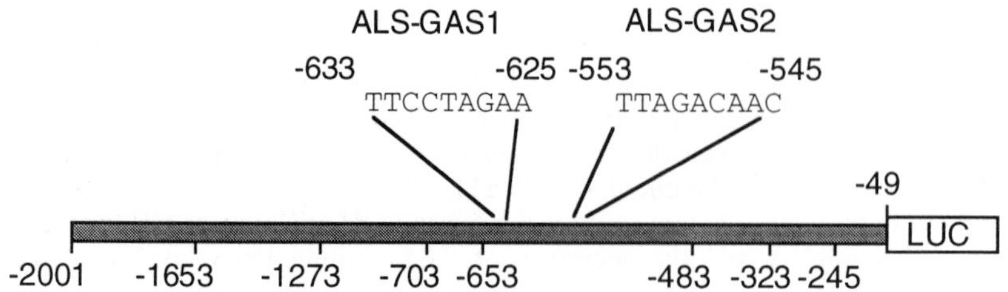

Fig. 1. Region of the mouse ALS promoter that is required for GH stimulation. Mouse ALS promoter fragments having 5' ends at nt -1653, -1273, -703, -653, -483, -323, or -245, and the same 3' end at nt -49 (relative to A_{+1}TG) were prepared by PCR amplification using Vent DNA polymerase (New England Biolabs, Inc, Beverly, MA). They were ligated upstream of a promoterless firefly luciferase gene (LUC) and transiently transfected into rat H4-II-E cell (12). Transfected cells were treated with 100 ng/ml recombinant human GH overnight, following which luciferase activity in the cell lysates was measured. An expression plasmid coding for secreted alkaline phosphatase was co-transfected to correct for differences in transfection efficiency.

Deletion of 5' end sequences up to nt -703 has no effect on GH stimulation of promoter activity, whereas deletion of the 5' end to nt -483 abolished the stimulation by GH. Sequences of the two GAS-like elements (ALS-GAS1 and ALS-GAS2) within the critical regulatory region (between nt -703 to nt -483) are shown.

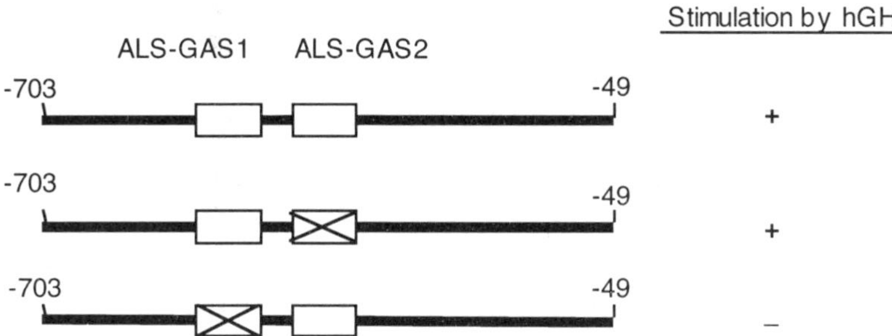

Fig. 2. Effect of site-mutagenesis of ALS-GAS1 and ALS-GAS2 in the mouse ALS promoter on stimulation of promoter activity by hGH. Block mutations overlapping the 9 bp GAS-like sequences were constructed in the nt -703 to nt -49 ALS promoter fragment. Native GAS-like elements are shown as open boxes, and mutated sites are indicated by an "X". Mutation of ALS-GAS1 abolished GH stimulation of ALS promoter activity whereas mutation of ALS-GAS2 had no effect.

2.2. ALS-GAS1 alone is sufficient to mediate GH-responsiveness to a heterologous promoter

The ALS-GAS1 element alone is sufficient to mediate the stimulation by GH. Three tandem copies of the 9 bp ALS-GAS1 element were introduced in a TK-LUC plasmid in which the luciferase reporter gene is driven by the minimal promoter of the heterologous thymidine kinase gene (16), designated TK-LUC-3GAS. In H4-II-E cells transfected with TK-LUC-3GAS, GH stimulated luciferase activity 3.6 ± 0.2 fold (Fig. 3). In contrast, GH did not significantly increase luciferase activity in cells transfected with TK-LUC plasmid alone, indicating that the 9 bp ALS-GAS1 is both necessary and sufficient to confer GH stimulation to the mouse ALS promoter.

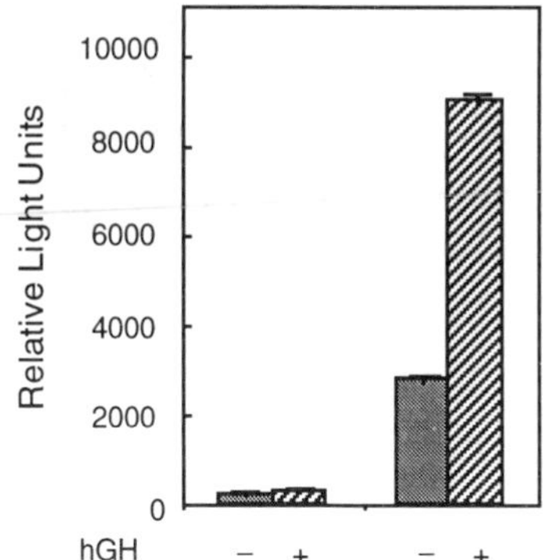

Fig. 3. ALS-GAS1 can confer GH responsiveness to a heterologous promoter. Three tandem copies of ALS-GAS1 were inserted upstream of the minimal promoter for the thymidine kinase gene (16) of TK-LUC plasmid to give TK-LUC-3GAS. These constructs were transfected into H4-II-E cells, and the transfected cells incubated with 100 ng/ml hGH overnight. Luciferase activity was measured in the cell lysates. Results (mean ± se) have been normalized against alkaline phosphatase for equivalent transfection efficiency, and were from duplicate transfection experiments.

3. STAT5 ISOFORMS BIND TO ALS-GAS1 IN A GH-DEPENDENT MANNER

The STATs are a family of 7 latent cytoplasmic proteins (Stat1-4, 5a, 5b and 6) which, upon tyrosine phosphorylation by the Janus family of kinases, will homo- or heterodimerize and translocate to the nucleus where they activate transcription by binding to target cis-regulatory elements such as GAS or ISRE (interferon-stimulated response element) (14). To determine whether STAT proteins bind to ALS-GAS1, we performed electrophoretic mobility shift assays using H4-II-E cell nuclear proteins. H4-II-E cells were maintained in serum-free media for 16 h, after which they were exposed to either 0 or 100 ng/ml of bovine GH for 15 min. Nuclear extracts prepared from these cells were incubated with an ALS-GAS1 (5'-AGGTGTTCCTAGAAGAGG-3') or ALS-GAS2 (5'-ACTGGGCCTTAGACAAACCCCT GGA-3') oligonucleotide probe. The prolactin response element (PRE) of the rat ß-casein gene (17-19) which binds to STAT5 isoforms (5'-GGACTTCTTGGAATTAAGGGA-3') also was used as a DNA probe.

When ALS-GAS1 was used as probe, a specific protein-DNA complex was detected only in cells treated with GH. This complex migrated with the same mobility as the protein-DNA complex formed with the PRE probe, suggesting that the complex may contain STAT5 proteins. This was confirmed by the immunoreactivity of the complex with STAT5 antibodies. Addition of STAT5a antibody prior to incubation with either the ALS-GAS 1 or PRE probes resulted in a 60% decrease in the abundance of the specific DNA-complex, and the concomitant formation of a supershifted complex. Inhibition of the DNA-complex, however, was complete when an antibody reacting with both STAT 5a and 5b was used instead. Antibodies against STAT1 and STAT3 (which also are known to be induced by GH) did not alter the formation or mobility of the GH-dependent DNA-complex In contrast, the ALS-GAS2 probe failed to form any GH-induced specific protein-DNA interactions with either GH-treated or non-GH-treated extracts. Thus, STAT5 isoforms can account for all of the observed binding to the ALS-GAS1 oligonucleotide probe.

The induction of STAT 5 binding to ALS-GAS1 in H4-II-E cells following GH-treatment is time-dependent. Gel shift assays were performed with nuclear extracts prepared from cells exposed to bovine GH for various times. The GH-dependent DNA-protein complex was detected 5 min following GH treatment, was maximal between 15 and 30 min, and slowly decreased over the next 24 h.

4. ACTIVATION OF STAT5 BUT NOT RAS IS SUFFICIENT FOR GH-STIMULATION OF ALS PROMOTER ACTIVITY

In addition to activating STATs (STATs1, -3, and -5a/b), GH can also activate molecules involved in the RAS-MAP kinase signaling pathway (15). To demonstrate that STAT5 itself is critical for the stimulation of ALS promoter activity by GH, wild-type and dominant-negative STAT5 expression plasmids were co-transfected into H4-II-E cells together with an ALS promoter plasmid. Dominant negative STAT5 isoforms are truncated at the carboxy-terminal domain; although they can be inducibly tyrosine phosphorylated, they form inactive heterodimers with wild-type STAT5 isoforms (20). In cells transfected only with the ALS promoter plasmid, GH-stimulated promoter activity 2.2 ± 0.4 fold (Fig. 4). Cells transfected with the wild-type STAT5a plasmid gave an enhanced stimulation by GH (3.3 ± 0.1 fold). In contrast, stimulation by GH was abolished in cells co-transfected with either dominant-negative STAT5a or dominant-negative STAT5b plasmids, whereas co-transfection with either dominant-negative RAS or constitutively active RAS expression plasmids (21,22) did not affect the GH-stimulation of ALS promoter activity. These results indicated that the GH signaling pathway leading to ALS gene transcription is dependent on the activation of STAT5 and independent of RAS activation.

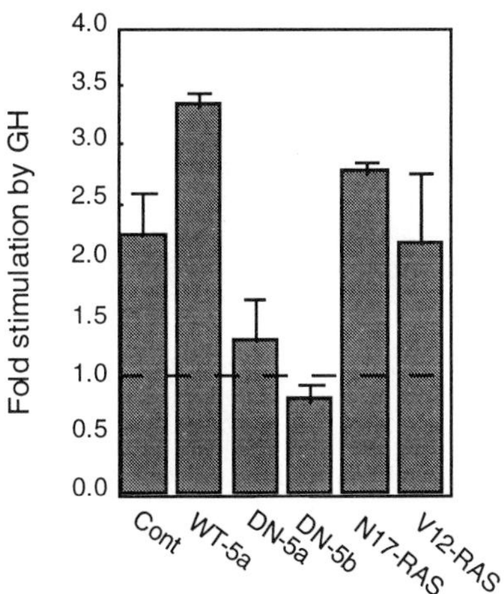

Fig. 4. STAT5 is critical for the stimulation of ALS promoter activity by GH. H4-II-E cells were co-transfected with a nt-703/-49 ALS promoter fragment and expression plasmids for either wild-type STAT5a (WT-5a), dominant negative STAT5a (DN-5a), dominant negative STAT5b (DN-5b) (J.N. Ihle, St Jude Children's Research Hospital, Memphis, TN), constituitively active RAS (V12-RAS), or dominant negative RAS (N17-RAS) (J.S. Gutkind, NIH, Bethesda, MD). Transfected cells were treated with hGH after which luciferase activity was measured in the cell lysates. Fold stimulation (+GH/-GH, mean ± se) from duplicate transfections are plotted. Dashed line indicates no stimulation by GH.

5. ACTIVATION OF ALS PROMOTER ACTIVITY DOES NOT DEPEND ON PULSATILE GH STIMULATION

Studies in rodents have indicated that some of the effects of GH on gene expression are sex-dependent, and are due to the pulsatile plasma GH secretion seen in male, but not in female, animals (23). This sex-dependent expression of some GH-regulated genes is attributed to the differential activation of STAT1, STAT3 and STAT5 by either intermittent pulses or constant levels of GH (17,24). Since STAT5 activation is crucial for the GH-stimulation of ALS promoter activity, we asked whether ALS promoter activity is differentially regulated by either steady-state or episodic exposure to GH. H4-II-E cells stably transfected with a nt -703/-49 ALS promoter fragment linked to a luciferase reporter gene (David F. Lee, unpublished results) were exposed to three intermittent pulses of GH (hGH at 100 ng/ml exposed for 1 h) separated by a 2.5 h incubation without hGH between pulses. Following the third GH treatment, the cells were lysed and the luciferase activity determined. This experimental regimen, which mimics physiological GH profiles in the plasma of adult male rats (24), however, did not increase the magnitude of GH-stimulation of ALS promoter activity over that seen with cells which had been exposed continuously to hGH for 8 h (Fig. 5). The fact that ALS gene expression is not differentially regulated by the pattern of GH exposure is consistent with the observation that ALS is equally expressed in the livers of both male and female rats (6).

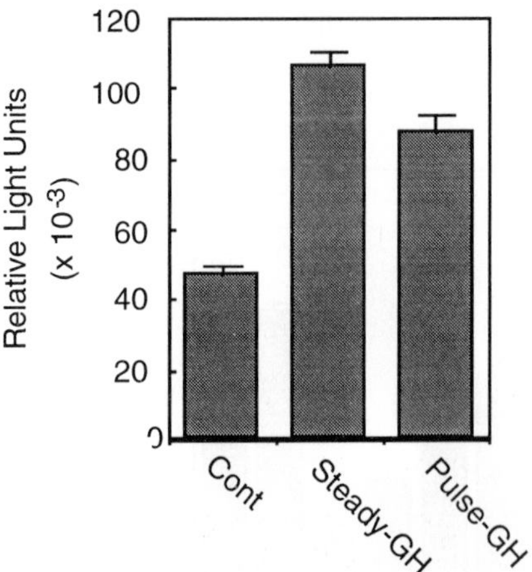

Fig. 5. ALS promoter activity is equally stimulated by either constant or pulsatile exposure to hGH. H4-II-E cells were stably transfected with a nt -703/-49 promoter fragment linked to a luciferase reporter gene using Lipofectamine (Life Technologies, Gaithersburg, MD). A stable clone containing the promoter fragment which retains responsiveness to GH was used. Confluent cells were conditioned in serum-free media overnight, and then exposed to three intermittent pulses of hGH (100ng/ml hGH for 1 h). Each pulse was separated by a 2.5 h incubation in GH-free media. Cells were lysed following the third GH exposure, and luciferase activity measured (Pulse-GH). For comparison, cells not exposed to GH (Control) or have been exposed continuously to 100 ng/ml hGH (Steady-GH) were included. Results (mean ± sd, n=6) plotted are representative from 2 separate experiments.

6. FUTURE PERSPECTIVES

Using the H4-II-E rat hepatoma cell model system, we have shown that the binding of activated STAT5 isoforms to a single GAS element in the mouse ALS promoter is sufficient to convey GH-responsiveness. Though ALS mRNA is not expressed in H4-II-E cells even when grown in the presence of fetal calf serum and/or GH [(11) and our unpublished results], we have obtained similar results in isolated primary rat hepatocytes (25). It remains however to be determined why the endogenous ALS gene is not expressed in H4-II-E cells or other continuous liver cell lines. It is likely that unusually high levels of methylation at CpG islands, the overwhelming majority of which occurs in the 5' ends of genes (26), in continuously cultured cells (27) lead to the loss of expression of tissue-specific genes such as ALS.

Cooperative binding of STATs to adjoining multiple GAS sites has been recently established, and this may be a mechanism whereby the affinity and selectivity of various STATs to low affinity GAS sites can be increased (28). In addition, dimers of STATs bound to DNA also can interact with other transcription factors (28). In the mouse ALS promoter, the second GAS-like element (ALS-GAS2) located downstream of ALS-GAS1 was not necessary for GH stimulation and did not bind any GH-induced nuclear proteins. It remains to be determined, however, whether other cis-elements (eg. liver-specific regulatory sites) are involved, or whether binding of STAT5 to ALS-GAS1 is regulated by protein-protein interaction with other transcription factors.

REFERENCES

1. Baxter RC 1988 Characterization of the acid-labile subunit of the growth hormone-dependent insulin-like growth factor binding protein complex. J Clin Endocrinol Metab 67:265-272
2. Rechler MM 1993 Insulin-like growth factor binding proteins. Vitamins & Hormones 47:1-114
3. Gargosky SE, Tapanainen P, Rosenfeld RG 1994 Administration of growth hormone (GH), but not insulin-like growth factor-I (IGF-I), by continuous infusion can induce the formation of the 150-kilodalton IGF-binding protein-3 complex in GH-deficient rats. Endocrinology 134:2267-2276
4. Zapf J, Hauri C, Futo E, Hussain M, Rutishauser J, Maack CA, Froesch ER 1995 Intravenously injected insulin-like growth factor (IGF) I/IGF binding protein-3 complex exerts insulin-like effects in hypophysectomized, but not in normal rats. J Clin Invest 95:179-186
5. Chin E, Zhou J, Dai J, Baxter RC, Bondy CA 1994 Cellular localization and regulation of gene expression for components of the insulin-like growth factor ternary binding protein complex. Endocrinology 134:2498-2504
6. Dai J, Baxter RC 1994 Regulation in vivo of the acid-labile subunit of the rat serum insulin-like growth factor-binding protein complex. Endocrinology 135:2335-2341
7. Baxter RC 1990 Circulating levels and molecular distribution of the acid-labile a subunit of the high molecular weight insulin-like growth factor-binding protein complex. J Clin Endocrinol Metab 70:1347-1353
8. Zapf J, Hauri C, Waldvogel M, Futo E, Hasler H, Binz K, Guler HP, Schmid C, Froesch ER 1989 Recombinant human insulin-like growth factor I induces its own specific carrier protein in hypophysectomized and diabetic rats. Proc Natl Acad Sci USA 86:3813-3817
9. Camacho-Hubner C, Clemmons DR, D'Ercole AJ 1991 Regulation of insulin-like growth factor (IGF) binding proteins in transgenic mice with altered expression of growth hormone and IGF-I. Endocrinology 129:1201-1206
10. Baxter RC, Dai J 1994 Purification and characterization of the acid-labile subunit of rat serum insulin-like growth factor binding protein complex. Endocrinology 134:848-852
11. Dai J, Scott CD, Baxter RC 1994 Regulation of the acid-labile subunit of the insulin-like growth factor complex in cultured rat hepatocytes. Endocrinology 135:1066-1072
12. Ooi GT, Cohen FJ, Tseng LY-H, Rechler MM, Boisclair YR 1997 Growth hormone stimulates transcription of the gene encoding the acid-labile subunit (ALS) of the circulating insulin-like growth factor- binding protein complex and ALS promoter activity in rat liver. Mol Endocrinol 11:997-1007
13. Boisclair YR, Seto D, Hsieh S, Hurst KR, Ooi GT 1996 Organization and Chromosomal Localization of the Gene Encoding the Mouse Acid Labile Subunit of the Insulin-Like Growth Factor Binding Complex. Proc Natl Acad Sci USA 93:10028-10033
14. Schindler C, Darnell JE,Jr. 1995 Transcriptional Responses to Polypeptide Ligands: The Jak-Stat Pathway. Ann Rev Biochem 64:621-651
15. Argetsinger LS, Carter-Su C 1996 Mechanism of signaling by growth hormone receptor. Physiol Rev 76:1089-1107
16. Nordeen SK 1988 Luciferase reporter gene vectors for analysis of promoters and enhancers. BioTechniques 6:454-458
17. Ram PA, Park S-H, Choi HK, Waxman DJ 1996 Growth hormone activation of Stat 1, Stat 3, and Stat 5 in rat liver. J Biol Chem 271:5929-5940
18. Wakao H, Schmitt-Ney M, Groner B 1992 Mammary gland-specific nuclear factor is present in lactating rodent and bovine mammary tissue and composed of a single polypeptide of 89 kDa. J Biol Chem 267:16365-16370

19. Liu X, Robinson GW, Gouilleux F, Groner B, Hennighausen L 1995 Cloning and expression of Stat5 and an additional homologue (Stat5b) involved in prolactin signal transduction in mouse mammary tissue. Proc Natl Acad Sci USA 92:8831-8835
20. Wang D, Stravopodis D, Teglund S, Kitazawa J, Ihle JN 1996 Naturally occurring dominant negative variants of Stat5. Mol Cell Biol 16:6141-6148
21. Crespo P, Xu N, Simonds WF, Gutkind JS 1994 Ras-dependent activation of MAP kinase pathway mediated by G-protein beta gamma subunits [see comments]. Nature 369:418-420
22. Feig LA, Cooper GM 1988 Inhibition of NIH 3T3 cell proliferation by a mutant ras protein with preferential affinity for GDP. Mol Cell Biol 8:3235-3243
23. Jansson J-O, Ekberg S, Isaksson O 1985 Sexual dimorphism in the control of growth hormone secretion. Endocr Rev 6:128-150
24. Gebert CA, Park S-H, Waxman DJ 1997 Regulation of signal transducer and activator of transcription (STAT) 5b activation by the temporal pattern of growth hormone stimulation. Mol Endocrinol 11:400-414
25. Berry MN, Edwards AM, Barritt GJ 1991 Isolated hepatocytes: preparation, properties and applications. Laboratory Techniques in Biochemistry and Molecular Biology, vol.21. Elsevier, New York, pp. 1-460
26. Bird Adrian P 1990 CpG-rich islands and the function of DNA methylation. Nature 321:209-213
27. Antequera F, Boyes J, Bird A 1990 High levels of de novo methylation and altered chromatin structure at CpG islands in cell lines. Cell 62:503-514
28. Darnell JE,Jr. 1997 Stats and Gene Regulation. Science 277:1630-1635

Molecular Mechanisms to Regulate the
Activities of Insulin-like Growth Factors
K. Takano, N. Hizuka and S-I. Takahashi (Editors)

Insulin-like Growth Factor Gene Targeting

Wesley Won and Lyn Powell-Braxton

Cardiovascular Research, Genentech Inc., 1 DNA Way, South San Francisco, CA 94080.

Targeted disruption of the genes in the insulin-like growth factor (IGF) signaling pathways has conclusively demonstrated an essential role for both IGF-I and IGF-II in prenatal growth and development. Postnatal mouse growth is also dependent on normal levels of IGF-I. Results obtained with these models emphasize the importance of this growth factor family in the normal development of many different tissues including muscle, bone and reproductive systems. The majority of IGF-I knockout animals (*Igf1*$^{-/-}$) die at birth due to runting and anoxia secondary to muscle insufficiency. However, the occasional surviving *Igf1*$^{-/-}$ animal has enabled us to study the role of IGF-I in tissues like the CNS where several neuronal populations develop post-natally. We have also undertaken physiological and hormone replacement studies on these *Igf1*$^{-/-}$ animals, heterozygous *Igf1* $^{+/-}$ animals which have 65% of wild type levels, and a third group of animals homozygous for a partial IGF-I reduction (*Igf1* $^{m/m}$), which have 30% of normal circulating levels providing a titration curve for the *in vivo* effects of IGF-I.

1. INTRODUCTION

The effects of targeted mutations abolishing *Igf1* or *Igf2* gene function on prenatal growth and development have been well described (1-3). As both peptides have classically been shown to promote cell proliferation it is not surprising that mice homozygous for either of these mutations are born approximately 60% of wild type litter mate size (1-4). However, knowledge of the embryonic mRNA levels of these two growth factors would not have predicted the result of gene targeting on development and viability. IGF-II peptide is found at much higher levels than IGF-I during embryonic development (4). *Igf2* $^{-/-}$ animals survive to maturity and are fertile, demonstrating that *Igf2* is not required for the correct differentiation of any organ systems (5). In contrast, the majority of the *Igf1*$^{-/-}$ mice do not survive birth (1-3). This dramatic reduction in viability appears to be due to the delayed development of musculature essential for respiration. In addition the reproductive systems of both sexes are significantly affected and the occasional survivors have not been fertile (6).

Gene targeting also allows some investigation of growth factor-receptor interaction. The effect of disruption of the tyrosine kinase IGF type I receptor (*Igf1r*) is absolute perinatal lethality (1). In addition, *Igf1r* null animals show a more profound effect on neonatal body weight (45% of wild type) indicating that both IGF-I and IGF-II exert their effects on cell proliferation via this pathway. The cation-independent mannose 6-phosphate receptors (CI-MPR), which is one of the receptors for the phosphomannosyl recognition marker of lysosomal hydrolases, also interacts with IGF-II (7). Referred to as the type 2 receptor (IGF2R/CI-MPR) it is believed to function mainly to regulate turnover of IGF-II by a clearance mechanism rather than signaling. In mice *Igf2r* targeted animals are also embryonic lethal. Supporting the turnover hypothesis these animals are larger than wild type at birth, potentially due to excess circulating IGF-II. Complex combinations of these mutations in mice has been carried out to clarify the various interactions of the IGF proteins with their different receptors (8). The daunting task of dissecting the multiple complex effects of IGF's interactions with their binding proteins has begun in several laboratories and will not be addressed further here.

We and others have studied the phenotype of the *Igf1* mutation on multiple organ systems both embryologically and, in some cases, in a small number of surviving animals. Efstratiadis and colleagues have extensively characterized both male and female reproductive systems (6) and we have collaborated on studies of the central nervous (9) and immunological systems (10). In addition we are continuing studies on muscle and bone structure which are both affected in the *Igf1$^{-/-}$* animals. In particular there is a delay in ossification of some but not all bones in the *Igf1$^{-/-}$* mouse. The most dramatic effect is seen in the sternum of the *Igf1$^{-/-}$*

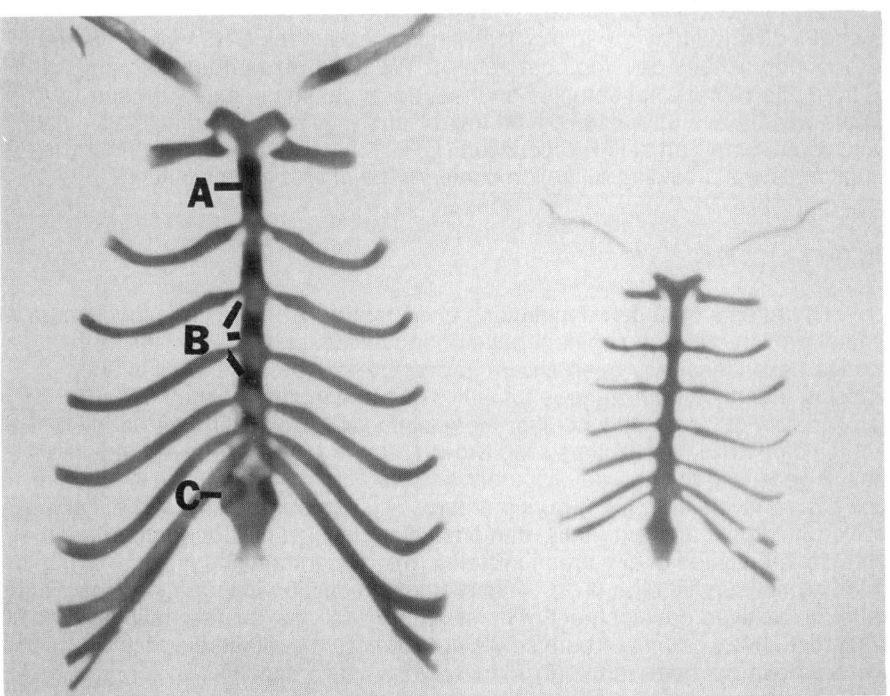

Figure 1. Sternums of neonatal wild type (left) and sibling *Igf1$^{-/-}$* (right) animals stained with Alcian (blue to show cartilage) and Alizarin (red to show calcified tissue). Although the clavicula bones are small in the *Igf1$^{-/-}$* animal, they are normally calcified. However, there is complete lack in ossification of the manubrium (A) the sternebra (B) and the xiphoid process (C) in the *Igf1$^{-/-}$* animal.

mice (Figure 1) which is barely ossified at birth. Thi is a 2.5 days delay over the normal initiation of ossification in this region. A similar delay is seen in areas of the skull and vertebral column. In contrast, the late calcifying areas of the phalanges are normally ossified (not shown).

2. *Igf1* $^{m/m}$ MICE

The perinatal lethality of the *Igf1* and type I receptor null animals makes it very difficult to study the ramifications of reducing IGF-I levels post natally. To this end we utilized a mouse model containing a site-specific insertion which created a leaky mutation in the *Igf1* gene (*Igf1* ^{m}) resulting in circulating levels of 30% of wild type in homozygous *Igf1* $^{m/m}$ animals (11). These animals are viable and fertile and 60 % of normal adult size enabling studies on various physiological systems. Confirming the homeostatic relationship between circulating IGF-I and growth hormone (GH) secretion, these animals have lower pituitary GH content and increased serum GH (11). Miniaturization of standard catheterization techniques enabled the measurement of blood pressure in the *Igf1* $^{m/m}$animals which was found to be elevated over age and sex matched siblings (11). We also found an elevation of cardiac contractility which may be related to the higher circulating levels of GH seen in these animals.

The slightly larger size of the *Igf1* $^{m/m}$ animals, compared to *Igf1*$^{-/-}$, permitted the investigation of nerve function by *in vivo* measurement of sciatic motor and sensory nerve conduction velocity as will be discussed below. We have also been able to carry out studies on the effects of replacement therapy on the peripheral nervous system (PNS) in these *Igf1* $^{m/m}$ mice by infusion of recombinant human IGF-I.

3. IGF-I AND THE PERIPHERAL NERVOUS SYSTEM

Having established that IGF-I deficiency lead to changes in selected nerve populations in the CNS (9) we turned our attention to the PNS (12). It appears that the optimal functioning of the PNS is very sensitive to circulating IGF-I levels as animals carrying only a single allele, with circulating levels of 65% wild type controls, exhibit a significant reduction in both motor nerve conductance velocity (MNCV) and sensory nerve conductance velocity (SNCV) when measured using electrophisiological recording of sciatic nerves in adult animals *in vivo* (12). In addition there is a significant reduction of the amplitude of the A-fiber response from isolated peroneal nerves *in vitro*. Similarly, *Igf1* $^{m/m}$ mice with 70% reduced serum levels, or *Igf1*$^{-/-}$ mice, totally lacking IGF-I, show concomitantly greater reductions in *in vivo* MNCV and SNCV measurements. Furthermore, *Igf1*$^{-/-}$ animals show a dramatic decrease in area and amplitude of A-fiber recordings and an impairment of synchronization of firings in isolated peroneal nerves. Histopathology of the peroneal nerves in *Igf1*$^{-/-}$ mice showed a marked reduction of nerve fiber area with an almost total loss of large diameter fibers. There was no noticeable decrease in the degree of myelination of the nerves in either the *Igf1* $^{m/m}$ or *Igf1*$^{-}$

/– mice. This is not inconsistent with the decreased myelination seen in certain areas of the brain (9). The effect of IGF-I on myelination is probably indirect and secondary to a reduction of nerve fiber diameter. In *Igf1* $^{m/m}$ animals replacement therapy with subcutaneously delivered infusions of exogenous recombinant human IGF-I restores both motor and sensory nerve conduction velocities (12 and see below). These findings demonstrate not only that IGF-I serves an important role in the growth and development of the PNS but also that systemic IGF-I treatment, as opposed to local replacement therapy, can enhance nerve function in mice.

4. IGF-I ADMINISTRATION TO *Igf1* $^{m/m}$ MICE

One of the advantages of the *Igf1* $^{m/m}$ mice is that they are sufficiently large to enable administration of therapeutics by osmotic mini-pump. In order to see if exogenous IGF-I was capable of restoring growth or physiological deficiencies seen in the *Igf1* $^{m/m}$ mice we undertook the following experiment. Cohorts of age and sex matched adult wild type and *Igf1* $^{m/m}$ mice were weighed, their motor and sensory nerve conduction velocities measured electrophysiologically, then implanted with osmotic mini-pumps administering 50 µg/mouse/day recombinant human IGF-I or excipient (ex.). Animals were weighed weekly, their IGF-I serum levels monitored and at the end of 4 weeks of treatment their conductance velocities were re-measured. As shown in Table 1 there is a significant increase in circulating IGF-I levels with treatment in both wild type and *Igf1* $^{m/m}$ animals. Although the *Igf1* $^{m/m}$ mice do show an increase in body weight with treatment, they are still significantly smaller than the wild type controls, however, both motor and sensory nerve conductance velocities are restored to near normal levels.

Table 1
Treatment of *Igf1* $^{m/m}$ and wild type mice with IGF-I

	Igf1 $^{+/+}$ + ex.	*Igf1* $^{+/+}$ + IGF-I	*Igf1* $^{m/m}$ + ex.	*Igf1* $^{m/m}$ + IGF-I
serum IGF (ng/ml)	196.9 ± 40.7	496.6 ± 39.1**	76.2 ± 37.4	610.4±139.4**
start body weight (g)	23.6 ± 1.2	23.0 ± 1.3	15.2 ± 2.6*	14.1 ± 1.4*
end body weight (g)	26.5 ± 1.8	27.7 ± 1.7	17.1 ± 2.1*	19.7 ± 1.1*
body weight change (g)	3.4 ± 1.5	4.1 ± 0.8	2.3 ± 1.6	5.4 ± 1.0
MNCV (m/sec)	54.61 ± 8.04	55.71 ± 8.44	35.62 ± 6.83*	48.69 ± 11.56
SNCV (m/sec)	54.07 ± 7.23	50.16 ± 7.61	31.79 ± 9.83*	42.67 ± 10.84

Igf1 $^{m/m}$ values compared with wild type values * $p < .005$
treatment compared with non treatment ** $p < .0001$

5. IGF-I and GH ADMINISTRATION TO *Igf1*$^{-/-}$ MICE

Although the small size of the *Igf1*$^{-/-}$ mice makes it difficult to administer IGF-I in the most optimal way, i.e.. by continuous infusion, it is possible to administer it by subcutaneous injection. We used this mode of delivery to treat surviving *Igf1*$^{-/-}$ mice with 10 μg/g of recombinant IGF-I twice daily for 14 days. We also administered the same 10 μg/g dose to control, sibling wild type animals. Animals were weighed daily and the dose adjusted accordingly, at the end of the treatment we harvested the tissues and compared treatment groups with excipient treated animals. The rates of growth of female animals are shown in Figure 2, expressed as the percentage change in body weight daily

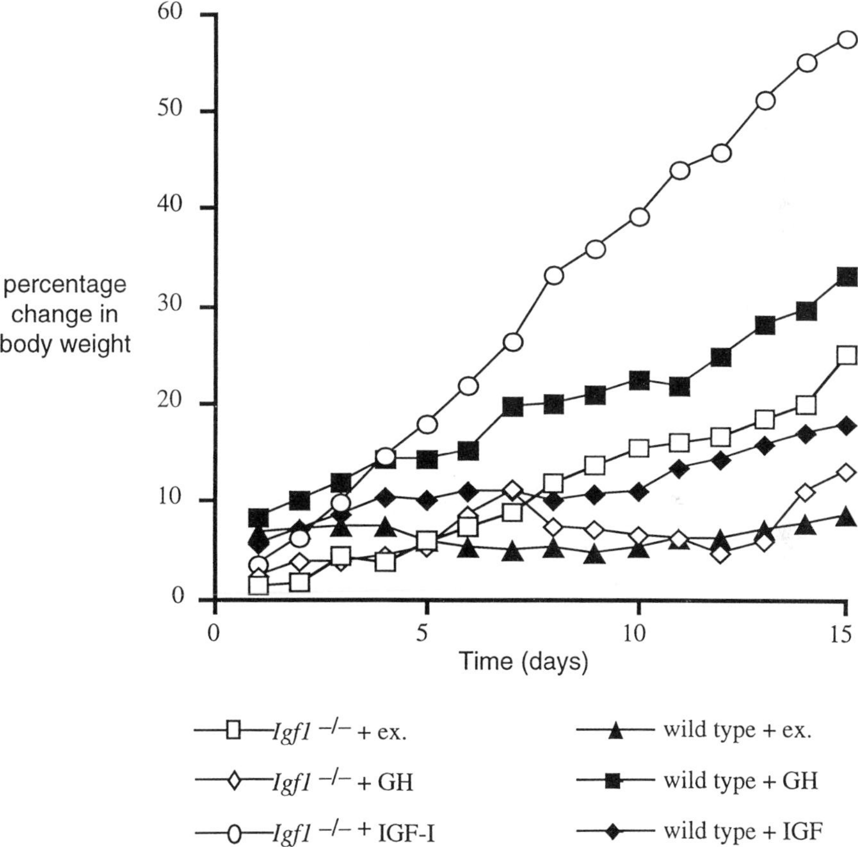

Figure 2. Change in body weight of wild type or *Igf1*$^{-/-}$ animals treated with twice daily injections of IGF-I (wild type n = 12; *Igf1*$^{-/-}$ n = 7), GH (wild type n = 3; *Igf1*$^{-/-}$ n = 3) or excipient (wild type n = 11; *Igf1*$^{-/-}$ n = 7) expressed as percentage change from day zero.

from the start of the experiment. We chose to analyze the data in this way to facilitate a direct comparison of wild type and *Igf1*$^{-/-}$ animals which are otherwise not easily comparable due to the large difference in body weight.

There is a clear difference in the responsiveness of the *Igf1*$^{-/-}$ and wild type animals to the different drug treatments. Both male and female animals were treated and there is no significant difference in response to the growth factors between the sexes. For the sake of clarity, there are no error bars shown and significant differences are ellaborated below. The wild type animals respond more to GH than IGF-I and the *Igf1*$^{-/-}$ animals show no response to GH but respond dramatically to IGF-I treatment. There was no difference between the means of the groups at the start of treatment. At day fifteen female wild type animals treated with excipient were 8.6% larger than on day zero. In response to IGF-I treatment, wild type females were 18% larger and GH treated females 33.1% larger. Differences were analyzed using multiple analysis of varience (ANOVA) with Scheffe F post hoc testing (ex. v. IGF-I $p = 0.016$; ex. v. GH $p < 0.0001$; IGF-I v. GH $p = 0.011$). In contrast there was no significant response of the *Igf1*$^{-/-}$ animals to GH. In fact, although it was not significantly different , the GH treated animals gained less weight than the excipient treated group $p=0.5287$). The smaller weight gain with GH is probably due to the smaller size of this cohort. As in the wild type groups, there was no difference in the means at the start of the experiment but after the treatment period *Igf1*$^{-/-}$ females treated with excipient were 25.1% larger, with IGF-I treatment 57.3% larger and with GH treatment only 12.9% larger (ex. v. IGF-I $p = 0.001$; ex. v. GH $p =0.5287$; IGF-I v. GH $p = 0.0005$). The response of the *Igf1*$^{-/-}$ animals to IGF-I and not to GH does not necessarily mean that all of GH's growth promoting actions are mediated by IGF-I but does suggest that if there are IGF-I independent effects they are not readily detectable during this period of treatment which is sufficient to give a robust response in the wild type animals. We have analyzed the tissues from these animals and do not detect any unusual differential growth responses between the *Igf1*$^{-/-}$ and wild type animals. All of the tissues that respond strongly to treatment with IGF-I in wild type animals also respond in the *Igf1*$^{-/-}$ animals, however, there is no response to GH treatment on particular tissues. One observation that should be emphasized is that the circulating levels of GH in *Igf1*$^{-/-}$ animals are higher than in wild type controls (11) which may mask any subtle effects on specific tissues in this present study.

6. SUMMARY

Targeted disruption of the insulin like growth factors and their receptors has made it possible to study the combined and separate effects of these genes on mouse growth and development. This has confirmed that both IGF-I and IGF-II function to promote growth via the type-I receptor *in vivo* and IGF-I has more dramatic effects on tissue differentiation and functioning in a number of systems including muscle, nerve and bone development. More detailed analysis of the rare surviving *Igf1*$^{-/-}$ mice and animals with reduced serum IGF-I levels has uncovered important effects of IGF-I in the optimal functioning of the peripheral nervous system. More recently the beginning of exogenous drug replacement therapy of these targeted animals has enabled us to begin addressing the effect of exogenous versus endogenous growth factor synthesis on growth and tissue function and incidentally, the role that growth hormone may have independent of IGF-I *in vivo*. Further studies remain to be carried out on the expression patterns in IGF-I deficient tissues versus normal tissues and the induction of gene expression in these tissues in response to exogenous drug delivery to see if exogenous drug replacement therapy can completely restore normal function to affected tissues and organs.

REFERENCES

1. J-P. Liu, *et al.*, Cell, 75 (1993) 59.
2. J. Baker, *et al.*, Cell 75 (1993) 73.
3. L. Powell-Braxton, et al., Genes & Develop. 7 (1993) 2609.
4. T. M. DeChiara, A. Efstratiadis and E. J. Robertson, Nature 345 (1990) 78.
5. W. H. Daughaday, *et al.*, Endocrinology 110 (1982) 575.
6. J. Baker, *et al.*, Mol. Endocrin. 10 (1996) 903.
7. S. Kornfeld. Annu. Rev. Biochem. 61 (1992) 307.
8. T. Ludwig, *et al.*, Dev. Biol. 177 (1996) 517.
9. K. Beck, *et al.*, Neuron 14 (1995) 717.
10. E. Montecino-Rodriguez, *et al.*, J. Immunol. 159 (1997) 2712.
11. G. Lembo, *et al.*, J. Clin. Invest., 98 (1996) 2648.
12. W-Q Gao, et al, J. Mol. Cell. Neurosci. submitted.

Molecular Mechanisms to Regulate the
Activities of Insulin-like Growth Factors
K. Takano, N. Hizuka and S-I. Takahashi (Editors)

Genetic disruption of IGF binding proteins

J. Pintar, A. Schuller, S. Bradshaw, J. Cerro, A. Grewal

Department of Neuroscience and Cell Biology, UMDNJ-Robert Wood Johnson Medical School, Piscataway, NJ 08854

Much recent work has begun to elucidate the individual contributions of IGF system legends and receptors to growth as well as specific differentiative events (1-3). In contrast, the required function(s) of IGFBPs, revealed by genetic approaches, continues to be relatively unexplored. The modest phenotype exhibited by IGFBP-2 null mice (4) is consistent with the possibility that compensation by other of the six "traditional" IGFBPs may mask required functions of the entire family. Moreover, additional proteins with IGF binding capacity have been recently identified (5) and potentially may complicate even further the analysis of individual IGFBP mutants. We here review our recent work that has lead to genetic disruption of two additional binding proteins, IGFBP-4 and IGFBP-6, and report initial results from analysis of homozygous mutant mice carrying these mutations.

I. Expression of IGFBP-4. IGF-II expression during fetal development in the rodent is widespread in mesoderm and its derivatives (6). Moreover, the required role for IGF-II normal prenatal growth in mice becomes apparent by embryonic day e11.5 (1). Since cell culture studies indicate that IGFBPs can modulate IGF action in multiple ways (7), IGFBPs could clearly be expected to influence IGF activity in vivo during critical times of IGF action if specific IGFBPs were indeed expressed during these times. To this end, we have have examined IGFBP expression within the developing embryo and fetus (8) as well as in the surrounding uterus during pregnancy (e.g. (9)) in an effort to identify which IGFBPs might be most likely to modulate IGF action.

Expression of several IGFBPs are readily detected soon after implantation by in situ hybridization. Of particular interest at early post-implantation stages has been the finding that IGFBP-4 expression exhibits a striking correlation with IGF-II expression. Thus expression of IGFBP-4 is widespread in mesoderm and its derivatives as is IGF-II, although with some additional sites, including cell groups in the CNS also express IGFBP-4. One example showing the extensive co-expression of IGFBP-4 and IGF-II at embryonic day 11 is shown in Figure 1.

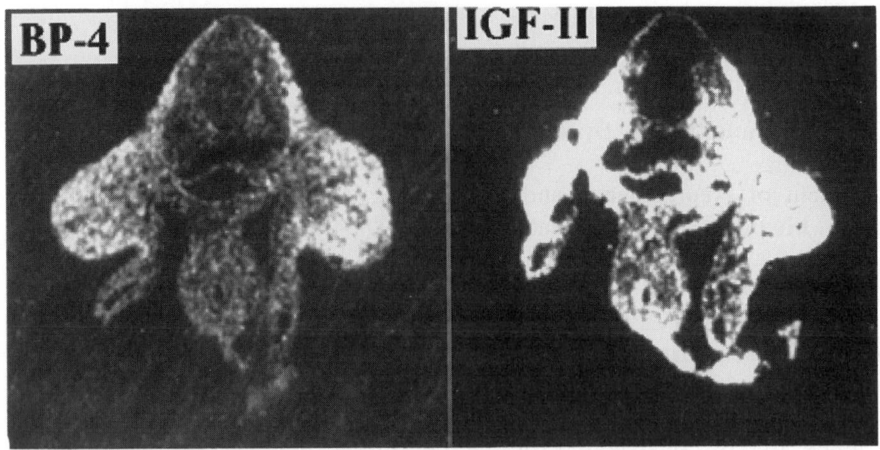

Figure 1. Expression of IGFBP-4 (a) and IGF-II (b) in the trunk of an e11 rat embryo. Note the extensive overlap in expression between these two genes.

Such an expression pattern suggests that IGFBP-4 would be likely to modify IGF-II action. The type of action that might be expected can be predicted from in vitro studies, which have consistently indicated

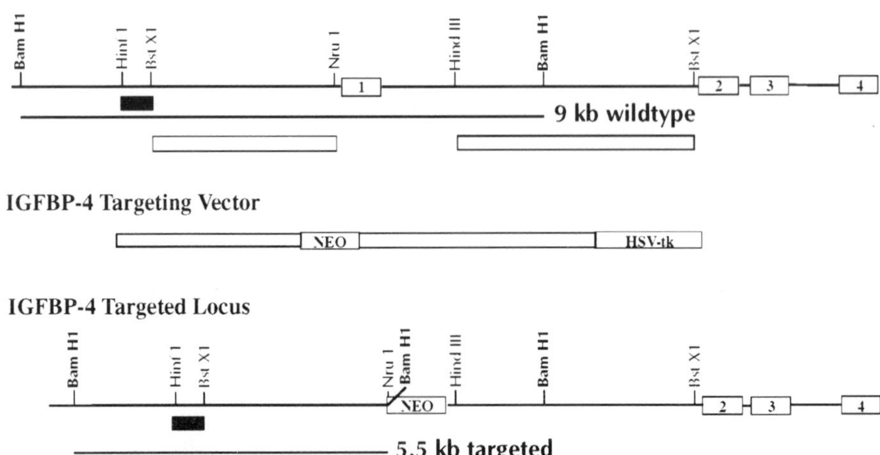

Figure 2. Strategy used for IGFBP-4 targeting, which leads to deletion of exon 1.

that IGFBP-4 is an inhibitory binding protein. IGFBP-4 was indeed originally isolated by virtue of its ability to inhibit growth of specific cell types in culture. For example, IGFBP-4 was isolated both from an osteosarcoma cell line by its ability to inhibit cartilage cell growth and independently isolated from follicular fluid by virtue of its ability to inhibit FSH-stimulated steroid production (10,11). Thus one might expect that inhibiting IGFBP-4 action by genetic ablation would enhance IGF-II action and thus increase growth as observed in mutations of the IGF-II receptor (3) and H19 loci.

II. Targeting of IGFBP-4. To test this prediction, gene-targeted mice were produced that lacked mouse IGFBP-4 exon 1 using the strategy diagrammed in Figure 2. A single targeted line was isolated and gave rise to germ-line trasmitting chimeric mice (Schuller and Pintar, in preparation). Offspring from the mating of heterozygous mice produce all three expected genotypes. Mice homozygous for the mutation are viable and are born in the expected Mendelian ratio. Thus far homozygous mutant mice have exhibited no obvious lethality compared to littermates. The growth characteristics of these mutant mice have been examined and compared to littermates. We find that homozygous mutant mice are clearly smaller than both wild-type and heterozygous littermates. A representative growth curve for a pair of +/+ and -/- mice is shown in Figure 3.

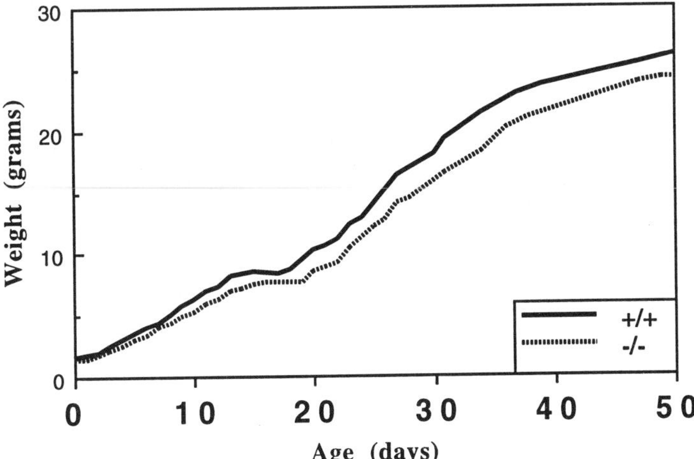

Figure 3. Post-natal growth curve from one representative wild-type (solid line) and BP-4 homozygous mutant embryo (-----).

These results are in marked contrast to the prediction that genetic ablation of IGFBP-4 would lead to enhanced growth. Instead the data are most consistent with the possibility that the IGFBP-4 is required for optimal action of IGF-II. The time at which the growth deficiency in IGFBP-4 mutants first becomes apparent is now being determined. If the deficit begins during prenatal ages, it will be of specific interest to determine whether IGFBP-4 mutations synergize with IGF-II mutants or whether the growth deficit of the double IGF-II/BP-4 mutant is no greater than IGF-II alone.

III. Expression of IGFBP-6.

We have also continued to characterize expression of IGFBP-6. This gene for this IGFBP, which was originally isolated from CSF by Binoux and colleagues (12), has been shown to be expressed in the brain, with particularly high levels in the pons (4). We have further clarified additional sites of BP-6 expression in both the adult and early post-natal brain. We find that BP-6 is expression in mitral cells of the olfactory bulb and, moreover, find expression outside the neural parenchyma in the meninges as well (Figure 4).

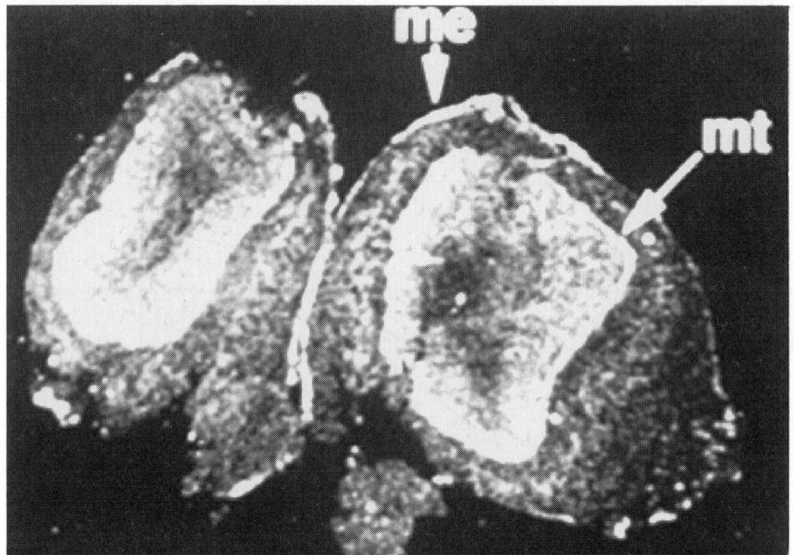

Figure 4. Expression of IGFBP-6 in the adult mitral cells (mt) and meninges (me).

Although BP-6 is expressed at very low levels in the fetal rodent brain, CNS expression is clearly evident by p11 in at least some regions that exhibit relatively high BP-6 expression in the adult. One such region in the pons (Fig. 5; Grewal and Pintar, in preparation).

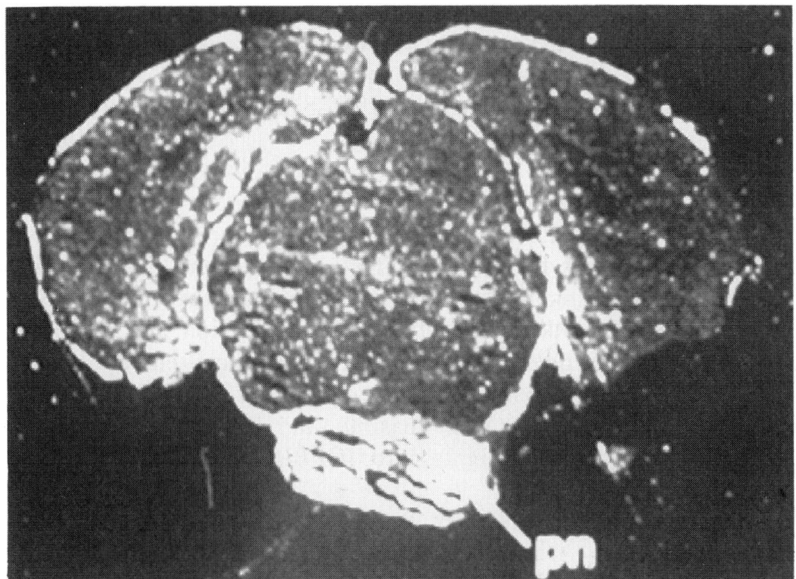

Figure 5. Expression of BP-6 in the p11 pons.

IV. Gene targeting of IGFBP-6.

To investigate required roles for IGFBP-6, a gene targeting strategy analogous to that used for IGFBP-4 was used to ablate exon 1 from the murine IGFBP-6 gene (Figure 6). Several targeted ES lines

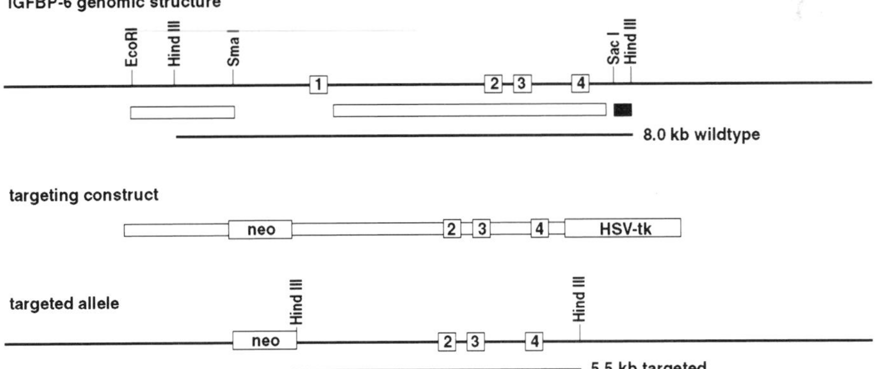

were identified (Bradshaw and Pintar, unpublished) and one line has thus far given rise to germ-line chimeric males. Matings of heterozygous mice have given rise to viable mice homozgyous for the

BP-6 mutation. Although the numbers of mice are low, it does not appear that prenatal lethality of BP-6 mutants occurs. IGFBP-6 mutants at present do not appear to exhibit gross deficits in growth from wild-type littermates, but detailed growth curves are not yet available. It is clear that subsequent analysis of BP-6 function in control of motor systems and reproductive capacity (see (9)) can now be approached genetically.

V. Conclusions. The production of the IGFBP-4 knock-out strain provides clear evidence that this IGFBP has a required role in growth that cannot be substituted by other IGFBPs. Moreover, the unexpected growth deficit demonstrated by these mice indicates that the in vivo role of IGFBP-4 is most likely to optimize IGF-II action and provides an another example of the unexpected findings that have emerged from analysis of the IGF system using genetic approaches. It is likely that combinatorial matings that are now feasible and in progress will provide additional surprising insights into IGFBP action in vivo.

1. Baker, J., Liu, J. P., Robertson, E. J., and Efstratiadis, A. (1993). *Cell* 75:73-82.
2. Baker, J., Hardy, M. P., Zhou, J., Bondy, C., Lupu, F., Bellve, A. R., and Efstratiadis, A. (1996). *Mol Endocrinol* 10: 903-918.
3. Ludwig, T., Eggenschwiler, J., Fisher, P., D'Ercole, A. J., Davenport, M. L., and Efstratiadis, A. (*1997*). *Develop. Biol* 177:517-535.
4. Pintar, J. E., Cerro, J. A., and Wood, T. L. (1996). *Horm Res* 45:172-177.
5. Oh, Y., Nagalia, S.R., Yamanaka, Y., Kim, H-S., Wilson, E., and Rosenfeld, R.G. (1996). *J. Biol. Chem.* 271:30322-3-325.
6. Stylianopoulou, F., Efstradiatis, A., Herbert, J. and Pintar, J.E. (1988). *Development* 103:497-507.
7. Jones, J. I., and Clemmons, D. R. (1995). *Endocrine Rev* 16:3-34.
8. Cerro, J. A., Grewal, A., Wood, T. L., and Pintar, J. E. (1993). *Regulatory Peptides* 48:189-198.
9. Cerro, J.A. and Pintar, J.E. (1997). *Develop. Biol.*184:278-295.
10. Mohan, S., Bautista C.M., Wergeal, J., and Baylink, D. (1989). *Proc. Natl. Acad Sci USA* 86:8338-8342.
11. Ui, M., Shimonaka, M., Shimasaki, S., and Ling, N. (1989). *Endocrinology* 125:912-916.
12. Hossenlopp, P, Seurin, D., Segovia-Quinson, B., and Binoux, M. (1986). FEBS Let. 208:439-444.

Molecular Mechanisms to Regulate the
Activities of Insulin-like Growth Factors
K. Takano, N. Hizuka and S-I. Takahashi (Editors)
1998 Elsevier Science B.V.

Targeted Mutations Of Insulin And IGF-1 Receptors In Mice.

Domenico Accili, Hiroko Kanno, Yoshiaki Kido, Davide Lauro,
and Kristina I. Rother

Unit on Genetics and Hormone Action,
Developmental Endocrinology Branch,
National Institute of Child Health and Human Development,
National Institutes of Health, Bethesda, MD, 20892

1. Summary

Mice with targeted mutations in the receptors for insulin (IR), IGF-1 (IGF-1R), IGF-2 (IGF-2R), as well as insulin receptor substrate-1 (IRS-1), an important substrate of the IR kinase, have provided important new information on the concerted roles of these proteins. A set of combined mutations in the IR and IGF-1R genes has shed new light onto the physiology of embryonic growth. The phenotype of mice with combined mutations in the insulin and IGF-1R genes suggests that both receptors play important, yet distinct roles to mediate embryonic growth. Interestingly, the growth-promoting actions of IRs during embryogenesis are mediated in response to IGF-2, rather than insulin.

2. Introduction

Targeted mutagenesis in mice has become an integral part of the scientific armamentarium (1-3). How profoundly the generation of targeted mouse mutants has impacted upon the field of insulin and IGFs action, including our understanding of the genetics of growth, is the subject of this review.

Insulin and IGF-1Rs play important and complementary roles in embryonic growth and development (4). These roles have been addressed by crossing mice with null mutations in both genes. The experiments indicate that both receptors are required for mouse embryonic growth. The roles of insulin and IGF-1Rs are developmentally distinct, so that IGF-1Rs are important for embryonic growth in mid-gestation, whereas IRs play a more important role in late gestation. Interestingly, IGF-2, and not insulin, is the ligand that mediates the growth-promoting actions of IRs in embryonic life (5).

3. The IR-deficient mouse

Mice lacking functional IRs have been generated in two laboratories with identical results (6,7). Mice nullizygous at the IR locus have normal features at birth; their intrauterine growth and development appear to be normal. However, upon closer inspection and careful measurements of embryonic weights, a small reduction (~10%) in size can be detected. Such reduction is statistically significant when large numbers of embryos are compared (5). The relevance of these findings to the genetics of growth is discussed in the following paragraphs.

Lack of IRs results in severe metabolic derangement. Within few days, mutant mice die of diabetic ketoacidosis (DKA). There are small differences in the exact survival of the two different strains, which probably reflect differences in their genetic make-up. However, the conclusion is that IRs are required to mediate the metabolic actions of insulin in post-natal life. Whether insulin exerts other, non-metabolic affects, and whether these effects are also mediated by IRs cannot be addressed by this model. Likewise, it is intriguing that fetal metabolism is unaffected by the lack of IRs. Two possible explanations come to mind: that fetal metabolism is insulin-independent, or that maternal metabolism can prevent the onset of metabolic abnormalities in utero. More work with inducible and/or tissue-specific mutants is required to address these questions. Interestingly, similar alterations are observed in humans with acute-onset insulin-dependent diabetes, with the notable exception that in humans with DKA plasma insulin levels are undetectable, whereas in mice lacking IRs plasma insulin levels are higher than in normal litter mates. The early post-natal death caused by inactivation of the IR gene is reminiscent of other targeted inactivations of genes important for fuel metabolism, such as glucokinase (8) and C/EBP-α (9).

4. Differences between mice and humans with homozygous nonsense mutations of the IR gene

The notion that mice can develop and be born without functional IRs had been predicted, based on the description of four cases of leprechaunism, a genetic syndrome of extreme insulin resistance, in which children were born with homozygous non-sense mutations or deletions of the IR gene (10-13). Nevertheless, there are striking differences between IR-deficient mice and humans with similar mutations. In humans, fasting hypoglycemia and retarded growth are the hallmarks of severe insulin resistance (14,15). The pathophysiologic basis of these differences remains speculative. It is especially difficult to reconcile the findings on glycemic control, since fasting hypoglycemia in children with leprechaunism is essentially a paradoxical finding. A simple explanation is that newborn mice, unlike newborn humans, are constantly nursed by their mothers. Thus, there exists a constant flow of nutrients that prevent hypoglycemia, while exacerbating hyperglycemia and ketogenesis. To partially support this view, it should be noted that children with leprechaunism do experience transient hyperglycemia and ketosis following meals. Furthermore, leprechauns present with dysmorphic and virilized features that are absent in IR knock-out mice.

5. Genetics of growth: role of IRs in embryonic development

IRs are expressed in the pre-implantation mouse embryo, and insulin has been shown to stimulate glucose utilization by isolated blastocysts (16). Furthermore, it is known that lack of IRs in humans is associated with severe growth retardation (14,17). The role of fetal IRs is unclear. One possibility is that IRs regulate growth by regulating fuel metabolism. In addition to the growth retardation observed in cases of extreme insulin resistance, macrosomia is a well recognized complication of fetal hyperinsulinemia, a metabolic consequence of the diabetic pregnancy (18,19). An alternative explanation to the role of IRs in development is that they mediate growth-promoting actions, either directly or through hybrid insulin/IGF-1Rs. We have shown that growth and development of mouse embryos are scarcely affected by the lack of IRs (6). The phenotype of nullizygous IR mice suggests that, during gestation, IRs can be replaced by other receptors. Because of the close structural and functional homology between insulin and IGF-1Rs, the IGF-1R is a prime candidate to compensate for functions normally exerted by IRs. Consistent with this hypothesis, we have observed increased expression of IGF-1Rs in mouse embryos lacking IRs (5).

Figure 1. The effects of IGF-2 are mediated by three different receptors. IGF-2 binds to the mannose-6-phosphate/IGF-2 receptors, the IGF-1R, and the IR. The former mediates clearance of IGF-2, while the two latter ones are responsible to mediate the growth-promoting effects of IGF-2. Starting at mid-gestation, the effects of IGF-2 are mostly exerted through the IGF-1R, so that 90% of the growth effect of IGF-2 goes through an IGF-1R-mediated mechanism. This interaction declines quantitatively to approximately 60% prior to birth. Contrariwise, the interaction of IGF-2 with the IR becomes more prominent during the last part of gestation, rising from 10 to 40%.

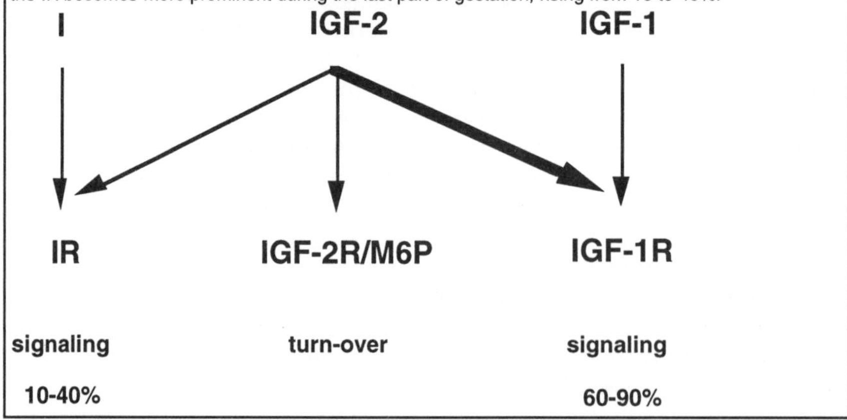

The role of IRs in mouse embryos has been examined in more detail by crossing mice with mutations in the IR, the IGF-1R, the IGF-2 receptor, and IGF-2 (5). Because of the complexity of the experimental approach, it is worthwhile summarizing the conclusions of these studies before dwelling on specific aspects of the data (Table 2). Mice lacking either IGF-1 or IGF-2 are small (~60% of the normal size) (20,21). Thus, both growth factors are required for embryo development. Remarkably, absence of both growth factors is not incompatible with embryo growth, although the latter is greatly impaired (30% of normal size). Thus, the actions of the two growth factors are in part synergistic and in part overlapping (21). Insulin, on the other hand, does not appear to play a major role during mouse embryo development (22). Which receptors mediate the growth-promoting actions of IGF-1 and IGF-2 during embryogenesis? IGF-1Rs are the main, and possibly the only mediators of IGF-1 actions. In contrast, IGF-2 acts through three receptors: insulin, IGF-1 and IGF-2 receptors. Insulin and IGF-1Rs mediate the growth-promoting actions of IGF-2 in mouse embryos, whereas IGF-2/mannose-6-phosphate receptors clear IGF-2 from the circulation and thus limit the ability of IGF-2 to act in a classic endocrine fashion (23) (fig. 1). It is likely that most of IGF-2's actions occur in a paracrine fashion (5). Thus, both insulin and IGF-1Rs are required to support mouse growth.

Let us now proceed more systematically through the evidence supporting the abovementioned conclusions. While knock-outs of the insulin genes and the IR gene are associated with modest (10-20%) growth retardation and lethal metabolic abnormalities (6,7,22), knock-outs of the IGF-

1 gene and its receptor are associated with severe intrauterine growth retardation (~45% of normal size) without metabolic abnormalities (20). These data can be construed to support the time-honored view that IGF-1Rs mediate growth, and IRs mediate metabolic responses. Genetic crosses between IR-deficient and IGF-1R-deficient mice suggest otherwise. Mice lacking both insulin and IGF-1Rs are more severely growth retarded than mice lacking either receptor alone (~30% of normal size) (5). Together with data on the up-regulation of IGF-1Rs in IR null mice, these findings indicate that IRs do affect embryonic growth and development, but in a qualitatively different way from IGF-1Rs. First, embryonic growth curves of single, double, and triple knock-out mice (in the latter, expression of both IR, IGF-1R, and IGF-2 have been abrogated) indicate that IGF-1Rs support embryonic growth in mid-gestation, whereas IRs support embryonic growth in late gestation (5). Second, it is likely that the growth promoting actions of the IR are mediated in response to IGF-2, and not to insulin. This conclusion is supported by the following data: ablation of IGF-1 has the same effect as ablation of IGF-1Rs, suggesting that IGF-1 interacts exclusively with the IGF-1R (20). Likewise, ablation of both insulin 1 and insulin 2 genes has the same effect as ablation of IRs (22). In contrast, inactivation of IGF-2 has the opposite effect compared to inactivation of the IGF-2 receptor. Thus, ablation of IGF-2 is associated with a growth deficient phenotype (24), whereas ablation of the IGF-2/mannose-6-phosphate receptor, is associated with an overgrowth syndrome reminiscent of Beckwith-Wiedemann syndrome in humans (25). Surprisingly, ablation of both IGF-1 and IGF-2 receptors is associated with a normal phenotype at birth . The likeliest explanation of these findings is that the IGF-1R mediates the growth-promoting actions of IGF-2, whereas the IGF-2 receptor serves to clear IGF-2 from the circulation. In the absence of IGF-2 receptors, IGF-2 is not cleared form the circulation and over-stimulates IGF-1Rs, resulting in a lethal phenotype. After ablation of IGF-1 and-2 receptors, excess IGF-2 cannot act through the IGF-1R, but can act at the IR, thus rescuing the phenotype (23). It is apparent that IRs are less potent in their ability to stimulate growth, since, in the absence of IGF-1Rs, IGF-2 is not lethal. These data are consistent with in vitro data indicating that IGF-2 binds to IGF-1Rs with slightly higher affinity than to IRs. Furthermore, Baserga's laboratory has recently shown that IGF-2 is more potent than insulin itself to stimulate cell growth through the IR (26). It remains to be determined whether the growth-promoting actions of the IR are mediated by holodimeric IRs (composed of two α- and two β-subunits) or by heterodimeric receptors composed by an IR α / β monomer and an IGF-1R α / β monomer.

6. Conclusions

Knock-out mice have rapidly become the gold standard to evaluate gene function. Intrinsic limitations to gene ablation strategies have not prevented dangerous generalizations. Obvious as this may sound, conclusions reached in a knock-out mouse may only be applicable to mice and not to humans. The best illustration of this aspect is the different developmental abnormalities observed in mice lacking IRs and in humans with similar mutations. Furthermore, when a given gene exerts a variety of complex functions, it is likely that only those functions required for survival of the mouse will be observed in a knock-out experiment, and that more adaptive functions will be overlooked. Again, the early demise of the IR deficient mice does not allow us to test the role of IRs in more arcane insulin functions, such as in the central nervous system, or in the hematopoietic and immune systems. Conversely, when a gene is a part of a family of genes, compensation by related genes may overshadow the real contribution of the gene of interest to the process under investigation. IRS-1 gene knock-out provides a compelling case for systematically studying all members of a gene family. Finally, it is becoming increasingly clear that genetic diversity among different mouse strains is important in the determination of the phenotype. The effects of modifier genes should not be overlooked, even though their role in human disease remains to be established (27,28). The development of tissue-specific or inducible knock-outs provides a novel strategy to address these issues (29,30).

References

1. Capecchi, M. R. (1989) *Science* **244**(4910), 1288-92
2. Capecchi, M. R. (1994) *Sci Am* **270**(3), 52-9
3. Accili, D., and Suzuki, Y. (1994) in *Molecular Endocrinology: Basic Concepts and Clinical Correlations* (Weintraub, B. D., ed), pp. 95-104, Raven Press, New York, NY
4. Efstratiadis, A. (1996) in *Exper. Clin. Endocrinol. Diabetes* Vol. 104, pp. 4-6
5. Louvi, A., Accili, D., and Efstratiadis, A. (1997) *Dev Biol*, in press
6. Accili, D., Drago, J., Lee, E. J., Johnson, M. D., Cool, M. H., Salvatore, P., Asico, L. D., Jose, P. A., Taylor, S. I., and Westphal, H. (1996) *Nat Genet* **12**(1), 106-109
7. Joshi, R. L., Lamothe, B., Cordonnier, N., Mesbah, K., Monthioux, E., Jami, J., and Bucchini, D. (1996) *EMBO J.* **15**(7), 1542-1547
8. Grupe, A., Hultgren, B., Ryan, A., Ma, Y. H., Bauer, M., and Stewart, T. A. (1995) *Cell* **83**(1), 69-78
9. Wang, N. D., Finegold, M. J., Bradley, A., Ou, C. N., Abdelsayed, S. V., Wilde, M. D., Taylor, L. R., Wilson, D. R., and Darlington, G. J. (1995) *Science* **269**(5227), 1108-12
10. Wertheimer, E., Lu, S. P., Backeljauw, P. F., Davenport, M. L., and Taylor, S. I. (1993) *Nat Genet* **5**(1), 71-73
11. Psiachou, H., Mitton, S., Alaghband, Z. J., Hone, J., Taylor, S. I., and Sinclair, L. (1993) *Lancet* **342**(8876)
12. Krook, A., Brueton, L., and O'Rahilly, S. (1993) *Lancet* **342**(8866), 277-8
13. Jospe, N., Kaplowitz, P. B., and Furlanetto, R. W. (1996) *Clin Endocrinol* **45**, 229-235
14. Accili, D. (1995) *Diabetes Metab Rev* **11**(1), 47-62
15. Taylor, S. I., Cama, A., Accili, D., Barbetti, F., Quon, M. J., de, la, Luz, Sierra, M, Suzuki, Y., Koller, E., Levy, T. R., Wertheimer, E., and et, a. l. (1992) *Endocr Rev* **13**(3), 566-595
16. Schultz, G. A., Hogan, A., Watson, A. J., Smith, R. M., and Heyner, S. (1992) *Reprod Fertil Dev* **4**(4), 361-71
17. Taylor, S. I. (1992) *Diabetes* **41**(11), 1473-1490
18. Tyrala, E. E. (1996) *Obstet Gynecol Clin North Am* **23**, 221-241
19. Naeye, R. L. (1965) *Pediatrics* **35**, 980-988
20. Liu, J. P., Baker, J., Perkins, A. S., Robertson, E. J., and Efstratiadis, A. (1993) *Cell* **75**(1), 59-72
21. Baker, J., Liu, J. P., Robertson, E. J., and Efstratiadis, A. (1993) *Cell* **75**(1), 73-82

22. Duvillie, B., Cordonnier, N., Deltour, L., Dandoy-Dron, F., Itier, J. M., Monthioux, E., Jami, J., Joshi, R. L., and Bucchini, D. (1997) *Proc Natl Acad Sci U S A* **94**(10), 5137-40
23. Ludwig, T., Eggenschwiler, J., Fisher, P., D'Ercole, A. J., Davenport, M. L., and Efstratiadis, A. (1996) *Dev Biol* **177**(2), 517-35
24. DeChiara, T. M., Efstratiadis, A., and Robertson, E. J. (1990) *Nature* **345**(6270), 78-80
25. Wang, Z. Q., Fung, M. R., Barlow, D. P., and Wagner, E. F. (1994) *Nature* **372**(6505), 464-7
26. Morrione, A., Valentinis, B., Xu, S. Q., Yumet, G., Louvi, A., Efstratiadis, A., and Baserga, R. (1997) *Proc Natl Acad Sci U S A* **94**(8), 3777-82
27. Dietrich, W. F., Lander, E. S., Smith, J. S., Moser, A. R., Gould, K. A., Luongo, C., Borenstein, N., and Dove, W. (1993) *Cell* **75**(4), 631-9
28. Bonyadi, M., Rusholme, S. A. B., Cousins, F. M., Su, H. C., Biron, C. A., Farrall, M., and Akhurst, R. J. (1997) *Nat Genet* **15,** 207-211
29. Gu, H., Zou, Y. R., and Rajewsky, K. (1993) *Cell* **73**(6), 1155-64
30. Kuhn, R., Schwenk, F., Aguet, M., and Rajewsky, K. (1995) *Science* **269**(5229), 1427-9

Molecular Mechanisms to Regulate the
Activities of Insulin-like Growth Factors
K. Takano, N. Hizuka and S-I. Takahashi (Editors)

METHODS TO DETECT AND ANALYSE MODIFIED INSULIN-LIKE GROWTH FACTOR BINDING PROTEINS

S. M. Firth

Kolling Institute of Medical Research, Royal North Shore Hospital, St. Leonards, New South Wales 2065, Australia

Abstract

In the last two decades, research into the functional roles of insulin-like growth factor binding proteins (IGFBPs) have advanced rapidly due to the burst of reproducible methodologies developed for identifying, detecting, analysing and purifying the proteins. All six IGFBPs undergo some form of post-translational modifications – phosphorylation, glycosylation or proteolytic modification. The post-translational modifications of the proteins are now recognised as another mechanism and level of regulating the biological actions of IGFBPs. While current methods used in detecting and analysing IGFBPs, like RIAs or SDS-PAGE followed by immuno and ligand blot analysis, could be used in studying post-translationally modified IGFBPs, these techniques may not identify changes in the post-translational modifications. In order to investigate the role of post-translational modifications in the activity of IGFBPs, it is necessary to compare the activities of modified and either "non-modified" or "un-modified" variants of the proteins. This workshop will discuss the current methods available to identify/detect the post-translational states of IGFBPs. "Un-modified" and "non-modified" variants of IGFBPs can be generated by enzymatic modification (dephosphorylation, deglycosylation and proteolysis) of the proteins and recombinant techniques including site-directed mutagenesis. A review of current techniques used in analysing the activity of the modified IGFBP will also be discussed.

1. INTRODUCTION

The insulin-like growth factors (IGF-I and IGF-II) play a role in the regulation of proliferation, differentiation, survival and specific functions of many cell-types [1]. The IGFs are regulated by their interaction with specific receptors and a family of binding proteins (IGFBPs). The IGFBP family comprises of 6 well-characterised and structurally homologous proteins but there have been recent reports of at least two new members to the family [2-5]. Originally, the IGFBPs were believed merely to be transporters since the majority of the IGFs present in circulation and extracellular space are bound to IGFBPs, and that this could prolong the half-lives of the IGFs.

With the advent of studies using cultured cell systems and the availability of purified proteins, it became apparent that the interplay between IGFs and IGFBPs was more complex than a simple ligand-binding protein relationship. The IGFBPs are potent inhibitors of IGF action due to the higher affinity of IGFBPs for IGFs compared to the receptors. In addition to their IGF-inhibitory function, the IGFBPs have been shown to potentiate IGF action (for review, see [1]). More recent studies have now shown that the IGFBPs (IGFBP-3, in particular) have IGF-independent actions [6]. The differential expression of IGFBPs by different cell-types are regulated by various factors including growth factors, hormones and development; this may in turn regulate the functions of the IGFBPs. Another level and mechanism of modulation of the IGF-dependent and -independent actions of IGFBPs may be afforded by the post-translational modifications (proteolysis, phosphorylation, glycosylation) or differential localisation (soluble, associated with the cell-surface or extracellular matrix) of the IGFBPs.

In order to study the effects of specific post-translational modifications on the structure-function of the IGFBPs, it is necessary to compare the activities of the modified and either un-modified or non-modified variants of the protein. Different forms of IGFBPs may be purified from natural sources like serum, other body fluids and cultured cell systems. However, the availability of recombinant proteins and pure enzymes has opened up a whole new spectrum of possibilities in generating customised modified and non-modified proteins. Recombinant DNA technology has enabled the expression of specific fragments to mimic proteolysis products and proteins with specific amino acid changes which disrupt consensus sequences for specific modifications. Commercially available phosphatases, kinases, endoglycosidases and proteases allow for the generation of various modified IGFBPs. This review of the methods currently available for the purification, detection and analysis of modified IGFBPs will focus on phosphorylation, glycosylation and proteolysis.

2. DETECTION AND ANALYSIS OF MODIFIED IGFBPS

The range of techniques that have been developed for detecting the native IGFBPs are in general applicable to modified forms of the proteins [7]. Techniques which are applicable to all forms of modified IGFBPs will be discussed followed by a review of techniques that are specific to each form of modified IGFBP.

2.1 Antibody-based techniques

All antibody-based techniques are reliant on firstly, that the modified forms of IGFBPs have retained the immunoreactivity and secondly, that both the native form used as the standard and the modified IGFBP react similarly to the antibody.

Radioimmunoassays are specific and sensitive for both detecting and quantifying IGFBPs [8]. However, it is necessary to determine that the test protein displays a parallel dilution curve to the standard protein in order to validate the quantitation. In addition, these assays will not differentiate between various forms of the IGFBP in a mixed sample, e.g. proteolysed and intact protein. It has also been reported that changes in the phosphorylation state of IGFBP-1 results in a significant change in its immunoreactivity [9]. Recently, comparison of IGFBP-1 ELISAs developed using a common capture antibody and three different detection antibodies indicated that there was only one assay which was unaffected by the various phosphoforms present in the sample [10]. The future development and use of monoclonal antibodies specific for epitopes on the core protein may alleviate problems with differences in immunoreactivity due to modifications on proteins. Despite the quantitation problems, immunoassays are quick and reliable in the detection of IGFBPs.

Western immunoblot is an important technique for the detection of modified IGFBPs (in particular, proteolysis products). This method involves the separation of proteins on SDS-PAGE [11] followed by the electrophoretic transfer of the proteins to nitrocellulose [12]. After transfer, the nitrocellulose blot is probed with antiserum raised against a specific IGFBP and the complexes detected by iodinated protein A [13, 14]. Significant advances in Western blotting detection systems (chemiluminescence and chromogenic methods) have allowed higher sensitivity and quicker detection of proteins as well as the advantage of being non-radioactive. As with radioactive detection methods, these methods are most commonly used in a purely qualitative sense since the strength of the signal may not be proportional to the amount of protein present on the blot. In addition, there is some evidence that the accumulation of signal is non-linear over time (J. Martin, personal communication).

2.2 Ligand-based techniques

The ligand blotting technique is useful in qualitative comparisons of IGF binding ability among various modified forms of IGFBPs. Following the transfer of proteins to nitrocellulose, the blot is probed with [^{125}I]IGF-I and/or [^{125}I]IGF-II. As with immunoblotting, it is now possible to prepare non-radioactive ligand probes with the added advantage of longer storage life, more sensitivity and quicker detection [15]. Another ligand which has been used successfully in the detection of IGFBP-3 by ligand blotting is [^{3}H]heparin [16]. A negative result from IGF ligand blotting must be interpreted with caution. For example, proteolysed IGFBP-3 in pregnancy serum is measurable by solution binding assays but is undetectable by ligand blotting [17, 18]. The proteolysed IGFBP-3 in pregnancy serum appears to bind native IGF-I but not iodo-IGF-I [17]. This raises the question of relying on ligand blotting as the only method in determining IGF binding ability in modified IGFBPs.

Another popular ligand-based technique used in detecting and characterising IGFBPs is affinity cross-linking. The IGFBP is incubated with radiolabelled IGF in the presence of a cross-linking agent like disucciminidyl suberate (DSS). The cross-linked complexes are then separated on SDS-PAGE and the radiolabelled bands are quantified by densitometry. The concentration of the proteins and cross-linking agent and time and temperature of the incubation has to be optimised in order to prevent cross-linking of non-specific interactions. This method can detect IGF-IGFBP interactions of low affinity which are not detected by other methods. For example, recombinant fragments of IGFBP-3 can be affinity-labelled and yet showed no binding to IGFs when assessed by ligand blotting and solution binding assays (S. Firth and R. Baxter, unpublished results).

A number of assays have been developed for measuring the binding of IGFBPs to IGFs in solution. Following incubation of radiolabelled IGF with IGFBP, free IGF is separated from IGF–IGFBP complexes by charcoal adsorption [19]. Alternatively, the complexes may be precipitated with IGFBP-specific antibodies [20] and polyethylene glycol [21]. In addition, lectins like concanavalin A or wheat germ agglutinin can be used if the IGFBP is glycosylated [20]. Another technique for analysing IGF-IGFBP interactions is gel filtration chromatography which provides high resolution separation of monomeric from dimeric molecules. Regardless of the method of separation used in distinguishing bound from free ligand, it should be noted that the binding assay in solution is affected by the ionic strength and pH of the buffer and the time and temperature of the incubation [22, 23]. Biosensors have been used in analysing the interaction between IGFs and IGFBP-3 [24, 25]. Besides being a non-radioactive method, biosensors have the advantage of yielding real-time kinetics and both association and dissociation constants of the interaction.

2.3 Detecting cell-surface or extracellular matrix –associated IGFBPs

The technique of affinity cross-linking has been applied to the detection of IGFBPs associated with the cell-surface or extracellular matrix. Radiolabelled IGF-I was incubated with cell monolayers pretreated with IGFBP-3 in the presence of DSS and the cell lysates were analysed by SDS-PAGE [26]. Alternatively, crude microsomal membranes have been affinity-labelled by [^{125}I]IGFs and analysed by SDS-PAGE to detect membrane-associated IGFBP-3 [6]. Immunohistochemistry has been used for the detection of cell-surface-associated IGFBP-3 [27]. Another immunological technique utilises the binding of [^{125}I]protein A to the IGFBP-specific antibody [28]. Cell-binding assays are used to demonstrate the binding ability of IGFBPs. Cell monolayers are incubated with radiolabelled IGFBP-3 and bound tracer can be determined by counting the radioactivity in the solubilised cell lysates [6].

In various studies, IGFBP-1 to -5 have all been shown to bind to cells [28-32]. Although both IGFBP-1 and -2 have a tripeptide RGD integrin recognition sequence in their carboxyterminal region, only IGFBP-1 has been shown to adhere to a cell-surface integrin,

$\alpha_5\beta_1$ [33]. IGFBP-3 and -5 have been shown to bind to cell-surfaces via a highly basic sequence in their respective carboxyterminal region since mutagenesis of the basic residues led to a reduction in cell-binding [34, 35]. The involvement of a heparin binding domain in the same vicinity has been tested with synthetic peptides [29].

3. GENERATION AND PURIFICATION OF MODIFIED IGFBP

The choice of purification protocols will depend on the source (natural or recombinant) of the modified IGFBP and the presence of either IGFs or other IGFBPs. It is also essential to have had some initial characterisation of the IGFBP. Purification can be achieved by IGF, immuno, heparin and lectin affinity chromatography. The choice of IGF (I or II) as the ligand may be important if the modified IGFBP has preferential affinity for one over the other; for example, IGFBP-6 has 10-100–fold higher affinity for IGF-II [36]. Heparin affinity chromatography was used in the purification of recombinant IGFBP-3 [37]. It can also be used to separate the IGFBPs due to their differential affinities for heparin; IGFBP-1 did not bind to heparin, IGFBP-2, -4 and -6 eluted with low NaCl concentration while IGFBP-3 and -5 eluted with high NaCl concentration [29].

The availability of kinases, phosphatases and glycosidases has enabled *in vitro* phosphorylation, dephosphorylation and deglycosylation of IGFBPs. Non-glycosylated IGFBP can be expressed in *Escherichia coli.* Several proteases have been identified for IGFBP-2 to -6 [38-40] and it is now possible to generate proteolysed fragments by incubating pure IGFBPs with either pure, commercially available proteases or biological samples containing known protease activity [17, 41-43]. If the cleavage site(s) of a protease is known or the N-terminal sequence of a proteolysed fragment has been determined, recombinant proteins can be generated which mimic specific proteolysis products [44] or specific regions of IGFBPs [34, 45].

In addition, recombinant DNA technology can be employed to introduce specific amino acid substitutions which disrupt potential phosphorylation [46, 47], glycosylation [48], proteolysis [49] and cell-binding [34] sites. For modified IGFBPs which have markedly reduced or lost the various affinities to enable purification, there are several expression systems which generate either N-terminal or C-terminal peptide fusions. Affinity resins specific to each peptide tag can then be used to purify the fusion protein and the peptide tag is usually removed by enzymatic cleavage. Specific antibodies to the peptide tag are also available for analytical or purification purposes.

4. SPECIFIC TECHNIQUES

4.1 Detection and analysis of phosphorylated IGFBPs

Phosphorylation is an important mechanism of acute and reversible regulation of protein function. Among the IGFBPs, phosphorylated IGFBP-1, -3 and -5 have been detected [50-52]. Although potential phosphorylation sites have been identified in the amino acid sequences of the other IGFBPs [53], there has been one preliminary report which indicated that IGFBP-6 is not phosphorylated [54].

Protein phosphorylation is usually studied by metabolic labelling of cells in culture with [^{32}P]orthophosphate in phosphate-free medium. The IGFBP of interest is then immunoprecipitated from the conditioned medium and analysed by either denaturing or non-denaturing PAGE [50, 55]. It should be noted that attempts to determine the number of IGFBP-3 phosphoforms by non-denaturing PAGE analysis have so far been unsuccessful ([47]; J. Coverley, personal communication). Alkaline phosphatase treatment of [^{35}S]methionine-labelled IGFBP-1 converted the 5 forms resolved on non-denaturing PAGE to a single band, clearly indicating that the 4 most negative forms were phosphovariants of IGFBP-1 [50].

IGFBP-1 phosphoforms in plasma and amniotic fluid have been detected by ligand blotting of immunoprecipitated protein separated on non-denaturing PAGE [9]. Alternatively, IGFBP-1 phosphoforms can be detected by immuno blotting of samples electrophoresed using n-octyl glucoside instead of SDS to permit resolution of phosphorylated and non-phosphorylated variants [56, 57]. Phosphoforms of IGFBP-1 have been separated and purified from decidual and amniotic fluid and serum by anion-exchange chromatography [30, 56, 57].

Most proteins are phosphorylated at serine, threonine or tyrosine residues and identification of the phosphoamino acid is accomplished by acid hydrolysis and two-dimensional thin-layer electrophoresis. Phosphoamino acid analysis identified phosphoserine in IGFBP-1 [50, 56] and IGFBP-3 [51], and both phosphoserine and phosphothreonine in IGFBP-5 [52]. Immuno blotting or immuno precipitating with commercially available phosphoserine-, phosphothreonine- and phosphotyrosine-specific antibodies is as yet an untried technique for detecting the type of phosphoacceptor sites in IGFBPs. The specific site(s) of phosphorylation in IGFBP-1 [46] and IGFBP-3 [47] were identified by different methods.

In IGFBP-1, the [^{32}P]orthophosphate-labelled protein was digested with trypsin and endoproteinase Glu-C and purified by HPLC. The recovered radiolabelled fractions were then subjected to radiosequencing; the concomitant release of radioactivity when an amino acid is cleaved from the peptide identified that amino acid as the site of phosphorylation [46]. In IGFBP-3, endoproteinase Lys-C digestion of [^{32}P]orthophosphate-labelled protein indicated that the majority of phosphorylation sites were located in the carboxyterminal 17kDa fragment. Instead of radiosequencing, the investigators performed several computational analyses of the amino acid sequence of the fragment and identified 2 serine residues which together with surrounding residues conformed to the consensus sequence of the casein kinase II phosphorylation site [47]. In both IGFBP-1 and IGFBP-3, the serine residues were changed to alanine by site-directed mutagenesis to examine the effect of blocking phosphorylation [46, 47]. While phosphorylation was markedly reduced in the site-directed mutants, it was not completely abolished suggesting that there are as yet unidentified minor phosphorylation sites in both proteins.

Phosphorylation of IGFBP-1 can be demonstrated *in vitro* by incubation with [γ-^{32}P]ATP and either protein kinase A or casein kinase II [50, 57] while phosphorylation of IGFBP-3 has been shown with casein kinase II, protein kinase A and protein kinase C (J. Coverley and R. Baxter, unpublished results; [53]). Dephosphorylation of IGFBP-1 by alkaline phosphatase digestion [9, 57] and of IGFBP-3 by either alkaline phosphatase or acid phosphatase digestion (J. Coverley and R. Baxter, unpublished results; [51]) has also been shown *in vitro*.

Comparison of phospho and either nonphospho or dephospho forms of IGFBP-1 (regardless of whether the sample was of natural, recombinant or *in vitro* origin) indicated that the state of phosphorylation has a major effect on the activity of the protein. The phosphoforms have higher affinities for IGF-I and inhibit their biological activity, whereas the nonphospho or dephospho forms have lower affinities for IGF-I and stimulate IGF-I action [46, 57]. Alterations in the phosphorylation form of IGFBP-1 in different states of health and disease may therefore influence the modulatory role of IGFBP-1 on IGF bioavailability [9, 56, 58, 59]. In contrast to IGFBP-1, the phosphorylation state of IGFBP-3 had no influence on its IGF-I affinity but dephosphorylated IGFBP-3 appears to have a slight increase in its affinity for ALS [47, 51]. Phosphorylation of IGFBP-3 appears to be stimulated by IGF-I and there is some evidence that the cell-associated form of IGFBP-3 may either be non-phosphorylated or is dephosphorylated when released from the cell-surface [55].

Of the remaining questions in the study of phosphorylated IGFBPs, the most challenging would be the determination of the location of the phosphorylation event. It remains to be shown as to whether phosphorylation occurs intracellularly before the protein is secreted, or extracellularly after the secretion process.

4.2 Detection and analysis of glycosylated proteins

Glycosylation has important functions in determining conformation, secretion, antigenicity and clearance of glycoproteins. Glycosylation occurs in IGFBP-3–6 and there appears to be more than one glycoform for each of the IGFBP [20, 36, 60-62]. There are three potential *N*-glycosylation sites in IGFBP-3 while IGFBP-4 has one; IGFBP-5 and -6 appear to be only *O*-glycosylated. The techniques used in determining the presence of carbohydrates on the IGFBPs include carbohydrate staining of proteins after SDS-PAGE using periodic acid-Schiff's reagent [20], deglycosylation by endoglycosidases [60, 62], lectin affinity [36] and metabolic labelling in the presence of tunicamycin, an inhibitor of *N*-glycosylation [63].

A variety of lectins of known specificity are readily available for the identification of specific oligosaccharide side-chains in glycoproteins. Specific glycosidases which can help in determining the oligosaccharide profile are also available; for example, neurominidase which specifically removes sialic acids. In the last few years, advances in methods development have resulted in several kits for determining whether proteins are glycosylated, for deglycosylating proteins and determining the extent of glycosylation, for obtaining *N*- or *O*-linked oligosaccharide profiles of glycoproteins, for sequencing *N*-linked oligosaccharides obtained from glycoproteins and for determining the monosaccharide composition of glycoproteins or oligosaccharides. While such fine analysis has not been carried out on any of the IGFBPs, there has been some evidence that desialylated ALS has decreased affinity for the IGF–IGFBP-3 binary complex (J. Janosi, P. Delhanty and R. Baxter, unpublished results).

Recombinant non-glycosylated IGFBP-3 has been expressed in *Escherichia coli* [24], while proteins with mutated *N*-glycosylation sites have been expressed in Chinese hamster ovary cells [48]. The panel of IGFBP-3 mutant proteins which has one, two or three *N*-glycosylation sites knocked out led to the identification that the process of glycosylation adds approximately 4, 5 and 6kDa of carbohydrate on ^{89}N, ^{109}N and ^{172}N, repectively. There is variable occupancy on ^{172}N which results in the characteristic doublet of 40-45kDa observed for IGFBP-3 [48].

Comparison of recombinant glycosylated and non-glycosylated IGFBP-3 suggests that glycosylation is not essential for IGF-I, ALS or cell-binding [24, 26, 48]. Non-glycosylated IGFBP-3 is as efficient as its glycosylated counterpart in modulating IGF-I activity [26]. Similarly, enzymatic deglycosylation of IGFBP-6 by neuraminidase, fucosidase and *O*-glycanase had no effect on its affinity for either IGF-I or IGF-II [62]. Interestingly, one study has provided evidence for *in vitro* non-enzymatic glycation of IGFBP-3 by incubating IGFBP-3–containing culture media with [U-^{14}C]glucose and analysing the radiolabelled proteins by SDS-PAGE [64]. Scatchard analysis of the competition assays for [^{125}I]IGF-I binding showed that both glycated and non-glycated IGFBP-3 had similar affinity constants. However, the glycated samples bound a larger amount of [^{125}I]IGF-I compared to control which was reflected as an increase in the number of binding sites and glycated IGFBP-3 appears to blunt the potentiation of IGF-I activity in fibroblasts [64]. It remains to be shown that this process occurs *in vivo*, which could have physiological significance in diseases like diabetes.

4.3 Detection and analysis of proteolysed proteins

Significant inroads have been made in the study of IGFBP proteolysis since the surge of reports in the early 1990s. Proteolysis has been shown to occur with IGFBP-2 to -6, and proteolytic activity has been reported in serum in various health and disease states as well as in specific tissues and cell culture systems. Limited proteolysis of IGFBPs has been proposed as a mechanism for modulating both IGF bioavailability in the circulation and IGF activity in the cellular environment. Recently, it has been shown that a proteolytic fragment of IGFBP-3 may have intrinsic biological activity of its own [65]. Several proteases have been identified and they include members of the serine protease, cysteine protease and metallo-protease families. Limited proteolysis results in stable, lower molecular weight fragments of IGFBPs which have either retained or lost IGF affinity and/or modulation of IGF activity. Proteolysis of IGFBPs can be regulated by IGFs and other IGFBPs. It is beyond the scope of this paper to review the extensive literature available but several reviews which discuss IGFBP proteolysis have been published [1, 7, 38, 66].

Modification to IGFBPs by limited proteolysis is usually observed as the disappearance of the intact IGFBP band from either immuno or ligand blots. This may be accompanied by the appearance of one or more smaller molecular weight forms. Studies usually involve the incubation of purified or recombinant IGFBP with serum or cell culture media prior to electrophoresis. Alternatively, IGFBPs in serum or endogenously produced IGFBPs in culture media can be used as substrates for proteolysis. Results from this fairly crude method may be hard to interpret due the presence of IGFs, IGFBPs, multiple proteases and protease inhibitors. Another important consideration would be the potential loss of IGF affinity or immuno affinity of the proteolysed fragment, which would then render detection by ligand or immuno blot difficult or useless. This can be overcome by the use of radiolabelled IGFBP as the substrate.

Once proteolytic activity is determined in a sample, the protease involved is identified by its characteristics - the pH dependence of activity, susceptibility to specific protease inhibitors and N-terminal sequencing of proteolytic products to determine cleavage site specificity. Another way of characterising the protease is by zymography, where the protease-containing sample is electrophoresed in a gel containing the IGFBP substrate compared to one containing a known substrate like gelatin [67]. If the IGFBP protease bears similarity to a known protease, this can then be confirmed by reproducing the proteolysis with pure protease [41, 68] or by specifically inhibiting the action of the protease by immunoadsorption [69].

Whilst there have been reports of cell-conditioned media or biological fluid (that is unpurified protease) which can proteolyse more than one IGFBP [70-72], most proteases identified to date are specific for each IGFBP. However, there are exceptions like prostate specific antigen which has been shown to proteolyse IGFBP-3 to -5 [73]. The protease cleavage sites are usually in the non-conserved central region which is unique to each IGFBP.

Studies examining the regulation of protease activity usually involved the use of IGFs and IGFBPs other than the substrate in the inhibition or stimulation of the protease activity [43, 74-77]. The use of IGF analogs which have reduced affinity for IGFBPs can yield useful information regarding the regulatory role of IGF [43, 78]. Synthetic peptides representing regions of the substrate IGFBP have also been used as inhibitors of proteolysis [70, 79].

Since the amino acid sequences of the IGFBPs are known, the mapping of the proteolysed fragments by N-terminal sequencing is most conclusive. However, this requires isolation and purification of the fragments and several of the fragments sequenced to date have yielded the natural N-terminal sequence [16, 67]. If the antigenic specificity of an IGFBP-antibody is known, and monoclonal antibodies are available [44, 80], it is possible to map the proteolysed fragment to a fairly specific domain of the IGFBP.

The effects of proteolysis are usually analysed by comparing the binding and/or biological activities of the proteolysed fragment to that of the intact IGFBP. Mutants of IGFBP-4 which are relatively resistant to proteolysis have been generated by altering the cleavage site sequence [49]. Proteolysis has resulted in IGF-binding fragments, non-IGF-binding fragments, fragments which still inhibit/potentiate IGF action, fragments which no longer inhibit/potentiate IGF action and an IGFBP-3 fragment which has intrinsic inhibitory effects on mitogenic signals [38, 65]. The physiological significance of each protease/IGFBP system has to be rigorously tested given that some of the proteolysis shown to date uses a large molar excess of the protease to the IGFBP substrate which may also lead to adventitious proteolysis.

5. SUMMARY

Post-translational modifications of the IGFBPs add greatly to the complexity of the IGF–IGFBP system. The development of a wide-range of techniques have greatly aided their characterisation and it is becoming clear that modifications to IGFBPs can modulate the way in which they affect IGF activity. Research into the effects of modifications to IGFBPs have concentrated mainly on proteolysis and, to a lesser extent, on phosphorylation and glycosylation. It remains a challenge to investigators to unravel the interplay among the IGFs, the IGF receptors and the IGFBPs and their variants in the regulation of growth and development.

REFERENCES

1. Jones JI, Clemmons DR Endocrine Rev 1995;16:3-34.
2. Swisshelm K, Ryan K, Tsuchiya K, Sager R Proc Natl Acad Sci USA 1995;92:4472-4476.
3. Oh YM, Nagalla SR, Yamanaka Y, *et al* J Biol Chem 1996;271:30322-30325.
4. Rosenfeld RG, Oh Y Proceedings, 79th Annual Meeting, The Endocrine Society, Minneapolis, Minnesota 1997;p59.
5. Kim HS, Nagalla SR, Oh Y, *et al* Proceedings, 79th Annual Meeting, The Endocrine Society, Minneapolis, Minnesota 1997;p352.
6. Oh Y, Muller HL, Lamson G, Rosenfeld RG J Biol Chem 1993;268:14964-14971.
7. Holly JMP, Martin JL Growth Regul 1994;4:20-30.
8. Blum WF, Breier BH Growth Regul 1994;4:11-19.
9. Westwood M, Gibson JM, Davies AJ, *et al* Clin Endocrinol Metab 1994;79:1735-1741.
10. Khosravi MJ, Diamandi A, Mistry J Clin Chem 1997;43:523-532.
11. Laemmli UK Nature 1970;227:680-685.
12. Towbin H, Staehelin T, Gordon J Proc Natl Acad Sci, USA 1979;76:4350-4354.
13. Burnette WN Anal Biochem 1981;112:195-203.
14. Hossenlopp P, Seurin D, Segovia-Quinson B, *et al* Anal Biochem 1986;154:138-143.
15. Fowlkes JL, Serra D Endocrinology 1996;137:5751-5754.
16. Booth BA, Boes M, Bar RS Am J Physiol 1996;271(Endocrinol & Metab 34):E465-E470.
17. Suikkari AM, Baxter RC J Clin Endocrinol Metab 1991;73:1377-1379.
18. Suikkari AM, Baxter RC J Clin Endocrinol Metab 1992;74:177-183.
19. Zapf J, Waldvogel M, Froesch ER Arch of Biochem and Biophys 1975;68:638-645.
20. Martin JL, Baxter RC J Biol Chem 1986;261:8754-8760.
21. Clemmons DR, Dehoff ML, Busby WH, *et al* Endocrinology 1992;131:890-895.
22. Bach LA, Rechler MM Biochim Biophys Acta 1996;1313:79-88.
23. Holman SR, Baxter RC Growth Regul 1996;6:42-47.
24. Sommer A, Spratt SK, Tatsuno GP, *et al* Growth Regul 1993;3:46-49.
25. Heding A, Gill R, Ogawa Y, *et al* J Biol Chem 1996;271:13948-13952.
26. Conover CA Endocrinology 1991;129:3259-3268.
27. Oh Y, Muller HL, Pham H, *et al* Endocrinology 1992;131:3123-3125.
28. Martin JL, Ballesteros M, Baxter RC Endocrinology 1992;131:1703-1710.
29. Booth BA, Boes M, Andress DL, *et al* Growth Regul 1995;5:1-17.
30. Busby WH, Klapper DG, Clemmons DR J Biol Chem 1988;63:14203-14210.
31. Russo VC, Bach LA, Werther GA Prog Growth Factor Res 1995;6:329-336.
32. Hasegawa T, Cohen P, Hasegawa Y, *et al* Growth Regul 1995;5:151-159.
33. Jones JI, Gockerman A, Busby WHJ, *et al* Proc Natl Acad Sci USA 1993;90:10553-10557.
34. Baxter RC, Firth SM Prog Growth Factor Res 1995;6:215-222.
35. Parker A, Clarke JB, Busby WHJ, Clemmons DR J Biol Chem 1996;271:13523-13529.
36. Martin JL, Willetts KE, Baxter RC J Biol Chem 1990;265:4124-4130.
37. Tressel TJ, Tatsuno GP, Spratt K, Sommer A Biochem Biophys Res Commun 1991;178:625-633.

38. Baxter RC Adv in Mol Cell Endocrinol 1997;1:123-159.
39. Claussen M, Buergisser D, Schuller AGP, *et al* Mol Endocrinol 1995;9:902-912.
40. Rajah R, Bhala A, Nunn SE, *et al* Endocrinology 1996;137:2676-2682.
41. Fielder PJ, Rosenfeld RG, Graves HCB, *et al* Growth Regul 1994;1:164-172.
42. Lalou C, Lassarre C, Binoux M Hormone Res 1996;45:156-159.
43. Salahifar H, Baxter RC, Martin JL Endocrinology 1997;138:1683-1690.
44. Vorwerk P, Oh Y, Lee PDK, *et al* Proceedings, 79th Annual Scientific Meeting, The Endocrine Society, Minneapolis, Minnesota 1997;p346.
45. Spencer EM, Chan K Prog Growth Factor Res 1995;6:209-214.
46. Jones JI, Busby WHJ, Wright G, *et al* J Biol Chem 1993;268:1125-1131.
47. Hoeck WG, Mukku VR J Cell Biochem 1994;56:262-273.
48. Firth SM, Baxter RC Prog Growth Factor Res 1995;6:223-229.
49. Conover CA, Durham SK, Zapf J, *et al* J Biol Chem 1995;270:4395-4400.
50. Frost RA, Tseng L J Biol Chem 1991;266:18082-18088.
51. Mukku VR, Chu H, Baxter R Proceedings, 2nd International IGF Symposium, San Francisco, California 1991;237.
52. Jones JI, Gockerman A, Clemmons DR Proceedings, 74th Annual Meeting, The Endocrine Society, San Antonio, Texas 1992;p372.
53. Coverley JA, Baxter RC Mol Cell Endocrinol 1997;128:1-5.
54. Marinaro JA, Bach LA Abstracts, 6th Biennial IGF Symposium, Sydney, Australia 1996;14.
55. Coverley JA, Baxter RC Endocrinology 1995;136:5778-5781.
56. Jones JI, D'Ercole AJ, Camacho-Hubner C, Clemmons DR Proc Natl Acad Sci USA 1991;88:7481-7485.
57. Koistinen R, Angervo M, Leinonen P, *et al* Clin Chim Acta 1993;215:189-199.
58. Frost RA, Berekat A, Wilson TA, *et al* J Clin Endocrinol Metab 1994;78:1533-1535.
59. Frost RA, Fuhrer J, Steigbigel R, *et al* Clin Endocrinol 1996;44:501-514.
60. Ceda GP, Fielder PJ, Henzel WJ, *et al* Endocrinology 1991;128:2815-2814.
61. Camacho-Hubner C, Busby WH, McCusker RH, *et al* J Biol Chem 1992;267:11949-11956.
62. Bach LA, Thotakura NR, Rechler MM Biochem Biophys Res Commun 1992;186:301-307.
63. Martin JL, Baxter RC Growth Regul 1992;2:88-99.
64. Cortizo AM, Gagliardino JJ J Endocrinol 1995;144:119-126.
65. Zadeh SM, Binoux M Endocrinology 1997;138:3069-3072.
66. Clemmons DR Adv Exp Med Biol 1993;343:245-253.
67. Fowlkes JL, Enghild JJ, Suzuki K, Nagase H J Biol Chem 1994;269:25742-25746.
68. Cohen P, Graves HCB, Peehl DM, *et al* J Clin Endocrinol Metab 1992;75:1046-1053.
69. Conover CA, Perry JE, Tindall DJ J Clin Endocrinol Metab 1995;80:987-993.
70. Chernausek SD, Smith CE, Duffin KL, *et al* J Biol Chem 1995;270:11377-11382.
71. Matsumoto T, Gargosky SE, Kelley K, Rosenfeld RG Growth Regul 1996;6:185-190.
72. Besnard N, Pisselet C, Monniaux D, Monget P Biol Reprod 1997;56:1050-1058.
73. Lee KO, Oh Y, Giudice LC, *et al* J Clin Endocrinol Metab 1994;79:1367-1372.
74. Fowlkes J, Freemark M Endocrinology 1992;131:2071-2076.
75. Conover CA, Kiefer MC, Zapf J J Clin Invest 1993;91:1129-1137.
76. Donnelly MJ, Holly JMP J Endocrinol 1996;149:R1-R7.
77. Nam TJ, Busby WHJ, Clemmons DR Endocrinology 1994;135:1385-1391.
78. Irwin JC, Dsupin BA, Giudice LC J Clin Endocrinol Metab 1995;80:619-626.
79. Fowlkes JL, Thrailkill KM, George-Nascimento C, *et al* Endocrinology 1997;138:2280-2285.
80. Schuller AGP, J. Lindenbergh-Kortleve DJ, de Boer WI, *et al* Growth Regul 1993;3:32-34.

Molecular Mechanisms to Regulate the
Activities of Insulin-like Growth Factors
K. Takano, N. Hizuka and S-I. Takahashi (Editors)

IGFBP Proteases - Physiology/Pathophysiology

Jeff M.P. Holly, Laura A. Maile, Sian C. Cwyfan Hughes, Janet K. Fernihough and Su Xu

Division of Surgery, University Department of Hospital Medicine, Bristol Royal Infirmary, Bristol BS2 8HW, UK.

1. INTRODUCTION

In addition to the ligands, cell receptors and IGFBPs it has become increasingly clear that IGFBP proteases form further integral components of the IGF system. The complex system of six high affinity IGFBPs are clearly not just there to sequester IGFs away from cell receptors since unbound IGFs are very rapidly cleared and would not accumulate to high levels without the IGFBPs. As soon as it became apparent that these IGFBPs maintained extracellular IGFs at concentrations far greater than that required for maximal cell stimulation, it was evident these complexes must be latent and that there had to be a mechanism for controlling the availability of IGF from these complexes. Estimates of affinity constants with which the IGFs bound to the IGFBPs and to the signalling type I IGF receptor confirmed that the equilibrium strongly favours the former. This implied that mechanisms must exist whereby the affinity of binding to the IGFBPs could be lowered to enable receptor activation. Since the initial observations of IGFBP proteolysis in the circulation of pregnant mothers [1,2] it has gradually become apparent that limited proteolysis of IGFBPs with a resultant decrease in affinity for IGFs is a general phenomenon and hence may be the mechanism whereby the latent stores of IGFs may be made available for biological actions in a controlled manner.

The study of *in vivo* IGFBP proteolysis has been hampered because the actual enzymes responsible have remained unidentified, despite considerable attention. This means that they have been studied by indirect means, with measurements of their activity or of their effects. Activity has generally been measured in terms of the relative amount of IGFBP substrate cleaved following a set incubation time. As activity may vary due to varying amounts of more than one protease, varying amounts of substrates and/or cofactors and varying amounts of inhibitors; such measures provide fairly limited information. The activity found also depends considerably upon the conditions of the assay, the nature of the substrate used and the length of time incubated.

The effects of proteolysis has generally been measured by assessments of relative levels of intact and fragmented IGFBP by Western blotting. In this respect ligand blotting has proven of limited value since it does not discriminate between changes in amount or changes in function of an IGFBP. Western immunoblotting has been informative since proteolytically modified IGFBPs run as lower molecular weight bands distinct from intact IGFBP; although prior to electrophoresis the fragments may well remain in association. The relative proportion of intact and fragmented forms of a particular IGFBP can therefore be assessed; although it should always be remembered that such measures are not precisely quantitative since the relative recognition of intact IGFBP and different fragments by antisera is unknown and hence the sensitivity of detection of the different forms may vary considerably. This approach is also limited since the relative presence of intact and fragmented IGFBP may vary due to differences in their relative clearance rates and fragments may be present *in vivo* without any protease activity if fragmented IGFBP is perfusing from another compartment. Decreases in protease activity may also be difficult to assess with such an approach since detection of cessation of proteolysis would depend upon the rate at which fragmented IGFBP was cleared and the rate at which new intact IGFBP is produced.

2. CIRCULATING IGFBP PROTEASES

2.1. Occurrence

Although originally described in the circulation in late pregnancy when virtually all the IGFBP-3 present is proteolysed; it soon became apparent that circulating IGFBP-3 proteases occur to varying extents in many different conditions. We have reported increased circulating IGFBP proteolysis in critical illness [3], following surgery [4], following trauma [5], in cancer patients [6], in growth hormone insensitivity [7] and in insulin-dependent diabetes [8]. Others have confirmed these findings and reported increases in other conditions, noticeably malnutrition [9]. In pregnancy all circulating IGFBPs with the exception of IGFBP-1 are cleaved. The activities directed against individual IGFBPs differ in their sensitivity to inactivation and inhibition and probably are due to individual specific enzymes [10]; although one enzyme could have preference for different substrates and respond in such a manner. However, in virtually all other conditions the activity appears to be primarily directed towards IGFBP-3, with the other IGFBPs relatively unaffected. The one exception to this that we have found was with bypass surgery when following the opening of the chest there was a transient appearance of an IGFBP-4 protease in the circulation, cleaving most of the IGFBP-4 present [4]. Such an IGFBP-4 protease was not observed following other surgical procedures [11]. This all implies that each IGFBP has a corresponding specific protease, although it remains feasible that such specificity could be conferred by another modification or a cofactor.

Like others we have found that a significant proportion of IGFBP-3 and other IGFBPs (particularly IGFBP-2 and -4) are generally present in the circulation in fragmented forms even in perfectly healthy individuals. Whilst IGFBP protease activity can be detected in such individuals following prolonged incubations, it would appear that physiologically such protease activity is probably small and the fragments may well be generated within other compartments.

2.2. Regulation

At present there is still only limited data regarding the regulation of circulating IGFBP proteases. The increase in IGFBP-3 protease activity in GH insensitivity [7] and a report that proteolysis of IGFBP-3 was inversely related to

GH status [12] suggested that there may be a feedback mechanism whereby IGFBP-3 proteolysis is altered according to IGF-I availability. However, many studies, including some from our group, have failed to find any marked alteration in IGFBP proteolysis following administration of exogenous IGF-I. These studies would appear to indicate that there is no direct feedback of IGF-I controlling its own availability by regulating IGFBP-3 proteolysis.

In critically ill patients we reported that IGFBP-3 protease activity increased during a 24 hour fast and decreased with refeeding [3]. In seriously malnourished subjects an increase in IGFBP-3 protease activity has also been reported to be decreased with refeeding [9]. These reports suggest that this protease is under some form of metabolic activity.

The start of insulin therapy in newly diagnosed children with insulin-dependent diabetes mellitus (IDDM) has been reported to result in a decrease in IGFBP-3 protease activity [13]. However in addition to being insulin-deficient these subjects were initially ketotic and presumably catabolic, both of which would have been corrected with insulin therapy and hence it was not possible to deduce a direct effect of insulin. We have recently re-addressed this question in hyperinsulinaemic clamp studies of subjects with IDDM. Subjects with good glycaemic control received insulin infusions to achieve hyperinsulinaemia throughout the night but with euglycaemia maintained by a continually adjusted glucose infusion. This resulted in a significant consistent decrease in IGFBP-3 protease activity. This indicates that insulin may directly affect IGFBP-3 proteolysis over a matter of hours. In our overnight study the affect of insulin was relatively rapid, within 6 hours activity was decreased by 40-50%, whilst in the previous report of newly diagnosed subjects IGFBP-3 protease was unaffected after one day but normalised after one week. The discrepancy in the time at which the effect became apparent probably relates to the other complicating factors in the newly diagnosed subjects.

The other situation in which we have been able to intervene and manipulate circulating IGFBP-3 protease activity has been in cancer patients. In women with breast cancer we found that therapy with the synthetic progestin, megestrol acetate led to a significant decrease in IGFBP-3 protease activity [14]. However megestrol acetate has well documented effects stimulating appetite and increasing insulin levels and the changes in IGFBP-3 protease activity may have been secondary to these affects on metabolism. We have also found that treatment of breast cancer patients with the anti-estrogen tamoxifen also resulted in a reduction of IGFBP-3 protease activity [15]. However the effect was only apparent in the patients showing a clinical response with a reduction in tumour burden. In patients with disease progression despite the tamoxifen treatment there was an increase in IGFBP-3 protease activity. This suggested that the change in IGFBP-3 protease may not have been a direct effect of the tamoxifen but may relate to tumour burden and be consequent to a tumour-host interaction.

2.3. Function

Although we initially showed that the proteolysis of IGFBP-3 in pregnancy serum could increase the IGF biological activity measured in a cell line based bioassay [8] and this was confirmed by others [16]; it was always apparent that this was a very artificial model. We have shown that proteolysis of IGFBP-3 does not disturb the distribution of IGFBP-3 in serum samples, the majority still remains in a ternary complex with the ALS. As there is a considerable concentration gradient for IGFs and IGFBP-3 between serum and extravascular fluids; this complex is relatively restricted to within the vasculature. Therefore IGFBP-3 that is proteolysed within this ternary complex would not normally directly result in increasing IGF availability to target tissue receptors as these are separated by the endothelial barrier. However, proteolysis of IGFBP-3 in the circulation has been reported to decrease the circulating half-life of IGF [17] indicating that it is more readily available for transport out of the circulating pool. Whilst much remains to be learned regarding the capillary transfer of IGF, these data imply a number of possible consequences of circulating IGFBP proteolysis. A decrease in affinity with which the IGFs are bound to IGFBP-3 may result in re-equilibration of the IGF to other IGFBPs (such as IGFBP-1, -2, etc.) or to endothelial cell receptors which may then transport the IGF into the tissues. Although present at much lower concentrations the other IGFBPs only form binary complexes with much shorter circulating half-lives and may therefore represent a significant flux of IGF between the tissues and the circulating reservoir. In addition there are a number of tissues in the body where blood perfusion is sinusoidal rather than by capillary bed perfusion, such as the liver, pancreas and pituitary. In such tissues there may be less of a barrier between blood and parenchymal cells such that proteolysis of IGFBP-3 may result in a direct increase in availability of IGFs to the target receptors.

3. EXTRA-VASCULAR IGFBP PROTEASES

3.1. Occurrence

Whilst many general extracellular proteases (including metalloproteinases, plasminogen activators, cathepsins and kallikreins) have been shown to be able to act on IGFBPs within cell models *in vitro*; the extent to which these contribute to IGFBP proteolysis within tissues has not been defined. Analysis of cathepsin D-deficient mice revealed no obvious changes in IGFBP proteolysis implying that if cathepsin D played any physiological role, this could be compensated for by other proteases [18]. This obviously does not rule out an important role for such proteases locally, particularly in pathologies such as tumours where levels and activation of such enzymes may be abnormally elevated. At present tools are not available to assess the action of individual proteases such as this *in vivo* and investigations of what occurs within tissues have been largely restricted to analysing tissue fluids using the same techniques employed for serum.

We have found that in normal subjects, in whom little IGFBP-proteolysis can be detected in the circulation, there is much more extensive proteolysis in a variety of extra-vascular fluids, including skin interstitial fluid, peritoneal fluid, ovarian follicular fluid and joint synovial fluid. Others have reported similar findings in lymph [19].

The ovarian follicle represents a very distinct structure in which very dynamic changes occur in cell metabolism, proliferation, differentiation and programmed cell death (PCD). We have show that within follicles IGFBP-2, IGFBP-3 and IGFBP-4 have been cleaved, each by a very distinct specific protease although there may be complex interactions between these [20]. For example, the proteolysis of IGFBP-4 can be inhibited by unsaturated IGFBP-3, although

IGFBP-3 is not cleaved by this activity and hence is not acting as a competitive substrate [20]. The proteolysis of the IGFBPs change according to the state of follicular maturation indicating a complex regulation system.
Within a more general tissue fluid, skin interstitial fluid, there appeared to be extensive fragmentation specifically of IGFBP-3 which was predominantly present in the fragmented form [21]. Local inflammation due to psoriasis was accompanied by a marked decrease in IGFBP-3 protease activity resulting in much more of the intact form being present [22]. This appeared to relate to the local production of an inhibitor. Initial observations revealed no marked alterations in other IGFBPs in this condition. We observed similar extensive fragmentation of IGFBP-3 in synovial fluid from normal knee joints [23]. This activity and the fragmentation of IGFBP-3 was markedly decreased in synovial fluid from knees affected by osteoarthritis or rheumatoid arthritis; in contrast there was an increase in fragmentation of IGFBP-2 in both conditions [23]. Again the decrease in IGFBP-3 protease activity appeared to relate to the appearance of a local inhibitor.

3.2. Regulation

The regulation of extravascular IGFBP proteases has yet to be addressed. From the preliminary observations in various conditions it is clear that the proteolysis of each IGFBP can be independently regulated and that each may be effected by changes in amounts or activation of specific enzymes or by changes in levels of specific inhibitors. In arthritic joints a decrease in IGFBP-3 protease activity and an increase in IGFBP-2 proteolysis occurs in a situation where widely reported increases occur in general extracellular matrix proteolysis indicating very specific regulation mechanisms.

3.3. Function

As indicated above, from the evidence originally gathered in relation to proteolysis of IGFBP-3 in the circulation, the inference has been that it represents a mechanism for decreasing the affinity with which IGF is bound and hence increasing IGF bio-availability. An extension of this inference is that in conditions where there is an increase in extracellular proteolysis this may result in increased IGF activity. This has obvious implications to tumour biology where enhanced extracellular proteolysis is well documented. Our findings in the ovary fit well within this model in that increased IGFBP proteolysis only occurs in maturing follicles where an increase in IGF activity could contribute to the enhance cellular activity and proliferation; whereas in atretic follicles restricted proteolysis would limit IGF actions [20]. However our findings that in general tissue fluids the majority of IGFBP-3 is normally in a cleaved form does not easily fit this model and suggests that there may be more complex mechanisms operating. If virtually all IGFBP-3 in the tissues is already cleaved then the significance of further proteolysis which may occur in situations such as around tumours is unclear. It may seem reasonable that the IGFs that are held on the main carrier IGFBP-3, should in the tissues generally be made available to the target cell receptors and their local activity may be modulated by other locally produced IGFBPs. In arthritic joints if the normal proteolysis is blocked, a consequent restriction in IGF activity may remove an anabolic signal resulting in a shift in tissue balance favouring catabolic joint destruction. However in psoriasis the observations are more difficult to reconcile with simple alterations in IGF availability; since a decrease in fragmentation of IGFBP-3 occurs in a condition with marked increases in cell proliferation. Recent findings that intact and fragmented forms of IGFBP-3 may have IGF-independent actions restricting cell growth [24] and enhancing PCD [25] indicate that tissue proteolysis of IGFBP-3 may serve a number of functions in addition to modulating IGF-availability.

4. CHARACTERISATION & COMPONENTS

Initial characterisation of the circulating IGFBP-3 protease using inhibitor profiles indicated that a cation-dependent serine protease was responsible. This was consistently found by a number of groups and in all the situations in which the protease activity was detected. Furthermore, the resultant IGFBP-3 fragmentation pattern was also consistent, but differed from that resultant from proteolysis of IGFBP-3 by a variety of characterised general proteases or media conditioned from various cultured cell lines. We have reported identical findings in our studies of extravascular fluids. All this suggests that the same uncharacterised enzyme(s) are responsible for the activity seen in vivo both within and outside the vasculature in all the different situations..

Further characterisation has proven problematic. The IGFBPs appear to be particularly sensitive to proteolysis and there are many general proteases present in the circulation, although these are normally inactive due to the presence of an excess of many forms of protease inhibitors. However whenever serum samples are fractionated in initial attempts to separate out and characterise IGFBP proteases, this can result in separation of inhibitors and activation of numerous proteases which are not normally active. Components of the plasminogen activator/plasminogen/plasmin system are clearly some of the most abundant proteases present and the most readily purified, but they are also the most tightly regulated. Thus whilst this system has clearly been shown to be present and have the capacity to cleave IGFBP-3 in a similar manner to the fragmentation observed *in* vivo, it is not necessarily responsible for the activity seen *in vivo* [26]. Similarly matrix metalloproteinases (MMPs) have been demonstrated to fragment IGFBP-3 and proposed to be responsible for the activity seen in serum [27]. The levels of MMPs increase throughout pregnancy whilst that of their natural inhibitors the tissue inhibitors of metalloproteinases (TIMPs) decreases.

We have recently carried out further characterisation of components of IGFBP-3 protease activity comparing fluids from the different compartments and different conditions in which variations in protease activity occurred. We have found that following size fractionation two regions of activity occur in serum from both normal controls and pregnant women; one at around 60-70 kDa and one at greater than 200 kDa. From inhibitor profiles it was apparent that the smaller sized region comprised a metal-dependent serine protease which could fragment IGFBP-3 and a metalloproteinase which could restrict this activity. In the higher molecular weight region there was a metal-dependent serine protease which could fragment IGFBP-3 and a metalloproteinase which could activate this protease. In the normal interstitial fluid and synovial fluid only the lower molecular weight activity was present. As these tissue fluids lacked the higher molecular weight activity but *in vivo* had exactly the same IGFBP-3 protease activity as serum samples; these results indicated that the higher molecular weight activity revealed following size fractionation of serum may not actually contribute to the *in vivo* effects. In the normal serum there was also fractions of around 150 kDa which could inhibit the IGFBP-3 protease activity seen in both size regions. This inhibitory activity was absent in pregnant serum indicating that the enhanced protease activity seen in pregnancy may be due to the degradation of an inhibitor rather than to the appearance of a protease. In rheumatoid arthritis the higher molecular weight protease activity and inhibitory activity appeared in synovial fluid possibly due to the associated local inflammation and increased capillary permeability. In contrast in psoriasis and osteoarthritis there appeared to be other local inhibitors present resulting in the decreased proteolysis seen *in vivo*. These results suggest that IGFBP-3 proteolysis is the result of complex interactions between a number of enzymes and inhibitors as shown schematically in Figure 1. The enzymes responsible for cleaving IGFBP-3 appear to be generally present and modulated in different conditions by the presence of inhibitors. Both the enzymes responsible and their inhibitors appear to be susceptible to modulation by proteolytic inactivation.

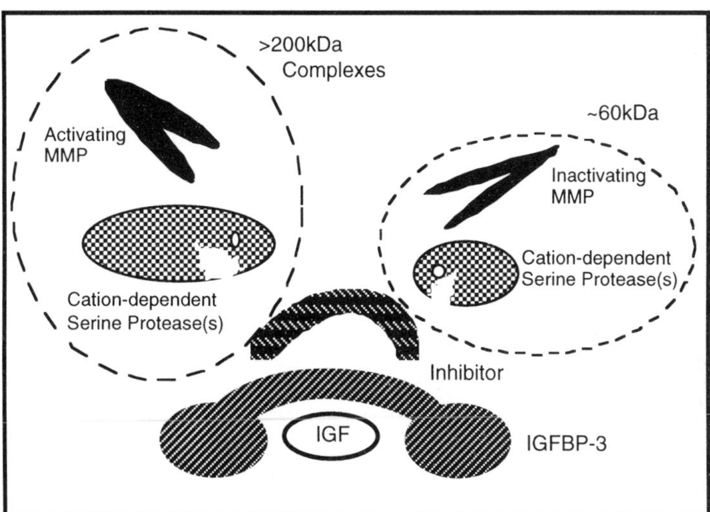

Figure 1. Schematic representing the different components present *in vivo* that are capable of contributing to IGFBP-3 proteolysis. The components of the >200 kDa complexes are present in serum but not in extravascular fluids despite apparent identical activity in the unfractionated samples; these components may therefore not contribute to the *in vivo* activity.

REFERENCES

1. Giudice LC, Farrell EM, Pham H, Lamson G & Rosenfeld RG. Insulin-like growth factor binding proteins (IGFBPs) in maternal serum throughout

gestation and in puerperium: Effects of pregnancy-associated serum protease activity. J. Clin. Endocrinol. Metab. 71 (1990) 806-816.

2. Hossenlopp P. Segovia B, Lassare C, Roghani M, Bredon M & Binoux M. Evidence of enzymatic degradation of insulin-like growth factor binding proteins in the 150K complex during pregnancy. J. Clin. Endocrinol. Metab. 71 (1990) 797-805.
3. Davies SC, Wass JAH, Ross RJM, Cotterill AM, Buchanan CR, Coulson VJ & Holly JMP. The induction of a specific protease for insulin-like growth factor binding protein-3 in the circulation during severe illness. J. Endocrinol. 130 (1991) 469-473.
4. Cwyfan-Hughes SC, Cotterill AM, Molloy AR, Cassell TB, Braude N, Hinds CJ, Wass JAH & Holly JMP. The induction of specific proteases for insulin-like growth factor binding proteins following major heart surgery. J. Endocrinol. 135 (1992) 135-145.
5. Timmins AC, Cotterill AM, Cwyfan Hughes SC, Holly JMP, Ross RJM, Blum W & Hinds CJ. Critical illness is associated with low circulating concentrations of insulin-like growth factor-I and -II, alterations in insulin-like growth factor binding proteins and induction of an insulin-like growth factor binding protein 3 protease. Crit. Care Med. 24 (1996) 1460-1466.
6. Frost VJ, Macaulay VM, Wass JAH & Holly JMP. Proteolytic modification of insulin-like growth factor binding proteins (IGFBPs): comparison of conditioned media from human cell lines, circulating proteases and characterised enzymes. J. Endocrinol. 138 (1993) 545-554.
7. Cotterill AM, Holly JMP, Taylor AM, Davies SC, Coulson VJ, Preece MA, Wass JAH & Savage MO. The insulin-like growth factor binding proteins and insulin-like growth factor-bioactivity in Laron-type dwarfism. J. Clin. Endocrinol. Metab. 74 (1992) 56-63.
8. Holly JMP, Claffey DCP, Cwyfan-Hughes SC, Frost VJ & Yateman ME. Proteases acting on IGFBPs; their occurrence and physiological significance. Growth Reg. 3 (1993) 88-91.
9. Pucilowska JB, Davenport ML, Kabir I, Clemmons DR, Thissen J-P, Butler T & Underwood LE. The effect of dietary protein supplementation on insulin-like growth factors (IGFs) and IGF-binding proteins in children with Shigellosis. J. Clin. Endocrinol. Metab. 77 (1993) 1516-1521.
10. Davies SC, Holly JMP, Coulson VJ, Cotterill AM, Abdulla AF, Whittaker PG, Chard T & Wass JAH. The presence of cation-dependent proteases for insulin-like growth factor binding proteins does not alter the size distribution of insulin-like growth factors in pregnancy. Clin. Endocrinol. 34 (1991) 501-506.
11. Cotterill AM, Mendel P, Holly JMP, Timmins AG, Camacho-Hubner C, Cwyfan Hughes S, Ross RMJ, Blum WF & Langford RM. The differential regulation of the circulating levels of the insulin-like growth factors and their binding proteins (IGFBP) 1, 2 and 3 after elective abnormal surgery. Clin. Endocrinol. 44 (1996) 91-101.
12. Lassarre C, Lalou L, Perin L & Binoux M. Protease-induced alteration of insulin-like growth factor binding protein-3 as detected by radioimmunoassay. Agreement with ligand blotting data. Growth Regul. 4 (1994) 48-55.
13. Bereket A, Lang CH, Blethen SL, Fan J, Frost RA & Wilson TA. Insulin-like growth factor binding protein-3 proteolysis in children with insulin-dependent

diabetes mellitus: possible role for insulin in the regulation of IGFBP-3 protease activity. J. Clin. Endocrinol. Metab. 80 (1995) 2282-2288.
14. Frost VJ, Helle SI, Lonning PE, van der Stappen JWJ & Holly JMP. Effects of treatment with megesterol acetate, aminoglutethimide or formestane on insulin-like growth factor-I and -II, IGF-binding proteins and insulin-like growth factor binding protein-3 protease status in patients with advanced breast cancer. J. Clin. Endocrinol. Metab. 81 (1996) 2216-2221.
15. Helle SI, Holly JMP, Tally M, Hall K, van der Stappen J & Lonning PE. Influence of treatment with Tamoxifen and change in tumour burden on the IGF-system in breast cancer patients. Int. J. Cancer. 69 (1996) 335-339.
16. Blat C, Villaudy J & Binoux M. In vivo proteolysis of serum insulin-like growth factor (IGF) binding protein-3 results in increased availability of IGF to target cells. J. Clin. Invest. 93 (1994) 2286-2290.
17. Davenport ML, Clemmons DR, Miles MV, Camacho-Hubner C, D'Ercole AJ & Underwood LE. Regulation of serum insulin-like growth factor-I (IGF-I) and IGF binding proteins during rat pregnancy. Endocrinol. 127 (1990) 1278-1286.
18. Braulke T, Claussen M, Saftig M, Weiland M, Neifer K, Schmidt B, Zapf J, von Figura K & Peters C. Proteolysis of IGFBPs by Cathepsin D in vitro and in cathepsin D-deficient mice. Progr. Growth Factor Res. 6 (1995) 265-271.
19. Lalou C & Binoux M. Evidence that limited proteolysis of insulin-like growth factor binding protein-3 (IGFBP-3) occurs in the normal state outside of the bloodstream. Regul. Peptides. 48 (1993) 179-188.
20. Cwyfan Hughes SC, Mason HD, Franks S & Holly JMP. Modulation of the insulin-like growth factor binding proteins by follicule size in the human ovary. J. Endocrinol. 154 (1997) 35-43.
21. Xu S, Cwyfan-Hughes SC, van der Stappen JWJ, Sansom J, Burton JL, Donnelly M & Holly JMP. Insulin-like growth factors (IGFs) and IGF binding proteins (IGFBPs) in human skin interstitial fluid. J. Clin. Endocrinol. Metab. 80 (1995) 2940-2945.
22. Xu S, Cwyfan Hughes SC, van der Stappen JWJ, Sansom J, Burton JL, Donnelly M & Holly JMP. Altered insulin-like growth factor-II (IGF-II) level and IGF binding protein-3 (IGFBP-3) protease activity in interstitial fluid taken from the skin lesion of psoriasis. J. Invest. Dermatol. 106 (1996) 109-112.
23. Fernihough JK, Billingham MEJ, Cwyfan Hughes SC & Holly JMP. Local disruption of the insulin-like growth factor (IGF) system in the arthritic joint. Arthritis Rheum. 39 (1996) 1556-1565.
24. Liu L, Delbe J, Blat C, Zapf J & Harel L. Insulin-like growth factor binding protein (IGFBP-3), an inhibitor of serum growth factors other than IGF-I and IGF-II. J. Cell. Physiol. 153 (1992) 15-21.
25. Gill Z, Perks C, Newcombe P & Holly JMP. Insulin-like growth factor binding protein-3 (IGFBP-3) predisposes breast cancer cells to programmed cell death in a non-IGF dependent manner. J. Biol. Chem. 272 (1997) 25602-25607.
26. Bang P & Fielder PJ. Human pregnancy serum contains at least two distinct proteolytic activities with the ability to degrade insulin-like growth factor binding protein-3. Endocrinol. 138 (1997) 3912-3917.
27. Fowlkes JL, Suzuki K, Nagase H & Thrailkill KM. Proteolysis of insulin-like growth factor binding protein-3 during pregnancy: a role for matrix metalloproteinases. Endocrinol. 135 (1994) 2810-2813.

Molecular Mechanisms to Regulate the
Activities of Insulin-like Growth Factors
K. Takano, N. Hizuka and S-I. Takahashi (Editors)

Proteolytic fragments of IGF binding protein-3 : physiological significance

M. Binoux, C. Lalou, S. Mohseni-Zadeh, P. Angelloz-Nicoud, C. Daubas and S. Babajko

Institut National de la Santé et de la Recherche Médicale U. 142, Hôpital Saint Antoine, 75571 Paris Cedex 12, France

INTRODUCTION

One of the concepts pivotal to the understanding of the roles played by the IGF binding proteins (IGFBPs) in modulating IGF action has come to light only over the past few years. This concerns the affinity shifts occurring as a result of structural alterations to the IGFBPs. Such alterations include phosphorylation, adherence to the cell surface or extra-cellular matrix and limited proteolysis by serine- and metallo-proteases on contact with the cell or in the extra-cellular medium. Phosphorylation enhances the affinity of IGFBP-1, thus promoting sequestration of IGFs in the extra-cellular environment. Adherence of IGFBP-1, -2, -3 and -5 to the cell surface or extra-cellular matrix means that IGFs are concentrated in the vicinity of their receptors and, owing to the diminished affinities of the IGFBPs, delivered slowly and continuously to the receptors. Limited proteolysis of IGFBP-2, -3, -4 and -5 results in a marked loss of affinity for IGFs, thereby accelerating dissociation and increasing the bioavailability of the IGFs. These functional changes at least in part account for the potentiation of IGF action described under certain experimental conditions [1]. Another discovery has been that IGFBP-3 is capable of inhibiting cell proliferation induced by IGFs or other growth factors via a mechanism divorced from its ability to bind IGFs [2, 3]. From this it could be inferred that IGFBP-3 may comprise more than one functional domain.

By way of confirmation, we first used an osteoblast cell line secreting IGFBP-3 and plasminogen activators to demonstrate that the fragments generated by plasmin exhibit the same electrophoretic migration as those resulting from *in vivo* proteolysis of IGFBP-3 in serum, with a major fragment of 30 kDa and one or two smaller products of 20 kDa and 16 kDa [4]. Thereafter we set out to reproduce this proteolysis *in vitro*, using plasmin and recombinant, non-glycosylated (*E. coli*) IGFBP-3 (generously provided by Celtrix Pharmaceuticals, Inc.). Our hypothesis was that by isolating these proteolytic fragments and investigating their properties, some progress might be made in identifying the different functional domains of IGFBP-3. Plasmin generated two major fragments [5]. The more abundant 22-25-kDa form had severely diminished affinity for IGFs, especially IGF-I, as does serum IGFBP-3 during pregnancy, the physiological condition with which maximal IGFBP-3 proteolysis is associated [6, 7]. This fragment was a weak antagonist of IGF action in a chick embryo fibroblast assay, whereas the intact protein was a powerful inhibitor. The other fragment, of 16 kDa, totally lacked affinity for IGF-I, but nevertheless blocked its mitogenic effects. This rather unexpected finding provided a new perspective in explaining some of the inhibitory effects previously described for IGFBP-3.

RESULTS and DISCUSSION

Amino acid sequences of the major proteolytic fragments of IGFBP-3

The N-terminal amino acids of proteolytic fragments of non-glycosylated rh IGFBP-3 have recently been sequenced. Based first on the primary structure of the protein deduced from its cDNA and secondly on sequencing of other proteolytic fragments with N-terminal amino acids in the middle and C-terminal parts of the protein, we concluded that the 22-25-kDa fragment comprises residues 1-160 and the 16-kDa fragment, residues 1-95 [8]. Two of the three glycosylation sites on the protein are located within the 1-160 sequence, one of which falls within the 1-95 sequence [9]. From analysis of the data taken together, it became clear that these 22-25-kDa and 16-kDa fragments correspond to the native 30- and 20-kDa fragments identified by western- and immuno-blotting in serum and cell culture media [4, 5]. Sequencing of the N-terminus of one of the other major fragments isolated by HPLC, but not detected by either ligand- or immuno-blotting, led us to identify this as residues 161-264 which are complementary to the 1-160 sequence generated by proteolytic cleavage. Figure 1 is a linear schematic representation of IGFBP-3, showing the cleavage sites observed under our experimental conditions.

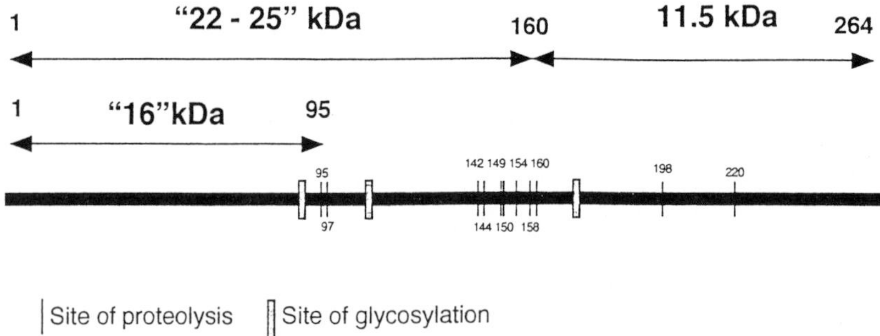

Figure 1 (from Ref. 8) : Schematic representation of the proteolytic cleavage sites of non-glycosylated rh IGFBP-3 by plasmin. The molecular masses indicated for the 22-25-kDa and 16-kDa fragments were calculated on the basis of electrophoretic migration in the absence of reducing agent.

Transfection of cDNA plasmids corresponding to the IGFBP-3 fragments

In order to confirm the above conclusions, we transfected cDNAs corresponding to intact IGFBP-3 and its proteolytic fragments (whose expression was under the control of the constitutive cytomegalovirus promoter) into a variety of cell types that are sensitive to stimulation by different mitogens, with the common characteristic of zero or very weak IGFBP-3 expression. Preliminary results obtained using two stably transfected lines (CCL 39 comprising hamster pulmonary fibroblasts and MCF7 derived from a breast cancer cell line) indicated that IGFBP-3 and its truncated fragments were expressed and secreted into the culture medium in different molecular forms, some glycosylated, some non-glycosylated. The electrophoretic migrations of the different forms were similar to those of the forms found in serum and conditioned media and therefore corroborated the fragment lengths deduced from N-terminal sequencing (Figure 2).

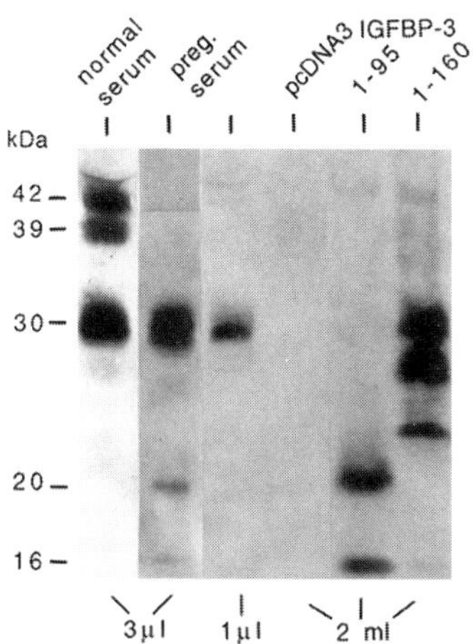

Figure 2 : Western blot analysis using anti-IGFBP-3 antibody of the culture media of CCL 39 cells transfected with the cDNAs corresponding to the proteolytic fragments of IGFBP-3 comprising residues 1-95 and 1-160. Controls were cells transfected with pcDNA-3 plasmid alone. Normal and pregnancy serum samples were analysed for comparison.

Physiological functions of the IGFBP-3 fragments

The fragment comprising residues 1-160 is always predominant and sometimes the only one detected by immunoblotting in biological fluids. At the time that this form was characterized in pregnancy serum, where the intact protein is totally or virtually absent [6], we found that its affinities for IGF-I and IGF-II were reduced 10- and 2-3-fold, respectively, as compared with those of intact IGFBP-3 [7]. The outcome of this weakened affinity is facilitated dissociation of the IGFs (especially IGF-I) bound to proteolysed IGFBP-3 within the 140-kDa complexes. The IGFs are thus redistributed among the serum pools of free or bound IGF, those in free form readily crossing the capillary endothelium and becoming available to the tissues [7, 10]. In the case of pregnancy, the physiological function of this proteolysis would therefore be to release IGFs in response to amplified metabolic needs and growth. IGFBP-3 proteolysis does in fact also occur in the normal state and our recent findings indicate that it is subject to nycthemeral variation. We have developed an immuno-functional assay for IGFBP-3, in which the intact and truncated forms in serum are bound to anti-IGFBP-3 monoclonal antibodies fixed to the test-tube wall, then incubated with radiolabelled IGF-I which binds only to intact IGFBP-3. A peak of intact IGFBP-3 is observed during the first half of the night, prior to the peak in IGFBP-1 (Figure 3). This would suggest that during nocturnal fasting there is co-operation between the two IGFBPs to limit the bioavailability of circulating IGF-I (by sequestration within the 140-kDa complexes) and curb its peripheral effects (prevented by IGFBP-1). This interpretation fits well with the notion of interaction between insulin and the IGF system in controlling substrate utilization.

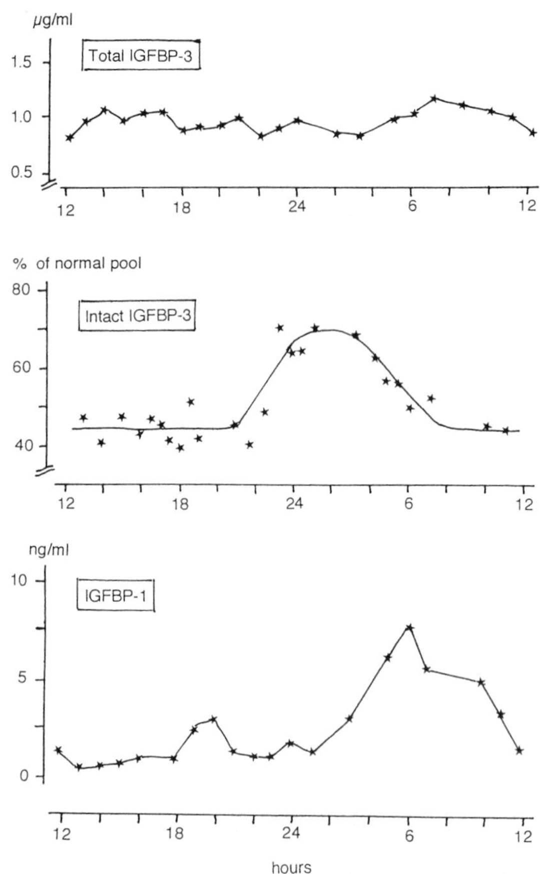

Figure 3: Nycthemeral variations in serum levels of total IGFBP-3, intact IGFBP-3 and IGFBP-1 in a child with partial GH deficiency. Total IGFBP-3 and IGFBP-1 were measured by immunoradiometric assay and intact IGFBP-3 by immunofunctional assay.

At cellular level, we previously showed that in PC-3 cells derived from a human prostate adenocarcinoma and capable of autocrine proliferation largely stimulated by the IGF-II that they secrete, cell growth was dependent upon the bioavailability of IGF-II, which in turn was regulated by the proteolytic state of the IGFBP-3 produced. Addition of IGF-II further induced proliferation, but the induction was weaker in the presence of protease inhibitors suppressing IGFBP-3 proteolysis and stronger in the presence of plasminogen which promotes it. This meant that potentiation of IGF-II action was directly related to IGFBP-3 proteolysis [11]. In this cell model, rh IGFBP-3 had a biphasic mitogenic effect : it was stimulatory at low concentrations and progressively inhibitory at higher concentrations. We were able to demonstrate that the biphasic effect reflected the changing ratios of proteolysed to intact IGFBP [12]. More recently, we have found that at the concentration at which rh IGFBP-3 causes slight stimulation of proliferation in these cells, IGFBP-3^{1-160} increases the stimulation by a factor of 5-7 [13] (Figure 4). This suggests that IGF binding weakly to the fragment dissociates readily so as to bind to its receptor. It is therefore conceivable that when the heparin binding site on the carboxy-terminal part of IGFBP-3 adheres to proteoglycans (HSP) on the cell surface, the protein is exposed to attack by proteases. IGFBP-3^{1-160} which contains essential IGF-binding sites would then be detached from the cell and release the IGFs associated with it.

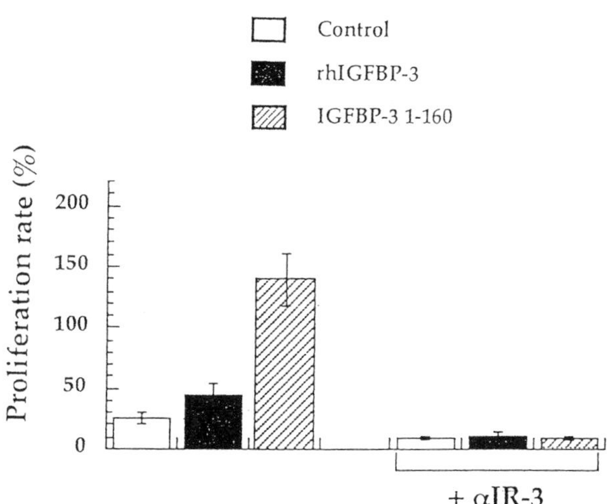

Figure 4 (from Ref. 13) : Effects of IGFBP-3 and its proteolytic fragment, IGFBP-3^{1-160} (5 nM), on autocrine proliferation in PC-3 cells. In this cell model, the mitogenic effect of IGFBP-3 is directly related to its limited proteolysis [12]. Stimulation of proliferation is IGF-dependent, as shown by its suppression in the presence of αIR-3 antibody which blocks the IGF receptor.

The proteolytic fragment comprising residues 161-264 remains something of a mystery, since it neither binds IGFs nor is recognized by any of our anti-IGFBP-3 antibodies. It contains several tyrosine residues and we have succeeded in ^{125}I-labelling it, noting that it binds with strong affinity to MCF-7 cells, most probably via its HSP-binding sites. There is also a nuclear localization signal (...K^{215}K...K.RK^{232}...) near the extremity of its carboxy-terminus and it will be interesting to learn if it is capable of internalization and translocation to the nucleus, as has recently been shown for intact IGFBP-3 [14, 15].

The fragment comprising residues 1-95 is irregularly detected in sera or culture media as a 20-kDa glycosylated form or 16-kDa non-glycosylated fragment. Being a product of secondary proteolysis of IGFBP-3^{1-160} [8], its appearance is related to the concentrations of IGFBP-3-degrading proteases and their inhibitors present in the medium. We have no data for its role *in vivo*, but from our findings for various cell lines, it would seem to be involved in the control of cell growth. Despite its lack of affinity for IGFs, this fragment inhibits the mitogenic effects of IGFs on chick embryo fibroblasts [5] and suppresses IGF-I- and oestradiol-induced proliferation in MCF cells (unpublished). Its relative specificity for the IGF receptor is noteworthy, particularly in comparison with that for the insulin receptor which is structurally the most closely related. IGFBP-3^{1-95} exerts 80-100% inhibition of the mitogenic effects of IGF-I and insulin (used at 100-fold concentrations), but, at most, 50% inhibition of insulin-induced glycogen synthesis in fetal rat hepatocytes (Figure 5).

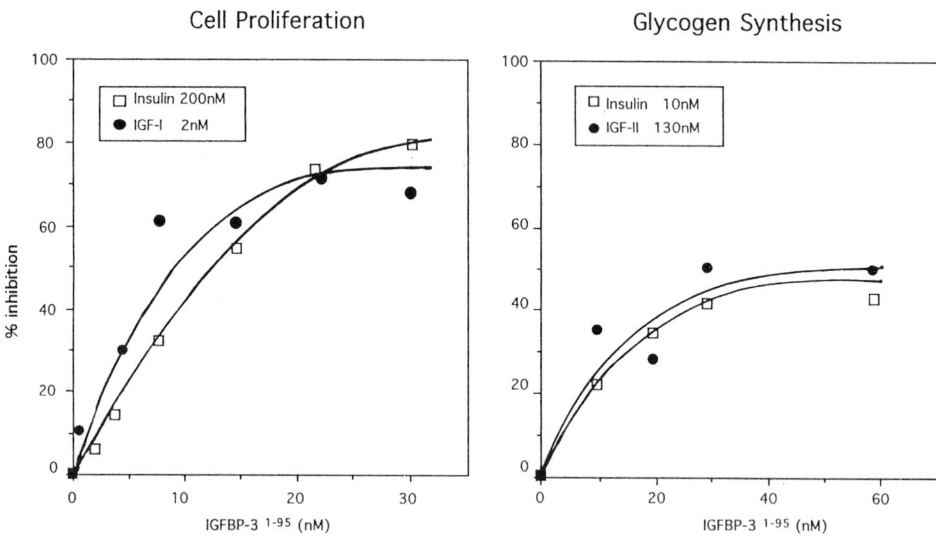

Figure 5 : Comparison of the effects of IGFBP-3^{1-95} on chick embryo fibroblast proliferation induced via activation of the IGF receptor and hepatic glycogen synthesis induced via activation of the insulin receptor.

However, it is capable of suppressing other signalling pathways for MAP kinase activation. We recently reported that IGFBP-3^{1-95} blocks the mitogenic action of bFGF on mouse fibroblasts with a targeted disruption of the type 1 IGF receptor gene [16]. The level(s) at which it interacts with these pathways remain unknown. Like the other IGFBP-3 fragments, it does not interfere with IGF binding to its receptor, but we discovered that ***intact IGFBP-3*** does interact with the type 1 IGF receptor, reducing the affinity of the receptor for its ligand [17]. This represents an alternative to the extra-cellular mechanism of sequestration and an additional means by which IGF action is regulated, although the molecular aspects of such control remain to be determined. To return to the mode(s) of action of IGFBP-3$^{1-95}$, interaction with the cell surface appears to be a necessary first step, possibly involving the IGFBP-3-specific binding site described by Oh *et al.* [18], or the type V TGF ß receptor very recently identified as the putative IGFBP-3 receptor [19], or some unknown site.

As regards the relationships between structure and activity of IGFBP-3 and its proteolytic fragments, we yet have much to learn.

ACKNOWLEDGEMENTS

This work was supported by INSERM, the Association de Recherche sur le Cancer and the Association pour la Recherche sur les Tumeurs de la Prostate. We are indebted to the colleagues who contributed to the study : S. Sawamura and Y. Ogawa (Celtrix Pharmaceuticals, Inc.) P. Menuelle (Université Paris VII), M.T. Tauber (Hôpital Purpan, Toulouse) ; and to B. Segovia for technical assistance.

REFERENCES

1. Jones JI, Clemmons DR (1995) Insulin-like growth factors and their binding proteins: biological actions. Endocrine Reviews 16:3-34.
2. Liu L, Delbé J, Blat C, Zapf J, Harel L (1992) Insulin-like growth factor binding protein-3 (IGFBP-3), an inhibitor of serum growth factors other than IGF-I and -II. J. Cell Physiol. 153:15-21.
3. Cohen P, Lamson G, Okajima T, Rosenfeld RG (1993) Transfection of the human insulin-like growth factor binding protein-3 gene into balb/c fibroblasts inhibits cellular growth. Mol. Endocrinol. 7:380-286.
4. Lalou C, Silve C, Rosato R, Segovia B, Binoux M (1994) Interactions between insulin-like growth factor-I (IGF-I) and the system of plasminogen activators and their inhibitors in the control of IGF binding protein-3 production and proteolysis in human osteosarcoma cells. Endocrinology 135:2318-2326.
5. Lalou C, Lassarre C, Binoux M (1996) A proteolytic fragment of insulin-like growth factor (IGF) binding protein-3 that fails to bind IGFs inhibits the mitogenic effects of IGF-I and insulin. Endocrinology 137:3206-3212.
6. Hossenlopp P, Segovia B, Lassarre C, Roghani M, Bredon M, Binoux M (1990) Evidence of enzymatic degradation of insulin-like growth factor binding proteins in the "150 K" complex during pregnancy. J. Clin. Endocrinol. Metab. 71:797-805.
7. Lassarre C, Binoux M (1994) Insulin-like growth factor binding protein-3 is functionally altered in pregnancy plasma. Endocrinology 134:1254-1262.
8. Lalou C, Sawamura S, Segovia B, Ogawa Y, Binoux M (1997) Proteolytic fragments of insulin-like growth factor binding protein-3 : N-terminal sequences and relationships between structure and biological activity. CR Acad Sci Paris/Life Sciences 320:621-628.
9. Firth SM, Baxter RC (1995.) The role of glycosylation in the action of IGFBP-3. Prog. in Growth Factor Res. 6:223-229.
10. Blat C, Villaudy J, Binoux M (1994) *In vivo* proteolysis of serum insulin-like growth factor (IGF) binding protein-3 results in increased availability of IGF to target cells. J. Clin. Invest. 93:2286-2290.
11. Angelloz-Nicoud P, Binoux M (1995.) Autocrine regulation of cell proliferation by the IGF and IGF binding protein-3 protease system in a human prostate carcinoma cell line (PC-3). Endocrinology 136:5485-5492.
12. Angelloz-Nicoud P, Harel L, Binoux M (1996) Recombinant human insulin-like growth

factor (IGF) binding protein-3 stimulates prostate carcinoma cell proliferation via an IGF-dependent mechanism. Role of serine proteases. Growth Regulation 6:130-136.

13. Angelloz-Nicoud P, Lalou C, Binoux M (1997) Prostate carcinoma (PC-3) cell proliferation is stimulated by the "22-25-kDa" proteolytic fragment (1-160) and inhibited by the "16-kDa" fragment (1-95) of recombinant human insulin-like growth factor binding protein-3. Growth Regulation (in press).
14. Li W, Fawcett J, Widmer HR, Fielder PJ, Rabkin R, Keller GA (1997) Nuclear transport of insulin-like growth factor-I and insulin-like growth factor binding protein-3 in opossum kidney cells. Endocrinology 138:1763-1766.
15. Jacques G, Noll K, Wegmann B, Witten S, Kogan E, Radulescu RT, Havemann K (1997) Nuclear localization of insulin-like growth factor binding protein-3 in a lung cancer cell line. Endocrinology 138:1767-1770.
16. Mohseni Zadeh S, Binoux M (1997) The 16-kDa proteolytic fragment of insulin-like growth factor (IGF) binding protein-3 inhibits the mitogenic action of fibroblast growth factor on mouse fibroblasts with a targeted disruption of the type 1 IGF receptor gene. Endocrinology 138:3069-3072.
17. Mohseni Zadeh S, Binoux M (1997) Insulin-like growth factor (IGF) binding protein-3 interacts with the type 1 IGF receptor, reducing the affinity of the receptor for its ligand : an alternative mechanism in the regulation of IGF action. Endocrinology (in press).
18. Oh Y, Müller HL, Pham H, Rosenfeld RG (1993) Demonstration of receptors for insulin-like growth factor binding protein-3 on Hs578T human breast cancer cells. J. Biol. Chem. 268:26045-26048.
19. Leal SM, Liu Q, Huang SS, Huang JS (1997) The type V transforming growth factor ß receptor is the putative insulin-like growth factor binding protein-3 receptor. J. Biol. Chem. 272:20572-20576.

Molecular Mechanisms to Regulate the
Activities of Insulin-like Growth Factors
K. Takano, N. Hizuka and S-I. Takahashi (Editors)

IGFBP REGULATION BY PROTEASES

Cheryl A. Conover

Endocrine Research Unit, Mayo Clinic and Mayo Foundation, 200 First Street SW, Rochester, MN 55905, USA

1. INTRODUCTION

IGFBP availability and bioactivity at the target cell is determined, not only by developmentally and hormonally directed IGFBP gene expression, but also by controlled proteolytic processing of the secreted protein. Indeed, IGFBP proteolysis may be the primary means of exerting acute and focal control over IGF cellular activity. Studies in 1990 reporting discrepancy between Western ligand blotting and radioimmunoassay measurements for IGFBP-3 in serum of pregnant women first introduced the notion of IGFBP proteases [1,2] Since then IGFBP-2 through IGFBP-5 have been shown to be physiological substrates for proteases produced by a variety of cells. Some are characterized enzymes that will cleave several IGFBPs: plasmin, prostate specific antigen (PSA), matrix metalloprotease (MMP), cathepsin D, nerve growth factor. Others are apparently novel proteases that are relatively specific for a single form of IGFBP. Control of IGFBP proteolysis is complex and can occur at the level of enzyme production/secretion, as well as by activation and inhibition of the secreted protease. The importance of IGFBP regulation by proteases is indicated by resultant modification in IGFBP affinity for IGFs or complete destruction of IGF binding potential. Moreover, IGFBP proteolysis may produce cleavage products with unique function, e.g., intracellular signaling independent of IGF binding. This article will focus on distinctive IGFBP proteases and their remarkable regulation by the IGFs themselves.

2. IGFBP-2 PROTEASE

Porcine vascular smooth muscle cells secrete a protease that cleaves IGFBP-2 (34 kD) into fragments of ~25 and 16 kD [3,4]. It is specific for IGFBP-2 and does not degrade other forms of IGFBPs. This IGFBP-2 proteolytic activity is inhibited by aprotinin, α_1-antichymotrypsin, and EDTA suggesting a cation-dependent serine protease. Little more is known about this enzyme. It binds weakly to heparin-Sepharose allowing it to be separated from a high affinity heparin binding IGFBP-5 protease also secreted by these cells (see below). The IGFBP-2 protease in this system appears to be secreted constitutively but is enhanced by serum-deprivation. It is further enhanced by addition of IGFs to the culture, with IGF-II being more effective than IGF-I. The mechanism of this IGF effect has not been elucidated, but it could not be reproduced with other peptide growth factors. It may be of relevance that the fragment sizes produced by IGFBP-2 proteolysis in porcine vascular smooth muscle cell conditioned medium are similar to those identified in the serum of newborn pigs fasted for 48 hours [5]. Since the fragments were not generated by an IGFBP-2 protease released into the serum, it can be speculated that proteolysis occurs on the endothelial surface by an enzyme such as that produced by the vascular smooth muscle cells *in vitro*.

3. IGFBP-3 PROTEASE

Circulating IGFBP-3 proteases were initially identified during pregnancy and then subsequently in association with a variety of catabolic states [1,2,6-8]. There is probably more than one protease responsible for the observed activity, but inhibitor profiles suggest predominantly serine proteases active at neutral pH. Cell-derived IGFBP-3 proteases have also been described for a number of normal and malignant cell types, and generally have been ascribed to known proteases: the serine proteases, PSA [9] and plasmin [10]; MMPs [11]; and the aspartic protease, cathepsin D [12]. Salahifar et al. [13] recently characterized proteolysis of IGFBP-3 by an enzyme secreted by MCF-7 human breast cancer cells that appears to be different from the above-mentioned proteases in three ways: pH optimum, inhibitor profile, and regulation by IGFs. MCF-7 serum-free conditioned medium was capable of proteolyzing exogenous IGFBP-3 during incubation under cell-free conditions, but only at pH 4.5-5.5. IGFBP-3 hydrolysis did not occur under conventional assay conditions, i.e., at neutral pH. Nor was there proteolysis of exogenous IGFBP-3 at pH < 4.5, which distinguishes it from cathepsin D previously reported as an acid-activated (pH 3) IGFBP-3 protease in MCF-7 media [12]. Proteolytic activity was inhibited with EDTA and leupeptin and not with aprotinin or benzamidine suggesting a cation-dependent cysteine protease. Gel chromatography indicated an approximate molecular mass for the proteolytic activity at 25-35 kD. Although the pH optimum, inhibitor profile and molecular size would be consistent with cathepsin B, there was no inhibition of IGFBP-3 proteolysis with cathepsin B inhibitors, and pure cathepsin B was not an effective IGFBP-3 protease *in vitro*. Moreover, MCF-7 cell-conditioned medium contained little or no immunoreactive cathepsin B. Thus, the exact biochemical nature of this enzyme activity remains to be determined.

The regulation of this IGFBP-3 protease by IGFs sparked particular interest. Salahifar et al. [13] showed that addition of IGF-I to the cell-free assay prevented IGFBP-3 proteolysis. Similar receptor-independent actions of IGF have been reported, and in most cases found to involve IGF binding to IGFBP-3 [14-16]. What was somewhat surprising in this account was that an IGF-I analog, [long Arg^3] ([LR^3]IGF-I), that does not bind IGFBP-3 also inhibited IGFBP-3 proteolysis in this cell-free system. These data suggest the possibility of a direct interaction between IGF and the protease. Grimes and Hammond [17] also noted that IGF-I and [LR^3]IGF-I could attenuate IGFBP-3 degradation during incubation in medium from cultured porcine ovarian granulosa cells. An IGF regulatory domain in a proteolytic enzyme has not been described in the literature, but other growth factor and growth-regulating peptide domains in enzymes have been recognized [18,19].

4. IGFBP-4 PROTEASE

We and others identified proteolytic activity secreted by normal human fibroblasts and osteoblasts in culture that cleaves the IGFBP-4 molecule into ~18 and 14 kD fragments [20-24]. The divalent cation chelator, EDTA, and the zinc metalloprotease inhibitor, 1,10 phenanthroline, were the most effective chemical inhibitors of IGFBP-4 proteolysis in these systems and their effects could be reversed with Ca^{++} and Zn^{++}, respectively. Serine protease inhibitors such as benzamidine and aprotinin produced variable effects, and there was no inhibition by TIMPs (tissue inhibitors of matrix metalloproteases). Based on these and other data, this IGFBP-4 protease appears to be a Ca^{++}-requiring metalloprotease (or metallo-serine protease), but evidently not a matrix metalloprotease. IGFBP-4 is the only IGFBP substrate (of the six known IGFBPs) for this protease, which is active in a broad pH range of 5.5-9. The signature feature of IGFBP-4 proteolysis in these systems is its strict dependence on IGFs for functional activity. Only very low concentrations of IGFs are needed, and, in general, IGF-II is more effective than IGF-I in activating IGFBP-4 proteolysis. A large number of other peptides (including structurally homologous insulin) and steroids have been shown to be incapable of inducing IGFBP-4 proteolysis during cell-free incubation. Even some of the apparently constitutive IGFBP-4 proteolysis as originally reported was subsequently determined to be due to endogenous IGFs [22,25-27]. These properties (metalloprotease, neutral pH, IGFBP-4 specificity, IGF-dependence) are also shared by IGFBP-4 proteases described in cultures of smooth muscle cells [28], endometrial stromal cells [29], decidual cells [30], granulosa cells [31], and certain malignant cell lines [26,32,33].

The exact mechanism underlying this IGF-dependency of IGFBP-4 proteolysis is unknown. It has been suggested that IGF binding to IGFBP-4 changes its conformation in such a way as to make the IGFBP-4 molecule more susceptible to proteolysis. Data of Irwin et al. [29] would support this mechanism as they showed that IGFBP-4 crosslinked with IGF-II was proteolyzed by endometrial stromal cell cultures in the absence of added IGF-II. Furthermore, a number of labs, including our own, have shown that insulin and IGF analogs that do not bind IGFBP-4 do not induce IGFBP-4 proteolysis, suggesting a need for IGF/IGFBP interaction. However, relative affinities of different IGFs for IGFBP-4 do not strictly correlate with ability to induce proteolysis [20], suggesting that IGF peptides may directly activate the IGFBP-4 protease. Recent data have come out in support of a third possible mechanism. Fowlkes et al. [34] showed that addition of IGFBP-3 can inhibit IGFBP-4 proteolysis in MC3T3-E1 cell cultures, and that IGF binding to IGFBP-3 reverses the inhibition. Thus, IGFBP-3 was suggested to be an inhibitor of the protease and to confer IGF-dependence on the system. Indeed, the ability of IGFs to induce IGFBP-4 proteolytic activity correlates with their affinity for IGFBP-3 (IGF-II>IGF-I>[des1-3]IGF-I>>>[LR^3]IGF-I, insulin [20,35]), but it is not clear whether this mechanism would operate in all systems, i.e., those that do not have IGFBP-3. Perhaps other IGFBPs serve this purpose in systems that do not express IGFBP-3 [36]. Purification of the IGFBP-4 protease will be required to resolve the issue of IGF-dependence.

There are a number of studies now emerging that further suggest regulation of IGFBP-4 proteolysis at the level of production/secretion of the enzyme. TGFβ treatment of normal human osteoblasts [37], dexamethasone treatment of B104 neuroblastoma cells [26], and FSH treatment of granulosa cells [31] have been shown to increase IGF-dependent IGFBP-4 proteolysis. None of these reagents act directly to activate IGFBP-4 proteolysis in cell-free assay. Rather, they appear to enhance the expression and/or secretion of the IGF-dependent IGFBP-4 protease. In addition, there is a preliminary report of an IGFBP-4 protease induced by vitamin C treatment in differentiating mouse MC3T3-E1 osteoblast cell cultures [38]. This proteolysis produced different sized IGFBP-4 fragments (~22 kD) and apparently is not regulated by IGFs. There are also inhibitors of IGFBP-4 proteolysis. Phorbol ester tumor promoters induce a cycloheximide-sensitive inhibitor of IGFBP-4 proteolysis in human fibroblasts and osteoblasts [39,40]. Estrogen treatment decreases IGFBP-4 proteolysis in estrogen-responsive bone cells, but it was not determined whether this inhibition was due to a decrease in protease and/or an increase in inhibitor [41].

5. IGFBP-5 PROTEASE

The existence of an IGFBP-5 protease was first indicated by the identification and purification by Andress and Birnbaum in 1991 [42] of a 23 kD IGFBP-5 fragment abundant in medium conditioned by U2 human osteosarcoma cells. We subsequently characterized the proteolytic activity in U2 conditioned medium [43]. Activity was specific for IGFBP-5; no proteolysis of IGFBP-1, -2, -3, -4, or -6 occurred during cell-free incubation in U2 conditioned medium, whereas 29 kD recombinant human IGFBP-5 was rapidly and completely degraded into immunoreactive fragments of ~16 kD. Endogenously generated IGFBP-5 fragments appeared as ~22, 17, and 16 kD immunoreactive forms reflecting the O-glycosylated nature of the native protein. IGFBP-5 proteolysis in U2 conditioned medium was inhibited most effectively by serine protease inhibitors (aprotinin, PMSF, DFP, TLCK), EDTA and 1,10 phenanthroline, indicating a member of the cation-dependent class of serine proteases as has been described for IGFBP-5-degrading proteases in conditioned media from human fibroblasts [44], osteoblasts [23], and rat articular chondrocytes [45]. IGFBP-5 proteolytic activity is acid-labile (no activity <pH 5.5) and optimal at pH 6-8. Protease inhibitor profile and pH optimum indicate a protease different than plasmin, MMP, PSA and cathepsin D, all of which have been shown to degrade IGFBP-5. Most of the biochemical characterization has come from the work of Clemmons and colleagues in human fibroblasts [44,46,47]. Taking advantage of the tight adherence of IGFBP-5 proteolytic activity to heparin-Sepharose, they have partially purified a 92 kD serine protease from human fibroblast conditioned medium. It is unclear at this time whether this protease is the same as that responsible for the IGFBP-5-specific degradation in U2 conditioned medium and other biological fluids.

Cellular exposure to IGFs results in markedly increased media levels of intact IGFBP-5. In U2 cells and human fibroblasts, this IGF-induced increase is not reflected in increased IGFBP-5 mRNA expression, and only those IGFs able to bind IGFBP-5 will stimulate IGFBP-5 accumulation [43,48]. In culture, it is obvious that IGFs act to retard IGFBP-5 proteolysis. Results in cell-free assays have been variable, calling into question whether the mechanism is solely by IGF binding and protecting the IGFBP-5 substrate. Kanzaki et al. [23] could find no effect of exogenous IGF-II on IGFBP-5 proteolysis during cell-free incubation in human osteoblast medium, and Nam et al. [44] found only a minimal effect in human fibroblast conditioned medium. These negative findings may be due in part to the assay conditions. We found that IGF-I altered the kinetics of proteolysis increasing the half-life of exogenous IGFBP-5 from 1.5 hours to 3.2 hours [24]; at >8 hours the protective effect of IGFs was less apparent. Matsumoto et al. [45] found that preincubation of [^{125}I]IGFBP-5 with IGF-I or IGF-II inhibited proteolysis during cell-free incubation, whereas simultaneous incubation of IGFs, conditioned medium and [^{125}I]IGFBP-5 did not.

IGFBP-5 proteolysis is also regulated by glycosaminoglycans in fibroblasts [47], FSH treatment of rat granulosa cells [49], and interleukin-6 treatment of rat fetal calvaria-derived osteoblasts [50].

6. SUMMARY

Researchers are just beginning to biochemically characterize the different IGFBP proteases, and to appreciate their role in IGF cellular physiology. We still have a long way to go with identifying and purifying the newly discovered IGFBP proteases, and then determining their regulation and biological actions. In this article we describe IGFBP proteolytic activity provisionally attributed to four novel proteases, and the complex regulation of these proteases, particularly by the IGFs themselves. Although it is well accepted that IGFBPs regulate the half-life of IGFs in the circulation, the converse seems to be true as well, i.e., that IGFs can alter the stability and local bioavailability of IGFBPs through the regulation of specific proteases. There is more than one protease in the different biological systems (see Table), and they are likely to be highly interactive in directing the temporal and spatial specificity needed for acute and focal growth responses.

IGFBP proteases in cell model systems

	HF	OB	GC	SMC	MCF-7	PEC
Known Proteases						
Plasmin		X				X
PSA						X
MMP	X	X				
Cathepsin D	X	X			X	X
Novel Proteases[a]						
IGFBP-2				X		
IGFBP-3			X		X	
IGFBP-4	X	X	X	X		
IGFBP-5	X	X	X	X		X

[a]As described in this article.

HF, human fibroblasts; OB, osteoblastic cells; GC, granulosa cells; SMC, smooth muscle cells; MCF-7, breast carcinoma cells; PEC, prostate epithelial cells.

7. ACKNOWLEDGMENTS

I thank my colleagues at Mayo (L. Bale, S. Durham, J. Lawrence, B.L. Riggs, R. Okazaki, M. Kassem) and elsewhere (D. Powell, D. DeLeon, M. Kiefer, J. Zapf) who have worked with me and who share my excitement over IGFBP proteases. This work was supported by NIH Grant DK-43258 and the Mayo Foundation.

REFERENCES

1. L.C. Giudice, E.M. Farrell, H. Pham, G. Lamson and R.G. Rosenfeld, J. Clin. Endocrinol. Metab., 71 (1990) 806.
2. P. Hossenlopp, B. Segovia, C. Lassarre, M. Roghani, M. Bredon and M. Binoux, J. Clin. Endocrinol. Metab., 71 (1990) 797.
3. W.S. Cohick, A. Gockerman and D.R. Clemmons, J. Cell. Physiol., 164 (1995) 164.
4. A. Gockerman and D.R. Clemmons, Circ. Res., 76 (1995) 514.
5. R.H. McCusker, W.S. Cohick, W.H. Busby and D.R. Clemmons, Endocrinology, 129 (1991) 2631.
6. S.C. Davies, J.A.H. Wass, R.J.M. Ross, A.M. Cotterill, C.R. Buchanan, V.J. Coulson and J.M.P. Holly, J. Endocrinol., 130 (1991) 469.

7. M.L. Davenport, W.L. Isley, J.B. Pucilowska, L.B. Pemberton, B. Lyman, L.E. Underwood and D.R. Clemmons, J. Clin. Endocrinol. Metab., 75 (1992) 590.
8. A. Bereket, C.H. Lang, S.L. Blethen, J. Fan, R.A. Frost and T.A. Wilson, J. Clin. Endocrinol. Metab., 80 (1995) 2282.
9. P. Cohen, H.C.B. Graves, D.M. Peehl, M. Kamarei, L.C. Giudice and R.G. Rosenfeld, J. Clin. Endocrinol. Metab., 75 (1992) 1046.
10. B.A. Booth, M. Boes and R.S. Bar, Am. J. Physiol., 271 (1996) E465.
11. J.L. Fowlkes, J.J. Enghild, K. Suzuki and H. Nagase, J. Biol. Chem., 269 (1994) 25742.
12. C.A. Conover and D.D. DeLeon, J. Biol. Chem., 269 (1994) 7076.
13. H. Salahifar, R.C. Baxter and J.L. Martin, Endocrinology, 138 (1997) 1683.
14. C.A. Conover, J. Clin. Invest., 88 (1991) 1354.
15. E.K. Neely and R.G. Rosenfeld, Endocrinology, 130 (1992) 985.
16. J.L. Martin, M. Ballesteros and R.C. Baxter, Endocrinology, 131 (1992) 1703.
17. R.W. Grimes and J.M. Hammond, Endocrinology, 134 (1994) 337.
18. J.S. Bond and R.J. Beynon, Protein Sci., 4 (1995) 1247.
19. L.M. Matrisian, Trends Genet., 6 (1990) 121.
20. C.A. Conover, M.C. Kiefer and J. Zapf, J. Clin. Invest., 91 (1993) 1129.
21. J. Fowlkes and M. Freemark, Endocrinology, 131 (1992) 2071.
22. S.K. Durham, M.C. Kiefer, B.L. Riggs and C.A. Conover, J. Bone Miner. Res., 9 (1994) 111.
23. S. Kanzaki, S. Hilliker, D.J. Baylink and S. Mohan, Endocrinology, 134 (1994) 392.
24. C.A. Conover, Prog. Growth Factor Res., 6 (1995) 301.
25. S.K. Durham, D.D. DeLeon, R. Okazaki, B.L. Riggs and C.A. Conover, J. Clin. Endocrinol. Metab., 80 (1995) 104).
26. P.-T. Cheung, J. Wu, W. Banach and S.D. Chernausek, Endocrinology, 135 (1994) 1328.
27. S.D. Chernausek, C.E. Smith, K.L. Duffin, W.h. Busby, G. Wright and D.R. Clemmons, J. Biol. Chem., 270 (1995) 11377.
28. A. Parker, A. Gockerman, W.H. Busby and D.R. Clemmons, Endocrinology, 136 (1995) 2470.
29. J.C. Irwin, B.A. Dsupin and L.C. Giudice, J. Clin. Endocrinol. Metab. 80 (1995) 619.
30. S.E. Myers, P.T. Cheung, S. Handwerger and S.D. Chernausek, Endocrinology, 133 (1993) 1525.
31. X.-J. Liu, M. Malkowski, Y. Gui, G.F. Erickson, S. Shimasaki and N. Ling, Endocrinology, 132 (1993) 1176.
32. W.A. Price, B.M. Moats-Staats and A.D. Stiles, Am. J. Respir. Cell Mol. Biol., 13 (1995) 466.
33. K. Noll, B.R. Wegmann, K. Havemann and G. Jaques, J. Clin. Endocrinol. Metab., 81 (1996) 2653.
34. J.L. Fowlkes, D.M. Serra, C.K. Rosenberg and K.M. Thrailkill, J. Biol. Chem., 270 (1995) 27481.
35. M.J. Donnelly and J.M.P. Holly, J. Endocrinol., 149 (1996) R1.

36. J.L. Fowlkes, K.M. Thrailkill, C. George-Nascimento, C.K. Rosenberg and D.M. Serra, Endocrinology, 138 (1997) 2280.
37. S.K. Durham, B.L. Riggs and C.A. Conover, J. Clin. Endocrinol. Metab., 79 (1994) 1752.
38. B. Bachrach, R.J. Wenstrup, E.P. Smith and S.D. Chernausek, Endocrine Society, Minneapolis, MN, June 11-14, 1997 (abstract).
39. C.A. Conover, J.T. Clarkson and L.K. Bale, Endocrinology, 133 (1993) 1347.
40. S.K. Durhan, B.L. Riggs, S.A. Harris and C.A. Conover, Endocrinology, 136 (1995) 1374.
41. M. Kassem, R. Okazaki, D. DeLeon, S.A. Harris, J.A. Robinson, T.C. Spelsberg, C.A. Conover and B.L. Riggs, Proc. Assoc. Am. Physicians, 108 (1996) 155.
42. D.L. Andress and R.S. Birnbaum, Biochem. Biophys. Res. Commun., 176 (1991) 213.
43. C.A. Conover and M.C. Kiefer, J. Clin. Endocrinol. Metab., 76 (1993) 1153.
44. T.J. Nam, W.H. Busby, Jr. and D.R. Clemmons, Endocrinology, 135 (1994) 1385.
45. T. Matsumoto, S.E. Gargosky, K. Kelley and R.G. Rosenfeld, Growth Regul., 6 (1996) 185.
46. T.J. Nam, W.H. Busby, Jr. and D.R. Clemmons, Endocrinology, 137 (1996) 5530.
47. T. Arai, A. Arai, W.H. Busby, Jr. and D.R. Clemmons, Endocrinology, 135 (1994) 2358.
48. C. Camacho-Hubner, W.H. Busby, Jr., R.H. McCusker, G. Wright and D.R. Clemmons, J. Biol. Chem., 267 (1991) 11949.
49. P.J. Fielder, H. Pham, E.Y. Adashi and R.G. Rosenfeld, Endocrinology, 133 (1993) 415.
50. N. Franchimont, D. Durant and E. Canalis, Endocrinology, 138 (1997) 3380.

Molecular Mechanisms to Regulate the
Activities of Insulin-like Growth Factors
K. Takano, N. Hizuka and S-I. Takahashi (Editors)

Regulation of Insulin-like Growth Factor I Actions by Insulin-like Growth Factor Binding Protein-5

David R. Clemmons, Yumi Imai, Bo Zheng, Jane Clarke, and Walker H. Busby, Jr.

Department of Medicine, University of North Carolina School of Medicine, Chapel Hill, NC 27599-7170, USA

Abstract

The insulin-like growth factors are present in the pericellular space complexed to insulin-like growth factor binding proteins. Since these binding proteins generally have a higher affinity for the ligand than the cell surface, type-I IGF receptor, they control the ability of IGF-I to access receptors, and thus indirectly regulate IGF actions. The variables that regulate the affinity of the binding proteins, and therefore their capacity to control receptor access, include proteolytic cleavage, binding to cell surfaces and to extracellular matrix, as well as the abundance of each protein in the pericellular environment and their affinities for IGF-I or IGF-II. We have determined that insulin-like growth factor binding protein-5 (IGFBP-5) associates with extracellular matrix (ECM) with high affinity. Furthermore, we have shown that several specific proteins within the ECM, such as plasminogen activator inhibitor-1 (PAI-1), bind to IGFBP-5. When associated with ECM, its affinity is reduced by 8 fold, and it is protected from proteolytic cleavage. In contrast, when IGFBP-5 is present in culture medium, it is rapidly cleaved to a non-IGF binding 22 kDa fragment. Specific basic amino acids between amino acids 201 and 218 regulate the ability of IGFBP-5 to bind to ECM. The residues Arginine 207 and 211 appear to be the most important mediators of binding. Mutagenesis of these residues results in decreased binding to extracellular matrix, and decreased cellular responsiveness to IGF-I. In contrast to ECM-associated IGFBP-5, the form of this protein that is present in the pericellular fluid has a high affinity for IGF-I. We determined its cleavage site and mutated it to create a protease resistant form. Addition of this form to smooth muscle cell cultures resulted in significant inhibition of replication. Therefore, ongoing proteolytic cleavage of IGFBP-5 may be necessary for full expression of IGF-I actions. In summary, several variables in the pericellular environment are capable of modulating the ability of IGFBPs to alter IGF actions. Understanding how these variables alter IGF actions could be an important key to understanding the mechanism by which this growth factor stimulates anabolic effects.

Introduction

The insulin-like growth factors are present in the pericellular environment bound to one or more of six forms of IGF binding proteins (1,2). These binding proteins have an affinity constant that is between 2 and 50 fold higher than the type I IGF receptor (3). Therefore, when present in their intact form, IGFBP's 1-6 are capable of controlling access of IGF-I or -II to receptors and are thus indirectly capable of controlling IGF stimulated actions. Generally, within the pericellular environment there is unsaturated binding protein, and therefore an increase in the IGF concentration may not necessarily result in a parallel increase in receptor saturation, and the degree to which this can occur is dependent upon the concentrations of intact, unsaturated binding protein forms that are also present and their affinities for IGF-I or IGF-II.

Several variables have been described that lower binding protein affinity. Three that have been studied in detail. They include (A) cell membrane or extracellular matrix (ECM) localization, which lowers the affinity principally of IGFBP-5, (B) proteolytic cleavage, which occurs with all binding proteins and drastically lowers their affinities, and, (C) phosphorylation of IGFBP-1 and -3 has been demonstrated, and for IGFBP-1 it has been shown to enhance affinity (3). In this paper, we will present a discussion of the variables that regulate insulin-like growth factor protein-5 affinity in the pericellular environment, principally proteolysis and adherence to ECM (4). IGFBP-5 is unique among the IGF binding proteins in that it has a very high affinity for ECM. This affinity is regulated both by the particular substances in the ECM that can bind to IGFBP-5 and by specific basic residues that are located between positions 201 and 218 within the molecule (5). To determine the IGFBP-5 regions that might be important for ECM binding, we prepared ECM from cultured human fibroblasts (6). This ECM was then incubated with IGFBP-5, and the capacity of 2 synthetic peptides that contained several basic amino acids corresponding to amino acids 131-141 (peptide B) and 201-218 (peptide A) (Figure 1) to compete for binding determined. As shown in figure 2, both peptides competed for binding; however, peptide A was the most potent inhibitor. In addition to inhibiting the binding of exogenously added IGFBP-5 to ECM, these peptides also inhibited the binding of endogenously synthesized peptide, thus showing that they are active even during the phase of matrix assembly (5). When ECM was prepared from other connective tissue cell types, for example chondrocytes, osteoblasts (7,8) and smooth muscle cells, similar findings were obtained. Because the amino acids within peptide A appeared to be more important for IGFBP-5 ECM association, we prepared mutants that had charged substitutions for these residues and determined the ability of the purified mutants to bind to ECM.

AMINO ACID SEQUENCE OF IGFBP-5

M V L L T A V L L L L A A Y A G P A Q S L 1

G S F V H C E P C D E K A L S M C P P S P 22

L G C E L V K E P G C G C C M T C A L A E 43

G Q S C G V Y T E R C A Q G L R C L P R Q 64

D E E K P L H A L L H G R G V C L N E K S 85

Y R E Q V K I E R D S R E H E E P T T S E 106

M A E E T Y S P K I F R P K H T R I S E L 127

Peptide B K A E A V K K D R R K K L T Q S K F V G G 148

A E N T A H P R I I S A P E M R Q E S E Q 169

G P C R R H M E A S L Q E L K A S P R M V 190

Peptide A P R A V Y L P N C D R K G F Y K R K Q C K 211

P S R G R K R G I C W C V D K Y G M K L P 232

G M E Y V D G D F Q C H T F D S S N V E

Figure 1. Amino acid sequence of IGFBP-5. The amino acid sequence is shown. The residues that are encoded by peptides A and B are underlined.

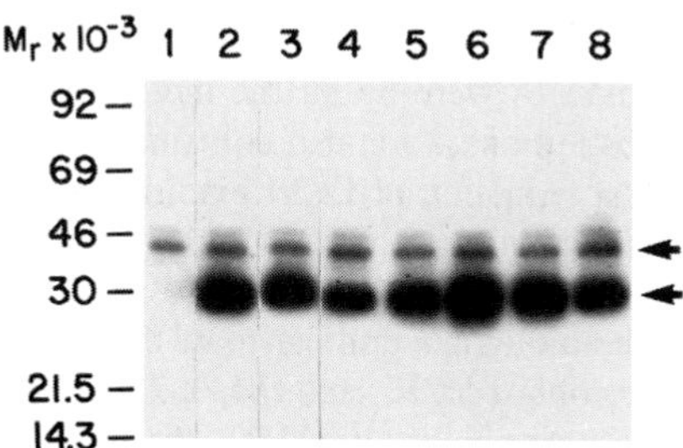

Figure 2: Ligand blot of IGFBP-5 in ECM in the presence of peptides A and B. Peptides A and B were incubated in increasing concentrations with native IGFBP-5 and ECM. Lane 1: ECM alone; lanes 2-8: ECM plus IGFBP-5. Lane 3: Peptide A (45 ng/ml); lane 4: 450 ng/ml; lane 5: 4.5 μg/ml; lane 6: Peptide B (45 ng/ml); lane 7: 450 ng/ml; lane 8: 4.5 μg/ml.

The basic amino acids at positions 202, 206, 207, 211, 214, 217, and 218 were targeted. CDNA's containing the substitutions were transfected into CHO cells, and the mutant proteins were expressed. Each mutant protein was purified to homogeneity and its protein concentration determined by amino acid composition analysis. Equimolar concentrations of each protein were incubated with fibroblast ECM, and the amount of each protein that associated with the ECM was determined by matrix extraction followed by Western ligand blotting. As shown in figure 3, the mutant containing substitutions for amino acids 211, 214, 217, and 218 had the greatest reduction in ECM association, followed by the mutant with 202, 206, 207 substitutions (5). Substitutions for B chain amino acids 134 and 136 had less of an effect, as did a single substitution for amino acid 211. Extension of these

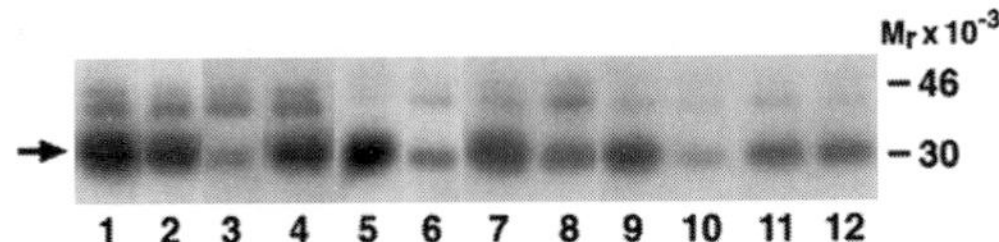

Figure 3. Ligand blot of IGFBP-5 mutants binding to ECM. Fibroblast ECM was prepared and incubated with 80 ng/ml of each mutant. The ECM was then extracted and IGFBP-5 binding determined by SDS PAGE with ligand blotting. Lane 1: IGFBP-5; lane 2: K211N; lane 3: K211N, R214A, K217A, R218N; lane 4: K134, R136A; lane 5: native IGFBP-5; lane 6: K202N, K206A, R207A; lane 7: IGFBP-5; lane 8: K134A, R136A, K211N; lane 9: IGFBP-5; lane 10: IGFBP-5 plus peptide A; lane 11: IGFBP-5 plus peptide A K211N; lane 12: K207A, K211N, peptide A.

findings to smooth muscle cell ECM and the use of additional single substitution mutants showed that amino acids 207 and 214 were by far the most important determinants of ECM binding. Since those substitutions were also contained in the 2 mutants tested on fibroblast ECM with the greatest reduction of ECM binding, it is likely that those specific residues account for much of the affinity of IGFBP-5 for fibroblast ECM. It should be noted that there is no alteration in affinity of any of these mutants for IGF-I or -II.

To further determine the biologic significance of these substitutions, the purified mutants were incubated with fibroblast ECM, and then cells were plated on top of the ECM containing mutant protein. As shown in table 1, the mutants that had reduced ECM binding resulted in the least potentiation of fibroblast growth, as compared to wild-type IGFBP-5, which caused a 2.2 fold increase in cell division in response to IGF-I (4). Extraction of the ECM at the end of the experiment confirmed that the 2 mutants that had previously been shown to have the lowest ECM binding were the least abundant. We concluded from these experiments that basic amino acids between positions 201 and 218 account for the ability of IGFBP-5 to bind to mesenchymal cell ECM and that Arg 207 and 214 may be the most important residues involved in binding. Lowering the amount of IGFBP-5 bound to the ECM by mutating these residues results in a marked attenuation in cell growth response to this growth factor.

Table 1
Binding of IGFBP-5 mutants to ECM and their effects on cell responses

Mutant	% reduction in binding	Fold increase in cell growth compared to IGF-I alone
Wild type IGFBP-5	0	2.2
K134A, R136A	26	1.9
K211N	21	1.9
R201A, K202N	23	2.0
K217N, R218A	36	1.7
R201A, K202N, K206A, R208N	21	1.8
R207A, K211N	49	1.6
R214A	51	1.6
K202N, K206A, R207A	76	1.2
K211N, R214A, K217A, R218N	91	1.0

To further analyze the importance of changes in ECM composition in determining the amount of IGFBP-5 that is associated with the ECM, and to determine if specific matrix substituents had high affinity for IGFBP-5, conditioned medium that contained soluble forms of matrix proteins from both smooth muscle cells and fibroblasts was chromatographed by IGFBP-5 affinity chromatography. Specific bands were eluted from the affinity column and their sequences determined by Edman degradation. Sequencing showed that 2 proteins that have been shown to be abundant substituents of the ECM of each these cell types were identified. These included plasminogen activator inhibitor-1 (PAI-1) and thrombospondin (9). To confirm that these proteins bound IGFBP-5 with high affinity, increasing concentrations of each protein were incubated with IGFBP-5 and their affinities for IGFBP-5 determined. The affinity for PAI-1 was 1.2 x 10^9M^{-1}, whereas the affinity of thrombospondin was higher at 2.1 x $10^{10}M^{-1}$. Incubation of PAI-1 with IGFBP-5 followed by coimmunoprecipitation using anti-PAI-1 antiserum showed that this protein could be specifically precipitated (figure 4). The significance of PAI-1 binding has not been defined, since it was shown that PAI-1 binding in solution did not alter the ability of IGFBP-5 to be proteolytically cleaved and did not alter its affinity for IGF-I. However, since PAI-1 is an abundant component of ECM, and IGFBP-5 affinity for IGF-I is lowered when it is bound to ECM; insoluble matrix associated, PAI-1 may participate in this reaction. In addition to these specific proteins, proteoglycans, such as tenascin, that are present in the ECM have also been shown to bind to IGFBP-5 (10). Specifically, proteoglycans that contain heparan sulfate that is O-sulfated in the 2 or 3 position of the iduronic acid ring will bind to IGFBP-5 and no doubt represent an important component of ECM binding.

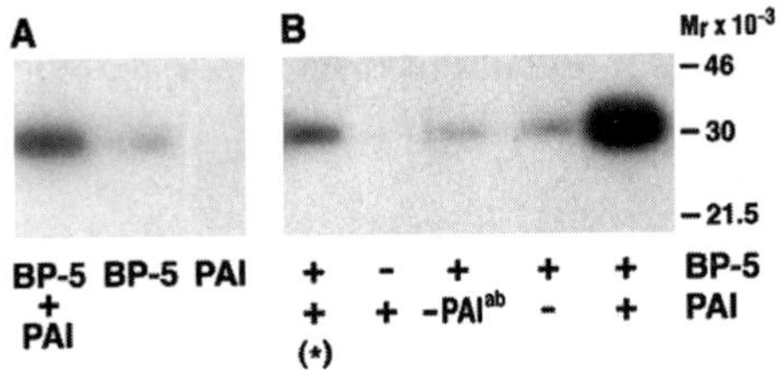

Figure 4: Panel A - IGFBP-5 (50ng/ml) was incubated with PAI-1 (100 ng/ml) and the complexes were immunoprecipitated with anti-PAI-1 then analyzed by ligand blotting using ^{125}I-IGF-I. Panel B - ^{125}I-IGFBP-5 (50000 cpm/ml) was incubated with PAI-1 (100 ng/ml) and anti-PAI-1 antibody. Following immunoprecipitation the proteins and the proteins were analyzed by SDS PAGE with autoradiography. The asterisk indicates an excess of unlabeled IGFBP-5.

Although IGFBP-5 that is associated with the ECM is present in the intact form when culture media from several cell types is analyzed there is often an abundant 22 kDa fragment. This fragment arises from proteolytic cleavage. To determine the functional significance of IGFBP-5 proteolysis in controlling IGF actions, we attempted to identify the cleavage site within this protein. The IGFBP-5 protease that is present in both human fibroblast and porcine aortic smooth muscle conditioned medium has been characterized. In both cases, the protease is a 95 kDa serine protease that cleaves IGFBP-5 into two fragments, the largest of which is 22 kDa. To confirm that this was a serine protease, we partially purified the activity from fibroblast conditioned media, then tested various types of protease inhibitors from activity. As shown in figure 5A, 3,4-DCI was a very potent inhibitor and TIMP-1 had weak activity (11). To further confirm that was a serine protease, the ability of the enzyme to immunoprecipitate ^{3}H-DFP was determined. As shown in figure 5B, a 45 kDa protein immunoprecipitated ^{3}H-DFP and zymography

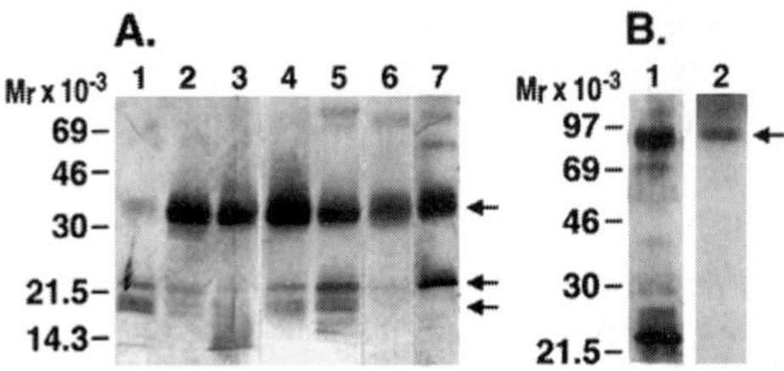

Figure 5. Panel A - Various protease inhibitors were incubated with the partially purified IGFBP-5 protease then the cleavage products analyzed by immunoblotting. The arrows denote the positions of intact IGFBP-5 and the

22 kDa fragment. Lane 1 - No protease inhibitor. Lane 2 - Alpha 1 antichymotrypsin. Lane 3 - BP-145 a serine protease inhibitor. Lane 4 - Heparin. Lane 5 - Heparin cofactor II. Lane 6 - 3,4-DCI. Lane 7 - Timp 1. Panel B - IGFBP-5 zymography Lane 1. ^{3}H-DFP incorporation into the protease Lane 2.The same band that accounts for zymographic activity incorporates ^{3}H-DFP incorporation.

showed that the protease activity had the same molecular weight. Thus, we confirmed by a very specific method that it was a serine protease. Both osteoblasts (13,14) and smooth muscle cells (15) produce a similar protease.

To determine the functional significance of proteolysis, we determined the cleavage site and prepared a protease resistant mutant. Transfection of the partially purified protease with native IGFBP-5 and isolation and sequencing of fragments showed that the 22 kDa fragment contained the N-terminus of the protein. In contrast, a 12 kDa fragment was isolated that contained the C-terminal sequence beginning at amino acid lysine 144. Based on observation, a mutant was prepared that mutagenized the serine 143 and lysine 144. This mutant was purified to homogeneity and incubated with the partially purified IGFBP-5 protease. In contrast to what was anticipated, the rate of cleavage by the IGFBP-5 protease was retarded but cleavage occurred. In order to determine the cleavage site in this mutant, the fragments were purified and again sequenced. A cleavage site was identified at lysine 138 and lysine 139. Since this was a di-basic site, it appeared to be more consistent with the fact that the IGFBP-5 protease is a serine protease (12). In order to confirm that this was the correct cleavage site, we mutagenized lysine 138 and 139 to asparaginines. When the mutant form was added with IGF-I to smooth muscle cell cultures, it completely blocked the ability of IGF-I to stimulated DNA synthesis, protein synthesis, and cell migration if a 6:1 molar excess was present (figure 6). In contrast,

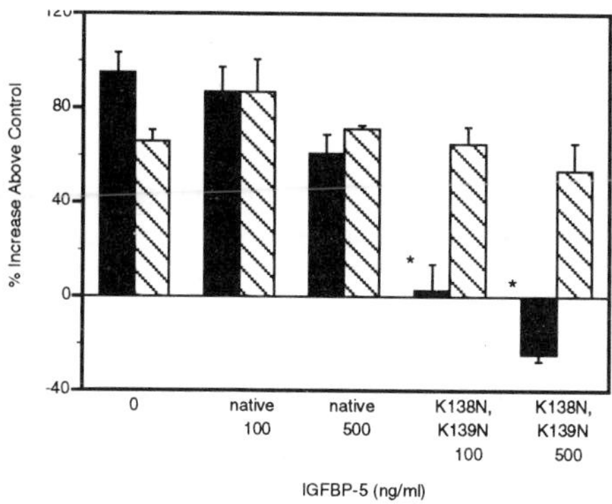

Figure 6: The purified protease resistant mutant and wild type IGFBP-5 were incubated with IGFBP-5 protease for the times shown. The products of the

reaction were detected by immunoblotting. After 14 hours all of the IGFBP-5 was degraded and none of the mutant was degraded.

when lower concentrations were added, it was relatively ineffective. Analysis of the conditioned media at the end of the experiments showed that the protein was not cleaved, even after a 72 hr incubation, indicating that the correct cleavage site is lysine 138-139 (figure 7). We surmised from these data that the initial cleavage occurs at that site, and

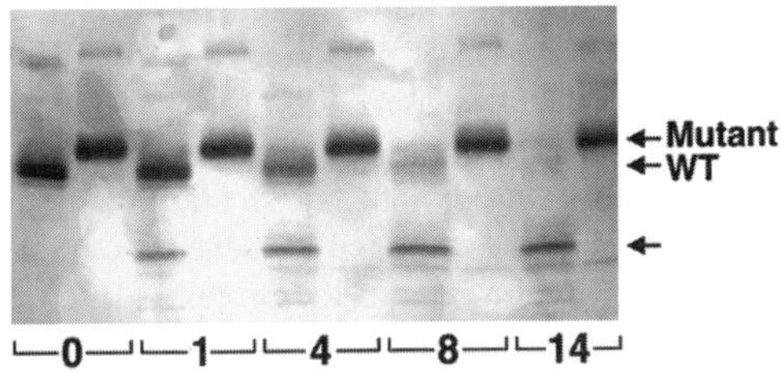

Figure 7: Increasing concentrations of native IGFBP-5 or IGFBP-5 mutant were incubated with 20 ng/ml of IGF-1 (solid bars) or des 1,3-IGF-I (hatched bars) and their capacities to inhibit ^{3}H-thymidine corporation into DNA determined. The difference between the two proteins when added at 500 ng/ml is statistically significant.

then an amino peptidase further cleaves the protein at the 143-144 site, but that the 138-139 site is the primary cleavage site. To further confirm the physiologic importance of IGFBP-5 cleavage, we transfected the protease-resistant mutant cDNA into smooth muscle cells and expressed the protein. The expressed protein was not cleaved and was easily detectable in its intact form. When these cells were tested for their ability to respond to IGF-I, a concentration of 5 ng/IGF-I had no effect on protein synthesis in the cells constituitively expressing the protease resistant mutant. We conclude from these findings that high affinity IGFBP-5 is a potent inhibitor of IGF-I action if added at molar excess of 6:1 over IGF-I. However, lower concentrations of intact IGFBP-5 are not effective inhibitors of these physiologic responses. This suggests that a large molar excess of intact IGFBP-5 is required and that nearly complete inhibition of proteolysis would be required to see inhibition of IGF-I action.

Summary

We have determined that IGFBP-5 adheres tightly to ECM when compared to other IGF binding proteins. This capacity to adhere to ECM is dependent upon the presence of basic residues in the region of 201-218, and residues 207 and 214 appear to be the most important for binding. Diminished binding results in altered cellular responsiveness. Our proposed mechanism by which IGFBP-5 is altering IGF-I mediated cell growth and anabolic responses is as follows. If IGFBP-5 synthesis is stimulated, such as is seen after exposure of smooth muscle cells to IGF-I (16), then the amount of intact IGFBP-5 that is deposited in ECM is also increased. This enhanced matrix content of a low affinity form of IGFBP-5 results in an increased amount of IGF-I being in a favorable equilibrium with the receptors. However, enhanced synthesis and release of IGFBP-5 into pericellular fluid does not result in inhibition of IGF-I action by high affinity IGFBP-5 because of ongoing proteolysis, that is, the rate of proteolysis is such that the intact IGFBP-5 concentration in medium is consistently less than the 6:1 molar excess over IGF-I, thus allowing the IGF-I to be preferentially associated with ECM associated IGFBP-5 and receptors. Since ECM associated IGFBP-5 is protected from proteolysis because its cleavage site is bound to ECM components, this allows the material to accumulate in the ECM and thus further enhance IGF-I actions. Factors that disrupt this equilibrium, such as potent protease inhibitors or soluble glycosaminoglycans, that lower the amount of ECM-associated IGFBP-5 might be expected to alter IGF-I actions. Future studies should be directed toward determining the *in vivo* significance of these variables.

References

1. Rechler, M. M. 1993. Insulin like growth factor binding proteins. *Vitamins and Hormones* 47:1-114.
2. Baxter, R. C. and J. L. Martin. 1989. Binding proteins for insulin-like growth factors: Structure, Regulation and Function. *Prog Growth Fact Res* 1:49-68.
3. Jones, J. I. and D. R. Clemmons. 1995. Insulin like growth factor and their binding proteins: biologic actions. *Endocrine. Rev.* 16:3-34.
4. Jones, J. I., A. Gockerman, W. H. Busby, C. Camacho-Hubner, and D. R. Clemmons. 1993. Extracellular matrix contains insulin-like growth factor binding protein-5: Potentiation of the effects of IGF-I. *J. Cell Biol.* 121:679-687.
5. Parker, A., W. H. Busby, and D. R. Clemmons. 1996. Identification of the extracellular matrix binding site for insulin like growth factor binding protein-5. *J. Biol. Chem.* 271:13523-13529.
6. Camacho-Hubner, C., W. H. Busby, R. H. McCusker, G. Wright, and D. R. Clemmons. 1992. Identification of the forms of insulin-like growth factor binding proteins produced by human fibroblasts and the mechanisms that regulate their secretion. *J. Biol. Chem.* 267:11949-11956.
7. Kanzaki, S., S. Hilliker, D. J. Baylink, and S. Mohan. 1994. Evidence that human bone cells in culture produce insulin-like growth factor bidning protein-4 and -5 proteases. *Endocrinology* 134(1):383-392.
8. Andress, D. L. 1995. Heparin modulates the binding of insulin-like growth factor binding protein-5 to a membrane protein in osteoblast cells. *J. Biol. Chem.* 270:28289-28297.
9. Nam, T. J., W. H. Busby, and D. R. Clemmons. 1997. Insulin-like growth factor binding protein-5 binds to plasminogen activator inhibitor-I. *Endocrinology* 138:2972-2998.

10. Arai, T., J. B. Clarke, A. Parker, W. Busby,Jr., T. J. Nam, and D. R. Clemmons. 1996. Substitution of specific amino acids in insulin-like growth factor-binding protein-5 alters heparin binding and its change in affinity for IGF-I in response to heparin. *J Biol Chem* 271:6099-6106.
11. Nam, T. J., W. H. Busby, and D. R. Clemmons. 1996. Insulin-like growth factor binding protein-5 bound to plasminogen activator inhibitor I. *Endocrinology* 137:5530-5536.
12. Imai, Y., W. H. Busby, C. E. Smith, J. B. Clark, G. D. Horvitz, C. Rees, and D. R. Clemmons. 1997. Protease resistant form of insulin-like growth factor binding protein-5 is an inhibitor of insulin-like growth factor-I actions in porcine smooth muscle cells in culture. *J Clin Invest* in press:
13. Conover, C. A., S. K. Durham, J. Zapf, F. R. Masiarz, and M. C. Keifer. 1995. Cleavage analysis of insulin-like growth factor (IGF) -dependent IGF-binding protein-4 proteolysis and expression of protease-resistant IGF-binding protein-4 mutants. *J Biol Chem* 270:4395-4400.
14. Thrailkill, K. M., P. Quarles, H. Nagase, K. Suzuki, D. M. Serra, and J. L. Fowles. 1995. Characterization of insulin like growth factor binding protein 5 degrading proteases produced throughout serine osteoblast differentiation. *Endocrinology* 136:3527-3533.
15. Gockerman, A. and D. R. Clemmons. 1995. Porcine aortic smooth muscle cells secrete a serine protease for insulin like growth factor binding protein-2. *Circ. Res.* 76:514-521.
16. Duan, C. M. and D. R. Clemmons. 1996. Insulin like growth factor I (IGF-I) regulates IGF binding protein-5 synthesis through transcriptional activation of the gene in aortic smooth muscle cells. *J Biol Chem* 271:4280-4288.

Molecular Mechanisms to Regulate the
Activities of Insulin-like Growth Factors
K. Takano, N. Hizuka and S-I. Takahashi (Editors)

IGF-independent actions of IGFBPs

Y. Oh, Y. Yamanaka, H.-S. Kim, P. Vorwerk, E. Wilson, V. Hwa, D.-H. Yang, A. Spagnoli, D. Wanek and R.G. Rosenfeld

Department of Pediatrics, School of Medicine, Oregon Health Sciences University, 3181 S.W. Sam Jackson Park Road, Portland, OR 97201, USA.

Abstract: The insulin-like growth factor (IGF) signaling system functions through interactions of the ligands (IGF-I, IGF-II and insulin) with IGF binding proteins (IGFBPs) and a family of transmembrane receptors (the insulin, type 1 IGF and type 2 IGF receptors). The current dogma is that the IGFBPs are modulators of the system, binding IGFs, but not insulin, with high affinity, thereby regulating the availability of IGFs to the receptors. The human IGFBP family consists of at least eight proteins, designated IGFBPs 1-8, which are found in physiological fluids and extracellular compartments and are produced by most cells. Although it was believed that these IGFBPs function as classical carrier proteins to transport IGFs and extend their half-lives in serum, several lines of recent evidence from various cell systems have suggested that IGFBPs may play more active roles (IGF-independent actions) in growth regulation. In support of this hypothesis, we have recently shown that IGFBP-3 binds specifically and with high affinity to the surface of various cell types and directly inhibits monolayer growth of these cells in an IGF-independent manner, presumably by specific interaction with cell membrane proteins that function as an IGFBP-3 receptor.

To extend our understanding of the IGF-independent action of IGFBPs, we have recently identified and characterized a new class of IGFBPs. These include IGFBP-7 (MAC25), IGFBP-8 (CTGF) and potentially IGFBP-9 (NOV) and IGFBP-10 (CYR61). These IGFBPs constitute low affinity members of the IGFBP family, and function primarily as modulators of cell growth in an IGF-independent manner, similar to the action observed with IGFBP-3 in breast cancer cells. Intriguingly, our recent studies have demonstrated that IGFBPs, especially IGFBP-7 and NH_2-terminal fragments of IGFBP-3, have the ability to bind insulin specifically, modulating insulin binding to its receptor, and subsequently inhibiting the early steps in insulin action such as autophosphorylation of the insulin receptor β subunit and phosphorylation of IRS-1. This indicates that the affinity of other IGFBPs for insulin can be enhanced by modifications which disrupt disulfide bonds or remove the conserved COOH-terminus (notably proteolysis of IGFBPs by specific proteases *in vivo*) and might contribute to the insulin resistance of pregnancy, type II diabetes mellitus, and other pathological conditions. These studies clearly indicate that investigations of the IGF-independent action of IGFBPs will augment our understanding of how the growth of normal or neoplastic cells can be modulated by the IGF/IGFBP system, how other growth factors or pharmacological agents can interface with this system, and how IGFBPs affect the pathophysiology of insulin resistance.

This work was supported by NIH grant # CA 58110 (to RGR) and US Army grant # DAMD 17-96-1-6204 (to YO).

1. INTRODUCTION

Since the existence of an IGF binding protein was initially postulated over 30 years ago (1), a total of six (IGFBPs 1-6) have been cloned and characterized (2-5). The relatively high sequence and structural similarity among the six IGFBPs indicate that their individual genes were derived from successive duplication of a common ancestor in a relatively short time frame during evolution. IGFBPs 1-6 bind IGF-I and IGF-II with high affinity and serve to transport the IGFs, prolong their half-lives, and modulate their proliferative and anabolic effects on target cells. The molecular mechanisms involved in the interaction of the IGFBPs with the IGFs and their receptors remain unclear, but these molecules appear, at least, to regulate the availability of free IGFs for interaction with IGF receptors (IGF-dependent actions of IGFBPs) (6, 7).

The complexity of IGF-IGFBP system was increased by the observation and identification of proteolytic fragments of IGFBPs *in vitro* and *in vivo* (8-12). These post-translationally modified IGFBPs appear to exhibit significantly reduced affinity for IGFs, thereby increasing the availability of IGFs to cell-membrane receptors. This suggests a possible mechanism for the liberation of IGFs from the IGF:IGFBP complex and delivery of IGFs to IGF receptor in targeted tissue (13).

The last five years have seen several significant leaps forward in our understanding of IGFBPs: (I) Some IGFBPs appear to exert biological action through an IGF-independent mechanism (14-30); (II) a new concept of an IGFBP superfamily emerges from identification of low-affinity binders of IGFs (table 1) (31-33); and (III) IGFBPs appear to bind insulin and, in some cases, modulate insulin receptor activation (34-35). This growing body of evidence thus indicates that each IGFBP might have its own, yet to be elucidated, biological actions beyond the ability to regulate IGF actions. Undoubtedly these findings bring with them a wealth of unanswered questions, providing clear and exciting directions for our future research on IGFBPs.

Table 1. Structural characteristics of the human IGFBPs

IGFBP	Molecular weight	Number of amino acids	Number of cysteines	N-linked glycosylation	Chromosomal localization	mRNA size (kb)
High affinity IGFBPs						
IGFBP-1	25,271	234	18	No	7p	1.6
IGFBP-2	31,355	289	18	No	2q	1.5
IGFBP-3	28,717	264	18	Yes	7p	2.4
IGFBP-4	25,957	237	20	Yes	17q	1.7
IGFBP-5	28,553	252	18	No	2q	1.7, 6.0
IGFBP-6	22,847	216	16	No	12	1.1
Low affinity IGFBPs						
IGFBP-7	?	251	18	Yes	4q	1.1
IGFBP-8	?	349 (pre)	39	Yes	6q (c-myb)	2.4
IGFBP-9	?	357 (pre)	41	? (No)	8q (c-myc)	2.4
IGFBP-10	?	379 (pre)	35	? (No)	?	2.4

2. IGF-INDEPENDENT ACTION OF IGFBPs

In addition to the structural and sequence homology among IGFBPs, some IGFBPs appear to possess unique characteristics: (I) IGFBP-1 and IGFBP-3 can be phosphorylated on their serine residues (36, 37); (II) IGFBP-1 and IGFBP-2 contain integrin recognition sequences (4); (III) IGFBP-3, -5 and -6 contain heparin binding motifs (3-5); and (IV) IGFBP-3 has a nuclear localization sequence and can be translocated to the nucleus (25, 26). At present, it remains unclear whether these characteristics are important for the ability of IGFBPs to modulate IGF actions or for the IGF-independent actions of IGFBPs. Studies of IGFBP-1, -3 and -5 suggest possible non-IGF-mediated actions of these binding proteins (5, 38, 39). IGFBP-3 is the most extensively characterized IGFBP species for IGF-independent actions; therefore, the following section will focus on IGFBP-3 studies in detail. Additionally, we will discuss the new concepts of an IGFBP superfamily and insulin binding, to provide insight into the IGF-independent actions of the high affinity IGFBPs.

2.1. IGF-independent anti-proliferative action of IGFBP-3

IGFBP-3 was initially identified by Hintz *et al.* as a large, growth hormone (GH)-dependent binding protein in human serum in 1977 (40). The complete primary structure of IGFBP-3 was provided by Wood *et al.*, following the cloning of its cDNA from an adult human liver library in 1988 (41). Subsequently, this GH-dependent protein has been extensively studied and identified as an IGF regulator and carrier protein, which forms part of the 150 kDa ternary complex with IGFs and an acid labile subunit (88 kDa glycoprotein) in circulation.

The first hint of IGF-independent action of IGFBP-3 came from outside the IGF research field in 1991: Villaudy *et al.* showed that a growth inhibitor, IDF45 (inhibitory diffusible factor of 45 kDa), purified from conditioned medium of Swiss 3T3 mouse fibroblasts and subsequently identified as mouse IGFBP-3 by N-terminal amino acid sequence analysis by Blat *et al.*, inhibited either serum- or fibroblast growth factor (FGF)-induced DNA synthesis in chick embryo fibroblasts (CEF) (14, 15). Furthermore, Liu *et al.* reported that the same protein inhibited the stimulation of CEF DNA synthesis by a serum fraction from which IGFs had been removed by acid gel-chromatography (16). These studies suggested that mouse IGFBP-3 could inhibit responsiveness to growth factors and serum components other than IGFs in chicken cells. Cohen *et al.* reported that overexpression of a transfected human IGFBP-3 cDNA in Balb c/3T3 mouse fibroblasts inhibited cell proliferation, with or without added IGF or insulin (17). They suggested that IGFBP-3 might exert a non-IGF-mediated action, based on the observation that insulin, which has mitogenic activity in these cells but does not bind to IGFBP-3, could not overcome IGFBP-3-induced cell growth inhibition.

Corroborating evidence showed that exogenous recombinant human IGFBP-3 potently inhibits human breast cancer cell growth in an IGF-independent manner by Oh *et. al.* (18). In particular, specific cell-surface binding of IGFBP-3 and an IGF-independent effect of IGFBP-3 on DNA synthesis and cell replication of breast cancer cells was demonstrated. For these experiments, IGF analogs which have differential affinities for IGF receptors or IGFBP-3 were employed, as was the Hs578T human breast cancer cell line, which lacks biological responsiveness to IGFs, despite the existence of IGF receptors. Further studies supported this new concept of IGF-independent action of IGFBP-3 by showing that: (I) IGFBP-3 binds to specific proteins on cell surfaces and this interaction is strongly correlated with the ability of

IGFBP-3 to inhibit cell growth (19); (II) the growth rate is significantly reduced in human IGFBP-3 transfected mouse fibroblast cells (R$^-$ cells) derived from mouse embryos homozygous for a targeted disruption of the type 1 IGF receptor gene (20); (III) IGFBP-3 mediates transforming growth factor-β (TGF-β)-, retinoic acid (RA)-, and antiestrogen-induced growth inhibition in breast cancer cells (21-23); (IV) regulation of IGFBP-3 gene expression has been shown to play a role in signaling by p53, a potent tumor-suppressor protein (24); (V) IGFBP-3 can be translocated into the nucleus (25, 26); (VI) exogenous IGFBP-3 induces programmed cell death in PC-3 human prostate cancer cells (27); (VII) IGFBP-3 fragments derived from limited proteolysis inhibit the stimulation of DNA synthesis induced either by IGF-I or insulin in CEFs or by FGF in R$^-$ cells (28, 29); (VIII) IGFBP-3 binds to the type V TGF-β receptor and competes with TGF-β1 for receptor binding (30); and (IX) the kinetics of IGFBP-3 binding to breast cancer cells is typical of ligand-receptor interactions, indicating the presence of an IGFBP-3 receptor in these cells (31).

Although these intensive efforts from various laboratories have supported the new concept of IGF-independent action of IGFBP-3, essential information is still lacking, such as definitive identification of the IGFBP-3 receptor and elucidation of the biological mechanisms of these actions.

2.2. Cell surface-interaction of IGFBP-3

The initial demonstration of IGFBP-3 binding directly to cells came from studies on CEFs by Delbe *et al.* (42). Their studies showed that mouse IGFBP-3 can bind to CEF membranes with low affinity. Binding of IGFBP-3 to the membranes was specific, and binding sites per cell were estimated at 60,000. Using an ER-negative human cell line, Hs578T, Oh *et. al.* showed that IGFBP-3 can be identified in membrane preparations (43), and subsequently demonstrated the existence of IGFBP-3 on monolayer cell surfaces, by use of immunoperoxidase staining with an IGFBP-3 specific antibody (αIGFBP-3g1). Immunoreactive IGFBP-3 was detectable on the cell surface of Hs578T cells treated with αIGFBP-3g1, but not on cells treated with nonimmune rabbit serum. Cell surface staining was clearly decreased by treatment with IGF-I, suggesting that IGF binds to cell surface-associated IGFBP-3 and causes it to dissociate from the cell membranes. Conversely, treatment with [Gln^3,Ala^4,Tyr^{15},Leu^{16}]IGF-I, an IGF-I analog with reduced affinity for IGFBPs, resulted in immunoreactive cell-surface staining equivalent to control, reflecting the analog's decreased affinity for IGFBP-3. These studies indicate that the cell-surface binding of IGFBP-3 is inhibited by the binding of IGF peptides.

Further studies showed that IGFBP-3 binding is not mediated through fibronectin receptors or through mechanisms responsible for IGFBP-1 association with the cell surface (18). Rather, the data supported the existence of specific proteins on Hs578T cell surface responsible for the cell-surface association of IGFBP-3.

2.3. Demonstration of putative IGFBP-3 receptors

Affinity cross-linking of Hs578T monolayers and cell lysates with [^{125}I]IGFBP-3$^{E.\ coli}$ has been employed to investigate possible mechanisms for the cell surface interaction of IGFBP-3 (19). One major radiolabeled band of apparent molecular mass 55 kDa was detected on monolayer cell surfaces by affinity cross-linking. Conversely, studies employing affinity cross-linking to Hs578T cell lysates suggested the presence of as many as three specific bands (55, 58 and 85 kDa). The additional two proteins may reflect either intracellular proteins contained within the whole cell lysates, components of an IGFBP-3 receptor complex, or precursor or degradation forms of the IGFBP-3 receptor. Nonetheless, these three bands appeared to be specific for IGFBP-3, as demonstrated by dose-dependent displacement with unlabeled IGFBP-3 and immunoprecipitation with IGFBP-3 specific antibody, but not with non-immune serum. In addition, IGF-II, but not IGF-II analogs with reduced affinity for IGFBP-3, prevented IGFBP-3 binding to these specific cell-surface association proteins, by forming [IGF-II:IGFBP-3] complexes, consistent with previous data employing monolayer binding assays (18). The specificity of IGFBP-3 binding to cell-surface association proteins was further supported by Western ligand blot (WLB) experiments with [^{125}I]IGFBP-3. WLB of Hs578T cell lysates revealed multiple bands, including 20, 23 and 50 kDa as major species. These bands correspond to those shown by cross-linking, after subtraction of the MW of non-glycosylated IGFBP-3 (19). Rajah *et al.* also reported that IGFBP-3 appears to bind to several cell-surface association proteins in PC-3 prostate cancer cells (27).

IGFBP-3 binding sites on breast cancer cell membranes were further evaluated through competitive binding studies with unlabeled IGFBPs 1-6 and various forms of IGFBP-3, including synthetic IGFBP-3 fragments (44). Scatchard analysis revealed the existence of high affinity sites for IGFBP-3 in estrogen receptor (ER)-negative Hs578T human breast cancer cells (Kd = $8.19 \pm 0.97 \times 10^{-9}$ M with $4.92 \pm 1.51 \times 10^{5}$ binding sites/cell) and 30-fold fewer receptors in ER-positive MCF-7 cells (Kd = $8.49 \pm 0.78 \times 10^{-9}$ M with $1.72 \pm 0.31 \times 10^{4}$ binding sites/cell), using a one-site model. These observations demonstrate binding characteristics of typical receptor-ligand interactions, providing evidence for specific, high affinity IGFBP-3 receptors on Hs578T cell membranes. These receptors have properties which support the hypothesis that they may mediate the IGF-independent inhibitory actions of IGFBP-3 in breast cancer cells.

Recently, Leal *et al.* postulated that the type V TGF-β receptor is the putative IGFBP-3 receptor (30). They demonstrated that IGFBP-3 binds to the type V TGF-β receptor specifically and can compete with TGF-β for receptor binding in mink lung epithelial cells, suggesting that IGFBP-3 is a functional ligand for the type V TGF-β receptor. However, at the present time, it is unclear whether this type V TGF-β receptor represents the primary IGFBP-3 receptor, whether it is identical to the protein observed in breast or prostate cancer cell systems, and whether it plays a role in the growth inhibitory action of IGFBP-3 in those cell systems. Nonetheless, this new finding provides further evidence for "IGF-independent anti-proliferative action of IGFBP-3".

3. CHARACTERIZATION OF NEW LOW-AFFINITY MEMBERS OF THE IGFBP FAMILY

Recently, the protein encoded by the mac25 cDNA and connective tissue growth factor (CTGF) have been identified as IGFBP-7 and -8, respectively, using baculovirus-expressed recombinant proteins and polyclonal antibodies specific for human MAC25 and CTGF (31-33). These findings lead to the proposal of the concept of an "IGFBP superfamily", which includes high affinity IGF binders (IGFBPs 1-6) and low affinity IGF binders [IGFBP-7 and -8 as well as putative IGFBP-9 (NOV) and IGFBP-10 (CYR61)] (Fig. 1). This new family of low affinity IGFBPs shows an overall homology of 30-45% compared with IGFBPs 1-6. Moreover, these proteins contain the common IGFBP motif (GCGCCXXC) in their NH_2-

termini, a region containing a cluster of 12 conserved cysteines in IGFBPs 1-6, of which 11-12 are found in IGFBPs 7-10.

Binding of ^{125}I-IGF-I and ^{125}I-IGF-II to rh-MAC25 (IGFBP-7) and rh-CTGF (IGFBP-8) were demonstrated by western ligand blotting after non-denaturing polyacrylamide gel electrophoresis, as well as by affinity cross-linking. In comparison with IGFBP-3, these proteins showed a 5-10 fold lower affinity for IGF-I and 20-25 fold lower affinity for IGF-II, indicating that IGFBP-7 and -8 constitute low affinity members of the IGFBP superfamily (31, 33).

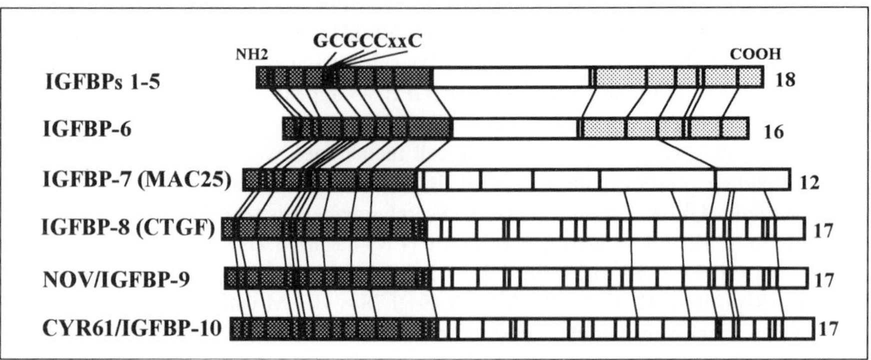

Figure 1. Comparison of total cysteine residues of IGFBP superfamily.
The conservation of the cysteine residues is shown by the vertical lines. The numbers represent the total conserved cysteines among IGFBPs. The shaded regions represent the conserved domains of IGFBPs in their NH_2- and COOH- termini.

To extend our understanding of the IGF-independent actions of IGFBPs, we have investigated the biological action of IGFBP-7 in human breast cancer cells. Using baculovirus-expressed IGFBP-7 (IGFBP-7^{bac}) and polyclonal antibodies specific for IGFBP-7, we found that expression of IGFBP-7 mRNA and protein is up-regulated by TGF-β in Hs578T breast cancer cells. Moreover, treatment with IGFBP-7^{bac} resulted in inhibition of DNA synthesis and cell proliferation in a dose-dependent manner in these cells, even in the absence of IGF peptides. Additionally, IGFBP-7 mRNA was down-regulated in human breast cancer cells, and up-regulated in normal, growing mammary epithelial cells by RA. It seems likely that, while IGFBP-7 constitutes a low affinity member of the IGFBP family, it primarily functions as a modulator of cell growth in an IGF-independent manner, similar to the action observed with IGFBP-3 in breast cancer cells.

IGFBP-8/CTGF, as well as highly related cDNAs (nov-oncogene, cyr61), have been identified in human tissues; the predicted proteins are products of "immediate-early genes" expressed after induction by serum, growth factors or certain oncogenes (45-47). CTGF/nov/cyr61 fall into the group of immediate-early genes that code for secreted, extracellular proteins which are needed for coordination of complex biological processes. Little is known about CTGF function and regulation, although partially purified CTGF from media conditioned by human umbilical vein endothelial cells demonstrated mitogenic and chemotactic activities (44). To date, information on the roles of CTGF has been inferred from studies of regulation of CTGF mRNA. For example, in human fibroblast cells, CTGF mRNA is up-regulated by TGF-β, a known growth stimulator for human fibroblasts, thus supporting the notion that CTGF is involved in regulating mitogenic activity (45, 48). Our preliminary studies indicate that mRNA and protein levels of IGFBP-8 are up-regulated by TGF-β and RA in ER-negative breast cancer cells, as well as in prostate cancer cells. These findings are consistent with the accumulated evidence that low affinity IGFBPs are important IGF-independent growth regulators.

We speculate that the IGFBP superfamily is derived from an ancestor gene/protein that was critically involved in the regulation of cell growth, probably in an IGF-independent manner. Over the course of evolution, the ability to bind IGF peptides with high or low affinity may have been a secondary role acquired by some of the IGFBPs, thereby conferring upon them the ability to influence cell growth (as in the breast) by both IGF-dependent and IGF-independent means (Fig. 2).

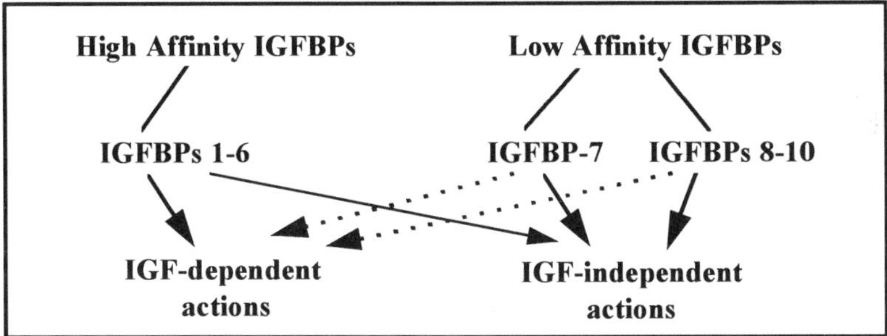

Figure 2. The potential biological actions of the IGFBP superfamily
The thick and thin arrows indicate proposed primary and secondary biological actions, respectively, while the dashed lines represent potential actions of IGFBPs.

4. CHARACTERIZATION OF IGFBPS AS INSULIN BINDING PROTEINS

We have recently identified an exciting, although surprising, finding that IGFBPs can bind insulin, as well as IGFs, subsequently modulating insulin receptor activation by inhibiting insulin binding to its receptor (34). Our studies demonstrated that, in particular, IGFBP-7, a newly identified member of low-affinity IGF binders, is a high-affinity insulin binding protein. IGFBP-7 not only binds insulin specifically but also modulates insulin binding to its receptor, subsequently inhibiting the early steps in insulin action, such as autophosphorylation of the insulin receptor β subunit and phosphorylation of IRS-1. These data raise the possibility that IGFBPs, and especially IGFBP-7, are functional insulin binding proteins. Indeed, the affinity of IGFBPs for insulin can be enhanced by modifications which disrupt disulfide bonds or remove the conserved COOH-terminus (Fig. 3). Like intact IGFBP-7, an NH_2-terminal fragment of IGFBP-3 derived from plasmin digestion, baculovirus-expressed IGFBP-3^{1-87} and IGFBP-3^{1-97}, and IGFBP-3 fragments in normal human urine, also bind insulin with high affinity and block its action (34, 35). These observations suggest that IGFBPs with enhanced affinity for insulin (especially following proteolysis of IGFBPs by specific proteases *in vivo*) might contribute to the insulin resistance of pregnancy, type II diabetes mellitus, and other pathological conditions.

These new findings are somewhat against the current dogma which proposes that a major difference between insulin and IGFs is that only IGFs have ability to bind IGFBPs 1-6 with high affinity. However, in retrospect, there is precedent for IGFBP interaction with insulin: (I) overexpression of a transfected human IGFBP-3 cDNA in Balb c/3T3 mouse fibroblasts inhibits insulin-induced cell proliferation (17); and (II) a 16 kDa proteolytic fragment of IGFBP-3 (IGFBP-3^{1-92}) inhibits insulin-stimulated DNA synthesis in CEFs (28). Whether these effects of IGFBP-3 and its fragments on insulin action can be explained by IGFBP-3 binding to insulin remain to be elucidated. However, it is unlikely to be the full explanation of IGF-independent action of IGFBP-3 in the light of the ability of IGFBP-3 fragments to inhibit FGF action (29), as well as the actions of insulin and IGF.

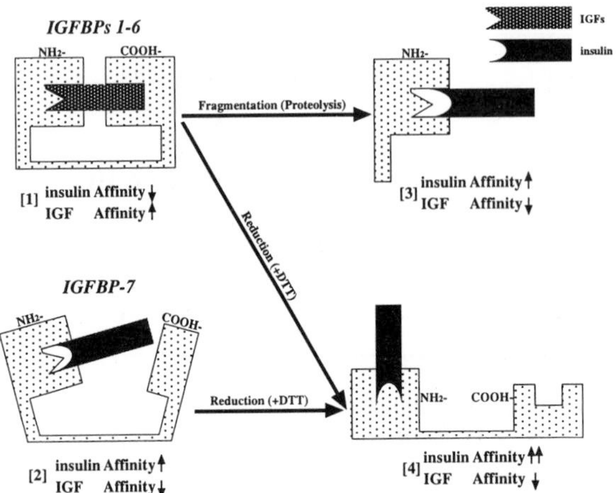

Figure 3. Schematic representation of IGFBP binding to insulin and IGFs

5. SUMMARY

A fuller understanding of the IGF-independent actions of IGFBPs will allow for a more complete understanding of how the growth of normal or neoplastic cells can be modulated by the IGF/IGFBP system, how other growth factors or pharmacological agents can interface with this system, and how IGFBPs affect the pathophysiology of insulin resistance. It is clear that we are beginning to unravel a mysterious, complex, and exciting story of the biological actions of insulin-like growth factor binding proteins and their significant implications on cell growth and proliferation.

References

1. W.H. Daughaday and D.M. Kipnis, Recent. Prog. Horm. Res., 22 (1966) 49.
2. R.C. Baxter and J.L. Martin, Prog. Growth Factor Res. 1 (1989) 49.

3. R.G. Rosenfeld, G.L. Lamson, H. Pham, *et. al.*, Recent. Prog. Horm. Res., 46 (1991) 159.
4. S. Shimasaki and N Ling, Prog. Growth Factor Res. 3 (1991) 243.
5. J.L. Jones and D.R. Clemmons, Endocrine Rev. 16 (1995) 3.
6. K.M. Kelley, Y. Oh, S.E. Gargosky, *et. al.*, Int. J. Biochem. Cell. Biol., 28 (1995) 619.
7. M.M. Rechler, Vitam. Horm., 47 (1993) 1.
8. P. Hossenlopp, B. Segovia, C. Lassare, *et. al.*, J. Clin. Endocrinol. Metab., 71 (1990) 797.
9. L.C. Giudice, E.M. Farrell, H. Pham, *et. al.*, J. Clin. Endocrinol. Metab., 71 (1990) 806.
10. C.A. Conover CA and D.D. De Leon, J. Biol. Chem. 269 (1994) 7076.
11. P. Cohen, H.C.B. Graves, *et. al..*, J. Clin. Endocrinol. Metab., 75 (1993) 1046.
12. D.Y. Lee, S.K. Park, P.D. Yorgin, *et. al.*, J. Clin. Endocrinol. Metab., 79 (1994) 1376.
13. P. Cohen, D.M. Peehl, G. Lamson, *et. al.*, J. Clin. Endocrinol. Metab., 73 (1991) 401.
14. J. Villaudy, J. Delbe, C. Blat, G. Desauty, *et. al.*, J. Cell. Physiol., 149 (1991) 492.
15. C. Blat, P. Bohlen, J. Villaudy, G. Chatelain, *et. al.*, J. Biol. Chem., 264 (1989) 6021.
16. L. Liu, J. Delbe, C. Blat, J. Zapf and L. Harel, J. Cell. Physiol., 153 (1992) 15.
17. P. Cohen, G. Lamson, T. Okajima and R.G. Rosenfeld, Mol. Endocrinol., 7 (1993) 380.
18. Y. Oh, H.L. Muller, G. Lamson and R.G. Rosenfeld, J. Biol. Chem., 268 (1993) 1496.
19 Y. Oh, H.L. Muller, H. Pham and R.G. Rosenfeld, J. Biol. Chem., 268 (1993) 26045.
20. B. Valentinis, A. Bhala, T. DeAngelis, R. Baserga, *et. al.*, Mol. Endocrinol., 9 (1995) 361.
21. Y. Oh, H.L. Muller, L. Ng and R.G. Rosenfeld, J. Biol. Chem., 270 (1995) 13589.
22. H. Huynh, X. Yang and M. Pollak, J. Biol. Chem. 271 (1996) 1016.
23. Z.S. Gucev, Y. Oh, K.M. Kelley and R.G. Rosenfeld, Cancer. Res. 56 (1996) 1545.
24. L. Buckbinder, R. Talbott, S. Velasco-Miguel, I. Takenaka, *et. al.*, Nature 377 (1995) 646.
25. W. Li, J. Fawcett, H.R. Widmer, P.J. Fielder, *et. al.*, Endocrinology, 138 (1997) 1763.
26. G. Jaques, K. Noll, B. Wegmann, S. Witten, *et. al.*, Endocrinology, 138 (1997) 1767.
27. R. Rajah, B. Valentinis and P. Cohen, J. Biol. Chem. 272 (1997) 12181.
28. C. Lalou, C. Lassarre and M. Binoux, Endocrinology, 137 (1996) 3206.
29. S.M. Zadeh and M. Binoux, Endocrinology, 138 (1997) 3069.
30. S.M. Leal, Q. Liu, S.S. Huang and J.S. Huang, J. Biol. Chem. 272 (1997) 20572.
31. Y. Oh, S.N. Nagalla, Y. Yamanaka, H.-S. Kim, *et. al.*, J. Biol. Chem., 271 (1996) 30322.
32. E. Wilson, Y. Oh and R.G. Rosenfeld, J. Clin. Endocrinol. Metab., 82 (1997) 1301.
33. H.-S. Kim, S.R. Nagalla, Y. Oh, *et. al.*, Proc. Natl. Acad. Sci. USA, in submission.
34. Y. Yamanaka, E. Wilson, R.G. Rosenfeld and Y. Oh, J. Biol. Chem., in press.
35. P. Vorwerk, Y. Yamanaka, A. Spagnoli, *et. al.*, J. Clin. Endocrinol. Metab., in submission.
36. R.A. Frost and L. Tseng, J. Biol. Chem., 266 (1991) 18082.
37. W.G. Hoeck and V.R. Mukku, J. Cell. Biochem., 56 (1994) 262.
38. J.I. Jones, A. Gockerman, *et. al.*, Proc. Natl. Acad. Sci. USA, 90 (1993) 10553.
39. D.L. Andress, J. Biol. Chem., 270 (1995) 28289.
40. R.L. Hintz and F. Liu, J. Clin. Endocrinol. Metab., 45 (1977) 988.
41. W.I. Wood, G. Cachianes, W.J. Henzel, *et. al.*, Mol. Endocrinol., 2 (1988) 1176.
42. J. Delbe, C. Blat, G. Desauty, *et. al.*, Biochem. Biophys. Res. Commun., 179 (1991) 495.
43. Y, Oh, H.L. Muller, H. Pham, G. Lamson, *et. al.*, Endocrinology 131 (1992) 3123.
44. Y. Yamanaka, J.L. Fowlkes, R.G. Rosenfeld and Y. Oh, in preparation.
45. D.M. Bradham, A. Igarashi, R.L. Potter, *et. al.*, J. Cell. Biol., 114 (1991) 1285.
46. M.L. Kireeva, F.E. Mo, G.P. Yan and L.F. Lau, Mol. Cell. Biol., 16 (1996) 1326.
47. C. Martinerie, E. Viegas-Pequignot, I. Guenard, *et. al.*, Oncogene, 7 (1992) 2529.
48. A. Igarashi, H. Okochi, D.M. Bradham, *et. al.*, Mol. Biol. Cell., 4 (1993) 637.

Molecular Mechanisms to Regulate the
Activities of Insulin-like Growth Factors
K. Takano, N. Hizuka and S-I. Takahashi (Editors)

The regulation and actions of ALS

P.J.D. Delhanty

The Kolling Institute of Medical Research, University of Sydney, St Leonards, NSW 2065, AUSTRALIA

1. INTRODUCTION

1.1. Control of IGF bioavailability

The combined concentration of the IGFs (IGF-I and IGF-II) in human serum (~800ng/mL) gives them a greater hypoglycaemic potential than circulating insulin [1]. The hypoglycaemic effect of circulating IGFs appears to be modulated by the IGF binding proteins (IGFBP-1 to -6), although there is evidence that binary IGFBP/IGF complexes can pass relatively freely between the blood and tissue compartments [2, 3]. Of particular importance in the regulation of the biological action of circulating IGFs is the acid-labile subunit (ALS). ALS has little affinity for the IGFs or IGFBP-3 alone, but has high affinity for the IGF-IGFBP-3 complex with which it forms a ternary complex in the blood [4]. In humans this ternary complex carries 75-90% of circulating IGFs [5, 6]. It is thought that the large size of this complex (~140 kDa) prevents or retards its passage from the circulation, thus contributing to the increased half-life of IGF–IGFBP-3 complexes in plasma from less than 1 hour in the free form to 12-15 hours in the ternary complex [7].

Ternary complex formation is thought to be the essential process which mediates regulation of the hypoglycaemic potential of IGFs in the blood. This is exemplified in patients with non-islet cell tumour hypoglycaemia (NICTH). Overexpression of IGF-II by (predominantly mesenchymal) tumours, and concurrent low levels of circulating ALS in these patients, contribute to their hypoglycaemia. In the short term, patients can be treated with growth hormone (GH) because of the marked GH dependence of ALS levels in the serum [8]. This treatment effectively raises the concentration of circulating ternary complex, and reduces the hypoglycaemic potential of the tumour derived IGF-II. Patients treated with GH show rapid return towards normal glucose homeostasis [9].

1.2. 140 kDa IGFBP complex formation

Purified human ALS is able to interact with binary IGF-IGFBP-3 complexes, but not with free IGFs [4, 10]. In these *in vitro* experiments formation of the IGF-IGFBP-3 complex was found to be a pre-requisite for ternary complex formation. In conditions designed to resemble physiological values of circulatory pH, temperature and ionic strength, ALS binds to binary IGF-I–IGFBP-3 and IGF-II–IGFBP-3 complexes with affinities of 2.5×10^{8} M^{-1} and 5.8×10^{7} M^{-1} respectively [11].

These values are 300 to 2000-fold lower than the binding constants for the formation of the respective binary complexes, evidence which supports the possible role for ALS in facilitating the passage of IGFs between the circulation and the tissues. Despite this relatively low affinity for the IGF–IGFBP-3 complex, recombinant human IGFBP-3 forms ~140kDa complexes within 2 minutes of being injected into rats [7]. From this experiment, it was calculated that a flux of approximately 700ng/mL of IGF-I would be required to occupy the exogenous IGFBP-3 and allow its incorporation into ternary complexes. Unlike human serum [1], rat serum may contain significant levels of free IGF-I (~75ng/mL) [12]. Thus the combination of readily dissociable IGF-I in the low molecular weight binding protein complex serum fraction where IGF-I is in rapid turn-over [5], and circulating free IGF-I may contribute to the apparent necessary flux of IGF, without appreciably affecting total circulating IGF levels. What is unclear is whether the kinetics of the IGF–IGFBP system in the circulation would support this flux.

Another interpretation of these results is that ALS is able to form binary complexes with free rhIGFBP-3, thus abrogating the need for a rapid movement of IGF into ternary complexes [13]. Recent evidence suggests that rat ALS can form binary complexes with recombinant human IGFBP-3, albeit at a 10-fold lower affinity than when forming ternary complexes [14]. Furthermore, cross-linking experiments suggest the possibility of interaction between human ALS and rhIGFBP-3 in the absence of IGF [15]. In addition, *in vivo* experiments indicate that infused IGFBP-3 is rapidly incorporated into 150 kDa complexes without increasing total circulating IGF content [16], presumably ruling out mobilisation of IGF from tissues. This has been interpreted as providing evidence for the formation of binary ALS–IGFBP-3 complexes in the circulation. However, evidence that the rapidly formed 140 kDa complexes consist of ternary or binary complexes remains equivocal [16]. A stable pool of binary complexes may occur in rat serum consisting of ALS and proteolytically nicked IGFBP-3 [13]. However, the 140 kDa complexes formed immediately after rhIGFBP-3 injection appear to be unstable, and the half-life of infused free IGFBP-3 is markedly reduced compared to IGFBP-3 infused in complexes with IGF-I [7, 16, 17].

The formation of 140 kDa complexes is clearly dependent on the kinetics of molecular interactions between components of the IGF–IGFBP–ALS system, not only within the circulation but in the organism as a whole. Of tremendous facility in examining the kinetics of these interactions are recent advances in biosensor technology, which enable measurements in real time without radio-labelled tracers [18].

1.3. Localisation of ALS and the 140 kDa IGFBP complex

Although ALS is mainly confined to the blood compartment, the ternary complex has been found outside the circulation (Table I), generally at much lower concentrations than in the blood. Its presence at certain of these sites, particularly in the ovary, skin and kidney, suggests potential extra-hepatic auto/paracrine roles for ALS in modulating IGF and IGFBP-3 activity which remain to be elucidated. The source of ALS at sites outside the circulation remains equivocal, mainly because of lack of evidence for extra-hepatic (and extra-renal) gene expression. Serum proteins, including the IGFBPs and hepatic derived acute phase proteins [19], are known to cross the capillary barrier. This makes it conceivable that the ALS at these sites is derived from the blood, perhaps having crossed the capillary endothelium in its free form. However, a large (~140 kDa) serum marker protein, ceruloplasmin, was not detected in skin interstitial fluid suggesting that ALS may be expressed locally in the skin [20].

Table I. Sites of ALS gene expression, and extracellular fluids which have been examined for the presence of ALS or ternary complex (*: radioimmunoassay for ALS, ‡: size exclusion chromatography).

Gene expression	**Ternary complex/ALS**	
	present	low abundance/undetectable
Liver (hepatocytes) [21, 22]	*Postnatal sera [8, 23]	*Cerebro-spinal fluid [8, 24]
Kidney [25]	‡Fetal sera [26]	‡Human lymph [27]
Fetal bone [25]	‡Ovine lymph [24]	*Seminal plasma [8]
	‡Follicular fluid [24]	*Amniotic fluid [8, 24]
	‡Peritoneal fluid [28]	‡Urine [24]
	‡Skin interstitial fluid [20]	‡Milk [24]
	‡Vitreous humour [24]	*Bone cells [29]

2. THE STRUCTURE AND GLYCOSYLATION OF ALS

2.1. The structure of ALS

The primary structure of ALS is highly conserved within primates (94% similarity between pre-peptides) and within rodents (92%). However, although there is conservation between the two groups (~76%) there are differences in the positioning of C-terminal N-glycosylation sites, and the position of the fifth cysteine residue in the mature peptide (Fig. 1). Rat ALS circulates in two variant forms, one of which has a four residue N-terminal extension relative to the human N-terminus and represents ~75% of the total population [23].

The main feature of ALS is its leucine-rich central domain which constitutes ~75% of this ~85kDa glycoprotein. This domain contains 20 tandem repeats of a leucine-rich motif common to a large family of proteins and glycoproteins which characteristically form protein-protein complexes [30]. Recently, the crystal structure of one of these leucine-rich repeat (LRR) proteins, porcine RNase inhibitor (RI), has been determined [31]. Subsequently, other members of the LRR group have been modelled on this structure [32, 33], and using the same approach we have developed a model for the central domain of ALS. In RI the LRR units consist of a short β-sheet and an α-helix lying approximately anti-parallel to each other. When placed in tandem these structures form a curved spring-like shape with the β-sheets on the inner surface and the α-helices on the outer surface, both perpendicular to the plane of the curve. Because of the relatively large number of LRR units in ALS, the structure curves round in a nearly complete circle, giving the molecule a doughnut shape.

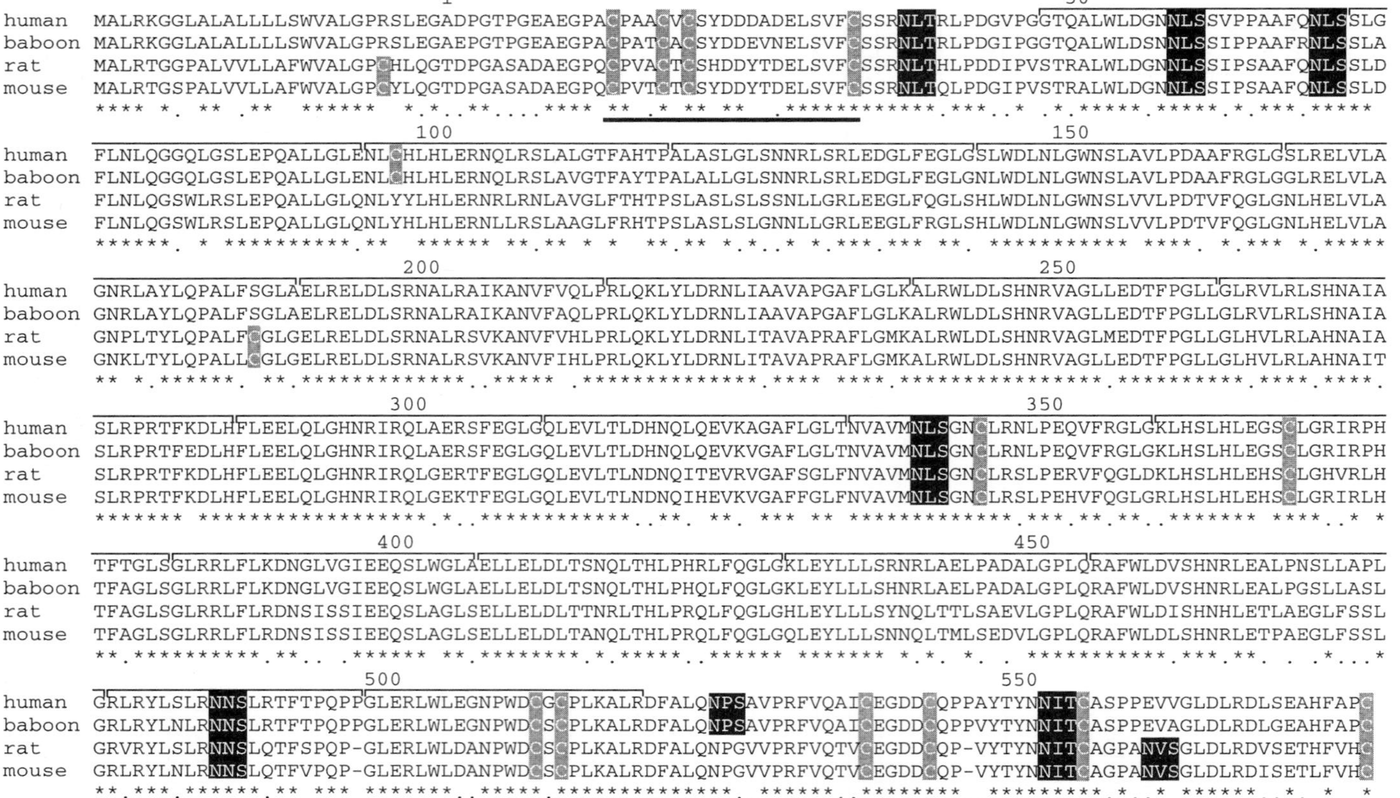

Fig 1. Comparison of deduced ALS primary amino-acid sequences from primates and rodents. Numbering of residues is relative to the first residue of the human mature peptide. Grey boxes highlight cysteine residues and black boxes indicate consensus N-linked glycan attachment sites. Horizontal brackets delineate LRR units, and black bars delineate the cysteine-rich elements characteristic of LRR proteins. *- identical residues and .- related residues.

Although not modelled specifically, the N- and C-terminal domains appear to extend in the same direction perpendicular to the plane of the "doughnut". Interestingly, the inner face of the "doughnut" contains a concentration of acidic residues near the N-terminus. Furthermore, the curved structure of the molecule brings the N-linked glycans, which are presumably sialylated (unpublished results) and negatively charged, into close apposition. It is possible that this structure may contribute to the acid-lability of the binding capacity of this protein

2.2. The role of ALS glycosylation in ternary complex formation

ALS occurs in at least two natural glycoforms in human circulation [34]. N-linked sugars contribute ~20 kDa to the molecular weight of ALS, and we now have evidence that these glycans are sialylated. Partial deglycosylation of serum ALS using endoglycosidase F generates at least six immunoreactive bands on electrophoresis [35], suggesting that most of the seven potential N-linked glycosylation sites are utilised. The glycosylation of several liver derived serum glycoproteins is affected by a number of factors including nutrition, age, diabetes, chronic inflammatory disease, cancer and cirrhosis [36]. Moreover, glycosylation can affect the half-lives of proteins in the circulation, and also can modulate their biological activity. Because of the potential biological effects of altered or aberrant glycosylation, recent studies in our laboratory have concentrated on the role of glycosylation of ALS. We have demonstrated that both glycosylation and sialylation of ALS play important roles in its ability to form the ternary complex. We found that although recombinant proteins lacking single glycans appear to form ternary complexes normally, complete enzymatic deglycosylation abolishes binding. Desialylation of ALS lowers its affinity for IGF–IGFBP-3 complexes by 2-fold or greater. In addition, sialylation appears to play a role in determining the preference of ALS for IGF-I–IGFBP-3 over IGF-II–IGFBP-3 binary complexes. Interestingly, the glycosylation state of IGFBP-3 appears to have little or no affect on its affinity for the IGFs or ALS [37].

3. ALS GENE STRUCTURE AND REGULATION

3.1. ALS gene structure

The ALS gene has been cloned from both rat and mouse, and appears to exist as a single copy in the rat genome [38, 39]. The relatively simple two exon structure of the gene is highly conserved between the two species. Unlike many other LRR proteins the exon structure of the ALS gene does not correspond to the leucine-rich repeat structure [30]. The gene lacks the usual consensus basal promoter elements such as TATA, CCAAT, or GC-boxes, or an *Inr* sequence. This property of the rodent genes may contribute to the occurrence of multiple transcriptional initiation sites.

3.2. ALS regulation by GH

The steady-state level of the ~2.2 kb ALS mRNA in hepatocytes is markedly GH-dependent, both *in vivo* and *in vitro* (see Table II), and recently it has been demonstrated that GH upregulates ALS gene transcription [40].

Furthermore, transiently transfected mouse ALS promoter-luciferase reporter gene constructs carrying ~2kb of 5′ flanking region are responsive to GH in both early passage H4-II-E rat hepatoma cells and primary hepatocytes. GH regulates gene transcription through members of a family of transcription factors known as the signal transducers and activators of transcription [41]. These bind to specific *cis*-elements related in sequence to the γ-interferon-activated site (GAS, TTNCNNNAA). Several sites homologous to the GAS are present in the 5′ flank of the ALS gene. Two with greatest homology to the GAS occur at -541 to -533 and -621 to -613 relative to the $A^{+1}TG$ in the rat gene. To examine the functionality of these sites we inserted deletion mutants containing bps -1 to -684 of the rat promoter upstream of the luciferase gene in a reporter construct. Recombinant human GH induced activity of this truncated promoter ~3-fold over basal levels in primary hepatocytes, suggesting that these elements are responsible for GH regulation of ALS transcription (unpublished observations). Similar results have been obtained with the mouse ALS promoter [40].

Growth hormone is not the only factor which regulates hepatic ALS expression (see Table II). However, many of these other regulatory factors may have indirect effects on ALS transcription by modulating components of the GH-signalling pathway. *In vivo* these factors may also affect systemic GH levels, thus having an indirect effect on hepatic ALS expression.

Table II. ALS expression is regulated *in vitro* in isolated (mostly rodent) hepatocytes, and *in vivo* in rodents and humans by a variety of factors.

Factor	*In vitro* (hepatocytes)		*In vivo*	
	gene expression	secretion	hepatic gene expression	serum
Growth hormone	↑[40, 42, 43]	↑ [21, 42]	↑ [40, 44]	↑ [9, 44-46]
Octreotide		↓ [47]		↓ [48]
T3		↑ [47]		
Insulin	↑ [43]	↑ [42]	(↑) [44]	(↑) [44, 49]
IGF-I	↑ [43]	→ [47]		↓ [50, 51]
IGF-II		→ [47]		
EGF	↓ [42]	↓ [42]		
TGFβ		↓ [47]		
Dexamethasone	↓ [42]	↓ [42]	↓ [44]	↓ [44]
Prednisolone				↑ [9]
cAMP	↓ [52]	↓ [52]		
Fasting (rodent)	—	—	→ [44]	↓ [44, 53]
Diabetes (rodent)	—	—	↓ [44]	↓ [44]
Cirrhosis	—	—	↓(unpublished)	↓ [54]
Burn patients	—	—		↓ [49]

3.3. Regulation of ALS expression by cAMP

Acutely fasted and chronically malnourished rats have significantly decreased levels of circulating ALS [44, 53]. A major mediator of the hepatic response to starvation is cAMP, which regulates the expression of a number of genes involved in this response [55-57].

To examine whether nutrition may modulate hepatic ALS output through cAMP, we have investigated the role of cAMP and its interaction with GH in the regulation of ALS expression in isolated rat hepatocytes. Plasma membrane permeable analogues of cAMP were found to suppress both ALS steady state mRNA levels, and secretion by primary hepatocytes, both in the absence and presence of 30ng/mL rhGH (Fig. 2). However, we found a consistently greater level of suppression of ALS mRNA levels by $(Bu)_2cAMP$ in cells treated with GH (~50%) than in cells cultured under basal conditions (~35%). This compared with a consistent ~60% suppression of secretion in the presence and absence of GH.

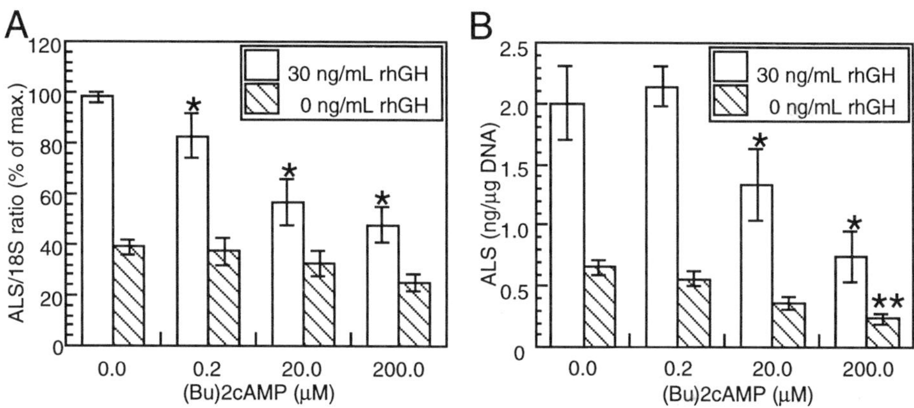

Fig. 2. Cyclic AMP suppresses ALS steady-state mRNA levels and secretion by isolated rat hepatocytes. **A** Dose response of $(Bu)_2cAMP$ on GH stimulated and basal ALS mRNA levels. **B** Dose response of $(Bu)_2cAMP$ on GH stimulated and basal ALS secretion. (*-P<0.05 relative to GH treated cells without $(Bu)_2cAMP$; **-P<0.05 relative to untreated cells). (Adapted from Delhanty & Baxter, 1998; [52]).

A similar effect was observed in cells treated with theophylline and cholera toxin, which increase endogenous cAMP levels. Cyclic AMP was found not to affect the stability of ALS mRNA in GH treated cells. Interferon-γ stimulation of Stat 1 and Stat 3 activity in mononuclear cells has been shown to be inhibited by cAMP [58, 59]. Similarly, we have preliminary evidence of cAMP mediated inhibition of Stat binding to its GAS-like element in the rat ALS promoter (unpublished observations). This evidence supports our hypothesis that cAMP has an indirect effect on ALS pre-translational regulation by acting on the GH signalling pathway.

In contrast to its effects on ALS transcription, the suppression of ALS secretion by cAMP appears to be independent of GH status. Cyclic AMP has been demonstrated to directly suppress translation by stimulating the binding of

PHAS-1 to the eukaryotic initiation factor (eIF) 4F [60]. We hypothesise that in the absence of GH, at least acutely, the effects of cAMP on ALS translation by hepatocytes becomes dissociated from those acting on transcription. This effect may be linked to our finding that although acutely fasted rats have significantly suppressed serum ALS levels, hepatic ALS mRNA levels remain unaffected in these relatively GH-insensitive animals.

4. DISCUSSION

This review aims to describe recent developments in our understanding of both the mechanisms by which the acid-labile subunit (ALS) interacts with the other components of the ternary complex, and the regulation of hepatic ALS expression. It is probably the combined effect of the ability of ALS to form the ternary complex, and the regulation of hepatic ALS output that contribute to the regulation of circulating IGF bioavailability. We have found that post-translational modifications of ALS, specifically its glycosylation state, can have a major impact on ternary complex formation. In addition, it is clear that GH has a central role in regulating both serum levels and hepatocyte gene expression and secretion. Indeed, other factors which control hepatocyte expression and circulating levels of ALS may well act indirectly through the GH signalling pathway, and *in vivo* by regulating GH levels. Furthermore, a number of disease states lead to GH resistance. The coincident reduction in circulating ALS may contribute to, and compound, the effects of lowered circulating levels of IGFs [54].

Although clinical investigations in man and *in vivo* experiments in rats have given us a somewhat indirect indication of the function of ALS, the current development of ALS knockout mice will hopefully give exciting new insights into the biological role of ALS throughout development [61].

ACKNOWLEDGMENTS

Much of the work on ALS glycosylation was performed by Jackie Janosi in our laboratory. We gratefully acknowledge the help of Paul Ramsland (University of Technology, Sydney) in developing the mathematical model of the LRR region of ALS. The author's studies are supported by a grant from the NHMRC, Australia

REFERENCES

1. Zapf, J. Eur. J. Endocrinol. 1997;**136**: 146
2. Bar, RS, et al. Endocrinology 1990;**127**: 497
3. Bar, RS, et al. Endocrinology 1990;**127**: 1078
4. Baxter, RC, et al. J. Biol. Chem. 1989;**264**: 11843
5. Jones, JI, and DR Clemmons. Endocrine Rev. 1995;**16**: 3
6. Baxter, RC, et al in *The insulin-like growth factors and their regulatory proteins.* (Baxter, R. C., Gluckman, P. D., and Rosenfeld, R. G., ed) 1994 p. 227, Elsevier Science B.V., Amsterdam
7. Lewitt, MS, et al. Endocrinology 1993;**133**: 1797
8. Baxter, RC. J. Clin. Endocrinol. Metab. 1990;**70**: 1347
9. Baxter, RC, et al. J. Clin. Endocrinol. Metab. 1995;**80**: 2700
10. Baxter, RC, et al. J. Biol. Chem. 1992;**267**: 60
11. Holman, SR, and RC Baxter. Growth Regul. 1996;**6**: 42

12. Frystyk, J, et al. Growth Regul. 1996;**6**: 48
13. Lee, CY, and MM Rechler. Endocrinology 1995;**136**: 668
14. Lee, CY, and MM Rechler. Endocrinology 1995;**136**: 4982
15. Barreca, A, et al. J. Clin. Endocrinol. Metab. 1995;**80**: 1318
16. Lee, CY, et al. Endocrinology 1997;**138**: 1649
17. Arany, E, et al. Growth Regul. 1996;**6**: 32
18. Heding, A, et al. J. Biol. Chem. 1996;**271**: 13948
19. Havenaar, EC, et al. Glycoconj. J. 1997;**14**: 457
20. Xu, S, et al. J. Clin. Endocrinol. Metab. 1995;**80**: 2940
21. Scott, CD, and RC Baxter. Biochem. J. 1991;**275**: 441
22. Leong, SR, et al. Mol. Endocrinol. 1992;**6**: 870
23. Baxter, RC, and J Dai. Endocrinology 1994;**134**: 848
24. Hodgkinson, SC, et al. J. Endocrinol. 1989;**120**: 429
25. Chin, E, et al. Endocrinology 1994;**134**: 2498
26. Bang, P, et al. Growth Regul. 1994;**4**: 68
27. Binoux, M, and P Hossenlopp. J. Clin. Endocrinol. Metab. 1988;**67**: 509
28. Bowsher, RR, et al. Endocrinology 1991;**128**: 805
29. Kanzaki, S, et al. J. Bone Miner. Res. 1995;**10**: 854
30. Kobe, B, and J Deisenhofer. TIBS 1994;**19**: 415
31. Kobe, B, and J Deisenhofer. Nature 1995;**374**: 183
32. Kajava, AV, et al. Structure 1995;**3**: 867
33. Weber, IT, et al. J. Biol. Chem. 1996;**271**: 31767
34. Baxter, RC, and JL Martin. Proc. Natl. Acad. Sci. U.S.A. 1989;**86**: 6898
35. Liu, F, et al. J. Clin. Endocrinol. Metab. 1994;**79**: 1883
36. Lis, H, and N Sharon. Eur. J. Biochem. 1993;**218**: 1
37. Firth, SM, and RC Baxter. Prog. Growth Factor Res. 1995;**6**: 223
38. Delhanty, PJD, and RC Baxter. J. Mol. Endocrinol. 1997;In press
39. Boisclair, YR, et al. Proc. Natl. Acad. Sci. U.S.A. 1996;**93**: 10028
40. Ooi, GT, et al. Mol. Endocrinol. 1997;**11**: 997
41. Ihle, JN. Nature 1995;**377**: 591
42. Dai, J, et al. Endocrinology 1994;**135**: 1066
43. Scharf, JG, et al. Hepatology 1996;**23**: 818
44. Dai, J, and RC Baxter. Endocrinology 1994;**135**: 2335
45. Fielder, PJ, et al. Endocrinology 1996;**137**: 1913
46. de Boer, H, et al. J. Clin. Endocrinol. Metab. 1996;**81**: 1371
47. Barreca, AM, et al. Eur. J. Endocrinol. 1997;**137**: 193
48. Barreca, A, et al. Clin. Endocrinol. 1995;**42**: 161
49. Bereket, A, et al. Clin. Endocrinol. 1996;**44**: 525
50. Baxter, RC, et al. Acta Endocrinol. 1993;**128**: 101
51. Kupfer, SR, et al. J. Clin. Invest. 1993;**91**: 391
52. Delhanty, PJD, and RC Baxter. Endocrinology 1998;In press
53. Oster, MH, et al. Am. J. Physiol. 1996;**270**: E646
54. Donaghy, AJ, and RC Baxter. Baillieres Clin. Endocrinol. Metab. 1996;**10**: 421
55. Beale, E, et al. Diabetes 1984;**33**: 328
56. Roesler, WJ, et al. J. Biol. Chem. 1996;**271**: 8068
57. Van Remmen, H, and WF Ward. Am. J. Physiol. 1994;**267**: G195
58. Ivashkiv, LB, et al. J. Immunol. 1996;**157**: 1415
59. Sengupta, TK, et al. Proc. Natl. Acad. Sci. U.S.A. 1996;**93**: 9499
60. Graves, LM, et al. Proc. Natl. Acad. Sci. U.S.A. 1995;**92**: 7222
61. Boisclair, YR, et al 1997; Proceedings, 79th meeting of the Endocrine Society, Minneapolis, MN, USA, p. 345

Molecular Mechanisms to Regulate the
Activities of Insulin-like Growth Factors
K. Takano, N. Hizuka and S-I. Takahashi (Editors)

INSULIN-LIKE GROWTH FACTORS IN THE DEVELOPMENT OF THE PANCREAS

D.J.Hill†§*, J. Petrik*†, E. Arany§, W. Reik‡and J.M. Pell‡

Lawson Research Institute, St. Joseph's Health Centre, London, Ontario, N6A 4V2, Canada
†Departments of Physiology, §Medicine, and *Paediatrics, University of Western Ontario, London, Ontario, N6A 5A5, Canada
‡The Babraham Institute, Babraham Hall, Cambridge CB2 4AT, UK

1. INTRODUCTION

In the rat fetus, the cellular area immunostained for insulin increases 2-fold over 2 days just prior to term, due to both β cell replication and recruitment and maturation of undifferentiated β cell precursors (1, 2). Endocrine cells develop from duct-like cells in the embryo, fetus and neonate, and form primitive islets in the mesenchyme adjacent to the ducts. Final differentiation into glucagon, somatostatin, pancreatic polypeptide or insulin-expressing cells likely depends on the expression of transcription factors such as Pdx-1 (3) and on the actions of local peptide growth factors within the surrounding mesenchyme (4). The population growth rate of all islet cells, including ß cells, slowed by postnatal days 3 and 4 and continued to decline thereafter (2). The rate of mitosis in adult pancreatic β cells is normally low (3% replication rate of β cells per day) (5).

The change from a fetal type of β cell population capable of expansion, to an adult β cell population which is not, may occur within a programmed ontological pathway that extends into postnatal life. The newborn rat made diabetic with streptozotocin has extensive β cell destruction, but at 14 days after birth demonstrates normal glycaemia with considerable mitotic activity being apparent in the pancreatic ductal epithelium, from which the precursor β cells derive (6). Similar β cell renewal occurs following alloxan treatment of the young rabbit (7). Pancreatectomy (90%) of the young rat is followed by regeneration of both exocrine and endocrine tissue (8). New β cells are derived by both neogenesis and cell replication within the remaining islets. Endocrine regeneration is also seen in association with duct epithelial cell proliferation in transgenic mice with β cells expressing IFN-γ (9). In these mice, β cell destruction leading to IDDM can be matched by new islets growing ectopically into duct lumens. Islet regeneration occurs in pancreas of young diabetic patients, and in some cases islet neogenesis is associated with centro-acinar and ductular cells leading to the formation of large islets consisting primarily of β cells (10). Islet cell regeneration in the pancreas of recent onset IDDM patients also occurs, particularly in infants and small children (11, 12). Recently it has been revealed that the ontogeny of islet cells in early life involves a balance between ß cell replication and neogenesis, and programmed ß cell death. A transient wave of apoptosis occurs in neonatal rat islets between 1-2 weeks of age (5, 13). However, β cell mass is not altered appreciably at this time, suggesting that a new population of β cells compensates for the loss. A similar episode of ß cell apoptosis has recently been described in the human fetal pancreas in third trimester (14). While the cellular triggers involved in the initiation of developmental ß cell apoptosis are not known, the apoptosis which mediates ß cell destruction in response to cytokine action during autoimmune diabetes involves an increased intracellular concentration of nitric oxide (NO), and increased expression of inducible nitric oxide synthase (iNOS) (15).

The mechanisms controlling developmental islet cell neogenesis or apoptosis are unknown, but these may be linked to the relative expression of peptide growth factors within the developing pancreas. There is considerable circumstantial evidence that the insulin-like growth factors (IGFs) are major contributors to ß cell growth, maturation and function, and are expressed by ß-cells throughout life. We previously reported that IGF-II mRNA is most abundant in the pancreas of the fetal rat, and declines during the neonatal period (16). Conversely, IGF-I mRNA levels were low but detectable in fetal life, and rose to adult levels within two weeks of birth. Others have reported that IGF-I mRNA abundance peaked transiently shortly after birth and then declined sharply (17), a pattern found by us for IGF-II but not IGF-I mRNA. Using in situ hybridization and immunohistochemistry IGF-I and -II mRNAs and peptides were shown to be present within islet cells throughout life, including ß-cells (18). Similarly, in the human fetus of mid-trimester IGF peptides and IGF binding proteins (IGFBPs) -1, -2 and -3 are localized by immunohistochemistry to the islets of Langerhans and ß-cells (19). We and others showed that isolated islets from the human (20) or rat fetus (21, 22) or neonate (23, 24) release both IGF-I and IGF-II; that exogenous IGF-I or -II promote increased islet cell DNA synthesis (21, 22, 25); and that isolated α and ß cells from rat islets contain the high affinity type 1 IGF signalling receptor (26), as do pancreatic ß cell lines (27). We also showed that isolated adult rat islets enriched in ß cells release IGF-I (28), while others have demonstrated a retention of IGF-II mRNA expression within the ß cells of the pancreas in both rat and man (29, 30), and high levels of IGF-II are expressed by the rat ß cell line, INS-1 (31). IGF binding proteins (IGFBPs) are also expressed in the developing pancreas (16). Both IGFBP-1 and -2 mRNAs transiently appear in the pancreas between postnatal week two and three and decline in the adult. IGFBP-3 and -4 mRNAs were detected in the pancreas throughout development, while IGFBPs -5 or -6 mRNAs were undetectable (16). We have shown that IGFBP-1 and -2 can potentiate the mitogenic actions of IGF-II on isolated fetal rat islets enriched in ß cells (21). The expression of IGFBP-2, but not IGF-II, in islets was potentiated by glucose.

The transient wave of developmental apoptosis responsible for a reduction in β cell number after 1-2 weeks of postnatal life in the rat (13) coincides temporally with our demonstration of a diminished pancreatic expression of IGF-II at this time, while pancreatic expression of IGF-I has not yet achieved adult values (16). A nadir in total IGF availability may therefore exist in pancreas when apoptosis is transiently high. IGFs inhibit apoptosis in mammary carcinoma cells, cerebellar granule neurons, ovarian pre-ovulatory follicles, human erythroid colony-forming cells and haematopoietic cells (32-35), and may act through a mechanism independent of Bcl-2, Bcl-x or Bax (36). However, IGF-I inhibits iNOS induction in some tissues (37), and may interfere with cytokine-stimulated NO synthesis. The ability of IGF-I to limit neuronal damage elicited by experimental hypoxia-ischaemia suggests an ability of IGF-I to limit free radical generation (38).

In these experiments we have examined the precise anatomical pattern of expression of IGF-I and -II and IGFBPs in the pancreas of the rat during early life, the contribution of IGF-II to ß cell growth in utero, the relationship between developmental ß cell apoptosis and the presence of iNOS, and present functional evidence that IGFs can influence islet ß cell survival in the neonatal rat.

2. MATERIALS AND METHODS

2.1 Animals and islet culture

Pregnant Wistar rats were killed at day 21 gestation, or were allowed to deliver and the offspring killed at postnatal days 4-29. The pancreas was immediately removed from each animal and placed in ice-cold Hank's buffered salts solution. Pancreata were either fixed for histology or processed immediately for islet isolation. For histology, tissues were fixed in ice-cold 0.2% glutaraldehyde, 4% paraformaldehyde buffered with 70 mM phosphate buffer for

16 h at 4°C.

The islet isolation technique using collagenase digestion was modified from that of Hellerstrom et al (39) and described by us previously (21). After an initial incubation of 48-72h islets were counted by eye under a dissecting microscope and transferred to non-tissue culture grade Petri dishes which did not permit cell attachment for the experimental period. A typical yield of islets from 10 pancreata was approximately 1500. At the beginning of the treatment period, the islets were distributed in equal batches (60-80 islets per plate) onto non-tissue culture grade Petri dishes containing glucose-free Dulbecco's Modified Eagle Medium (DMEM), supplemented with 2 mM glutamine and 8.7 mM glucose. Medium was further supplemented as required with recombinant human IGF-I or IGF-II (100 ng/ml), a monoclonal antibody against rat IGF-II (20 μg/ml; Amano Int.), recombinant human interleukin-1ß (IL-1ß) (2.5 ng/ml), tumour necrosis factor-α (TNF-α) (10 ng/ml), interferon-γ (IFN-γ) (10 ng/ml); alone or in combination. Islets were incubated for either 24, 48 or 72h. In some experiments [methyl-^{3}H] thymidine (20 Ci/mM, 5 μCi/ml) was added for the final 24h of culture. At the end of the treatment period conditioned medium was removed and the islets were washed in phosphate-buffered saline (PBS) and assessed either for viability or for DNA synthetic rate. In some experiments islets were plated into 8-well chamber slides (20 islets per well) to allow histological assessment.

2.2 Assessment of islet viability, DNA synthesis and insulin release.

To assess islet cell viability following incubation in test culture media the islets from each plate were resuspended in PBS containing 5mg/ml trypan blue. All islets were examined immediately under a dissecting microscope and any islet containing one or more cells which had taken up trypan blue was considered to be non-viable. At the time of viability assessment the recovery of islets was 92±5% (Mean±S.D.) of those initially added to each culture dish, and did not differ between control cultures or those which had contained cytokines. The within batch coefficient of variation on assessment of islet viability following repeated measures was 3%. Islets were suspended in PBS and solubilized by ultrasonication. DNA content of islets was measured by fluorometry using Hoechst fluorochrome 33258. To measure the rate of DNA synthesis, [^{3}H] thymidine incorporation (dpm/μg DNA) was measured by liquid scintillation counting. The insulin content of conditioned culture medium was measured by radioimmunoassay (RIA) in a modification of the method of Hales & Randle (40) as modified by Herbert et al. (41), and described by us previously (42). Rat insulin was used for the standard curve.

2.3 Immunohistochemistry and visualization of apoptosis

Histological sections of pancreas (5 μm) were subjected to immunohistochemistry to localize IGF-I or -II, IGFBPs, insulin, or iNOS within islets by a modified avidin-biotin peroxidase method (43). Tissue sections were counterstained with Carazzi's haematoxylin. Dual staining for iNOS and insulin involved first performing immunohistochemistry for insulin using diaminobenzidine as the chromagen, followed by alkaline phosphatase (blue) as the chromagen for iNOS. Immunocytochemistry was performed to localize apoptotic nuclei within either tissue sections or isolated islets (44) using the Apoptag in situ apoptosis detection kit (Oncor Inc.).

2.4 Northern blot analysis and in situ hybridization

Total RNA was extracted from isolated islets as previously described (16). For Northern blot hybridization, between 15 and 20 μg of total RNA was used. The blots were hybridized at 42°C overnight with 2 x 10^6 c.p.m./ml radiolabelled cDNA probe for IGF-I or -II, or 1 x 10^6 c.p.m./ml for a cDNA encoding 18S ribosomal RNA. Blots were exposed to X ray film at -70°C with intensifying screens for up to 10 days before developing. Complimentary DNA probes used for hybridization of Northern blots and for in situ hybridization were kindly provided by the following investigators: a 500 bp rat IGF-I cDNA in pGEM Blue (Promega) encoding exon 3 and part of exon 4 of the rIGF-I gene was provided by Dr. L. Murphy,

University of Manitoba, Winnipeg, MT (46); and a 807bp mouse IGF-II in pGEM 4z by Dr. G. Bell, University of Chicago, Chicago, IL.

For in situ hybridization, sections of pancreas were rehydrated before permeabilization with 10 μg/ml proteinase K. Hybridization with cRNA probes was in a humidified chamber at 50°C for 16 h. Kodak NTB-3 photoemulsion was applied subsequently to all sections and exposed for up to 14 days at 4°C prior to development. The restriction enzymes (BRL, Burlington, ON) and RNA polymerases (Promega) used to linearize the plasmids containing these cDNAs and to generate [^{35}S]-radiolabelled riboprobes are listed as follows: antisense rat IGF I, Hind III/T7; sense rat IGF I, Pvu II/SP6; antisense mouse IGF-II, Hind III/TSP6; sense mouse IGF-II EcoR1/T7.

2.5 IGF-II transgenic mouse

Transgenic constructs in which mouse Igf2 gene was expressed from its own promotors were introduced into embryonic stem (ES) cells (49). ES cell clones shown to over-express IGF-II were introduced into host blastocysts to make chimeras. Increased fetal weight gain was found in animals with the greatest percent chimerism from day 13 of gestation and these mice were 50-60% larger at birth. Multiple organ overgrowth occurred affecting predominantly the heart, kidney, liver and tongue. Most high level chimeras died within 24h of birth. Serum levels of IGF-II were elevated 2-3 fold at birth compared to age-matched controls.

3. RESULTS

Islets from mice transgenic for IGF-II demonstrated a 4-5-fold increase in islet cell area on the day of birth in animals which demonstrated the greatest degree of transgenic chimerism (transgenic 5.99±0.42% of pancreas, control 1.40±0.25%). Islet cell number was not significantly altered compared to control animals. Immunohistochemistry showed that ß cells were abundant in the large islets, but that staining was diffuse and insulin-filled storage granules were less apparent than in control mice. Circulating insulin levels and blood glucose levels were not significantly different between IGF-II transgenic animals and controls. The results suggest that the elevated IGF-II expression in these animals gives rise to a hyperplasia of ß cells within existing islets.

Table 1 Incidence of apoptotic cells in rat pancreatic islets (% of islet cells demonstrating apoptotic nuclei) in animals from 21 days gestation until postnatal day 22.

	Age	% islet cells
Fetal	21 days	1.99±0.25
Postnatal	4 days	5.90±1.23
	6 days	6.46±0.61
	12 days	9.17±0.47
	14 days	13.18±3.40
	18 days	8.14±1.25
	22 days	2.36±0.91
Adult		0.23±0.04

Figures represent mean values ± sem (n=3-4).

Sections of pancreas were stained for the presence of apoptotic cells determined by DNA breakage to establish the ontogeny of islet cell apoptosis. Apoptotic nuclei within condensed islet cells with little cytoplasm were seen at all ages examined. The incidence of apoptotic nuclei in endocrine islet cells of fetuses at 21 days gestation was less than 2%, but this increased postnatally and was maximal at 13% on postnatal day 14 (Table 1) and subsequently declined. The majority of apoptotic cells were located centrally within the islets, suggesting that they were ß cells. However, duel staining immunohistochemistry for both insulin and apoptosis failed to demonstrate insulin still remained in the condensed, apoptosing cells.

Table 2 Incidence (% area immunopositive cells) of cells in rat pancreatic islets demonstrating the presence of immunoreactive iNOS in animals from 21 days gestation until postnatal day 22.

	Age	% islet cells
Fetal	21 days	3.80±0.25
Postnatal	4 days	5.28±0.78
	6 days	5.22±0.91
	12 days	9.49±2.29
	14 days	4.56±0.29
	18 days	5.66±2.18
	22 days	3.43±1.05
Adult		0.10±0.10

Figures represent mean values ± sem (n=3-4).

Since the induction of apoptosis is pancreatic ß cells by cytokines has been shown to be mediated, in part, by an increase in intracellular nitric oxide; we investigated the presence of iNOS during developmental apoptosis in islets. Using immunohistochemistry a minority of islet cells, predominantly within the central area, were found to contain cytoplasmic staining for iNOS at all ages of animal. An occasional presence of iNOS was also seen in ductal acinar cells. Dual staining immunohistochemistry showed that at each age studied 95% of islet cells immunopositive for iNOS also contained insulin. When pancreata from various ages of rat were compared the percentage of islet cells containing immunoreactive iNOS was greatest at almost 10% on postnatal day 12, and declined thereafter to negligible levels by weaning (Table 2).

We localized IGF mRNAs in pancreatic sections using in situ hybridization, and the distribution of the peptides by immunohistochemistry. In situ hybridization for IGF-II mRNA showed a relatively high signal intensity within endocrine cells of the islets at 7 days postnatal with little signal in the ductal epithelial cells. By postnatal day 29 no specific hybridization signal for IGF-II mRNA remained within pancreatic islets. Immunohistochemistry showed a widespread distribution of IGF-II peptide in all islet cells in late gestation and postnatal day 7, but this was less intense on postnatal days 12 and 14, and was absent by postnatal day 22. Conversely, no specific hybridization signal for IGF-I mRNA was seen in the fetal or neonatal pancreas and only a low level of mRNA was detected in islets after weaning. However, at postnatal day 29 IGF-I mRNA was seen within ductal epithelial cells. Following immunohistochemistry for IGF-I peptide little immunoreactivity was seen in the fetal or neonatal islets or acinar tissue. These studies confirmed that IGF-II expression in the pancreas declined substantially in neonatal life, and showed that the major sites of IGF-II expression were the pancreatic islets. Immunohistochemistry for IGFBPs showed that IGFBPs-1-5 were all present on islet cells in late fetal life. The number of islet cells immunopositive for IGFBPs-3, -4 and -5 increased postnatally, while the percentage of cells staining for IGFBP-2 did not change, and those positive for IGFBP-1 declined abruptly postnatally. Since we showed previously that IGFBP-1 and -2 mRNA is barely detectable in the fetal and neonatal pancreas, the immunoreactive peptide observed on islets is most likely to be sequestrated from the circulation.

We determined if either IGF-I or IGF-II could protect islets from cell death induced by exposure to cytokines. Islets isolated from rats at postnatal day 20-22 days, or from adult female rats, were incubated in the presence of single concentrations of either IL-1ß (2.5 ng/ml), TNF-α (10 ng/ml) or IFN-γ (10 ng/ml), with or without IGF-I or -II (100 ng/ml), for either 24h or 48h. After 24h incubation the mean viability of islets in the presence of each cytokine alone was significantly lower than in control cultures for TNF-α and IFN-γ, but not for IL-1ß. Following incubation for 48h in the presence of IL-1ß, TNF-α or IFN-γ each cytokine alone caused between a 60% and 90%, significant reduction in islet cell viability after 48h (Table 3). Co-incubation with either IGF-I or IGF-II significantly reduced cell mortality in the presence of IL-1ß or TNF-α, but not in the presence of IFN-γ. Exposure to IL-1ß, TNF-α, or IFN-γ alone each caused a significant reduction in [^{3}H] thymidine incorporation per µg DNA. Co-incubation with either IGF-I or -II (100 ng/ml) significantly reversed the decrease in radio-thymidine incorporation seen with IL-1ß, or TNF-α, but not IFN-γ. The release of insulin into conditioned culture medium was substantially reduced in response to each of the cytokines alone, and this was not reversed by the presence of IGF-I or -II. Islets from rats of 20-22 days age were grown in chamber slides and exposed to IL-1ß for 48h, with or without IGF-I or -II (100 ng/ml). Islets were then fixed and apoptotic nuclei visualized by the staining for DNA breakage. Exposure to IL-1ß increased the incidence of apoptotic cells, and this was significantly reduced following co-culture with IGFs. This demonstrated that the cytotoxic effects of IL-1ß were likely to be mediated, at least in part, by apoptosis, and that this could be limited by IGF-I or -II.

Table 3 Viability of isolated rat pancreatic islets (% of islets containing only viable cells) from rats of 4-6 days or 20-22 days age following exposure for 24h or 48h to IL-1ß (2.5 ng/ml), TNF-α (10 ng/ml) or IFN-γ (10 ng/ml) without or with IGF-I or -II (100 ng/ml).

		No cytokine	IL-1ß	TNF-α	IFN-γ
4-6 days					
24h	Control	96±1	90±2	89±2	88±3
	IGF-I	94±2	89±1	88±2	83±2
	IGF-II	94±2	90±1	86±1	87±3
48h	Control	94±1	88±1	85±1	83±2
	IGF-I	95±1	89±1	92±2	89±2
	IGF-II	90±2	86±2	86±1	96±1
20-22 days					
24h	Control	92±3	65±5	47±4†	35±3†
	IGF-I	94±2	82±7	66±3	44±5†
	IGF-II	90±3	84±6	62±3	42±3†
48h	Control	89±2	38±2††	24±2††	11±4††
	IGF-I	89±2	71±4*	51±3†**	18±1††
	IGF-II	87±1	62±2*	48±8†**	14±3††

Figures represent mean values ± sem (n=3-4). † p<0.05, †† p<0.001 vs no cytokine in the absence of IGF -I or -II; *p<0.05, **p<0.01 vs incubation with cytokine but without IGF-I or -II.

Experiments were repeated with islets isolated from rats on days 4-6 of postnatal life. Exposure to cytokines alone at the same concentrations as stated above did not significantly alter either islet cell viability or DNA synthetic rate compared to control incubations after 24h or 48h (Tables 3). Exposure of neonatal islets did cause a significant reduction in insulin release, but this was not as great as that seen for islets taken at 22 days gestation. Insulin release from neonatal islets was not altered with the further addition of IGFs. The resistance of islets isolated from neonatal rats to cytokine-induced cell death may be related to the relatively greater expression of IGF-II seen in the neonatal rat pancreas than at older ages. To test this we first established that isolated islets from neonatal rats continued to express IGF-II after incubation in vitro for 48h. Total mRNA was isolated from approximately 500 islets derived from rats either 1-4 days or 20-22 days old and IGF-II mRNA visualized by Northern blot hybridization. A major transcript of IGF-II mRNA of 4.8 kb was seen for neonatal islets, which was no longer present at 20-22 days age. The presence of endogenous IGF-II released from neonatal islets was then functionally removed by incubation with an IGF-II antiserum, and the viability of the islets assessed following exposure to cytokines. Islets were isolated from neonatal rats of 5 days age and incubated for 48h in the presence of either IL-1ß (2.5 ng/ml), TNF-α (10 ng/ml) or IFN-γ (10 ng/ml), with or without antiserum against IGF-II or a non-specific antibody raised against human placental lactogen. Exposure to each cytokines alone did not alter islet cell viability (Table 4). A significant reduction in viability was seen in response to IL-1ß and TNF-α following co-incubation with IGF-II antiserum, but not in the presence of the control antiserum. These results suggest that an endogenous release of IGF-II contributed to the resistance of neonatal islets to cytokine-induced cell death in vitro.

Table 4 Viability of isolated rat pancreatic islets (% of islets containing only viable cells) from rats of 5 days age following exposure for 48h to IL-1ß (2.5 ng/ml), TNF-α (10 ng/ml) or IFN-γ (10 ng/ml) without or with antiserum against IGF-II or control antiserum.

	no cytokine	anti-IGF-II Ab	control Ab
Control	90.7±2.1	90.9±1.3	91.7±1.5
IL-1ß	94.4±5.6	68.4±0.8**	91.2±0.3
TNF-α	91.1±4.8	58.5±3.3**	97.4±2.6
IFN-γ	87.5±6.7	68.2±7.0*	94.4±2.8

Figures represent mean values ± sem (n=3-4). * p<0.05, **p<0.001 vs cytokine but no antibody, or vs cytokine and control antibody.

4. DISCUSSION

The fetal and neonatal endocrine pancreas in rat has a high degree of plasticity as ß cell mass increases in proportion to the rapid growth rate of the animal. While IGF-II has been shown to increase DNA synthesis in isolated fetal islets (21) we now show that in transgenic mice over-expressing IGF-II there is an increased islet size with hyperplasia of the ß cells. Islet number per pancreas was not altered, suggesting that IGF-II did not induce islet neogenesis from the pancreatic ductal tissue, but acted to promote ß cell replication within existing islets. The increased ß cell volume did not result in hyperinsulinaemia or hypoglycaemia. This may be due to a down-regulation of insulin release by IGF-II.

Only recently has it been appreciated that substantial remodelling of the endocrine pancreas occurs in the neonatal life of the rat in which ß cells are deleted by apoptosis, and ß cell mass maintained predominantly by the generation of new islets from the ductal epithelium (5, 13). Our studies confirm those of Scaglia et al. (13) in showing a peak of cell apoptosis in the rat islets of Langerhans at 13-17 days after birth. Since the incidence of apoptosis increases steadily from postnatal day 6, and continues to fall after day 14, at least a third of islet cells may be lost in this way in neonatal life.

Nitric oxide production, driven by increased expression of inducible NOS, has been implicated in the autoimmune destruction of ß cells by cytokines such as IL-1ß (14, 47).

Immunohistochemistry showed the presence of iNOS in pancreatic islets, with a mean of 10% of cells having immunopositive cytoplasm at postnatal day 12. The peak incidence of iNOS presence in islet cells was 2 days before the maximal incidence of apoptosis, suggesting a possible mechanistic relationship. Dual staining for insulin and iNOS showed that almost all iNOS-positive cells were ß cells, located within the central areas of the islets. The appearance of iNOS, and the occurrence of apoptosis, was seen in almost all islets at 12-14 days after birth. The events which lead to an increased distribution of iNOS and increased incidence of apoptosis at this time are not known. However, it is possible that these events also require the removal of a survival factor for ß cells, and this may be the local availability of IGF-II.

Exogenous IGFs have been demonstrated to prevent induced apoptosis in a variety of differing cell types (32-36, 48), and have been implicated in the modulation of iNOS activity (37). We have previously shown that there is an abrupt decline in the pancreatic expression of IGF-II mRNA 2-3 weeks after birth in the rat (16) which would correlate temporally with the increased incidence of apoptosis. In the present experiments he have found, by in situ hybridization and immunohistochemistry, that IGF-II mRNA and peptide is predominantly associated with the islets in fetal and neonatal life, and disappears with the same ontogeny as total pancreatic IGF-II mRNA. While IGF-I mRNA did appear postnatally by day 29, it was predominantly associated with the ductal epithelium of the pancreas rather than islet cells. IGF-I immunoreactivity was, however, associated with the endocrine pancreas. This may relate to the sequestration of IGF-I by IGFBPs. We found a progressive increase in the distribution of immunoreactive IGFBPs-3, -4, and -5 between birth and weaning.

IGF-I or -II were both able to reduce the islet cell death induced by autoimmune cytokines such as IL-1ß and TNF-α in islets from rats at 20-22 days age which no longer expressed IGF-II mRNA, and that this is likely to involve a protection against the induction of apoptosis. Exposure to cytokines also caused a reduction in DNA synthetic rate of islets, and a decreased release of insulin. This would be expected in part because of the induction of cell death. However, DNA synthetic rate was decreased 5-fold within 24h exposure to cytokines which could not be accounted for simply by a reduced cell viability at this time. IGFs were also able to protect islets from a decrease in DNA synthetic rate in response to IL-1ß or TNF-α, but did not reverse the decrease in insulin release. Islets from rats of 4-5 days age, which expressed IGF-II mRNA, showed no reduction in viability in response to the same concentrations of cytokines, but were susceptible once endogenous IGF-II had been immunoneutralized. This provides evidence that endogenous IGF-II can protect neonatal islet cells against cytokine-induced apoptosis at least in vitro. While it is not known if the developmental apoptosis seen in islet cells in vivo in the rat following birth involves cytokines, the experiments at least provide proof of principle that IGF-II has the capability to act as a survival factor.

6. REFERENCES

1. Hill DJ, Hogg J 1991 In: Clinical Endocrinology and Metabolism. Herington A (Ed). London: Bailliere Tindall pp 689-698.
2. Kaung HL 1994 Dev. Dyn. 200: 163-175.
3. Fernandes A, King LC, Guz Y, Stein R, Wright CVE, Teitelman G 1997 Endocrinology 138: 1750-1762.
4. Dudek RW, Lawrence IE 1988 Diabetes 37: 891-900.
5. Finegood DT, Scaglia L, Bonner-Weir S 1995 Diabetes 44: 249-256.
6. Cantenys D, Portha B, Dutrillaux MC, Hollande E, Roze C, Picon L 1981 Virchows Arch (Cell Pathol.) 35: 109-122.
7. Bencosme SA 1955 Am. J. Pathol. 31: 1149-1164.
8. Brockenbrough JS, Weir GC, Bonner-Weir S 1988 Diabetes 37: 232-236.
9. Gu D, Sarvetnick N 1993 Development 118: 33-46.
10. Cecil RL 1911 J. Exp. Med. 14: 500-519.
11. Gepts W, de Mey J 1978 Diabetes 27 (suppl. 1): 251-261.

12. Volk BW, Wellman KF 1985 In: Volk BW, Arquilla ER (Eds) The Diabetic Pancreas. 2nd ed. New York: Plenum Medical p353.
13. Scaglia L, Cahill CJ, Finegood DT, Bonner-Weir S 1997 Endocrinology 138: 1736-1741.
14. Tornehave D, Larsson L-I 1997 Exp. Clin. Endocrinol. Diabetes 105: A27.
15. Corbett JA, McDaniel ML 1995 J. Exp. Med. 181: 559-568.
16. Hogg J, Hill DJ, Han VKM 1994 J. Mol. Endocrinol. 13: 49-58.
17. Calvo EL, Bernatchez G, Pelletier G, Iovanna JL, Morisset J 1997 J Mol. Endocrinol. 18: 233-242.
18. Hill DJ, Hogg J 1992 In: Pancreatic Islet Cell Regeneration and Growth. Vinik A (Ed), New York: Plenum Medical pp. 113-120.
19. Hill DJ, Clemmons DR 1992 Growth Factors. 6: 315-326.
20. Hill DJ, Frazer A, Swenne I, Wirdnam PK, Milner RDG 1987 Diabetes 36: 465-471.
21. Hogg J, Han VKM, Clemmons DR, Hill DJ 1993 J. Endocrinol. 138: 401-412.
22. Swenne I, Hill DJ, Strain AJ, Milner RDG 1987 Diabetes. 36: 288-294.
23. Scharfmann R, Corvol M, Czernichow P 1989 Diabetes. 38: 686-690.
24. Romanus JA, Rabinovitch A, Rechler MM 1985 Diabetes 34: 696-702.
25. Rabinovitch A, Quigley C, Russel T, Patel Y, Mintz DH 1982 Diabetes. 31: 160-164.
26. Van Schravendijk CF, Foriers A, Van Den Brande JL, Pipeleers DG 1987 Endocrinology. 121: 1784-1788.
27. Fehmann HC, Jehle P, Markus U, Goke B 1996 Metabolism 45: 759-766.
28. Swenne I, Hill DJ 1989 Diabetologia. 32: 191-197.
29. Hoog A, Grimelius L, Falkmer S, Sara VR 1993 Regul. Pept. 47: 275-283.
30. Maake C, Reinecke M 1993 Cell Tissue Res. 273: 249-259.
31. Asfari M, Wei D, Noel M, Holthuizen PE, Czernichow P 1995 Diabetologia 38:927-935.
32. Geier A, Haimshon M, Beery R, Lunenfeld B 1992 In Vitro Cell Dev. Biol.-Animal 28A: 725-729.
33. Galli C, Meucci O, Scorziello A, Werge TM, Calissano P, Schettini G 1995 J. Neurosci. 15: 1172-1179.
34. Chun SY, Billig H, Tilly JL, Furuta I, Tsafriri A, Hsueh AJW 1994 Endocrinology 135: 1845-1853.
35. Muta K, Krantz SB 1993 J. Cell Physiol. 156: 264-271.
36. Jung Y, Miura M, Yuan J 1996 J. Biol. Chem. 271: 5112-5117.
37. Schini VB, Catovsky S, Schray-Utz B, Busse R, Vanhoutte PM 1994 Circ. Res. 74: 24-32.
38. Guan J, Williams C, Gunning M, Mallard C, Gluckman P 1993 J. Cerebr. Blood Flow Metab. 13: 609-616.
39. Hellerstrom C. Lewis NJ, Borg H, Johnson R, Freinkel N 1979 Diabetes 28: 769-776.
40. Hales CN, Randle PJ 1963 Biochem. J. 88: 137-146.
41. Herbert V, Lau K, Gottlieb CW, Bleicher SJ 1965 J Clin. Endocrinol. 25: 1375-1384.
42. Schnuerer EM, Rokaeus A, Carlquist M, Bergman T, Dupré J, McDonald TJ 1990 Pancreas 5: 70-74.
43. Hsu SM, Raine L, Fanger H 1981 J. Histochem. Cytochem. 29: 577-580.
44. Wijsman JH, Jonker RR, Keijzer R, Van de Velde CJ, Cornelisse CJ, Van Dierendonck JH 1993 J. Histochem. Cytochem. 41: 7-12.
45. Southern EM 1975 J. Mol. Biol. 98: 503-507.
46. Murphy LJ, Tachibana K, Friesen HG 1988 Endocrinology 122: 2027-2033.
47 Kenato H, Fujii J, Seo HG, Suzuki K et al. 1995 Diabetes 44: 733-738.
48. Stewart CE, Rotwein P 1996 J. Biol. Chem. 271: 11330-11338.
49. Sun F-L, Dean WL, Kelsey G, Allen ND, Reik W 1997 Nature (Genetics) in press.

Molecular Mechanisms to Regulate the
Activities of Insulin-like Growth Factors
K. Takano, N. Hizuka and S-I. Takahashi (Editors)

The early embryonic neuroretina: a CNS site of production and action of (pro)insulin and IGF-I

F. de Pablo*, B. Díaz, M. García-de Lacoba, E. Vega and E.J. de la Rosa

Department of Cell and Developmental Biology, Centro de Investigaciones Biológicas, C.S.I.C, Velázquez 144, Madrid 28006, Spain.

The chick embryo neuroretina is an avascular and easily accesible part of the central nervous system (CNS) from the early stages of organogenesis that provides the unique opportunity to study the autocrine/paracrine role of the insulin family polypeptides in neurogenesis. In embryonic stages day 3 (E3) to E7 the neuroretina actively proliferates, with apoptotic cell death ocurring as part of normal development *in vivo*. Progressively until E12, the neuroepithelial cells withdraw from the cell cycle, differentiate in neuronal and glial cell types, migrate to their definitive layer location and initiate synaptogenesis (1,2). By studying the transition between proliferative and differentiative stages of neuroretina, we aim at understanding the multiple, possibly complementary, roles of insulin and IGF-I, including protection from apoptosis, in the development of the CNS.

1. PROINSULIN mRNA EXPRESSION PREDOMINATES IN THE PROLIFERATIVE STAGES AND IGF-I mRNA EXPRESSION PREDOMINATES IN THE DIFFERENTIATIVE STAGES OF NEUROGENESIS

Proinsulin and IGF-I mRNAs have been studied in neuroretina by in situ hybridation and reverse-transcriptase-polymerase chain reaction (RT-PCR) from E3 to E20. In the E3-E6, neuroretina proinsulin mRNA is widespread, while IGF-I mRNA is less abundant and preferentially localized in the pigmented neuroretina and cilliary processes. After E8-E10 and through the end of embryogenesis, IGF-I mRNA expression predominates in the neuroretina (3,4 and B. Díaz *et al.* in preparation). Immunoactivity with the HPLC mobility of proinsulin is increasingly accumulated between E6 and E8 in the vitreous humor, in close contact with the neuroretina and cilliary processes (5). In E8 embryos its concentration (~55 ng/ml) is higher than IGF-I immunoactivity (<0.1 ng/ml), the latter increasing markedly in vitreous humor between E6 and E15 (5,6) in close parallelism with retinal IGF-I

*For correspondence, E-mail: cibfp1f@fresno.csic.es. We acknowledge funding for our research from the Dirección General de Investigación Científica y Técnica (PM 94-0052 to F. de P. and PM 96-0003 to E.J. de la R.), Spain

mRNA. It is very likely that this vitreal (pro)insulin is of local origin since in plasma the level of insulin immunoreactivity in E8 is below 0.1 ng/ml, while IGF-I is 1 ng/ml and the retina secretes (pro)insulin in culture (5). It is not known what up-regulates IGF-I gene expression and supresses proinsulin gene expression in developing neuroretina. Circulating growth hormone is very low in chick embryos until late stages (7) what makes it an unlikely candidate.

Although pancreatic expression of proinsulin mRNA begins coincident with the development of a circulatory system in the chick embryo (E2.5-E3), there is low immunoreactive insulin in blood until E12 (0.6 ng/ml). At this age, there is expression of low levels of proinsulin mRNA in chick embryo liver, brain, as well as the eye and high levels in the pancreas (8). Thus, it is possible that both pancreatic and extrapancreatic (pro)insulin are implicated in the modulation of growth and metabolism of several tissues during embryogenesis. In agreement with this, recently, mice null mutants for both insulin genes, have been shown to have embryonic growth retardation (9).

2. EXPRESSION OF FUNCTIONAL INSULIN AND IGF-I RECEPTORS IS REGULATED POSTRANSCRIPTIONALLY IN EMBRYONIC NEURORETINA

Characteristically, mRNA expression for both the insulin receptor and the IGF-I receptor (the only receptor that binds IGFs in chicken) is widespread in the developing retina and shows little temporal regulation between E6 and postnatal day 1 (3,4,10). However, there is high regulation at the protein level with marked changes in receptors affinity and especificity during proliferative vs. differentiative stages. Both [^{125}I]IGF-I and [^{125}I]insulin binding are higher in E6 than in E12 neuroretina.

The receptors present in E6 neuroretina are two distinct classes. The IGF-I receptor has the classic high affinity for IGF-I, lower for insulin and very little for proinsulin (Table 1). In contrast, the insulin binding sites are atypical, showing high affinity for both insulin and IGF-I, and significant affinity for proinsulin (Table 1). This 'low discriminating receptor' may correspond in part to hybrid receptors, the predominant form which binds [^{125}I]insulin at this stage (unpublished results). By E12, specific binding of both [^{125}I]IGF-I and [^{125}I]insulin decrease several fold in neuroretina and the two types of receptors have more typical specificities.

The insulin and IGF-I receptors are highly phosphorylated in the basal state during proliferative stages of neurogenesis, although they respond to high concentrations of ligands with further autophosphorylation. The overall phosphorylation of cellular proteins in whole mounts of E5 neuroretina is more responsive to IGF-I than insulin (Figure 1). Intriguingly, there is an apparent central-peripheral spatial gradient in this ligand-dependent phosphorylation that may correspond to differential expression of receptors and/or substrates which should be investigated.

Table 1.
Half-maximal inhibition of tracer binding by unlabelled peptides (M) derived from binding-competition curves in E6 chick embryo neuroretina.

	Competitor		
[125I]Insulin	Insulin	IGF-I	Proinsulin
	5.8×10^{-10}	7.0×10^{-10}	1.1×10^{-8}
[125I]IGF-I	Insulin	IGF-I	Proinsulin
	6.4×10^{-9}	1.0×10^{-10}	$>10^{-7}$

Figure 1. Ligand-estimulated phosphorylation of E5 neuroretinas *in toto*. Freshly isolated neuroretinas from chick embryos were placed over nitrocellulose filters and preincubated 1 hour in RPMI 1640 medium (M) free of phosphates suplemented with complete medium (9:1). Then, they were labelled for 1 hour with 100 μCi of $H_3{}^{33}PO_4$ in 1 ml of medium at 37C. During the last 15 min. either insulin or IGF-I at 10^{-7}M were added. The reaction was stoped by cold TCA precipitation (5%) and the dried retinas were exposed to a PhosphorImager screen.

3. MULTIPLE EFFECTS OF INSULIN AND IGF-I IN NEUROGENESIS

The presence in E6 neuroretina of a high-affinity, low-discriminating receptor for insulin, proinsulin and IGF-I, together with high affinity typical IGF-I receptors, fits well with the potency of these factors to estimulate different cellular processes in neuroretina. In organoculture of E5-E6 neuroretina, IGF-I is slightly more potent stimulating DNA and protein synthesis. However, the maximal effect, at 10^{-8} M, is comparable for IGF-I, proinsulin and insulin (5). Although proliferation and differentiation are mutually exclusive events in neuronal cells, these peptides also increase the differentiation marker G4 (Ng-CAM) in organoculture (5). This dual action of (pro)insulin and IGF-I could be due to effects on different cell populations of the neuroretina or be in part the result of a survival effect of the factors in both proliferating neuroepithelial cells and postmitotic neuroblasts. Indeed, all three IGF-I, proinsulin and insulin are very potent inhibitors of apoptosis in E5 neuroretina (Figure 2) when is placed in a growth factor deprived organoculture.

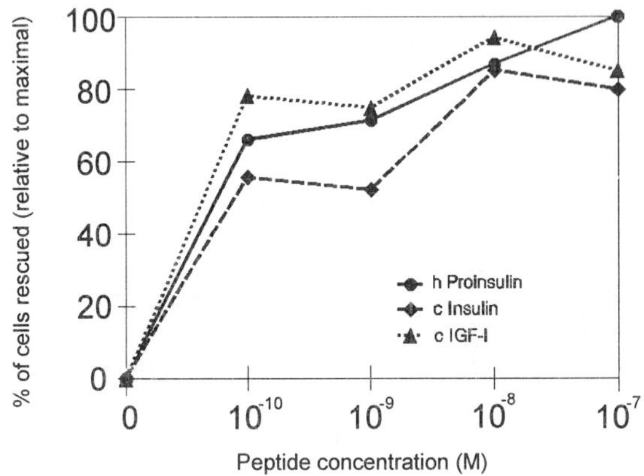

Figure 2: Protection from apoptosis in retina organoculture. E5 neuroretinas (four per experimental point) were placed for 6 hours in organoculture in defined medium alone or in the presence of chicken IGF-I, chicken insulin or human proinsulin at the concentrations indicated. Then, individual cells were obtained by trypsin dissociation of the retina and preparaded by cytospin for DAPI staining on a glass slide (for details see Ref. 11). The percentage of apoptotic nuclei were counted in at least 1000 cells from two cytospins in each point. The mean values of cells rescued relative to maximal effect (obtained with proinsulin at 10^{-7}M) are represented.

Studies on cell-cycle and proliferative fraction revealed that insulin apparently shortens the cell-cycle duration, thus, the cell flow in E5 neuroretina increases from 4.4% to 8.3% of the cells transversing S phase per hour in the presence of insulin (B. Díaz, *et al.* in preparation). In parallel, a reduction of apoptosis was observed by multiple criteria including presence of pyknotic nuclei, TUNEL staining and ELISA determination of free nucleosomes. Our present interpretation is that the overall developmental program of retinal neurogenesis is advanced forward in large part because the (pro)insulin found in the microenvironment of proliferating neuroepithelial cells protects them from apoptosis. We speculate that the up-regulation in IGF-I expression in neuroretina after E6-E8 could be involved in further stimulation of neuronal differentiation. The intrinsic developmental stage and potentiality of the cells can in part dictate the response, proliferative or differentiative to a growth factor, as shown for IGF-I in neuroblastoma cells (12).

4. INSULIN AND IGF-I EFFECTS ON THE CHAPERONE HSC-70. A MECHANISM TO PROTECT CELLS FROM APOPTOSIS?

When DNA synthesis of dissociated E5 retinal cells was studied in culture, we observed that under those conditions (more distant from physiological than the organotypic cultures) the subpopulation of cells sensitive to insulin and IGF-I were the neuroepithelial cells expressing the antigen PM1 (recognized by the monoclonal antibody termed Precursor Marker 1) (Figure 3, Ref. 13).

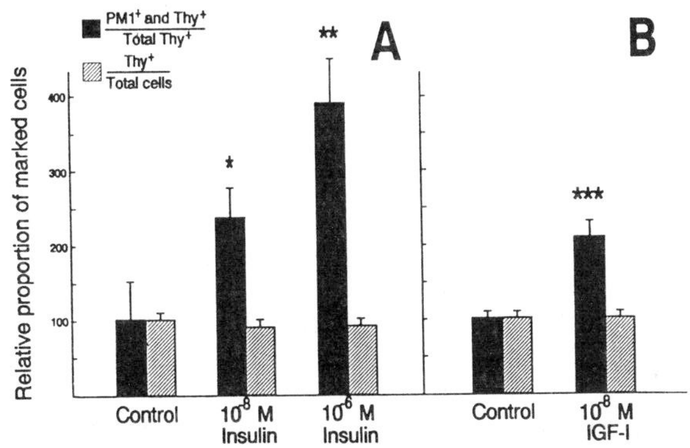

Figure 3 Effects of insulin and IGF-I on PM1-positive neuroepithelial cells. E5 retinal dissociated cell cultures were grown for 20 hours in control medium or in the presence of insulin (A) or IGF-I (B). During the last hour cells were labelled with [^{3}H]methylthymidine and immediately fixed, stained with monoclonal antibody PM1 and processed for autoradiography. The hatched bars represent the relative proportion of total cells that had incorporated [^{3}H]methylthymidine. The solid bars represent the relative proportion of [^{3}H]methylthymidine-labelled cells double labelled with PM1. (Reproduced from Ref. 3)

The PM1 antigen was present in nearly all cells in E4 neuroretina and decreased progressively until E12 (13). We had suggested that its corresponding protein might be involved in fate selection. After partial purification of the protein and using an RT-PCR cloning strategy we have identified PM1 as the heat shock cognate 70 protein (Hsc70). This is the first chaperone to be found regulated by insulin and IGF-I. Its role and modulation during development appear complex and its relation to cell cycle, proliferation and differentiation will require detailed studies. Presently, there is indirect evidence suggestive of an implication of Hsc70 in the mechanism by which insulin and IGF-I promote survival in early development. During neurulation, inhibition of endogenous embryonic (pro)insulin by antisense oligodeoxynucleotides increases apoptosis (14) and decreases the levels of Hsc-70 (A. Morales, *et. al.* manuscript in preparation). Exogenous insulin, in contrast, protects the embryo from apoptosis caused by growth factor deprivation and increases the percent of cells expressing high levels of Hsc70.

In conclusion, insulin, proinsulin and IGF-I, at concentrations possibly found in the embryonic CNS, have multifunctional influence in neural development. Their complementary expression from early stages in neuroretina, together with the high affinity receptors, is suggestive of a coordinated autocrine-paracrine role in all essential cellular processes that lead to the correct building of a vertebrate CNS: proliferation, differentiation and cell survival. Obviously, the members of the insulin family will have to interact and cooperate with other extracellular signals in a dynamic network of modulators of cellular activities.

REFERENCES

1. A.J. Khan. Dev. Biol. 38 (1974) 30.
2. C. Prada, J. Puga, L. Pérez-Méndez, R. López and G. Ramirez. Eur. J. Neurosci. 3 (1991) 559.
3. E.J. de la Rosa, C.A. Bondy, C. Hernández-Sánchez, X.Wu, J. Zou, A. López-Carranza, L.M. Scavo, and F. De Pablo. Eur. J. Neurosci. 6 (1994)1801
4. F. De Pablo, C. Alarcón, B. Díaz, M. García de Lacoba, A.López-Carranza, A.V. Morales, B. Pimentel, J. Serna, E.J. De la Rosa. Int. J. Dev. Biol. 40: Suppl. 1(1996)109S.
5. C. Hernández-Sánchez, A. López-Carranza, C. Alarcón, E.J. de la Rosa and F. De Pablo. Proc. Natl. Acad. Sci. USA 92 (1995) 9834.
6. Y.W-H.Yang, D.R. Brown, H.L. Robcis, M.M. Rechler and F. de Pablo. Reg. Peptides 48 (1993) 145.
7. K. Kikuchi, F.C. Buonomo, Y. Kajimoto and P. Rotwein. Endocrinology 128 (1991) 1323.
8. J. Serna, P. Gonzalez-Guerrero, C.G. Scanes, M. Prati, G. Morreale and F. de Pablo. Growth Reg. 6 (1996) 73.
9. B. Duvillié, N. Cordonnier, L. Deltour, F. Dandoy-Dron, J.M. Itier, E. Monthioux, J. Jami, R.L. Joshi and D. Bucchini. Proc. Natl. Acad. Sci. USA 94 (1997) 5137.
10. F. de Pablo and E.J. de la Rosa. Trends Neurosci 18 (1995) 143.
11. E.J. de la Rosa, B. Díaz and F. de Pablo. In: Cellular and Molecular Procedures in Developmental Biology (series: Current Topics in

Developmental Biology) F. de Pablo, A. Ferrús and C. D. Stern, Eds. Academic Press, Inc., San Diego, Vol. 36,1997, p.133.
12. S. Pahlam, G. Meyerson, E. Lindgren, M. Schalling and I. Johansson. Proc. Natl. Acad. Sci. USA 88 (1991) 9994.
13. C. Hernández-Sánchez, J.M. Frade and E.J. de la Rosa. Eur. J. Neurosci. 6 (1994) 105.
14. A.V. Morales, J. Serna, C. Alarcón, E.J. de la Rosa and F. de Pablo. Endocrinology 138 (1997) 3967.

Molecular Mechanisms to Regulate the
Activities of Insulin-like Growth Factors
K. Takano, N. Hizuka and S-I. Takahashi (Editors)
1998 Elsevier Science B.V.

IGF-I and uterine growth

O. O. Adesanya, J. Zhou, C. Samathanam, L. Powell-Braxton* and C.A. Bondy

Developmental Endocrinology Branch, NICHD, NIH, Bethesda, MD, 20892

* Cardiovascular Research, Genentech, San Francisco, CA

Estradiol has potent mitogenic effects upon uterine tissue in all species. The mechanism(s) of estradiol (E2) induced uterine growth have interested investigators for nearly fifty years. Early experiments showed that extracts from E2-treated uterine tissue contained factors capable of stimulating the growth of tumor cells, suggesting that E2 induced the expression of soluble growth-promoting factors ('estromedins') by uterine tissue (1). More recently, insulin like growth factor I (IGF-I) has been identified as a potential mediator of E2's growth promoting effects in the uterus (2). E2 induces IGF-I synthesis in rodent (3-5), farm animal (6 & 7) and primate (8-11) uterine tissue (Fig. 1).

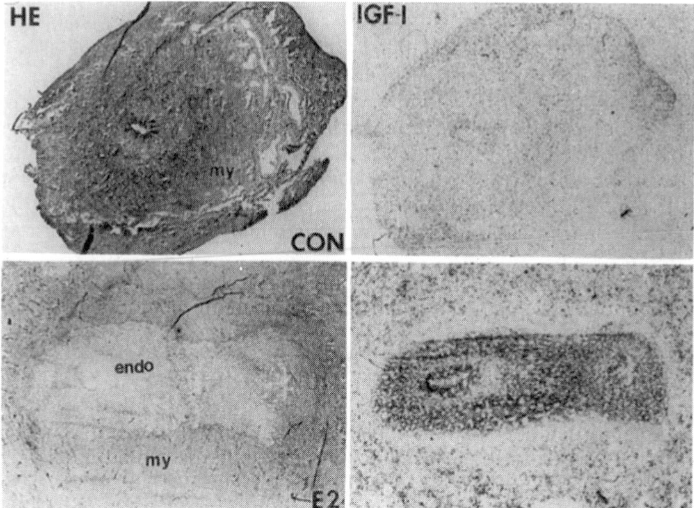

Fig. 1 Effects of estrogen on IGF-I gene expression in the primate uterus. Ovariectomized rhesus monkeys were treated with placebo (CON) or E2-containing pellets. Representative tissues sections stained with hematoxylin & eosin are shown on the right and matching film autoradiograms showing IGF-I mRNA signal on the left. IGF-I mRNA is barely detected in the ovariectomized animal but is abundant in the E2-treated endometrium.

Furthermore, we have recently shown that local IGF-I expression shows a strong positive correlation with spatiotemporal patterns of E2-induced uterine cell proliferation as revealed by the human cell proliferation marker, Ki67 antigen (11 and Figs. 2&3).

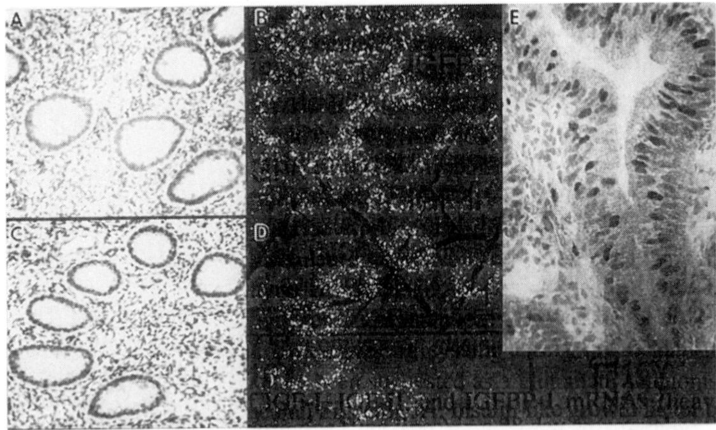

Fig. 2 Cellular patterns of IGF-I (A&B)and IGF-I receptor (C&D) gene expression and comparison with cell proliferation patterns (E) in the E2 treated rhesus uterus. These are paired bright and dark field photomicrographs showing *in situ* mRNA hybridization signal as white grains in the dark field. IGF-I mRNA is localized in the endometrial stroma while the receptor mRNA is concentrated in the epithelium (A-D. The human cell proliferation marker, Ki67, is detected immunohistochemically in the epithelium, suggesting a paracrine mode of IGF-I action.

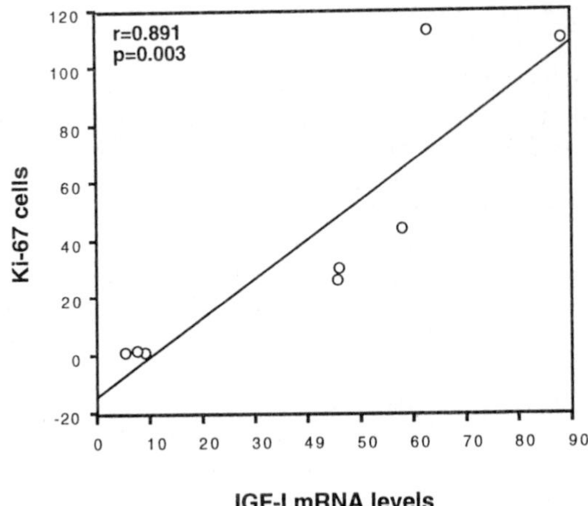

Fig. 3 Correlation between local IGF-I mRNA concentration and cell proliferation (Ki67 positive cells) in the primate uterus.

Very powerful evidence for IGF-I's role in uterine growth is provided by the development of IGF-I targeted gene deletion mice, which demonstrate a reduction in uterine size which is more severe than their generalized dwarfism (12). The interpretation of this latter observation is complicated by the fact that IGF-I nullizygous mice also demonstrates deficient estrogen production (12,13). Hence, it is not clear whether the reduction in uterine growth in the IGF-I null mouse is due to inadequate estrogen production or to inadequate estrogen effect upon the uterus secondary to absent local IGF-I.

To clarify this issue and to further investigate the role of IGF-I in mediating E2's mitogenic effects, we evaluated the effects of exogenous E2 treatment on uterine growth parameters in IGF-I null (KO) and wild-type (WT) littermate mice. The generation of the IGF-I targeted gene deletion mouse line has been described (14). Genotyping was carried out by Southern analysis or PCR, as previously described (14). Groups of female mice were studied at P20 and P40. The animals received an intraperitoneal (ip) injection of 17-beta-estradiol (1 μg/g) or diluent (95% normal saline and 5% ethanol) at time zero and ip injections of 3H-thymidine (2μCi/g) and colchicine (50 mg/g) one hour before sacrifice at 24 hours. Uteri were fixed in 10% formalin, embedded in paraffin, and cut in 10 μm (P40) or 6 μm (P20) sections. Deparaffinized sections were dipped in photographic emulsion (Kodak NTB2) for 3 weeks, developed and stained with HE. A set of sections were also stained with DAPI (4,6-diamidino-2-phenylindole, Sigma, St. Louis, MO).

Labeling index was determined by counting tritium-positive nuclei (>5 grains/nucleus) per 100 nuclei of a given cell type (epithelial, stromal or myometrial) under direct brightfield microscopy at a magnification of 200x. Mitotic index was determined by counting the number of mitotic figures per 100 nuclei for each cell type in HE-stained thin sections at a magnification of 400x. DAPI stained mitotic figures were counted in a similar manner under fluorescent illumination. 300-400 cells in each category were scored in each tissue section and data from 2-4 uterine sections were meaned for each animal. Six to 12 animals were in each group. Group means were statistically compared by ANOVA and significant differences among means were determined by Fischer's least significant difference test.

The uteri of nullizygous IGF-I mice (KOs) were 2.4-fold smaller in cross-sectional area compared with littermate wild-type (WT) mice uteri at P20 (prepubertal). By P40 the discrepancy between KO and WT uteri had increased greatly. P40 KO uteri showed no increase in size compared with P20 KO uteri, while WT uteri increased in size by 5-fold compared with P20 WT uteri. Thus, KO uteri were almost 10-fold smaller than littermate WTs at P40, indicating that the onset of puberty had zero effect on uterine growth in the KO mice. Despite their small size, the anatomical features of KO uteri appear normal, except for a disproportionate reduction of the size of the myometrium relative to the stromal and epithelial compartments.

To evaluate E2's ability to stimulate proliferation in KO uteri, animals were treated with E2 or diluent and killed the next day, one hour after tritiated thymidine injection. At baseline, i.e., in the absence of E2 treatment, the labeling index (LI) in uterine epithelium and stroma and myometrium is equal in WT and KO mice. Unexpectedly, E2 treatment resulted in equal, robust increases in LI in uterine epithelium and stroma in both WTs and KOs. The increase in myometrial LI in response to E2, however, was slightly but significantly less in KOs.

The finding of equal numbers of cells entering S-phase— as shown by DNA synthesis— in the IGF-I null uterus was unexpected. The mitotic index was determined in two ways—using DAPI-staining examined under fluorescent microscopy and HE staining examined under light microscopy. Both methods showed a highly significant decrease in the number of mitotic figures in all three uterine cell compartments for KO compared with WT uteri (e.g., for P20 mice: epithelium- 3.5-fold lower; stroma- 6.9-fold lower and myometrium- 7.4-fold lower).

Rethinking the Role of IGF-I in Cell Cycle Regulation

In vitro studies using the BALB/c3T3 cell line established the view that growth factors function primarily in the early stages of the cell cycle. Peptides such as PDGF, FGF and EGF were found to stimulate serum-starved cells to exit G0 and enter the G1-phase of the cell cycle, while IGF-I was found to promote progression of these cells through G1 to S-phase, and thus was deemed a "G1-progression factor" (15-17). The present study is, to our knowledge, the first analysis of the effects of IGF-I deletion on cell cycle dynamics in vivo.

Our data show that at baseline, i.e., without E2 treatment, the percent of cells in S-phase is equal in IGF-I null and WT uteri, indicating no block to G1 progression in the absence of IGF-I. Furthermore, the percent of uterine cells in S-phase increases dramatically in response to E2 treatment in both WT and KO mice, with the response being equal in the epithelium and stroma and mildly retarded in the KO myometrium. These findings show that IGF-I is not essential for the progression of these uterine cell types through the G1 phase of the cell cycle, although it may serve to enhance the rate of myometrial cell's transit through G1 into S-phase. IGF-I is, however, critically involved in the progression of all uterine cell types from S-phase through G2/M phase, since there is a profound decrease in cells in mitosis in all regions of the IGF-I null uterus. The fact that the myometrium seems to depend upon IGF-I for both the rate of G1 progression and the transition from S- to M-phase may explain the profound growth retardation demonstrated by this region of the uterus in the IGF-I KO mouse.

Analysis of cell cycle kinetics in IGF-I receptor null fibroblasts provides support for the view that IGF-I plays an important role in later phases of the cell cycle (18). A fibroblast cell line derived from IGF-I receptor null embryos was found to have a G2/M phase duration which was 4-fold longer than a comparable WT cell line, and an S-phase duration almost 3-fold longer than WT, while the G1-phase was prolonged only 2-fold. This approximately 7-fold prolongation of the time it takes from starting S-phase to mitosis in the absence of IGF-I effect fits quite well with the ~7-fold reduction in mitoses seen in KO compared with WT uterine tissues in the present study. The question arises as to whether this cell cycle delay merely slows growth, such that the mitotic index in KO uteri would 'catch-up' with the WT seven days after E2 treatment. Based on the complete absence of growth of KO uteri from P20 to P40, it appears that this is not the case. In any case, current studies are aimed at evaluating the fate of cells

In addition to demonstrating that IGF-I has a critical role in S- to M-phase transition of the cell cycle, this data also provides definitive proof that IGF-I is a critical mediator of E2's growth promoting effects in the uterus. Together with previous work showing that E2 induces IGF-I gene expression in murine (3) as well as primate (11) uteri, the present findings show that E2 stimulates uterine growth in two phases, with early cycle events to the initiation of S-phase independent of IGF-I while later events culminating in mitosis are dependent on E2-induced IGF-I synthesis and action.

REFERENCES

1. Sirbasku DA Estrogen induction of growth factors specific for hormone-responsive mammary, pituitary, and kidney tumor cells. Proc Natl Acad Sci U S A 1978 75:3786-90
2. Murphy LJ; Ghahary A Uterine insulin-like growth factor-1: regulation of expression and its role in estrogen-induced uterine proliferation. Endocr Rev 1990 11:443-53
3. Murphy LJ, Murphy LC, and Freisen HG. 1987 Estrogen induces insulin-like growth factor-I expression in the rat uterus. Mol Endocrinol. 1:445-450.
4. Norstedt G, Levinovitz A, and Erikson H. 1989 Regulation of uterine insulin-like growth factor I mRNA and insulin-like growth factor II mRNA by estrogen in the rat. Acta Endocrinol (Copenh). 120:466-472.
5. Kapur S, Tamada H, Dey SK, and Andrews GK. 1992 Expression of insulin-like growth factor-I (IGF-I) and its receptor in the peri-implantation mouse uterus, and cell-specific regulation of IGF-I gene expression by estradiol and progesterone. Biol Reprod. 46:208-219.
6. Simmen RCM, Simmen FA, Hofig A, Farmer SJ, and Bazer FW. 1990 Hormonal regulation of insulin-like growth factor gene expression in pig uterus. Endocrinology. 127:2166-2174.
7. Stevenson KR; Gilmour RS; Wathes DC Localization of insulin-like growth factor-I (IGF-I) and -II messenger ribonucleic acid and type 1 IGF receptors in the ovine uterus during the estrous cycle and early pregnancy. Endocrinology 1994 134:1655-64
8. Giudice LC, Dsupin BA, Jin IH, Vu TH, and Hoffman AR. 1993 Differential expression of messenger ribonucleic acids encoding insulin-like growth factors and their receptors in human uterine endometrium and decidua. J Clin Endocrinol Metab. 76:1115-1122.
9. Vollenhoven BJ, Herington AC, and Healy DL. 1993 Messenger ribonucleic acid expression of the insulin-like growth factors and their binding proteins in uterine fibroids and myometrium. J Clin Endocrinol Metab. 76:1106-1110.
10. Zhou J, Dsupin BA, Giudice LC, and Bondy CA. 1994 Insulin-like growth factor system gene expression in human endometrium during the menstrual cycle. J Clin Endocrinol Metab. 79:1723-1734.
11. Oluyemisi O. Adesanya, Jian Zhou, Carolyn A. Bondy (1996) Sex Steroid Regulation Of IGF System Gene Expression And Proliferation In Primate Myometrium J Clin Endo Metab. 81:1967-1974.
12. Baker J, Hardy MP, Zhou J, Bondy CA, Lupu F, Bellve AR, Efstratiadis A. 1996 Effects of an IgfI gene null mutation on mouse reproduction. Molec Endocrinol 10:903-918.
13. Zhou J and Bondy CA Insulin-Like Growth Factor I Regulates Gonadotropin Responsiveness In The Murine Ovary, Molec Endocrinol, in press.
14. Powell-Braxton L; Hollingshead P; Warburton C; Dowd M; Pitts-Meek S; Dalton D; Gillett N; Stewart TA IGF-I is required for normal embryonic growth in mice. Genes Dev 1993 7:2609-17
15. Rubin R; Baserga R Insulin-like growth factor-I receptor. Its role in cell proliferation, apoptosis, and tumorigenicity. Lab Invest 1995 73:311-31
16. Pardee AB 1989 G1 events and the regulation of cell proliferation. Science 246:603-608.

17. Stiles CD; Capone GT; Scher CD; Antoniades HN; Van Wyk JJ; Pledger WJ Dual control of cell growth by somatomedins and platelet-derived growth factor. Proc Natl Acad Sci U S A 1979 76:1279-83
18. Sell C; Dumenil G; Deveaud C; Miura M; Coppola D; DeAngelis T; Rubin R; Efstratiadis A; Baserga R Effect of a null mutation of the insulin-like growth factor I receptor gene on growth and transformation of mouse embryo fibroblasts. Mol Cell Biol 1994 14:3604-13.

Molecular Mechanisms to Regulate the
Activities of Insulin-like Growth Factors
K. Takano, N. Hizuka and S-I. Takahashi (Editors)

Bone Morphogenetic Proteins and IGF System

S. Mohan and D.J. Baylink

Departments of Medicine, Biochemistry and Physiology, Mineral Metabolism, Jerry L Pettis VA Medical Center, 11201 Benton Street, Loma Linda, CA 92357, USA

1. INTRODUCTION

The bone morphogenetic proteins (BMPs), which belong to TGFβ superfamily, represent a unique set of growth factors, because they are the only proteins that can stimulate mesenchymal stem cells to differentiate into osteoblasts, and subsequently induce newly formed osteoblasts to deposit bone matrix and promote new bone formation (1-4). Unlike any other known bone growth factors, BMPs are also unique in their ability to induce bone formation at non-bony sites (1-4). Although a single BMP, in combination with an appropriate carrier, is capable of inducing new bone formation at ectopic sites, little is known about how a single messenger molecule can initiate and perpetuate this entire cascade of events.

In our studies on the mechanism of BMP action, we proposed the hypothesis that it is likely that, in addition to direct action on responding cells, BMPs may promote their effects on osteoblast line of cells by stimulating local production of paracrine and autocrine factors. To test this hypothesis, we chose to examine the effects of BMP-7, also known as osteogenic protein-1, on local growth factor production using human osteoblasts. In previous studies using this model system, we found that treatment of human osteoblasts with BMP-7 increased both cell proliferation, and alkaline phosphatase activity, a marker of differentiation in human osteoblasts (5).

To evaluate if BMPs mediate their effects by recruiting other growth factors, we chose the IGF system for the following reasons: 1) IGFs are the most abundant mitogens produced by osteoblasts and stored in bone; 2) IGFs contribute to 40% of basal osteoblast cell proliferation; 3) IGFs stimulate bone formation parameters, in vitro and in vivo; 4) bone growth is significantly impaired in IGF knockout mice; and 5) IGF actions are controlled by both local and systemic osteoregulatory agents (6-9). Our findings on the effect of BMP-7 on the production of IGF system components, as discussed below, are consistent with this hypothesis that BMPs may mediate their effects on target cells in part via regulating production and/or actions of local growth factors. In this paper, we describe the effects of BMPs on IGF regulatory system and discuss the potential molecular mechanisms by which BMPs may mediate their effects on IGF system in osteoblasts.

2. REGULATION OF PRODUCTION OF IGF SYSTEM COMPONENTS BY BMP-7 IN HUMAN OSTEOBLASTS

To evaluate the effects of BMP-7 on the production of IGF system components, dose and time effects of BMP-7 on the protein and mRNA levels of various IGF system components were evaluated using serum-free cultures of human osteosarcoma cells (TE85 and SaOS-2) with osteoblastic characteristics as model systems (10). Table 1 summarizes the effects of BMP-7 on the production of various IGF system components in SaOS-2 and TE85 human osteosarcoma cells. After 48 h of treatment, BMP-7 increased the level of IGF-II (3- and 2-fold, respectively) in the conditioned medium of SaOS-2 and TE85 cells, whereas IGF-I levels were low to undetectable in the conditioned medium of either cell type. BMP-7 caused a dose-dependent increase in the level of stimulatory IGFBP-3 and IGFBP-5, as measured by specific radioimmunoassay, in the conditioned medium of both SaOS-2 and TE85 cells. In contrast, the level of inhibitory IGFBP-4 was decreased by more than 50% in the conditioned medium of SaOS-2 and TE85 cells.

Because the concentration of IGFBPs may be regulated both at the level of production and degradation (11), and because human osteoblasts have been shown to produce proteases capable of degrading IGFBPs (12-14), we determined if BMP-7 mediates its effects in part by regulating IGFBP proteolysis. To evaluate IGFBP proteolysis, [^{125}I]IGFBP tracer was incubated with conditioned medium collected from control or BMP-7 treated cultures for 24 h at 37 C prior to separation of intact IGFBP from fragments by polyacrylamide gel electrophoresis in the presence of sodium dodecylsulfate. Our data revealed that BMP-7 treatment decreased the amount of proteolysis of [^{125}I]IGFBP-3 by 29% compared with vehicle-treated controls. In addition, BMP-7 treatment decreased the amount of proteolysis of [^{125}I]IGFBP-5 by 71% as compared with vehicle treated controls. In contrast, BMP-7 treatment had no significant effect on the amount of [^{125}I]IGFBP-4 proteolysis. These data suggest that BMP-7-induced changes in the levels of stimulatory IGFBP-3 and IGFBP-5 protein levels may be mediated in part by changes in proteolysis of these binding proteins. The mechanism(s) by which BMP-7 treatment induces changes in the proteolysis of IGFBP-3 and IGFBP-5 remains to be established.

To evaluate if BMP-7-induced changes in IGFBP protein levels were mediated via corresponding changes in the synthesis of these proteins, the effects of BMP-7 on IGFBP mRNA levels were evaluated by Northern blot analysis. BMP-7 treatment increased steady state levels of IGFBP-3 mRNA by more than 10-fold at 24 h compared to vehicle treated controls, in SaOS-2 and TE85 cells. In addition, BMP-7 treatment increased the mRNA level of stimulatory IGFBP-5 by more than 5-fold in TE85 and SaOS-2 cells. In contrast to stimulatory IGFBP-3 and IGFBP-5, mRNA level for inhibitory IGFBP-4 was decreased by more than 50% after 24 h of treatment with BMP-7 in TE85 and SaOS-2 cells. These data suggest that IGFBP synthesis is a major control point at which the level of IGFBP-3, IGFBP-4 and IGFBP-5 are regulated by BMP-7 in TE85 and SaOS-2 human osteoblast cell model systems.

Because IGF actions could be modulated not only at the level of the IGFs and the IGF binding proteins but also by their receptors, we next determined the effects of BMP-7 on the mRNA level for type I and type II receptors in TE85 cells. We found that BMP-7 treatment had no significant effect on the mRNA levels for either type I or type II IGF receptor (10).

These data suggest that BMP-7 treatment modulates some but not all of the IGF system components in human osteoblasts.

Based on the findings of this study, we have proposed a model to explain the mechanism by which BMP-7 may in part modulate its effects on the proliferation and differentiation of osteoblast line cells (Figure 1). According to this model, BMP-7 treatment can act locally by modulating the IGF regulatory system. Since IGFs have been demonstrated to increase both proliferation and differentiation human osteoblasts, our findings are consistent with the hypothesis that the mitogenic/differentiative effect of BMP-7 on human osteoblasts may be mediated in part via IGF-II by increasing its synthesis, and by regulating the balance between the stimulatory (e.g. IGFBP-5) and inhibitory (e.g. IGFBP-4) classes of IGFBPs both at the level of production (mRNA) and at the level of degradation but not by up-regulating the IGF receptor. Further studies utilizing antisense or blocking antibody technologies against the various IGF system components are needed to examine the extent of the potential contribution of the IGF system in mediating the effects of BMP-7 on human osteoblasts.

Recent findings on the effects of BMP-7 on the expression of various IGF system components in rat osteoblasts revealed that BMP-7 treatment increased expression of IGF-I and IGF-II but decreased expression of IGFBP-4, IGFBP-5 and IGFBP-6 in primary cultures of fetal rat calvaria cells (15). In addition, BMP-2 treatment has also been shown to decrease expression of IGFBP-5 by a mechanism involving transcriptional regulation (16). Thus, BMP-7 treatment increased expression of stimulatory IGFBP-5 in human osteoblasts while both BMP-2 and BMP-7 treatments decreased expression of stimulatory IGFBP-5 in rat osteoblasts. The apparent reasons for this difference in BMP-7 regulation of IGFBP-5 expression between rat and human osteoblasts and the molecular mechanism by which BMP-7 differentially regulates IGFBP-5 expression between rat and human osteoblasts remains unknown at this time.

In addition to the effects on the production of various IGF system components, we also found evidence that BMP-7 treatment increased expression of BMP-6 and decreased expression of BMP-2 and BMP-4 in human osteosarcoma cells (17). These data suggest that BMP-7 actions on osteoblast line cells may involve: 1) complex regulation (both increases and decreases) in the expression of IGF system components; and 2) regulation of expression of members of the BMP family. Further studies are needed to establish the extent to which the various growth factor systems contribute to promoting the effects of BMP-7 on osteoblast cell proliferation and/or differentiation.

3. MOLECULAR MECHANISM BY WHICH BMP-7 INCREASES IGFBP-3 EXPRESSION IN HUMAN OSTEOBLASTS

In studies on how BMP-7 effects various members of the IGF system, we chose IGFBP-3 expression as an end point because BMP-7 consistently produced more than 10-fold increase in IGFBP-3 mRNA and protein. To begin to delineate the mechanism by which BMP-7 regulates steady state levels of IGFBP-3 mRNA, time course effects of BMP-7 were evaluated on IGFBP-3 mRNA level in SaOS-2 cells. These studies revealed that BMP-7 treatment caused a significant increase in IGFBP-3 mRNA levels with in 6 h of BMP-7 treatment. A maximal increase in IGFBP-3 mRNA of more than 10-fold was seen at 24 h. To investigate whether BMP-7 treatment increased IGFBP-3 expression by a mechanism that

Table 1
Effect of BMP-7 on the production of IGF system components in SaOS-2 and TE85 cells

IGF system component	Function	Change versus control
IGF-II	Stimulatory	>2-fold ↑
IGFBP-3	Stimulatory	>10-fold ↑
IGFBP-4	Inhibitory	>50% ↓
IGFBP-5	Stimulatory	>5-fold ↑
Type I IGF receptor	Stimulatory	No effect

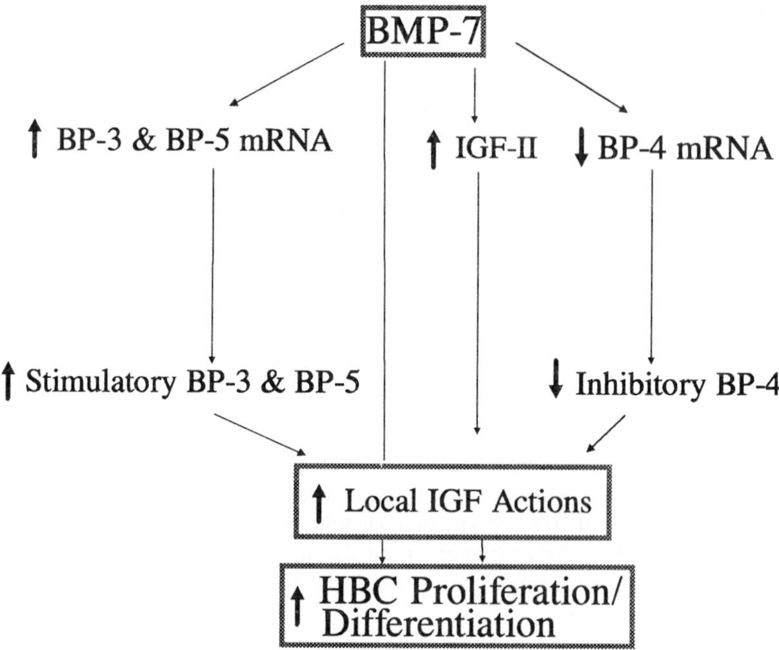

Figure 1: Regulation of IGF system components by BMP-7. Multifactorial model to explain the mechanism by which BMP-7 may in part modulate its effects on the proliferation and/or differentiation of osteoblast line cells.

involved cytosolic transcript stabilization, SaOS-2 cells were treated with 5,6-dichloro-1-β-D-ribofuranosylbenzimidazole (inhibitor of RNA polymerase-II) for measurement of the decay of IGFBP-3 mRNA over a 24 h period. In these experiments, the SaOS-2 cells were pretreated for 24 h with BMP-7 prior to DRB administration based on the time course study which showed that the maximal effect of BMP-7 on IGFBP-3 mRNA level occurred at 24 h. We found that BMP-7 treatment did not produce the anticipated increase in stabilization of IGFBP-3 mRNA but, in fact, decreased the half-life of IGFBP-3 mRNA. This effect of BMP-7 on the half-life of cytosolic IGFBP-3 transcript appears to be specific since BMP-7 had no effect on the half-life of IGFBP-4 mRNA (18). Because BMP-7 caused an acute increase in IGFBP-3 mRNA level and because BMP-7 did not cause an increase in the half-life of IGFBP-3 mRNA, we predicted that BMP-7 effects on IGFBP-3 mRNA level was due to increased transcription. To examine this, we evaluated the effects of BMP-7 on IGFBP-3 nuclear transcripts by using RT-PCR. After separation of RT-PCR products by Southern blot, IGFBP-3 transcripts were detected by ^{32}P labeled IGFBP-3 specific oligonucleotide probes. These results demonstrated that BMP-7 treatment increased IGFBP-3 pre-mRNA level by 6-fold within 6 h of treatment (18). To examine if de novo protein synthesis is required to promote stimulatory effects of BMP-7 on IGFBP-3 nuclear transcript and mRNA production, SaOS-2 cells were pretreated with cycloheximide prior to administering BMP-7. It was found that cycloheximide pretreatment completely abolished the stimulatory effect of BMP-7 on IGFBP-3 nuclear transcript level. In contrast to IGFBP-3, pre mRNA levels of β-actin control gene remained unaltered. These findings are consistent with the hypothesis that BMP-7 promotes IGFBP-3 production by a transcriptional mechanism that requires synthesis of new proteins.

To examine whether putative BMP-7 responsive cis-elements may reside within the promoter domain of the IGFBP-3 gene, we evaluated the effects of BMP-7 on IGFBP-3 promoter activity using CAT reporter vectors that contained either 0.4- or 0.8 kb of the 5'-flanking region. In these experiments, BMP-7 did not stimulate IGFBP-3 proximal promoter activity in either cell line, thus suggesting that BMP-7 responsive domains are located either beyond the currently established 5'-flanking region, or within internal exon/intron regions of the IGFBP-3 gene. Further studies are needed to identify the BMP-7 responsive domain(s) within the IGFBP-3 gene that mediate the transcriptional effects of BMP-7 on IGFBP-3 expression in human osteoblasts.

4. EVALUATION OF SIGNALING MECHANISM BY WHICH BMP-7 MEDIATES ITS EFFECTS ON IGFBP-3 EXPRESSION IN HUMAN OSTEOBLASTS

To identify the potential second messenger signal pathway by which BMP-7 modulates its effects on osteoblasts, we focused our efforts on two major signaling pathways which have been shown to be utilized by several known growth factors (19,20). They are: 1) protein tyrosine kinases and phosphatases; and 2) protein serine/threonine kinases and phosphatases. To evaluate the involvement of these 2 pathways, we determined the effects of BMP-7 in the presence of known inhibitors of kinases and phosphatases. In our studies on BMP-7 signaling mechanism, we compared the effect of BMP-7 with TGF-β1 using MG63 cell line, a cell line which we found to be responsive to both of these growth factors.

Table 2
Effects of inhibitors and activators of signal transduction events on IGFBP-3 production in MG63 cells

Inhibitor	Signal pathway	BMP-7 effect	TGFβ effect
Phorbol myrystyl acetate	PKC activator	Decrease	Increase
Staurosporine	PKC inhibitor	Increase	Decrease
Genistein	Tyrosine kinase inhibitor	No effect	No effect
Okadoic acid	Serine/threonine phosphatase inhibitor	Increase	Decrease
Vanadate	Phosphotyrosine protein phosphatase inhibitor	Increase	Decrease

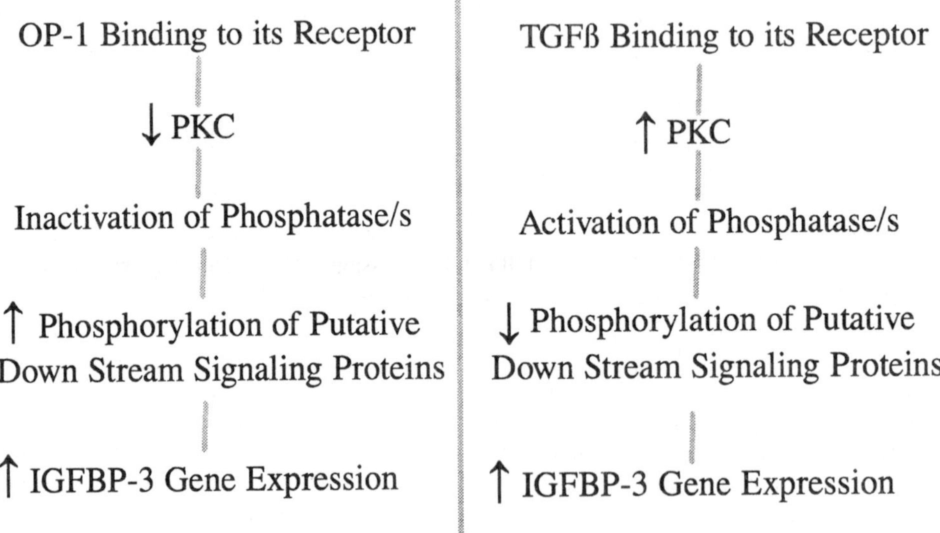

Figure 2: Models of BMP-7 and TGFβ1 signal transduction pathways. Hypothetical models to explain how BMP-7 and TGFβ may mediate their effects on IGFBP-3 gene expression (see text for details).

Since the signal transduction pathway utilized by both, BMP-7 and TGFβ1, involves activation of a receptor containing a serine/threonine kinase, we examined whether inhibitors or activators of PKC would alter BMP-7-stimulated increase in IGFBP-3 expression. We found that treatment of MG63 cells with staurosporine, an inhibitor of PKC, potentiated the stimulatory effect of BMP-7 on IGFBP-3 expression while it completely blocked the stimulatory effect of TGFβ1. Staurosporine treatment had no significant effect on IGFBP-3 expression in control cultures (i.e. not treated with BMP-7 or TGFβ1). Similar results were obtained at the mRNA level (21). Consistent with the staurosporine effect, pretreatment of MG63 cells with PMA, an activator of PKC, inhibited BMP-7 induced increase in IGFBP-3 expression while it potentiated TGFβ1 induced increase in IGFBP-3 expression. Our data with the inhibitors and activators of PKC suggest that the effect of BMP-7 and TGFβ1 on IGFBP-3 expression involve differences in PKC signaling processes.

To determine if BMP-7 and TGFβ1 mediate their effects on IGFBP-3 expression, in part, via regulating serine/threonine phosphatases, we investigated the effect of okadoic acid, a serine/threonine phosphatase inhibitor. Okadoic treatment potentiated the BMP-7 effect but blocked the inhibitory effect of TGFβ1 on IGFBP-3 expression (21). These data suggest that BMP-7 and TGFβ1 may mediate their effects on IGFBP-3 expression by regulating both serine/threonine kinase and phosphatase.

In contrast to PKC inhibition, treatment of MG63 cells with genistein, an inhibitor of tyrosine kinase, had no significant effect on BMP-7 or TGFβ1-induced increase in IGFBP-3 expression. These data suggest that BMP-7 and TGFβ1 effect on IGFBP-3 expression may not involve modulation of tyrosine kinase activity. On the other hand, pretreatment of MG3 cells with vanadate, an inhibitor of phosphotyrosine protein phosphatase, potentiated BMP-7 effect but inhibited TGFβ1 effect on IGFBP-3 expression. These data suggest that BMP-7 and TGFβ1 effect on IGFBP-3 expression may involve modulation of protein tyrosine phosphatase activity.

Table 2 summarizes the effects of inhibitors and activators of signal transduction events on IGFBP-3 expression. Two major conclusions can be drawn from these data: 1) those agents that increase the BMP-7 effect on IGFBP-3 expression decrease the TGFβ1 effect, while those agents that inhibit the BMP-7 effect increase the TGFβ1 effect on IGFBP-3 expression; and 2) BMP-7/TGFβ1-induced increase in IGFBP-3 expression may involve modulation of not only serine/threonine kinase but also serine/threonine phosphatase and protein tyrosine phosphatase activity.

Based on our findings, we have proposed the following hypothetical models (Figure 2) to explain how BMP-7 and TGFβ1 may mediate their effects on IGFBP-3 expression. We propose that the binding of BMP-7 to its receptors leads to inactivation of receptor kinase and/or upstream signaling serine/threonine kinases. The signaling cascade initiated by BMP-7 leads to an increase in the phosphorylation of serine/threonine and tyrosine containing down stream putative signaling proteins (via inhibiting phosphatases) resulting in the activation of IGFBP-3 gene. In contrast, we propose that the binding of TGFβ1 to its receptors leads to activation of receptor kinase and/or upstream signaling serine/threonine kinases. We also propose that phosphorylation of TGFβ1 receptors activates a signaling cascade which leads to activation of both serine/threonine phosphatase and protein tyrosine phosphatase, leading to dephosphosphorylation of one or more phosphorylated down stream signaling proteins ultimately resulting in the activation of the IGFBP-3 gene. Further studies would verify whether the newly identified Mothers against Drosophila Decapentaplegic (MAD) and TGFβ-associated kinase (TAK) proteins are involved in mediating BMP-7/TGFβ1 effects on IGFBP-3 expression (19,20,22).

5. CONCLUSION

Studies in our laboratory and other laboratories provide evidence that BMPs regulate production of multiple components of the IGF regulatory system in human osteoblasts. In addition, we have found by indirect evidence that the signaling pathway by which BMP-7 mediates its effects on the expression of IGFBP-3 is different from that of TGFβ1. Because BMP-7 and TGFβ1 are multifunctional factors that regulate many aspects of cellular events including cell migration, attachment, proliferation, differentiation and apoptosis, it is not surprising that the signal transduction pathways for these two growth factor systems are complex. It is anticipated that future studies on the identification of the components of the intracellular BMP signaling pathway may provide clues to how BMP-7 acts to regulate transcription of various genes encoding the IGF system components.

ACKNOWLEDGMENTS

This work was supported by funds from the National Institutes of Health (NIH AR31062) and the Veterans Administration.

REFERENCES

1. Bilezikian, L.G. Raisz and G.A. Rodan (eds.), Principles of Bone Biology, Academic Press, San Diego, 1996.
2. Reddi. Clin. Orth. Rel. Res., 313 (1994) 115-119.
3. Centrella, M.C. Horowitz, J.M. Wozney and T.L. McCarthy. Endocr. Rev., 15 (1994) 27-39.
4. Baylink, R.D. Finkelman and S. Mohan. J Bone Miner. Res., 8 (1993) S565-S572.
5. Knutsen, J.E. Wergedal, T.K. Sampath, D.J. Baylink, and S. Mohan. Biochemical & Biophysical Research Communications, 194 (1993) 1352-1358.
6. Mohan and D.J. Baylink. Horm. Res., 45 (1996) 59-62.
7. Bilezikian and G.A. Rodan (eds.), Principles of Bone Biology, Academic Press, San Diego, 1996.
8. Canalis and D. Agnusdei. Calcif. Tissue Int., 58 (1996) 133-134.
9. Rosen, L.R. Donahue and S.J. Hunter. Insulin-like growth factors and bone. Proc. Soc. Exp. Biol. Med., 206 (1994) 83-102.
10. Knutsen, Y. Honda, D.D. Strong, T.K. Sampath, D.J. Baylink and S. Mohan. Endocrinology, 136 (1995) 857-865.
11. Baxter, P.D. Gluckman and R.G. Rosenfeld (eds.), The Insulin-Like Growth Factors and Their Regulatory Proteins, Exerpta Medica, Amsterdam, 1994.
12. Kanzaki, S. Hilliker, D.J. Baylink and S. Mohan. Endocrinology, 134 (1994) 383-392.
13. Conover. Progress in Growth Factor Research, 6 (1995) 301-309.
14. McCarthy, S. Casinghino, M. Centrella and E. Canalis. J Cell. Physiol., 160 (1994) 163-175.

15. Yeh, M.L. Adamo, A.M. Kitten, M.S. Olson and J.C. Lee. Endocrinology, 137 (1996) 1921-1931.
16. Gabbitas and E. Canalis. Endocrinology, 136 (1995) 2397-2403.
17. Honda, R. Knutsen, D.D. Strong, T.K. Sampath, D.J. Baylink and S. Mohan. Calcif. Tissue Int., 60 (1997) 297-301.
18. Hayden, D.D. Strong, D.J. Baylink, D.R. Powell, T.K. Sampath and S. Mohan. Endocrinology, (in press), 1997.
19. Massagué. Cell, 85 (1996) 947-950.
20. Yamashita, P. Ten Dijke, C.H. Heldin and K. Miyazano. Bone, 19 (1996) 569-574.
21. Srinivasan, D.J. Baylink, T.K. Sampath and S. Mohan. J Cell. Physiol., (in press), 1997.
22. Hoodless, T. Haerry, S. Abdollah, M. Stapleton, M.B. O'Connor, L. Attisano J.L. Wrana. Cell, 85 (1996) 489-500.

Molecular Mechanisms to Regulate the
Activities of Insulin-like Growth Factors
K. Takano, N. Hizuka and S-I. Takahashi (Editors)

Regulation of IGF and IGFBP Gene Expression in Bone

Ryo Okazaki

Third Department of Medicine, Teikyo University School of Medicine,
3426-3 Anesaki, Ichihara, Chiba 299-01, Japan

1. BONE CELLS

Bone cells consist of bone-forming osteoblastic cells and bone-resorbing osteoclastic cells. Little is known about the regulation of IGF and IGFBP gene expression in osteoclastic cells. Thus this article deals with IGF and IGFBP expression in osteoblastic cells.

Osteoblasts derive from bone marrow mesenchymal stromal cells. Chondrocytes, myoblasts, fibroblasts, and adipocytes are also believed to be derived from the same stem cells. The exact mechanisms how the stem cells commit to a specific cell lineage are currently unknown. Commitment to a specific cell linage is followed by a sequential differentiation process. In osteoblastic linage, committed progenitor cells, or pre-osteoblasts, which are rapidly proliferating and elongated in shape, differentiate into spherical mature osteoblasts actively secreting matrix proteins such as type I collagen. As matrix proteins are calcified, osteoblasts begin to secrete a new set of proteins such as osteocalcin. Finally, approximately 15% of osteoblasts are encased in calcified bone and become osteocytes.

This whole commitment-differentiation process is tightly regulated by many systemic and local factors. IGFs are certainly among such factors; targeted disruption of IGF-I or type 1 IGF-receptor delays bone development in mice [1]. Most osteoblastic cells in culture are able to reproduce in vivo differentiation process to a certain degree. In these osteoblastic cultures, both IGF-I and -II stimulate cell replication and type I collagen synthesis.

2. BASAL EXPRESSION OF IGF AND IGFBP

2.1 . IGF Expression in Osteoblastic Cells

Studies using in situ hybridization demonstrated that osteoblasts in vivo express both IGF-I and -II in rats, mice, and humans [2-4]. Osteoblastic cells in vitro also express both IGF-I and -II. We studied various human osteoblastic cell models and reported that basal expression of IGFs varies depending on the cell models [5]. As shown in the Table 1., primary human osteoblast-like (hOB) cells express both IGF-I and -II, but none of the other human osteoblastic cells express both IGFs consistently.

Table 1
Expression of IGF-I and -II in various human osteoblastic cell models

	hOB	HOBIT	SaOS-2	MG-63	TE-85	U-2
IGF-I	+	−	±	−	−	±
IGF-II	++	++	±	−	+	++

When we reported the above observations we speculated transformation might be related to the difference in the IGF expression. Another possibility is that this may reflect difference in the differentiation stage of each cell model. Indeed, using in situ hybridization, Middleton et al. [4] reported that some alkaline phosphatase positive human osteoblasts do not express IGFs and that signal for IGFs are generally more intense in cuboidal active osteoblasts compared to flat lining osteoblastic cells. Furthermore, very little or no signal is detected for IGF mRNAs in bone marrow cells or osteocytes. These in vivo observations are in good agreement with in vitro findings. In primary rat osteoblast-like (ROB) cells and a mouse osteoblast-like cell line, MC3T3-E1, IGF-I and -II mRNA expression increases for a certain period of time which corresponds proliferation and early matrix maturation stages [6, 7]. Increase in IGF expression is followed by a gradual decrease as extending cultures, corresponding late matrix maturation and calcification stages. Interestingly, chondrocytes do not express IGF-I and express only IGF-II in vivo at least in rats and mice despite these cells are derived from same stromal cells as osteoblasts: though human chondrocytes express both IGF-I and -II [2-4].

These results suggest that expression of IGFs are regulated by differentiation/maturation process of osteoblastic cells. Among defined factors that promote osteoblastic differentiation are bone morphogenetic proteins (BMPs). BMP-2 and BMP-7 (osteogenic protein -1) increases both IGF-I and -II expression in ROB [8, 9]. In the latter study, BMP-7 induced alkaline phosphatase activity is inhibited by anti-sense RNA for IGF-I or IGF-II, suggesting BMP-7 effects on differentiation of ROB cells are in part mediated by IGFs.

2.2. IGFBP Expression in Osteoblastic Cells

Osteoblastic cells express multiple IGFBPs but the type of IGFBP expressed is different among species and osteoblastic cell models. ROB cells express IGFBP-2, -3, -4,-5, and -6 but not IGFBP-1, whereas hOB cells express IGFBP-1, -3, -4, -5, and -6 but not IGFBP-2 [10, 11]. In situ hybridization study reveals that mouse osteoblasts express IGFBP-2, -4, -5, and -6 but not IGFBP-1, or -3 [3]. Expression of IGFBP-7 has not been documented in osteoblastic cells so far.

IGFBP expression in osteoblastic cells are also regulated by differentiation process. In ROB, IGFBP-2 and -5 mRNA expression is maximal in proliferating preosteoblasts, whereas IGFBP-3, -4, and -6 expression is maximal in mature osteoblasts [10]. In MC3T3-E1 cells, IGFBP-2,-4, and -5 mRNA expression increases time-dependently [7]. However, these changes in mRNA expression do not always reflect protein level. For example, in MC3T3-E1 cells, medium IGFBP-5 level is increased up to 12 days of culture along with its mRNA level; the protein level is decreased thereafter despite continuous increase in mRNA level due to increased IGFBP-5 protease activity [12].

3. REGULATION OF IGF AND IGFBP

IGF and IGFBP expression in osteoblastic cells are regulated many defined factors. Table 2. is a partial list of ever expanding information available on the subject.

Table 2
Regulation of IGF and IGFBP in osteoblastic cells

	cAMP	PK-C	E2	GC	VD	BMPs	TGF-ß
IGF-I	↑	↓	→~↑	↓	→	↑	↓↑#
IGF-II	→	→	→	→	→	↑	↓→
IGFBP-1				↑			
IGFBP-2		↑				→*	
IGFBP-3	↑		→	↓		↑*	
IGFBP-4	↑		↑	↓	↑	↓	↓
IGFBP-5	↑			↓	→	↓↑*#	↓
IGFBP-6				→~↑		↓*	↓

VD: $1,25(OH)_2D$
#:opposite responses are reported depending on cell models tested
*: reported only for BMP-7 (OP-1)

Among defined factors that affect IGF/IGFBP gene expression, some have consistent effects regardless of the cell models (e.g. cAMP upregulation of IGF- I expression). However, in some cases opposite responses have been reported (e.g. BMP-7 effect on IGFBP-5 expression; TGF-ß effect on IGF-I expression) [9,13-15]. The reasons for these variations are not known; differences in the species and/or the maturation stage of each cell model should be in part accounted for.

3.1. cAMP pathway

Activation of protein kinase A (PKA) and subsequent increase in intracellular cAMP mediates many hormonal stimuli. In osteoblastic cells, two most important and well studied biological agents that activate cAMP pathway are parathyroid hormone (PTH) and prostaglandin E_2 (PGE_2). Activation of cAMP pathway consistently increases IGF-I mRNA level without affecting IGF-II mRNA expression in ROB, hOB, and MC3T3-E1 cells [15-17]. PKA activation leads to an increase in both type 1 and type 2 IGF-I transcripts [15, 18]. Thomas et al. [19] identified a cAMP response element (CRE) within exon 1 of rat IGF-I gene.

This CRE has no sequence homology with classical CRE. Whether this element is responsible for an increase in type 2 IGF-I transcripts in response to cAMP is not known.

PKA activation also affects IGFBP expression. It increases IGFBP-3, -4, and -5 mRNA expression in various types of osteoblastic cells [20,21].

3.2. Protein-kinase C (PK-C) pathway

PK-C pathway is another important pathway that mediates various hormonal action. PK-C activation by TPA or PGF2α causes a decrease in IGF-I mRNA level without affecting IGF-II expression in ROB and MC3T3-E1 cells [16, 17, 22]. TPA increases IGFBP-2 mRNA level in MC3T3-E1 cells [17].

3.3. Estrogen

Estrogen is one of the most important hormones in the maintenance of bone metabolism. 17ß-estradiol (E_2) increases IGF-I mRNA level by 2-fold in neonatal rat osteoblast-like cells [23]. However, this is not observed in fetal rat osteoblast-like cells or hOB cells [24, 25]. Because this difference may be due to low number of estrogen receptor (ER) in osteoblastic cells, Kassem et al. [26] tested estrogen and various anti-estrogen effects on IGF-I expression in a human fetal osteoblast-like cell line that express high number of ER (FOB/ER9). Assessed with semi-quantitative RT-PCR, E_2 increases IGF-I mRNA level up to 3-fold. Pure (type 2) anti-estrogens, ICI 182,780 and ICI 164,384, completely blocked E_2 induction of IGF-I expression.

Although consensus estrogen responsive element (ERE) has not been identified in any of the IGF-I genes so far cloned, Umayahara et al. [27] identified an ERE within upstream of exon 1 of chicken IGF-I gene, using HepG2 human hepatoma cell line. This element has no sequence homology with classical consensus ERE; the AP-1 like motif therein is essential to confer estrogen response. Whether or not this ERE is also active in osteoblastic cells is not known. Recently, McCarthy et al. [24] demonstrated that E_2 potently suppresses cAMP induced , but not basal, rat IGF-I promoter activity using the same system in which CRE was identified. However, whether or not E_2 suppresses cAMP- induced increase in IGF-I mRNA level is not known. In FOB/ER9 cells, E_2 does not suppress cAMP induction of IGF-I gene expression [26].

In contrast to the complex regulation of IGF-I by E_2, estrogen or anti-estrogen has no effect on IGF-II expression in osteoblastic cells.

Estrogen also affects IGFBP expression in osteoblastic cells. In FOB/ER9 cells, E_2 increases IGFBP-4 mRNA level without affecting IGFBP-3 mRNA expression [28]. As in the case of IGF-I, E_2 induction of IGFBP-4 mRNA is completely blocked by type 2 anti-estrogens but not by progesterone.

3.4. Glucocorticoid (GC)

Glucocorticoid has profound effects on bone cell metabolism. In osteoblastic cells, GC appears to promote osteoblastic maturation at an earlier stage as exemplified by alkaline phosphatase induction whereas inhibit it at a later stage by suppressing type 1 collagen and osteocalcin production [29]. GC has also been shown to potentiate IGF action in several osteoblastic cells including hOB cells [30].

In hOB cells, GC causes a decrease in IGF-I mRNA level without affecting IGF-II expression [31]; GC decreases IGFBP-3, -4, and -5 mRNA levels , increases IGFBP-1 level and has little effect on IGFBP-6 mRNA level [11]. In ROB cells, Gabbitas et al. [32] reported that GC increases IGFBP-6 expression through a transcriptional mechanism.

4. SUMMARY

Osteoblastic cells are one of the most extensively studied cell systems for the regulation of IGF and IGFBP gene expression. Information available is so vast and sometimes conflicting, that it is difficult to draw an overall scheme how it relates to bone cell physiology. However, most agents that have some effects on osteoblastic cells also regulate IGF/IGFBP gene expression in the same cells. This suggests that orchestrated regulation of IGF/IGFBP gene expression, and their action, play a major role in bone. Conflicting results may lead to an elucidation of transcriptional (co-)factors that determine differentiation stages of osteoblasts. Further characterization of complex IGF/IGFBP gene regulation in each bone cell system would help our understanding of bone cell physiology as well as pathophysiology.

REFERENCES

1. Baker, J., et al. Cell, 75,(1993) 73 .
2. Shinar, D.M., et al. Endocrinology, 132,(1993) 1158 .
3. Wang, E., et al. Endocrinology, 136,(1995) 2741 .
4. Middleton, J., et al. Bone, 16,(1995) 287 .
5. Okazaki, R., et al. J. Bone Miner. Res., 10,(1995) 788 .
6. Birnbaum, R.S., et al. J. Endocrinol., 144,(1995) 251 .
7. Thrailkill, K.M., et al. Bone, 17,(1995) 307 .
8. Canalis, E.and Gabbitas, B. J. Bone Miner. Res,. 9,(1994) 1999 .
9. Yeh, L.C., et al. Endocrinology, 137,(1996) 1921 .
10. Birnbaum, R.S.and Wiren, K.M. Endocrinology, 135,(1994) 223 .
11. Okazaki, R., et al. Endocrinology, 134,(1994) 126 .
12. Thrailkill, K.M., et al. Endocrinology, 136,(1995) 3527 .
13. Knutsen, R., et al. Endocrinology,136,(1995)857.
14. Canalis, E., et al. Endocrinology, 133,(1993) 33 .
15. Okazaki, R., et al. Biochem. Biophys. Res. Commun., 207,(1995) 963 .
16. McCarthy, T.L., et al. J. Biol. Chem., 265,(1990) 15353 .

17. Hakeda, Y., et al. J. Cell. Physiol., 158,(1994) 444 .
18. Pash, J.M., et al. Endocrinology, 136,(1995) 33 .
19. Thomas, M.J., et al. J. Biol. Chem., 271,(1996) 21835 .
20. LaTour, D., et al. Mol. Endocrinol., 4,(1990) 1806 .
21. Pash, J.M.and Canalis, E. Endocrinology, 137,(1996) 2375 .
22. Hakeda, Y., et al. J. Biol. Chem., 266,(1991) 21044 .
23. Ernst, M.and Rodan, G.A. Mol. Endocrinol., 5,(1991) 1081 .
24. McCarthy, T.L., et al. J.Biol. Chem., 272,(1997) 18132 .
25. Okazaki, R. Unpublished observations.
26. Kassem, M., et al. Calcif. Tissue. Int., (in press) .
27. Umayahara, Y. J. Biol. Chem., 269,(1994) 16433 .
28. Kassem, M., et al. Proc. Assoc. Am. Physicians., 108,(1996) 155 .
29. Subramaniam, M., et al. J. Cell. Biochem., 50,(1992) 411 .
30. Kream, B.E., et al. Endocrinology, 126,(1990) 1576 .
31. Swolin, D., et al. J. Endocrinol., 149,(1996) 397 .
32. Gabbitas, B.and Canalis, E. Endocrinology, 137,(1996) 1687 .

Molecular Mechanisms to Regulate the
Activities of Insulin-like Growth Factors
K. Takano, N. Hizuka and S-I. Takahashi (Editors)

The IGF System in the Ovary

G. F. Erickson, T. Kubo, D. Li, H. Kim and S. Shimasaki

Department of Reproductive Medicine, University of California at San Diego, La Jolla, California

STATEMENT OF THE PROBLEM

In mammals, the ovaries are responsible for the production of dominant follicles (DF) which ovulate eggs into the oviduct to be fertilized (1). After ovulation, the DF transforms into the corpus luteum which is responsible for producing hormones which maintain pregnancy. In this capacity, the development of the DF is the basis for fertility in the female mammal. Given the importance of this event, understanding how a DF develops in the ovaries is a major goal of reproductive research. Follicle stimulating hormone (FSH) is obligatory for DF formation and no other ligand, by itself, can serve in this regulatory capacity. Consequently, to understand the control of DF development, it is necessary to understand the mechanisms involved in the regulation of FSH action in target granulosa cells (GC).

There is now substantial evidence that the IGF growth factor family, replete with ligands, binding proteins (IGFBPs) and proteases plays a central role in determining FSH action in the murine ovary (2-4). Research done in the last few years has provided compelling evidence to support the concept that IGF-I is an essential ligand for FSH to exert its inductive effect on DF formation in the ovary (5, 6). There is also compelling evidence that FSH-induced DF formation requires the expression of inhibin, specifically the expression of the Inh-α gene (7). Collectively, this evidence argues that the stimulation of DF development by FSH requires the action of two growth factors, IGF-I and inhibin.

The question considered in this chapter concerns to what extent the endogenous IGF-I signaling pathway in GC might play a role in the regulated production of Inh-α. These experiments show that IGF-I signaling is absolutely necessary for ligand-dependent increases in Inh-α production and that the increases are completely blocked by IGFBP-4 and -5. Based on these results, we suggest that the endogenous IGF-I/IGFBP system in GC may be at or near the apex in the regulatory mechanisms that determine whether a follicle develops or dies by atresia in the murine ovary.

BACKGROUND INFORMATION

There is compelling evidence from gene knockout experiments that IGF-I and Inh-α are essential for DF formation and fertility in the mouse. IGF-I or inhibin nullizygotic female mice survive to adulthood, but are infertile (5-7). Although the ovaries of these animals are filled with large numbers of secondary preantral follicles, there are no DF, and hence, no estrous cycles (6, 7). A striking feature of the preantral follicles in the IGF-I and inhibin deficient animals is that they are resistant to stimulation with super physiological doses of FSH (6, 7). Consequently, the concept has emerged that IGF-I and Inh-α are critically important in the mechanism by which FSH stimulates DF formation and GC cytodifferentiation.

There have been important advances in our understanding of the role of IGF-I follicle development. Evidence from studies with rat GC has demonstrated the existence of an autocrine/paracrine IGF-I loop which is necessary for FSH action. Throughout the development of the DF, the GC express IGF-I and receptor proteins (8, 9). This intrinsic autocrine/paracrine pathway acts broadly to increase FSH-dependent functions in GC (10). Insight into the modulation of this IGF-I autocrine/paracrine loop came from the discovery that rat GC destined to die by apoptosis express two IGF binding proteins, IGFBP-4 and -5, (2). The significance of this finding was emphasized by the fact these IGFBPs inhibit FSH-dependent steroidogenesis in rat GC by neutralizing endogenously produced IGF-I (11). This observation implied that the IGF-I autocrine/paracrine loop is essential for FSH-induced steroidogenesis, and that IGFBP-4 and -5 produced by the GC themselves can sequester IGF-I, thereby, blocking FSH action.

The synthesis of Inh-α is also regulated by FSH and IGF-I (12). Thus, an interesting question arises: To what extent is the endogenous IGF-I signaling pathway obligatory for FSH-stimulated Inh-α gene activity? The purpose of this study is to investigate this question.

THE OBLIGATORY ROLE OF ENDOGENOUS IGF-I IN INH-α EXPRESSION

As a way of understanding this problem, a series of *in vitro* experiments was performed with primary cultures of rat GC grown in serum free medium. The GC were treated for 48 h with FSH, IGF-I, and/or a variety of anti-IGF-I molecules, after which Inh-α protein and mRNA levels were measured in both conditioned medium (CM) and cells by Western immunoblotting and Northern analysis respectively.

In our initial experiments, we found that FSH stimulated basal levels of 18 KDa Inh-α protein in a dose-dependent manner (ED_{50} = 0.4 ± 0.1 ng/ml) with maximal increases occurring at 3 ng/ml FSH. Inh-α was also stimulated in a dose-dependent fashion by treatment with IGF-I (ED_{50} = 10 ± 2 nM). The stimulatory effects of FSH and IGF-I were time-dependent with maximal increases occurring at 48 to 72 h. These results extend previous findings (12) and demonstrate that FSH and IGF-I exert a stimulatory effect on the synthesis of Inh-α by cultured rat GC.

To address the role of endogenously produced IGF-I in this response, we tested the effects of rat IGFBP-4 and 5 on FSH-stimulated Inh-α production. As shown in Fig 1A, Inh-α stimulation by FSH was inhibited by IGFBP-4 (IC_{50} =14 ± 2 nM) and IGFBP-5 (IC_{50} = 6 ± 2 nM). Concentrations of 30 and 100 nM IGFBPs produced over a 90% inhibition of Inh-α production. We also examined whether IGF-I could reverse these inhibitory effects. As shown in Fig. 1B, IGF-I produced a dose-dependent reversal of the inhibitory effects of IGFBP-4 and -5 in which > 30 nM IGF-I caused a near 100% reversal.

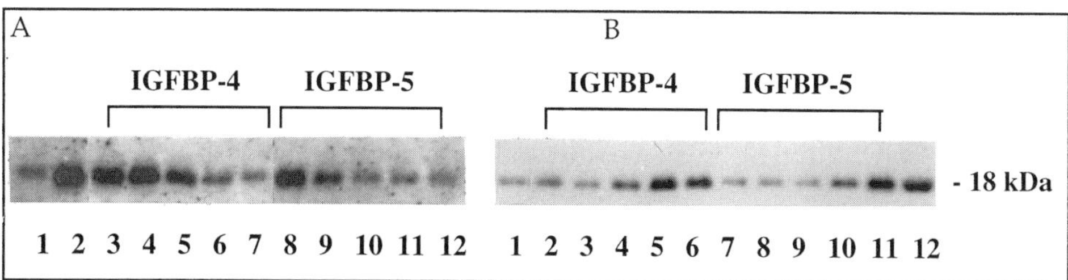

Figure 1. Inhibition of FSH-induced Inh-α production by IGFBP-4 and -5. Panel A) Granulosa cells were cultured for 48 h with a saturating dose of FSH and the indicated concentrations of rat IGFBP-4 or -5 and/or IGF-I. After culture, Inh-α in cell-free supernatants was analyzed by Western immunoblotting under reducing conditions. Panel A) Granulosa cells were cultured for 48 h with a saturating dose of FSH (3 ng/ml), with either IGFBP-4 or IGFBP-5. 1 (control); 2 (FSH); 3-7 (FSH + IGFBP-4, 1, 3, 10, 30, 100 nM); 8-12 [FSH + IGFBP-5 (1, 3, 10, 30, 100 nM)]; Panel B) effects of IGF-I on IGFBP inhibition. 1 (control); 2-6 [FSH + 30 nM IGFBP-4 + IGF-I (1, 3, 10, 30, 100 nM)]; 7-11 [FSH + 30 nM IGFBP-5 + IGF-I (1, 3, 10, 30, 100)]; 12 (FSH).

We next investigated the possible role of endogenous IGF-I in the mechanisms by which IGFBPs inhibit Inh-α production. This was accomplished using an anti-IGF-I antibody (Fig. 2). Treatment with increasing doses of IGF-I antiserum reduced the ability of a saturating dose of FSH to stimulate Inh-α production, with 30 μl causing complete inhibition. By contrast, treatment with 30 μl normal rabbit serum had no effect indicating that the inhibition is not due to endogenous IGFBP activity in the serum. These results with IGFBPs and anti-IGF-I antibody support an obligatory function of endogenously produced IGF-I in FSH dependent Inh-α production.

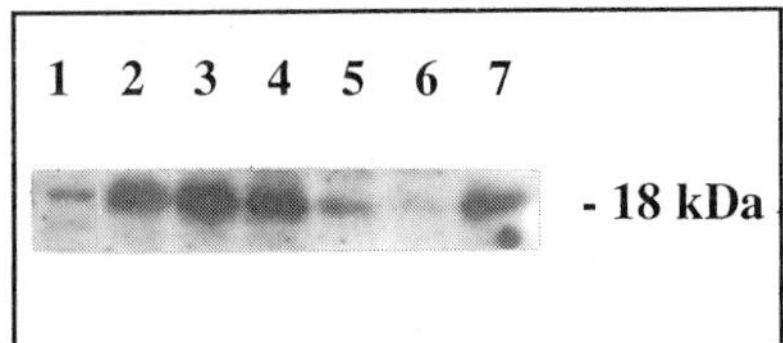

Figure 2. Inhibitory effects of anti-IGF-I antiserum on Inh-α production. Granulosa cells were cultured for 48 h with 3 ng/ml of FSH alone or together with the indicated concentrations of anti-IGF-I polyclonal antibody or control normal rabbit serum. After culture, the levels of 18 KDa Inh-α were determined by Western immunoblotting. 1 (control); 2 (3 ng/ml FSH) ; 3-6 [FSH + anti-IGF-I antibody (1, 3, 10, 30 μl)]; 7 (normal rabbit serum).

To investigate the possible role of endogenous IGF-I in regulating Inh-α gene activity, cells were treated either alone or together with FSH, IGF-I, IGFBP-4 or -5. After culture the amount of Inh-α mRNA was measured by Northern blotting. As shown in Fig. 3, a low basal level of expression of Inh-α mRNA was detected in control (untreated) cells, suggesting that a small amount of Inh-α is produced constitutively by the control (untreated) cells. FSH and IGF-I caused a significant stimulation of Inh-α mRNA levels when compared control cells. Interestingly, the FSH-induced increases in Inh-α mRNA were blocked by co-treatment with a maximally effective dose (30 nM) of IGFBP-4 or -5. These inhibitory effects were completely reversed by a super saturating dose (100 nM) of IGF-I. Therefore, it appears that FSH may act via IGF-I to achieve its stimulatory effect on Inh-α gene activity, and that the addition of IGFBP-4 or -5 results in an inhibition of FSH activity.

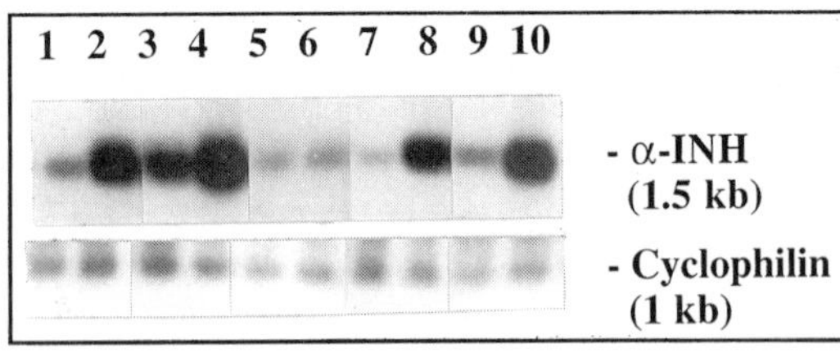

Figure 3. Effects of FSH, IGF-I, IGFBP-4 and IGFBP-5 on Inh-α mRNA steady state levels. Granulosa cells were cultured for 48 h with FSH, IGF-I, IGFBP-4 and/or IGFBP-5, after which total RNA was isolated and analyzed by Northern blotting. Upper panel; typical Northern blot of the 1.5 kb Inh-α mRNA. Lower panel; 1.0 kb cyclophilin mRNA. 1 (control); 2 (3 ng/ml FSH); 3 (100 nM IGF-I); 4 (FSH + IGF-I); 5 (30 nM IGFBP-4); 6 (30 nM IGFBP-5); 7 (FSH + IGFBP-4); 8 (FSH + IGFBP-4 + IGF-I); 9 (FSH + IGFBP-5); 10 (FSH + IGFBP-5 + IGF-I).

To rule out the possibility that the inhibitory effects of IGFBP-4 and -5 are caused by changes in FSH binding to its receptor, we tested the effects of the IGFBPs on Inh-α production by cells treated with forskolin, cholera toxin and 8-Br-cAMP instead of FSH. Each of these molecules substantially increased Inh-α production and the increases were completely abolished by the presence of 30 nM IGFBP-4 or -5 (data not shown). We suggest, therefore, that the IGFBPs must exert these inhibitory effects on FSH signaling at a step downstream from cyclic AMP (cAMP).

To further explore the mechanism of the IGF-I signaling in the stimulation of Inh-α production, cells were treated with various stimulatory molecules in the presence and/or absence of a protein tyrosine kinase (PTK) inhibitor, tyrphostin A23 (13, 14). As shown in Fig. 4, a complete inhibition of the stimulation of Inh-α production by FSH (3 ng/ml) and IGF-I (30 nM) occurred following treatment with A23, whereas the inactive isomer, tyrphostin A63, was without effect. Furthermore, the A23 also prevented the stimulation of Inh-α production by forskolin, cholera toxin or 8-Br-cAMP, while A63 had no effect. These observations fit the prediction that PTK activity, perhaps at the level of IGF-I receptor autophosphorylation, is obligatory for FSH- and cAMP-dependent increases in Inh-α production by rat GC.

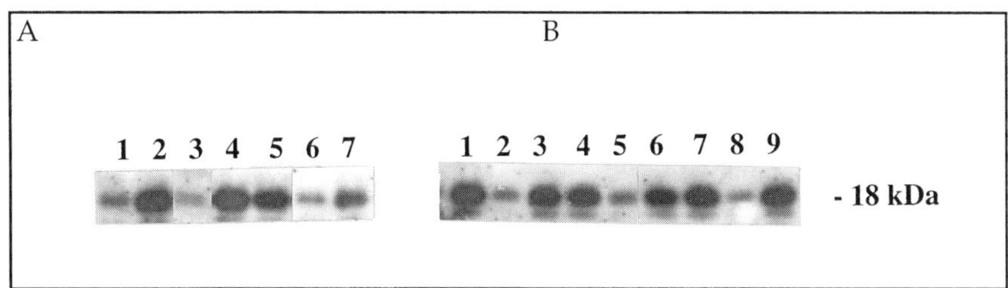

Figure 4. Inhibitory effects of tyrphostin A23 on Inh-α. Panel (A) Granulosa cells were cultured for 48 h with 3 ng/ml of FSH or 30 nM IGF-I with or without 30 μM of the protein tyrosine kinase inhibitor A23 or its inactive isomer A63. 1 (control); 2 (FSH); 3 (FSH + A23); 4 (FSH + A63); 5 (IGF-I); 6 (IGF-I + A23); 7 (IGF-I + A63). Panel (B) Granulosa cells were cultured for 48 h with 3 μM forskolin (FK) or 0.3 nM cholera toxin (CT) or 1 mM 8-Br-cAMP (8-Br) either alone or together with 30 μM A23 or A63. 1 (FK); 2 (FK + A23); 3 (FK + A63); 4 (CT); 5 (CT + A23); 6 (CT + A63); 7 (8-Br); 8 (8-Br + A23); 9 (8-Br + A63). Data are typical Western immunoblots.

Like FSH, there is evidence that activin can also exert stimulatory effects on Inh-α production (15). Therefore, we investigated whether an IGF-I autocrine/paracrine mechanism might also underlie the activin effects on Inh-α production. Using our *in vitro* GC model, we found that treatment with recombinant activin A stimulated Inh-α expression, and the effects were dose-dependent (ED_{50} = 48.3 ± 0.5 nM). As with FSH, co-treatment with A23, completely abolished the activin-dependent Inh-α expression, whereas the inactive isomer, A63, had no effect. The stimulatory effect of activin was also blocked in a dose-dependent manner by added IGFBP-4 (IC_{50} = 3.5 ± 0.5 nM) or IGFBP-5 (IC_{50} = 5.7 ± 1.9 nM), and the effects were reversed by IGF-I (Fig. 5). Finally, treatment with an anti-IGF-I antibody mimicked the inhibitory effects the IGFBPs on activin-stimulated Inh-α expression (data not shown). Thus, these results support the conclusion that an endogenous IGF-I plays an essential role in activin stimulation of Inh-α gene activity.

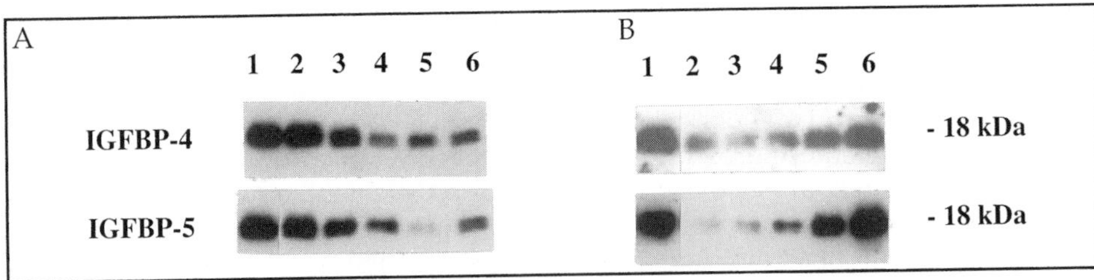

Figure 5. Inhibitory effects of IGFBPs on Inh-α production stimulated by recombinant human activin A (300 ng/ml). Granulosa cells were cultured for 48 h with the indicated proteins and the amount of 18 KDa Inh-α in conditioned medium was measured by Western immunoblotting. (A) Effects of graded doses of IGFBP-4 and -5. 1 (activin); 2-5 (activin + IGFBPs, 1, 3, 10, 30 nM); 6 (control). (B) Reversibility of the IGFBP-4 and -5 (30 nM) inhibitory effects on Inh-α production by increasing doses of IGF-I. 1 (activin); 2-6 (activin + IGFBP + IGF-I, 3, 10, 30, 100 nM).

STATEMENT OF CONCLUSION

Two major conclusions emerge from these studies: 1) the IGF-I receptor signaling in rat GC is obligatory for activin- and FSH-dependent increases in Inh-α expression, and 2) IGFBP-4 and -5 are potent inhibitors of Inh-α production by FSH and activin. These findings lead us to propose that the IGF-I receptor signaling pathway is the major route of increased Inh-α gene activity in rat GC, while FSH and activin signaling serve to enhance or amplify the synthesis induced by IGF-I. Given the fact that IGF-I, FSH and Inh-α are critical determinants of DF development, our observations support the idea that the regulated expression of the intrinsic IGF-I/IGFBP system may be placed at or near the apex in the control pathway governing DF development in the murine ovary.

A schematic model outlining this hypothesis is shown in Fig. 6. Three ligands (at least) have a marked stimulatory effect on Inh-α gene expression, IGF-I, activin and FSH. We propose that there exists a hierarchy in the ligand regulated Inh-α production in which IGF-I receptor signaling appears dominant. This proposal is supported by the evidence that the FSH- and activin-stimulation of Inh-α can be completely abolished by the presence of IGFBP-4, IGFBP-5, anti-IGF-I antibody, or the PTK receptor inhibitor, A23. These observations argue that an IGF-I autocrine/paracrine loop is absolutely necessary for activin and FSH-stimulated Inh-α gene activity. Perhaps the simplest explanation of these findings is that the activin and FSH signaling pathways do not act directly to induce Inh-α production, but rather act to amplify the signals elicited by IGF-I signaling. If true, then FSH and activin would be amplifiers rather than inducers of inhibin production. It follows from this reasoning that IGF-I, but not FSH or activin, is the primary regulatory mechanism responsible for causing the activation of Inh-α gene expression in rat GC.

There are two possible mechanisms that could explain our results. One is that the activin and FSH could increase the stimulatory effects of IGF on Inh-α production by increasing the levels of endogenously produced IGF-I and/or up-regulate the number of IGF-I receptors. Another possibility is that the FSH and activin signaling could directly potentiate the IGF-I signaling pathway that eventually leads to the stimulation of Inh-α gene activity. Regardless of the mechanism of the IGF-I, activin, and FSH cross-talk, an important modulatory role of IGFBP-4 and -5 in controling Inh-α production is clear: these proteins appear to act by preventing IGF-I from binding to its receptor and thereby preventing the IGF-I stimulation of a PTK cascade. Thus, we propose that the levels of IGFBP-4 and -5 in follicular fluid could be a critical determinant of inhibin production, which in turn could serve to determine follicle development.

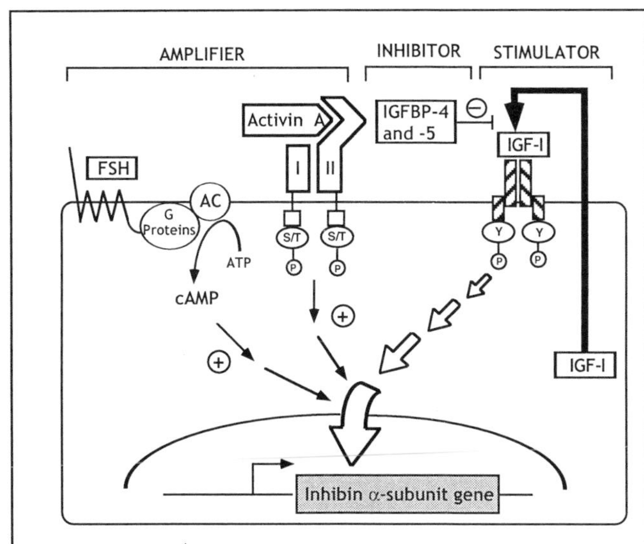

Figure 6. Schematic model of the regulation of Inh-α expression in rat GC by IGF-I, Activin, and FSH. Endogenous IGF-I interacts with its receptor activating PTK which phosphorylates (P) specific tyrosine residues (Y), in turn initiating signal transduction pathways leading to activation or induction of Inh-α gene transcription. Termination of the signal occurs when the inhibitors, IGFBP-4 and -5, neutralize bioavailable IGF-I. Activin and FSH interact with transmembrane receptors increasing Serine/Threonine (S/T) kinase-dependent and cAMP-dependent signaling pathways, respectively which lead to increased Inh-α gene activity. Evidence that IGFBP-4 and -5 totally abolish these increases argues that IGF-I signaling is a dominant pathway inducing Inh-α expression while the main function of FSH and activin are to amplify the IGF-I induction pathway.

PERSPECTIVES

What might be the physiological relevance of our findings? Understanding the regulation of Inh-α production is physiologically significant because of the key role that inhibin plays in DF development and tumor prevention in murine ovaries (7). Of the three signaling mechanisms regulating Inh-α production (IGF-I, FSH, activin), the most critical appears to be IGF-I. To understand the potential physiological implications of these findings, it is important to consider the fact that IGFBPs are potent inhibitors of FSH and activin stimulation of Inh-α expression. It has been demonstrated that the GC in atretic but not healthy follicles produce IGFBP-4 and -5 (2). This observation leads us to propose the concept that these IGFBPs are physiologically important in the process of follicle death.

There is evidence that IGF-I and Inh-α are expressed in GC of newly recruited preantral follicles, demonstrating that these genes are turned on in GC very early in folliculogenesis (8, 15). An attractive hypothesis is that the expression of Inh-α gene activity might be activated by endogenously produced IGF-I. When activin and/or FSH enter follicular microenvironment, one might expect that the level of Inh-α expression will be enhanced. Because IGF-I is expressed in GC of healthy but not atretic follicles, the progressive increase in inhibin production could be an important step in the process by which FSH evokes a DF. Indeed evidence has been obtained which supports this hypothesis. For example, atretogenic ligands which trigger programmed cell death also cause dramatic increases in IGFBP-4 and -5 production (16, 17). One would expect this stimulation of IGFBP-4 and -5 to be intrinsically destructive because it would lead to a loss of IGF-I signaling, and thereby a loss of FSH and activin-dependent stimulation of GC cytodifferentiation. In the absence of stimulation by these trophic ligands, the follicle will die.

ACKNOWLEDGMENTS

The authors gratefully thank Melissa Huang for technical help and Andrea Hartgrove for typing the manuscript.

REFERENCES

1. P. Felig, J. D. Baxter, A. E. Broadus and L. A. Frohman, (eds.), Endocrinology and Metabolism (Third Edition), McGraw-Hill, New York, 1995 .
2. J. K. Findlay, (eds.), Molecular Biology of the Female Reproductive System, Academic Press, New York, 1994 .
3. G. F. Erickson and D. R. Danforth, Am J Obstet Gynecol, 172 (1995) 736-747.
4. E. Y. Adashi, J Soc Gynecol Invest, 2 (1995) 721-726.

5. J. Baker, M. P. Hardy, J. Zhou, C. Bondy, F. Lupu, A. R. Bellvé and A. Efstratiadis, Mol Endocrinol, 10 (1996) 903-918.
6. D. LeRoith, (eds.), Frontiers in Endocrinology, Serono Symposia, 1996 .
7. M. M. Matzuk, T. R. Kumar, W. Shou, K. A. Coerver, A. L. Lau, R. R. Behringer and M. J. Finegold, Recent Prog Horm Res, 51 (1996) 123-157.
8. J. E. Oliver, T. J. Aitman and J. F. Powell, Endocrinology, 124 (1989) 2671-2679.
9. J. Zhou, E. Chin and C. Bondy, Endocrinology, 129 (1991) 3281-3288.
10. E. Y. Adashi, C. E. Resnick, A. J. D'Ercole, M. E. Svoboda and J. J. Van Wyk, Endocr Rev, 6 (1985) 400-420.
11. T. A. Bicsak, M. Shimonaka, M. Malkowski and N. Ling, Endocrinology, 126 (1990) 2184-2189.
12. P. S. LaPolt, G. N. Piquette, D. Soto, C. Sincich and A. J. Hsueh, Endocrinology, 127 (1990) 823-831.
13. A. Gazit, P. Yaish, C. Gilon and A. Levitzki, J Med Chem, 32 (1989) 2344-2352.
14. M. Párrizas, A. Gazit, A. Levitzki, E. Wertheimer and D. LeRoith, Endocrinology, 138 (1997) 1427-1433.
15. J. K. Findlay, Biol Reprod, 48 (1993) 15-23.
16. G. F. Erickson, D. Li, R. Sadrkhanloo, X.-J. Liu, S. Shimasaki and N. Ling, Endocrinology, 134 (1994) 1365-1372.
17. N. Onoda, D. Li, G. Mickey, G. Erickson and S. Shimasaki, Mol Cell Endocrinol, 110 (1995) 17-25.

Molecular Mechanisms to Regulate the
Activities of Insulin-like Growth Factors
K. Takano, N. Hizuka and S-I. Takahashi (Editors)

The Roles of IGFs and IGFBP-1 in Non-pregnant Human Endometrium and at the Maternal:Placental Interface During Human Pregnancy

LC Giudice, MD, PhD

Division of Reproductive Endocrinology & Infertility, Department of Gynecology and Obstetrics, Stanford University Medical Center, Stanford, CA 94305-5317 USA

INTRODUCTION

Human endometrium is a dynamic tissue. It undergoes cycle-dependent changes in response to fluctuating levels of circulating ovarian-derived steroids which act primarily or through mediators, including growth factors and related peptides (1). During the first half of the cycle, which is estradiol (E_2)-dominant and named the "proliferative phase", endometrial cellular components undergo intense proliferation. During the secretory phase (second-half of the cycle and progesterone (P)-dominant), cellular differentiation prevails, heralded by glandular secretion and stromal decidualization. Both proliferative and differentiative processes are regulated by growth factors and cytokines preferentially expressed in the proliferative and secretory phases.

Appropriate endometrial development is essential for successful blastocyst implantation, which is temporally restricted. The IGF system is one of several growth factor systems important in cyclic endometrial development and in the process of implantation (1). Interest in this growth factor system in the uterus arose from two landmark studies. The first study revealed high levels of IGF-I mRNA expressed in rat uterus (about two-thirds in the myometrium and about one third in the endometrium). In addition, expression of IGF-I mRNA in this tissue was shown to be E_2-dependent, increasing nearly 20-fold after E_2 administration (2). The second landmark study revealed that a major secretory protein of decidualized human endometrium was an IGF binding protein, IGFBP-1 (3-5). This monograph describes the expression, regulation, and putative roles of IGF-I, IGF-II, and IGFBP-1 in human endometrium during the normal menstrual cycle and in pregnancy.

THE IGF FAMILY IN CYCLING ENDOMETRIUM.

Members of the IGF family are expressed in human endometrium and undergo unique changes throughout the menstrual cycle (6-13). IGF-I mRNA is expressed primarily in mid-late proliferative and early secretory endometrium, and IGF-II mRNA is preferentially expressed in secretory, compared to proliferative, endometrium(**Figure 1**). Because of its temporal expression and E_2-dependence, IGF-I is believed to mediate the mitotic actions of E_2 in this tissue. IGF-I is stimulated by E_2 in endometrium of other species (*vida supra*) and has been implicated in E_2-dependent neoplasms of human endometrium(14). Thus, IGF-I is believed to be one of several estromedins during the proliferative phase, acting as a mitogenic stimulus to effect rapid endometrial growth. IGF-II, on the other hand, is expressed abundantly in mid-late secretory endometrium and may be a mediator of progesterone (P) action. Frost, Tseng, and colleagues found that medroxyprogesterone (MPA)-induced stromal mitosis was inhibited by increasing amounts of IGFBP-1 added to stromal cell cultures (15). Since decidualized stromal cells synthesize IGF-II, this response was attributable to MPA-induction of IGF-II.

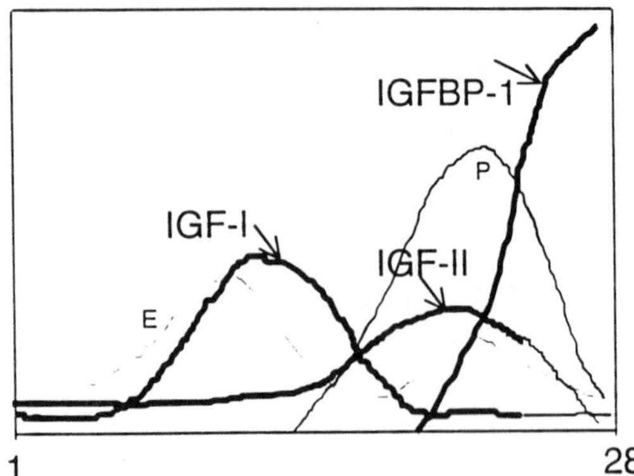

Figure 1. Relative expression of IGF-I, IGF-II, and IGFBP-1 mRNAs (heavy lines) and circulating E=estradiol and P=progesterone (fine lines) in a 28 day menstrual cycle.

With regard to IGF receptors, the Type I IGF receptor is expressed primarily in glandular epithelium and lower amounts are expressed in stroma. Since the IGF peptides are stromally-derived, they are believed to participate, by autocrine mechanisms in stromal proliferation, and, by paracrine mechanisms, in epithelial proliferation. IGF-I and IGF-II are mitogenic to *in vitro* cultured human endometrial cells and regulate secretory functions of stromal cells (15,16), supporting roles for these peptides in endometrium *in vivo*.

Messenger RNAs encoding IGFBP-1-6 are expressed in human endometrium, primarily in stroma. IGFBP-5 is the only IGFBP that is preferentially expressed in proliferative phase endometrium (12). It may facilitate transport of the IGFs from their sites of synthesis to their sites of action in proliferative phase. This conjectural function of IGFBP-5 is based on the observation that IGFBP-5 is found in many tissues in the extracellular matrix and proteolytic mechanisms exist for "off-loading" IGFs from this IGFBP (17). IGFBP-1, -2, -3, -4, and -6 are differentially expressed in secretory endometrium, with IGFBP-1 being by far the most abundantly expressed. It is likely that IGFBP-2, -3, and -4 regulate IGF actions in both phases of the menstrual cycle. In contrast, IGFBP-1 is exclusively expressed in the secretory, but not proliferative, phase and likely plays a major role in regulating IGF availability to receptors on both glandular epithelium and stroma during this phase of the cycle. Recently, an IGFBP-1 transgenic mouse model was developed and used to explore the hypothesis that IGF-I is a mediator of E_2 action in the rodent uterus. In the IGFBP-1 transgenic mice, IGFBP-1 mRNA and protein were expressed in the glandular epithelium. Impaired E_2 action (decreased DNA synthesis) in the uterus was noted, compared to wild type controls, suggesting that IGF-I is a mediator of E_2 action, and an IGFBP can regulate its actions (18). The temporal and spatial relationships of IGF-I, IGF-II, and IGFBP-1 expression suggest that these members of the IGF family play major roles in endometrium during the menstrual cycle and in early pregnancy (*vide infra*).

THE IGF FAMILY AT THE DECIDUAL:TROPHOBLAST INTERFACE

In addition to their roles in nonpregnant endometrial development, the IGFs and IGFBP-1 are believed to play a major role in the process of implantation. As in cycling endometrium, IGF-II mRNA is expressed in the endometrium of pregnancy (decidua), but only when the conceptus is extrauterine (eg. tubal ectopic pregnancy) (10). With an intra-uterine gestation, IGF-II is exclusively expressed in the placenta (13). The most abundant expression of IGF-II is in the columns of the intermediate (invading) trophoblasts in the anchoring villi **Figure 2** (13), suggesting a role for it in trophoblast invasion. In addition, there is a gradient of IGF-II mRNA expression in the cytotrophoblast columns, with the greatest levels expressed at the invading front, further suggesting a role for IGF-II in trophoblast invasion (13) (**Figure 2**). IGF receptor mRNAs are expressed in placental trophoblasts (13), suggesting these cells are targets for IGF actions. With regard to IGFBPs, trophoblasts do not express IGFBP mRNAs (except for weak expression of IGFBP-3 mRNA). This is in contrast to secretory endometrium and decidua in which IGFBP-1 is the major IGFBP expressed and is a major protein product of these tissues (11-13,19). The spatial pattern and relative abundance of IGFBP-1 in decidua suggest it interacts with the IGF-II-expressing, invading cytotrophoblast. In human implantation, invasion of the trophoblast is aggressive, and there is intermingling of the trophoblasts within the maternal decidua. There is likely extensive maternal:placental cross-talk at the molecular level, and these two members of the IGF family are well situated to conduct such a dialogue. *In vitro* studies support modulatory roles for IGFBP-1 in trophoblast cellular function. For example, IGFBP-1 inhibits the binding and biological activity of IGF-I on choriocarcinoma cells, suggesting that abundant decidual IGFBP-1 may regulate trophoblast-derived IGF auto/

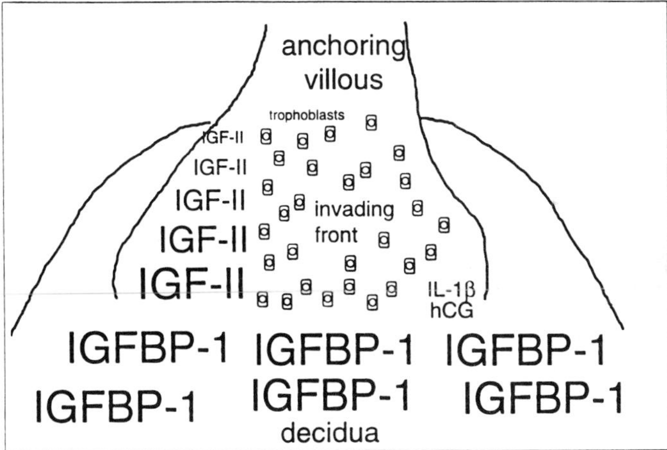

Figure 2. Schematic of invasion of the IGF-II producing trophoblasts into the IGFBP-1 producing maternal decidua. IL-1β and hCG are also shown.

paracrine actions (20). IGFBP-1 also has IGF-independent actions, binding to cell membranes and altering cellular motility (21). Recently, IGFBP-1 has been shown to inhibit trophoblast invasion into decidualized endometrial stromal cultures (22), suggesting that this IGFBP is a maternal "restraint" on trophoblast invasion (*vide infra*). Whether it does this as an IGF binding protein or via IGF-independent effects at the maternal-fetal interface is yet to be determined.

PRODUCTION AND REGULATION OF IGFBP-1 IN ENDOMETRIUM AND DECIDUA

Production

IGFBP-1, formerly known as placental protein-12 and α_1-progesterone-dependent endometrial globulin (α_1-PEG), is made in few tissues of the adult human, including liver, granulosa-luteal cells, secretory endometrium, and decidua. It is exclusively made by the decidua and is not a placental protein at all. It is made in endometrium of a variety of species (23,24, reviews). In humans IGFBP-1 is a major product of late secretory endometrium (16 μg/g protein), and is abundantly produced by maternal decidua, reaching levels of 1224 μg/g protein by mid-gestation (19,25-27). IGFBP-1 has been immunolocalized to the extracellular matrix and stromal cells of decidualizing endometrium, in the periarteriolar regions, and on the villous trophoblast, but not on placental fibroblasts (28-33).

Regulation

Progesterone. *In vivo* human endometrial stromal cells undergo proliferation and differentiation in response to progesterone, after E_2-priming. Isolated from human endometrium, stromal cells can be decidualized *in vitro* with progesterone or progestins, or stimulators of cyclic AMP (cAMP), and E_2 or epidermal growth factor (EGF) (4,34,35). This *in* vitro model has provided an opportunity to investigate production and regulation of IGFBP-1, whose mRNA is expressed primarily in this cell type in endometrium *in vivo*. IGFBP production in endometrium is dependent on stromal differentiation (9,16), and while decidualized stromal cells increase their production of all IGFBPs upon decidualization *in vitro*, most striking are the markedly high levels of IGFBP-1 produced by these cells upon decidualization (25.3 + 3.2 μg/day per 10^6cells). This is in contrast, eg., to another decidual protein, prolactin (40 ng/10^6 cells/day) (9,16,36). Furthermore, progesterone regulates IGFBP-1 protein and mRNA expression in decidualized endometrial stromal cells, and the progesterone receptor antagonist RU486 is inhibitory (9,36). The IGFBP-1 gene has a glucocorticoid response element, and it is likely that progesterone, via this promoter element, exerts its effects on IGFBP-1 gene expression (24, review).

Insulin, IGFs. There are likely autocrine and paracrine regulatory mechanisms in microenvironments of the endometrium that influence stromal IGFBP production. The net result would affect stromal and glandular function, as well as trophoblast-stromal interactions during the implantation process. With regard to implantation, the placental cytotrophoblast expresses high levels of IGF-II mRNA (13) and protein (Irwin and Giudice, unpublished), while the adjacent decidua expresses IGFBP-1. The close proximity of the IGF-II producing trophoblast and the IGFBP-1 producing decidua suggests either an effect of IGFBP-1 on trophoblast function and/or an effect of IGF-II on decidual and/or trophoblast processes. Thus, whether insulin-like peptides regulate endometrial stromal IGFBP-1 is important to ascertain in this paradigm. Insulin and IGFs are known inhibitors of IGFBP-1 mRNA and protein expression in liver and HepG2 cells (24). Similarly, in endometrium, insulin, IGF-I, and IGF-II inhibit decidualizing stromal cell IGFBP-1 secretion into conditioned media (CM) (16), in a dose dependent fashion, with ED_{50}'s consistent with actions through their respective receptors. Insulin and IGFs are also inhibitory to IGFBP-1 secretion by endometrial cells decidualized *in vivo* (37). The physiologic relevance of IGF and insulin regulation of IGFBP-1, a major product of secretory phase endometrium and decidua, likely rests in the need to regulate IGF bioavailability to IGF receptors in target cells, including trophoblast and/or decidua and endometrial glandular epithelium. In addition, these regulatory mechanisms may affect direct interactions of this IGFBP with the invading trophoblast (*vide infra*).

Other modulators. The close proximity of the invading trophoblast and the IGFBP-1-producing decidua suggests that growth factors and cytokines at the decidua:trophoblast interface may regulate IGFBP-1 production in the maternal compartment (**Figure 2**). A recent study has investigated the effects of interleukin-1β (IL-1β), transforming growth factor-β (TGF-β), stem cell factor (SCF), colony stimulating factor-1 (CSF-1), leukemia inhibitory factor (LIF), and IGF-II on IGFBP-1 production by decidualized endometrial stromal IGFBP-1 production *in vitro* (38). Only IGF-II and IL-1β had an effect on IGFBP-1 secretion by decidualized endometrial stromal cells in culture. They were both inhibitory, with ED_{50}'s consistent with actions through their respective receptors. IL-1β inhibits the process of decidualization of endometrial stromal cells (39,40). Also, since IL-1β stimulates production of trophoblast matrix metalloproteinase-9 (MMP-9) which promotes the invasive trophoblast phenotype (41), it is likely that inhibition of IGFBP-1 in the decidua by this cytokine further promotes trophoblast invasion, by inhibiting this decidual "restraint protein" (*vide infra*).

Human chorionic gonadotropin (hCG), another trophoblast product, has been reported to stimulate stromal IGFBP-1 production, likely due to its effects on promoting stromal decidualization (42.43). Similar results have been reported with free hCG-α subunit (44), although the mechanisms for free α-subunit action alone are not well understood at this time. HCG has no appreciable effect on stromal cell products, once the cells are decidualized *in vivo* (45). While the physiologic relevance of hCG action in endometrium is not well established, premature ovulation or elevated luteinizing hormone (LH) (which cross reacts with hCG at the LH/hCG receptor) in the follicular phase may lead to premature decidualization of the endometrium, and elevated LH has been associated with increased risk of spontaneous abortion (46). Whether elevated endometrial IGFBP-1 predisposes to poor implantation and miscarriage has yet to be determined. Since IL-1β, IGF-II, and hCG are trophoblast-derived, these observations cumulatively support roles for trophoblast-derived growth factors and cytokines in the regulation of decidualization of the maternal endometrium and of decidual IGFBP-1 production at the trophoblast: decidual interface (**Figure 2**).

IGFBP-1 AND IMPLANTATION

During the invasive phase of implantation, the "intermediate" trophoblast of the anchoring villous produces large amounts of IGF-II and invades into the maternal decidua which is producing large amounts of IGFBP-1 (13,47) (**Figure 2**). Trophoblast invasion into the maternal decidua occurs by processes similar to those accompanying tumor invasion, including attachment to the extracellular matrix, local matrix proteolysis, cellular migration, and inhibition of these processes (48). IGF-II and IGFBP-1 are temporally and spatially positioned to participate in the regulation of trophoblast invasion into the maternal endometrium. IGFBP-1 is relatively resistant to proteolysis, being intact in fluids where proteolysis of other IGFBPs occurs (49,50). There is a gradient of IGF-II mRNA expression in the trophoblast columns, with greatest levels expressed at the invading front, strongly suggesting a role for IGF-II in invasive trophoblast invasion (13). The spatial pattern and relative abundance of IGFBP-1 in decidua suggest it interacts with the IGF-II-expressing, protease-secreting, invading trophoblast. The abundance of IGF-II at the invading front and the proximity of decidual cells expressing IGFBP-1 are suggestive of a role for this growth factor and its inhibitor in invasion. IGFBP-1 has been shown to have inhibitory effects on IGF binding and IGF actions on choriocarcinoma cells in culture (20), and may have similar effects on IGF-II on normal, invading cytotrophoblasts at the maternal:fetal interface *in vivo*.

Besides its IGF binding properties, IGFBP-1 also has IGF-independent actions, binding to cell membranes and altering cellular motility (17). IGFBP-1 contains the tripeptide motif, Arg-Gly-Asp (RGD), which is a recognition site for several cell adhesion molecules, including the $\alpha_5\beta_1$ integrin. In Chinese Hamster Ovary (CHO) cells, IGFBP-1 binds to the $\alpha_5\beta_1$ integrin, presumably via its RGD sequence, and stimulates motility of these cells *in vitro* (21). The invading trophoblast at the maternal:fetal interface in humans uniquely expresses the $\alpha_5\beta_1$ integrin, among all trophoblast phenotypes (51). Recent studies have demonstrated that IGFBP-1 specifically binds to human trophoblast and to the $\alpha_5\beta_1$ integrin in trophoblast. Furthermore, it inhibits trophoblast attachment to fibronectin, another RGD ligand found in the placental bed (22). Cytotrophoblast interaction with fibronectin, through the $\alpha_5\beta_1$ integrin, restrains invasion (52). Using a co-culture model, we have found that when human cytotrophoblasts are co-cultured with human endometrial stromal cells decidualized in vitro and producing high amounts of IGFBP-1, trophoblast invasion into the stromal cell multilayers is inhibited (22). Furthermore, when IGFBP-1 production is inhibited by insulin, co-cultured trophoblasts invade into the stromal multilayers. The invasion is inhibited by the addition of exogenous IGFBP-1 into the co-culture system, in a dose-dependent fashion (22). These studies cumulatively suggest that IGFBP-1 is a maternal "restraint" on trophoblast invasion. Whether it does this via direct interactions with the trophoblast (as an $\alpha_5\beta_1$ ligand or a ligand for other trophoblast cell surface proteins) or by inhibiting IGF-II actions on the trophoblast is not known at this time. Signaling events can occur upon perturbing membrane integrins, however, how IGFBP-1 (and fibronectin) interacting with cytotrophoblast integrins results in inhibition of invasion remains uncertain. IGFBP-1 has also been shown to stimulate migration of passaged human trophoblasts on a plastic surface (53). These seemingly opposing observations of IGFBP-1's effects on trophoblast behavior may reflect differences in *in vitro* culture conditions and cell populations or intermediate mediators that are present in co-cultures versus singly cultured cells. Cumulatively, the data suggest a role for IGFBP-1, either as a direct modulator of trophoblast invasion or as an IGF-II binding protein, at the maternal:fetal interface in humans.

IGFBP-1 IN PRE-ECLAMPSIA

Pre-eclampsia is a pregnancy-specific disorder, occurring in 5-10% of pregnant women. In its severe form, in first world countries, it is a major cause of fetal and maternal morbidity and mortality (54). Clinically, it is usually detected in the second and early third trimesters, and if left undiagnosed or untreated, it can progress to maternal multi-organ failure, coagulopathy, seizures, and maternal and fetal death. In this disorder there is generalized hypoxia within the placenta (55) and abnormally shallow cytotrophoblast invasion into the decidua (56,57). The latter may result from abnormal trophoblast adhesion molecules (58) and/or from elevated decidual levels of IGFBP-1 preventing deeper placental invasion (33). In support of the latter hypothesis is the clinical finding that in pregnancies complicated by severe pre-eclampsia, maternal serum IGFBP-1 levels in the second and early third trimesters are about 6-fold higher (33) and at term are about 2-fold higher (59,60 and **Table 1** ((33) and Giudice et al, unpublished) than in normal pregnancies. This is not a non-specific elevation of hepatic-derived proteins, as often occurs in this disorder, since maternal serum levels of

IGFBP-3 (also of hepatic origin) are normal and IGFBP-3 is not a useful predictor of pre-eclampsia (61). In addition, hepatic-derived IGF-I levels are about half of those in the circulation of controls ((33) and **Table 1**). A significant positive correlation was observed

Table 1.

IGFs and IGFBP-1 (in ng/mL ± sem) in the circulation of women with severe pre-eclampsia (SPE) in the second/third trimester and at term, compared to gestational age-matched controls[#]

	IGF-I	IGF-II	IGFBP-1
Second/Third Trimester			
Control (n=29)	**179.4 ± 28.2**	**1015.8 ± 134.1**	**76.6 ± 11.8**
SPE (n=16)	**80.9 ± 17.2^**	**1113.7 ± 156.2**	**428.3 ± 85.9***
Term			
Control (n=23)	**291.2 ± 31.4**	**765.3 ± 59.1 +**	**121.2 ± 24.2**
SPE (n=21)	**228.4 ± 33.8**	**741.8 ± 81.2**	**220.3 ± 41.5[+]**

^P=0.0007, compared to controls; *P=0.0001, compared to controls; [+]P=0.041, compared to controls

from (33) and Martina and Giudice, unpublished.

between maternal diastolic blood pressure, aspartate transcarbamylase and IGFBP-1, suggesting that IGFBP-1 reflects severity of pre-eclampsia and hepatic involvement (33). It is likely that elevated levels of IGFBP-1 in the circulation of women with severe pre-eclampsia are derived from both the decidua and the liver. In decidua IGFBP-1 is found in the periarteriolar region. With enhanced vascular permeability and vasospasm, common in pre-eclampsia, escape from the decidua into the maternal circulation may occur, contributing to the observed elevated circulating levels of IGFBP-1. Higher levels of immunoreactive IGFBP-1 were observed at the decidual:placental interface in pregnancies complicated by severe pre-eclampsia, compared to controls (33). This may contribute to the shallow placentation observed in this disorder, although decidual origin of the elevated levels of IGFBP-1 at the decidual: placental interface in severe pre-eclampsia remains to be determined. Recently, a longitudinal study revealed that IGFBP-1 levels were decreased in maternal serum in midgestation in women who subsequently developed mild pre-eclampsia (62). It is likely that mild and severe pre-eclampsia are different disorders, and controversy still surrounds whether circulating IGFBP-1 early in gestation will be a predictor of developing severe pre-eclampsia later in gestation.

SUMMARY

IGF-I, IGF-II, and IGFBP-1 are important in endometrial development during the menstrual cycle and in the process of implantation. The mitogenic, differentiative, and anti-apoptotic properties of these growth factors, as well as their spatial and temporal expression in cycling endometrium, suggest that they may participate in endometrial growth, differentiation, inhibition of apoptosis, and perhaps angiogenesis. IGFBP-1, a major protein product of secretory endometrium and decidua, is an IGF-binding protein and a trophoblast integrin ligand. These properties suggest that it may have multifaceted roles in endometrial development and in interactions between the decidua and the integrin-expressing, invading trophoblast. Elucidation of the mechanisms underlying IGF and IGFBP-1 action at the decidual:trophoblast interface in early pregnancy await further investigation and will likely provide valuable insight into disorders of placentation, intrauterine growth retardation, implantation failure, and infertility. The potential predictive value of IGFBP-1 in serum for these disorders should unfold in the near future.

REFERENCES

1. L.C. Giudice, Fertil Steril 61 (1994) 1-17.
2. L.J. Murphy, L.C. Murphy, H.G.Friesen, Trans Assoc Amer Physicians 99 (1987) 204-14.
3. R. Koistinen, N. Kalkinnen, M-L. Huhtala, M. Seppala, H. Bohn, E-M. Rutanen, Endocrinology 118 (1986) 1375-8.
4. S.C. Bell, J.A. Jackson, J. Ashore, H.H. Zhu, L. Tseng, J Clin Endocrinol Metab 72 (1991) 1014-1019.
5. M. Julkunen, R. Koistinen, K. Aalto-Setala, M. Seppala, O.A. Janne, K. Kontula, FEBS Letters 236 (1988) 295-301.
6. M. Julkunen, R. Koistinen, A-M. Suikkari, M. Seppala, O.A. Janne, Molecular Endocrinol 4 (1990) 700-7.
7. K.D. Boehm, M. Daimon, I.G. Gorodeski, L.A. Sheean, W.H. Utian, J. Ilan, Mol Reprod Devel 27 (1990) 93-101.
8. L.C. Giudice, D.A. Milkowski, G. Lamson, R.G. Rosenfeld, J.C. Irwin JC, J Clin Endocrinol Metab 72 (1991) 779-87.
9. L.C. Giudice, B.A. Dsupin, J.C. Irwin JC, J Clin Endocrinol Metab 75 (1992) 1235-41.
10. L.C. Giudice, B.A. Dsupin, I.H. Jin, T.H. Vu, A.R. Hoffman, J Clin Endocrinol Metab 76 (1993) 1115-1122.
11. L.C. Giudice, J.C. Irwin, B.A. Dsupin, L. de las Fuentes, I.H. Jin, T.H. Vu, A.R. Hoffman, In: The Insulin-like Growth Factors and Their Regulatory Proteins. R.C. Baxter, P.D. Gluckmann, R.G. Rosenfeld. Excerpta Medica International Congress Series 1056, Amsterdam, 1994, pp 351-361.
12. J. Zhou, B.A. Dsupin, L.C. Giudice, C.A. Bondy, J Clin Endocrinol Metab 79 (1994) 1723-1734.
13. V.K.M. Han, N. Bassett, J. Walton, J.R.G. Challis, J Clin Endocrinol Metab 81 (1996) 2680-2693
14. E-M. Rutanen, T. Nyman, P. Lehtovirta, M. Ammala, F. Pekonen, Int J Cancer 59 (1994) 307-312.
15. R.A Frost, J. Mazella, Tseng, Biol Reprod 49 (1993) 104-11.
16. J.C. Irwin, L. de las Fuentes, B.A. Dsupin, L.C. Giudice, Regulatory Peptides 48 (1993) 165-77.
17. J.I. Jones, D.R. Clemmons, Endo Reviews 18 (1994) 1-31.
18. K. Rajkumar, T. Dheen, M. Krsek, L.J. Murphy, Endocrinology 137 (1996) 1258-1264.
19. E-M. Rutanen, R. Koistinen, T. Wahlstrom, H. Bohn, T. Ranta, M. Seppala, Endocrinology 116 (1985) 1304-9.

20. O. Ritvos, T. Ranta, J. Jalkanen, A-M. Suikkari, R. Voutilainen, H. Bohn, M. Seppala, Endocrinology 122 (1989) 2150-7.
21. J.I. Jones, A. Gockerman, W.H. Busby, D.R. Clemmons, Proc Natl Acad Sci USA 90 (1993) 10553
22. J.C. Irwin, L.C. Giudice, Growth Regulation, in press (1997).
23. A.T. Fazleabas, S. Hild-Pertito, H.G. Verhage, Sem Repro Endocrinol 13 (1995) 120-132.
24. P.D.K. Lee, D.R. Powell, C.A. Conover, L.C. Giudice, Proc Soc Expt Med Biol , in press, (1997).
25. E-M. Rutanen, R. Koistinen, T. Wahlstrom, J. Sjoberg, U-H. Stenman, M. Seppala, Br J Obstet Gynecol 91 (1984) 377-81.
26. E-M. Rutanen, M. Menabawey, K. Isaka, H. Bohn, T. Chard, J.G. Grudzinskas, J Clin Endocrinol Metab 63 (1986) 675-679.
27. E-M. Rutanen, R. Koistinen, J. Sjoberg, M. Julkunen, T. Wahlstrom, H. Bohn, M. Seppala, Endocrinology 118 (1986) 1067-71.
28. G.T. Waites, R.F.L. James, S.C. Bell, J Endocrinol Metab 67 (1988) 1100-1104.
29. G.T. Waites, R.F.L. James, S.C. Bell, J Endocrinol 120 (1989) 351-357.
30. S.C. Bell, S.R. Patel, J.A. Jackson, G.T. Waites, J Endocrinol 118 (1988) 317-28.
31. E-M. Rutanen, E. Gonzalez, J. Said, G.D. Braunstein, Endocr Path 2 (1991) 132-8.
32. G.D. Bryant-Greenwood, E-M. Rutanen, S. Partanen, T.K. Coelho, S.Y. Yamamoto, Molecular and Cellular Endocrinology 95 (1993) 23-29.
33. L.C. Giudice, N.A. Martina, L. de Las Fuentes, R.A. Crystal, M.L. Druzin, Am J Obstet Gynecol, 176 (1997) 751-757.
34. J.C. Irwin, D. Kirk, R.J.B. King, M.M. Quigley, R.B.L. Gwatkin, Fertil Steril 52 (1989) 761-768.
35. J.C. Irwin, W.H. Utian, R.L. Eckert, Endocrinology 129 (1991) 2385-92.
36. M. Rosenberg, J. Mazella, L. Tseng, Annal NY Acad Sci 622 (1991) 138-44.
37. K.M. Thrailkill, D.R. Clemmons, W.H. Busby Jr, S. Handwerger, J Clin Invest 86 (1990) 878-883.
38. S.P. Mark, N.A. Martina, J.C. Irwin, L.C. Giudice, Am Soc Reprod Med Annual Meeting, Boston, November, p 84, Abstract No. O-007. (1996).
39. M. Kariya, H. Kanzaki, K. Takakura, K. Imai, N. Okamoto, N. Emi, Y. Kariya, T. Mori, J Clin Endocrinol Metab 73 (1991) 1170-1174.
40. G.R. Frank, A.K. Brar, H. Jikihara, M.I. Cedars, S. Handwerger, Biol Reprod 52 (1995)184-191.
41. J. Librach, S.L. Feigenbaum, K.E. Bass, K-Y Cui, N. Versastas, Y. Sadovsky, J.P. Quigley, D.L. Franch, S.J. Fisher, J Biol Chem 269 (1994) 17125-17131.
42. B. Tang, E. Gurpide, J Steroid Biochem Molec Biol 47 (1993) 115-121.
43. S.W. Han, Z.M Lei, J.S. Sanfilippo, Ch.V. Rao, 77th Ann Mtg US Endocrine Soc, Wash DC, Abstr P2-84. (1995).
44. E. Moy, L.M. Kimzey, L.M. Nelson, D.L. Blithe, Endocrinology 137 (1996) 1332-1339.
45. S.G. Ren, G.D. Braunstein, J Clin Endocrinol Metab 70 (1990) 983-989.
46. T.C. Li, E. Serle, M.A. Warren, I.D. Cooke, Human Reprod 8 (1993) 1021-1024.
47. J. Zhou, C. Bondy, Endocrinology 131 (1992) 1230-1240.
48. W.G. Stetler-Stevenson, S. Aznavoorian, L.A. Liotta, Ann Rev Cell Biol 9 (1993) 541-573.
49. P. Hossenlopp, B. Segovia, C. Lassarre, M. Roghani, M. Bredon, M. Binoux, J Clin Endocrinol Metab 71 (1990) 797-805.
50. L.C. Giudice, E.M. Farrell, H. Pham, R.G. Rosenfeld, J Clin Endocrinol Metab 71 (1990) 806-816.

51. S.J. Fisher, C.H. Damsky, Cell Biol 4 (1993) 183-188.
52. C.H. Damsky, C. Librach, K-H. Lim, M.L. Fitzgerald, M.T. McMaster, M. Janatpour, Y. Zhou, S.K. Logan, S.J. Fisher, Development (1994) 3657-3666.
53. J.A. Irving, P.K. Lala, Exp Cell Res 217 (1985) 419-429.
54. S.A. Friedman, R.N. Taylor, J.M. Roberts, Clinics Perinatol 4 (1991) 661-682.
55. E.W. Page, Obstet Gynecol Survey 3 (1984) 615-628.
56. T.Y. Khong, F. DeWolf, B. Robertson, I. Brosens, Br J Obstet Gynecol 93 (1986) 1049-1059.
57. J. Moodley, R. Ramsaroop, S Afr Med J. 75 (1989) 376-378.
58. Y. Zhou, C.H. Damsky, K. Chiu, J.M. Roberts, S.J. Fisher, J Clin Invest 91 (1993) 950-960.
59. G.N. Than, I.F. Csaba, D.G. Szabo, A.A. Arany, Z.J. Bognar, H. Bohn, Arch Gynecol 236 (1984) 41-45.
60. K. Iino, J. Sjoberg, M. Seppala, Obstet Gynecol 68 (1986) 58-60.
61. M. Varma, C.J.M. de Groot, S. Lanyi, R.N. Taylor, Am J Obstet Gynecol. 169 (1993) 995-999.
62. C.J.M. de Groot, T.H. O'Brien, R.N. Taylor, Amer J Obstet Gynecol 175 (1996) 24-29.

Molecular Mechanisms to Regulate the
Activities of Insulin-like Growth Factors
K. Takano, N. Hizuka and S-I. Takahashi (Editors)

The Prostatic IGF System: New Levels of Complexity

A. Grimberg, R. Rajah, H. Zhao and P. Cohen

Division of Endocrinology, Dept. of Pediatrics, The Children's Hospital of Philadelphia, University of Pennsylvania School of Medicine, 3400 Civic Center Blvd. Philadelphia, PA, 19104 USA. Supported in part by NIH grant 2RO1 DK 47591

INTRODUCTION

The insulin-like growth factor (IGF) system is a multi-layered network of molecules involved in the regulation of cell growth and death in a variety of tissues, both normal and pathologic. The prostate gland has proven a fertile organ for study of the IGF system; not only does the human prostate contain the many components of the IGF system, but disturbances in this carefully tuned regulatory cascade have been found in benign prostatic hypertrophy (BPH) and prostatic carcinoma (CaP), two conditions of aberrant growth which import major societal clinical significance. The IGF system centers around the IGFs themselves, their receptors and their binding proteins (IGFBPs), and all three participate in controlling prostatic growth. Recent research has not only sought to clarify the actions of these three elements, but also to discern their roles in the larger regulatory scheme of cell growth and death. To this end, several modulators of the IGF axis including IGFBP proteases, upstream regulators such as cytokines and transcription factors, and downstream response molecules have been elucidated. We will review what is currently known about the many levels of the IGF system as they pertain to both normal and abnormal prostatic growth.

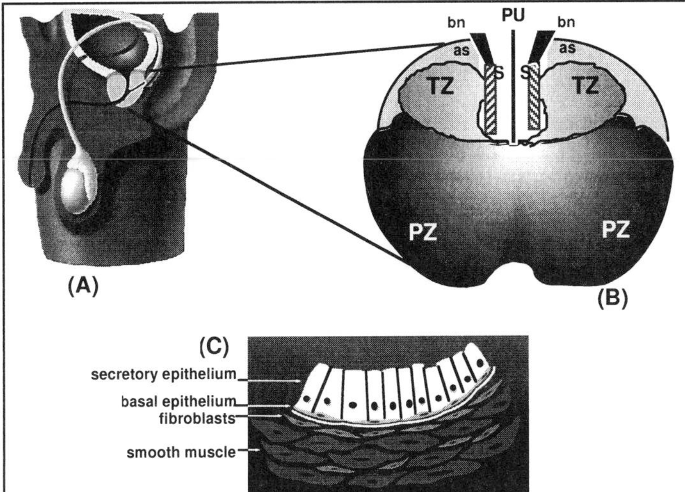

Figure 1. Prostate anatomy. (A) Gross anatomic position of the prostate gland. (B) Schematic of the prostatic zones. TZ=transition zone, origin of BPH. PZ=peripheral zone, origin of most CaP. PU=proximal urethra, bn=bladder neck, s=preprostatic sphincter, as=anterior fibromuscular stroma. (C) Representation of the four cell types found within the prostate. Secretory and basal epithelium = PC-E (prostate cells of epithelium), and fibroblasts and smooth muscle = PC-S (prostate cells of stroma).

PROSTATE ANATOMY

Within a common capsule, the human prostate contains both anteromedial nonglandular tissue and posterolateral glandular segments [1]. The glandular tissue is divided into four distinct zones, based on their origin from the prostatic urethra (Figure 1A). The distal urethra relates to the largest zone, the peripheral zone, comprising about 70 percent of the mass of the glandular prostate, as well as the central zone which constitutes another 25 percent [2]. The remaining 5 to 10 percent of the glandular prostate relates to the proximal urethra; most of this is considered the transition zone, with a minority of tissue referred to as the periurethral zone [3]. The peripheral zone is the region most susceptible to inflammation and malignant transformation, giving rise to roughly 70 percent of all prostate cancers [4]. In contrast, BPH exclusively originates in the transition and periurethral zones [3].(Figure 1B)

Besides the four gross zones which seem to behave differently, the prostate gland can also be subdivided histologically into different tissues with distinct growth patterns. The epithelial cells (PC-E) are the source of CaP while the stromal cells (PC-S) harbor abnormalities in BPH [1]. (Figure 1C)

IGFs, IGF RECEPTORS, AND IGFBPs IN THE PROSTATE

Although human seminal fluid contains both IGF-I and IGF-II, cultured PC-S produce readily detectable levels of IGF-II but not IGF-I [5], and cultured PC-E do not synthesize or secrete significant amounts of either IGF-I or IGF-II [6]. Nonetheless, both PC-S and PC-E express type 1-IGF receptors and respond to the mitogenic effect of the IGFs [6,7]. In fact, the relative mitogenic potency of growth factors added to primary PC-E cultures reflects their relative affinities for the type 1-IGF receptor: IGF-I >IGF-II > insulin [6]. The type 2-IGF receptor, on the other hand, is not found in substantial amounts in either PC-E or PC-S [6].

The IGF-IGF receptor interaction is modulated by the IGFBPs, of which at least seven have been characterized in humans. The IGFBPs not only alter the availability of free IGFs for interaction with the IGF receptor, but also exhibit some IGF-independent growth-mediating effects [8]. The IGFBPs may be growth-enhancing, especially IGFBP-5, [9] or growth-inhibitory, such as IGFBP-3 which also participates in apoptosis [8,10]. In the prostate, IGFBP-1 peptide has not been detected in the cultured medium of either normal PC-E or PC-S, and its mRNA was absent on Northern blots of both cell types [5,6]. IGFBP-2 and IGFBP-4, the major IGFBPs in seminal fluid [11], are both synthesized and secreted predominantly by normal PC-E [6] and also by PC-S [5]. IGFBP-3, however, is definitively made by PC-S [5,12] and possibly by PC-E [6,12] as well as by several CaP cell lines. IGFBP-5 and -6 are made in cultured prostatic cells and are detected in histologic section of prostate tissues [13]. MAC-25 (IGFBP-7) is expressed in PC-S [Cohen et al., personal communication]. IGFBP production by PC-E appears to be constitutive, as treatment of PC-E with various growth regulators and media additives did not change IGFBP levels; PC-S IGFBP production, on the other hand, is influenced by a variety of cytokines (Table 2).

IGFBP PROTEASES: PROSTATIC GROWTH MODULATORS

Although IGFBP-3 is produced by prostate cells, its detection in seminal plasma can be achieved only by radioimmunoassay and not by Western ligand blotting; IGFBP-3 proteolysis by seminal plasma, even in azospermic samples, has been confirmed using labeled IGFBP-3 in a protease assay [14] and immunoblot assays [15]. Prostate-specific antigen (PSA), produced by the luminal PC-E, was further identified as a seminal IGFBP-3 protease [14]. A significant inverse correlation was found between PSA and non-digested IGFBP-3 in prostatic massage fluid [16].

The role of PSA as a prostatic growth modulator is directly linked to its IGFBP-3 protease function. Cleavage of IGFBP-3 by PSA caused a ten- to hundred-fold decrease in the affinity of IGFBP-3 fragments for the IGFs [10]. Exposure of PC-E in serum-free culture conditions to IGFs stimulated a doubling of cell number; this stimulation was blocked by addition of IGFBP-3 and subsequently re-instated by further addition of PSA [10]. Thus, PSA functions as a co-mitogen with the IGFs, by freeing them from IGFBP binding and thereby increasing their potential interaction with the IGF receptors. This effect may be compounded by the fact that PSA reduces levels of IGFBP-3, which in and of itself participates in apoptosis.

PSA is not the only IGFBP protease active in the prostate, nor is IGFBP-3 the only IGFBP cleaved [15]. The prostatic IGFBP proteases can be divided into three categories: kallikrein enzymes (like PSA), matrix metalloproteinases (MMPs), and the cathepsins [17-19]. Whereas PSA is functional at neutral pH, the cathepsins, lysosomal enzymes implicated in malignant processes, are acid-activated [20]. All of these proteins appear to be secreted into seminal plasma, but their role there is not yet known. The prostatic IGFBP proteases are summarized in Table 1.

Table 1. Prostatic IGFBP proteases and their characteristics. hK-2 = human kallikrein 2, TIMP = tissue inhibitor of metalloproteinase, PCI = protease C inhibitor, PAI = plasminogen activator inhibitor.

Protease	Class	IGFBPs cleaved	other substrates	known inhibitor
PSA (HK-3)	kallikrein	3, 5	fibronectin, laminin	alpha$_1$-ACT
HK-2	kallikrein	1-5	fibronectin	PCI, kallistatin
urokinase	serine	3	fibronectin	PAI
plasmin	serine	1-6	fibrin	anti-plasmin
cathepsin D	aspartyl	1-5	T-kininogen, fibronectin	pepstatin
MMP-2	metallo	3, 5	collagen	TIMPs

UPSTREAM PROSTATIC IGF AXIS REGULATORS

The prostatic IGF system is regulated by androgens, cytokines and several transcription factors and oncogenes. Androgens are the key circulating factors which promote prostate growth, and androgen ablation therapy is the hallmark of non-surgical approaches to prostatic disease [21-23]. Paradoxically, the incidence of prostate disease increases with age, as androgen levels are waning, and neoplastic prostatic tissues often become androgen-independent. Two-month treatment with a gonadotropin-releasing hormone analog, producing sub-castration serum levels of testosterone in men with BPH, significantly increased their prostatic IGF-receptor binding capacity, both low and high affinity, when compared to that of non-androgen-ablated BPH prostates; furthermore, expression of the type 1-IGF receptors was found not only in the basal epithelial cells, as in the untreated controls, but also extending to the glandular epithelium [24]. When a constitutively active androgen receptor was stably transfected into PC-3 cells, an androgen receptor-negative prostate carcinoma line, IGFBP-3 transcription and protein levels decreased [25]. Further studies are needed to elucidate the effects of androgens on the different IGF system components in the normal prostate.

Cytokines, particularly transforming growth factor-beta (TGF-ß) and tumor necrosis factor-alpha (TNF-a), have been found by our lab to also regulate the prostatic IGF system. Both TGF-ß and TNF-a modulated the production of IGFBP-3 by normal PC-S cells. TGF-ß also inhibited normal PC-S growth to 30% of baseline, and it seems that IGFBP-3-induced apoptosis is involved in the growth inhibition [26].

Most recently, transcription factors, like the tumor suppressors Wilms' tumor gene product (WT-1) and p53, have been linked to the prostatic IGF system. WT-1 represses transcription of both the type 1-IGF receptor [27] and IGF-II [28]. p53, on the other hand, induces IGFBP-3 gene expression [29], and mutant p53s from a variety of non-prostate cancers which fail to activate IGFBP-3 demonstrate impaired apoptosis [30,31]. Peehl and colleagues found that exposing normal, BPH or CaP tissues to ionizing radiation, a form of genotoxic stress, does not increase p53 in PC-E but does produce a 3- to 9-fold augmentation of p53 in PC-S; in parallel, the PC-E cells did not undergo cell cycle arrest as did the PC-S cells [32]. We are now investigating the role IGFBP-3 plays in mediating p53 effects in a prostate cell model.

PROSTATIC IGFs AND IGFBPs: EFFECTS AND DOWNSTREAM RESPONSE MOLECULES

Little is currently known about the IGFs' downstream response molecules in the prostate. The IGFs are primarily mitogenic, acting through the IGF-receptor. The type 1-IGF receptor is a tyrosine kinase that activates the mitogen activated protein (MAP) kinase pathway [33]. The IGFBPs not only influence the amount of free IGF available for receptor interaction, but also produce growth effects via IGF-independent pathways. For example, IGFBP-3 promotes apoptosis. Apoptosis-inducing agents are reported to mediate increases in the bax family or reductions in the bcl-2 family of proteins. These opposing molecules form dimers, and the balance between the two (mainly through the levels of free bax or bcl-2) determines the degree of apoptosis incurred during specific conditions [34-36]. Bax, in turn, activates the interleukin-1ß-converting enzyme (ICE) family of cysteine proteases which cleave numerous proteins and lead to apoptosis [37]. Our lab demonstrated that IGFBP-3 mediates apoptosis by inducing bcl-2 serine phosphorylation, and hence, ICE activation. Research investigating the role of the MAP kinase pathway or of bcl-2 induction in mediating the mitogenic effects of the IGFs in the prostate is yet to be undertaken. Finally, tyrosylphosphorylation of c-jun, an IGF nuclear target, is believed to participate in IGF-mitogenicity [38]. Sequential strains of SV40-immortalized PC-E cells demonstrated an increased expression of c-jun that paralleled the cells' incremental proliferative response to IGF [39].

THE IGF SYSTEM IN BPH

BPH, the most commonly occurring benign proliferative abnormality found in any internal organ, has been described as an abnormal proliferation of prostatic epithelial ducts and acini, probably preceded by hyperproliferation of the stroma in the prostatic transition zone [40]. This histologic observation has borne out in molecular studies. No abnormalities in the IGF axis were detected in PC-E from BPH prostates [5]. BPH PC-S, however, demonstrated a ten-fold increase in IGF-II transcription as well as a tripling in type 1-IGF receptor transcription; these findings are most likely the result of a significant diminution of WT-1 expression to 0-20% of normal [41], as it is well described that WT-1 is a tumor suppressor which inhibits transcription of multiple growth-promoting genes including IGF-II and type 1-IGF receptor. In BPH PC-S, IGFBP-2 expression was also markedly reduced and was replaced by IGFBP-5 expression, as revealed by both mRNA and peptide analyses [5].

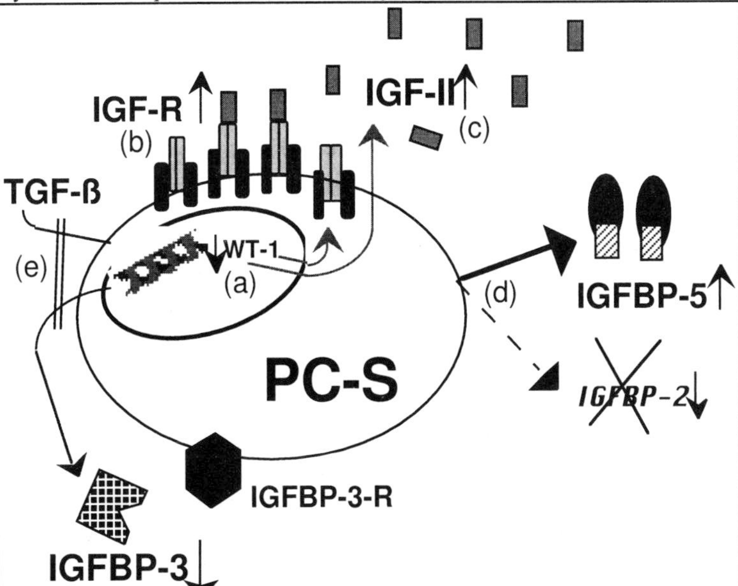

Figure 2. IGF axis disturbances in BPH. (a) Decreased WT-1 expression appears to lead to increased IGF-R (b) and IGF-II (c) expression which may result in accelerated proliferation of BPH PC-S. Additionally, (d) loss of IGFBP-2 but gain of IGFBP-5 secreted from PC-S influence the mitogenic stimulus. The blunted TGF-ß induction of IGFBP-3 associated with a decreased TGF-ß induced growth inhibition (e) in PC-S from BPH may also contribute to the hyperplastic phenotype and may involve binding of IGFBP-3 to its receptors (IGFBP-3-R).

The genes for IGFBP-2 and -5 are tightly linked on chromosome 2q33-34 and may follow coordinated regulation [42]. Thus, it is tempting to postulate that a genetic event leads to diminished expression of IGFBP-2, an IGF-inhibitory molecule, and augmented expression of IGFBP-5, an IGF-enhancing peptide. In addition, IGFBP-3, another growth-inhibitory IGFBP, was found by our lab to not only be four-fold lower at baseline in BPH versus normal PC-S, but to manifest a blunted induction by TGF-ß in BPH PC-S; this phenomenon was associated with failed TGF-ß-induced growth inhibition of BPH cells [43]. This IGFBP imbalance, coupled with the IGF-II overproduction, constitutes an autocrine stimulus for stromal proliferation and a paracrine signal for epithelial growth. The IGF system disturbances in BPH are illustrated in

THE IGF SYSTEM IN CaP

CaP is the most commonly diagnosed cancer in American men, and the second most lethal. Each of the IGF system levels delineated above has been reported affected in CaP. CaP cells are responsive to the mitogenic effects of the IGF/IGF-receptor signal. For example, adding exogenous IGF-I or IGF-II to DU 145, an androgen-independent CaP cell line that expresses IGF-receptors, stimulates monolayer and anchorage-independent cell growth [44]. Whereas normal PC-E do not synthesize or secrete significant amounts of either IGF-I or IGF-

II, IGF-I secretion has been found in the established metastatic cancer cell lines PC-3, LNCaP and DU 145 [45], but this may reflect a late genetic change since IGF-I production has not been confirmed in any PC-E from CaP tissues. IGF-II production has been detected in DU 145 [44], PC-3 [46], SV-40 immortalized PC-E [39], and CaP tissues [47]. Furthermore, monoclonal anti-IGF-antibodies and anti-type 1-IGF receptor-antibodies could inhibit PC-3 cell growth dose-dependently up to 80% [46]. Reduction in type 1-IGF receptor has been reported in SV-40 immortalized cell lines [39] and CaP tissues [47], but this down-regulation may not be a primary event in neoplastic transformation; it may result from a downstream mutation, making the cells IGF-receptor-independent.

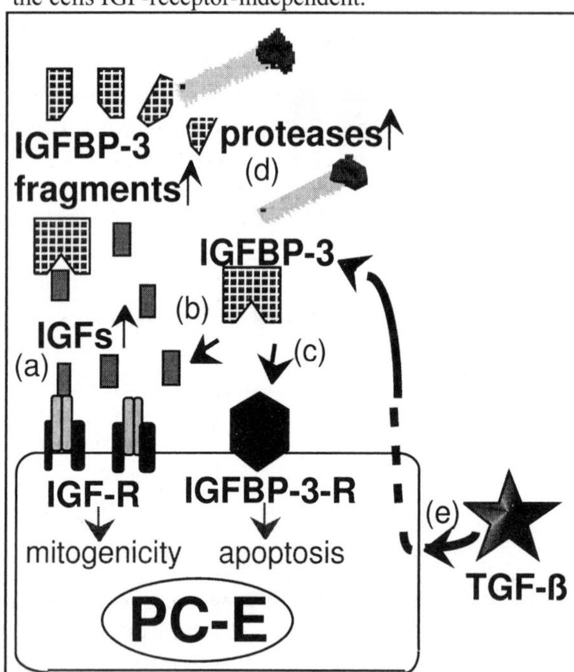

Figure 3. IGF axis disturbances in CaP. (a) Elevated IGF levels may lead to increased mitogenicity of CaP cells by binding to IGF-R. (b) IGFBPs, which block IGF action, may be reduced in CaP. (c) IGFBP-3 can also inhibit cell growth and induce apoptosis by binding its own receptor; it may be reduced in p53 mutant CaP. (d)Augmented protease activity in CaP increases IGFBP-3 fragmentation and diminishes its functioning. (e) TGF-ß induction of IGFBP-3 activates cell cycle arrest and apoptosis; this, too, may be blunted in CaP.

Changes in the IGFBP profile have also been associated with CaP. In one study comparing benign PC-E with prostate intraepithelial neoplasia (PIN) and prostatic adenocracinoma, IGFBP-2 mRNA and protein increased with progressive malignancy; IGFBP-3 mRNA levels did not differ, but protein immunoreactivity changed PIN > benign > adenocarcinoma [48]. IGFBP-4 and IGFBP-5 also increased with progressive malignancy, and there was no change in IGFBP-6 levels [13]. A direct relation of these findings to the progression of CaP has not been demonstrated, and they may represent secondary phenomena. Our lab has shown that IGFBP-3 dose-dependently induces apoptosis through an IGF/IGF receptor-independent pathway in PC-3 cells [26], suggesting that decreased IGFBP-3-induced apoptosis might be involved in prostate cancer progression.

IGFBP alterations have been linked to protease perturbations. The volume of prostate cancer is correlated to the elevation in serum PSA level, which has also been correlated with increased serum IGFBP-2 [49-51] and decreased intact IGFBP-3 [50,51]. Since the pattern of IGFBP-3 cleavage fragments in serum differs from that created by seminal PSA [14,49], and serum PSA is inactivated by inhibitors [52], PSA unlikely causes of the proteolysis of serum IGFBP-3. PSA is active, however, in seminal fluid and presumably in the prostate as well. Whereas PSA is normally sequestered in the prostatic lumens, malignancy may disrupt the acinar architecture and thereby lead to PSA leakage into the surrounding stroma and eventually, the bloodstream. Thus, IGFBP-3 proteolysis by PSA is more likely a local effect, within the prostate or metastatic foci, whose consequences contribute to the local propagation of neoplasia or metastasis. Cathepsin D may also contribute to the local and metastatic progression of malignancy by similar mechanisms. Although Cathepsin D secretion did not differ between cancerous and normal PC-E in cell culture, early malignant changes including increased sialic acid content in surface glycoproteins and exaggerated proton release by membrane-bound proteins, may create an acidic microenvironment which would activate Cathepsin D protease activity [20]. Urokinase and plasminogen-activator proteolysis of IGFBP-3 were found to impact growth of cultured PC-3 cells [46].

Cytokines, particularly transforming growth factor-ß1 (TGF-ß1) and tumor necrosis factor-a (TNF-a) which are known to induce apoptosis also induce IGFBP-3 expression in the p53 negative prostate cancer cell line, PC-3 [24]. This effect of TGF-ß1 was prevented by co-treatment with IGFBP-3 neutralizing antibodies or IGFBP-3 specific antisense thiolated oligonucleotides, suggesting the role of IGFBP-3 in mediating TGF-ß1 induced apoptosis in PC-3 cells [26]. Our current understanding of the IGF system in CaP is illustrated in Figure 3.

Table 2. Summary of the known prostatic IGF system components.

IGF molecule	PC-E	PC-S	Prostate cell lines	Circulation
IGFs	none	IGF-II	IGF-II (in PC-3)	IGF-I, IGF-II
IGF action	proliferation	unknown	proliferation (all); survival (PC-3)	unknown
IGF receptors	IGF-1R	IGF-1R	IGF-1R (all lines)	none
IGFBPs	IGFBP-2,4,5	IGFBP-2,3,4,5	IGFBP-2 -6	IGFBP-3, -2
IGFBP action	IGF-inhibitory (IGFBP-2, -3)	unknown	apoptosis: IGFBP-3 in PC-3	IGF-inhibitory
IGF axis regulators	none found	TGFß, WT-1, androgens	TGFß, (PC-3), DHT (LNCaP)	prostate cancer age
putative IGFBP receptors	unknown	35,50, & 60 kD	20, 60, & 300 kD	currently being examined
IGFBP proteases	PSA, hK-2, Cathepsin D	MMP-1	cathepsins, MMPs, urokinase	MMPs, others
protease action	block IGFBP inhibition	degrade IGFBP-5	promote growth	cleave IGFBPs

CONCLUSION

As our understanding of the prostatic IGF system increases, so too does our awareness of its complexity (Table 2). This carefully balanced network of stimulatory and inhibitory molecules keeps cellular growth in check. When the IGF system is disturbed, aberrant growth such as BPH and cancer may ensue. Cell cycle control points lie not only with the IGFs themselves, their receptors and binding proteins. IGFBP proteases, upstream regulators and downstream response molecules all provide additional sites for IGF-related growth modulation. Further research will likely unravel yet unknown components and more tangled interactions. The prostate is an ideal model for IGF investigation, because it not only contains the many parts of the IGF axis, but this axis may hold the key for the successful treatment of BPH and CaP, two clinical conditions with tremendous societal impact.

REFERENCES

1. J.E.McNeal, Prostate, in S.S.Sternberg (ed.), Histology for Pathologists, Raven Press, Ltd., New York, 1992.
2. J.E.McNeal, Am. J. Clin. Pathol., 49 (1968) 347.
3. J.E.McNeal, Invest. Urol., 15 (1978) 340.
4. J.E.McNeal, Cancer, 23 (1969) 24.
5. P.Cohen et al., J. Clin. Endocrin. Metab., 79 (1994) 1410.
6. P.Cohen et al., J. Clin. Endocrin. Metab., 73 (1991) 401.
7. M.Iwamura et al., Prostate, 22 (1993) 243.
8. B.Valentinis et al., Molec. Endocrinol., 9 (1995) 361.
9. C.A.Conover, Endocrinol., 130 (1992) 3191.
10. P.Cohen et al., J. Endocrinol., 142 (1994) 407.
11. R.Rosenfeld et al., J. Clin. Endocrinol. Metab., 70 (1990) 551.
12. R.S.Birnbaum et al., J. Endocrinol., 141 (1994) 535.
13. M.K.Tennant et al., J. Clin. Endocrinol. Metab., 81 (1996) 3783.
14. P.Cohen et al., J. Clin. Endocrinol. Metab., 75 (1992) 1046.
15. K.O.Lee et al., J. Clin. Endocrinol. Metab., 79 (1994) 1367.
16. S.R.Plymate et al., J. Clin. Endocrinol. Metab., 81 (1996) 618.
17. G.Frenette et al., Internat. J. Cancer, 71 (1997) 897.
18. D.Deperthes et al., J. Andrology, 17 (1996) 659.
19. C.Lalou et al., Endocrinol., 135 (1994) 2318.
20. S.E.Nunn et al., J. Cell. Physiol., 171 (1997) 196.
21. J.H.Kim et al., Endocrinol., 137 (1996) 991.
22. D.P. Sells et al., Cell Growth Differen, 5 (1994) 457.
23. A.C.Levine, Trends Endocrinol. Metab., 6 (1995) 128.
24. G.Fiorelli et al., J. Clin. Endocrinol. Metab., 72 (1991) 740.
25. M.Marcelli et al., Endocrinol., 136 (1995) 1040.
26. R.Rajah et al., J. Biol. Chem., 272 (1997) 12181.
27. H.Werner et al., J. Biol. Chem., 269 (1994) 12577.
28. I.A.Drummond et al., Science, 257 (1992) 674.
29. L.Buckbinder et al., Nature, 377 (1995) 646.
30. R.L.Ludwig et al., Molec. Cell. Bio., 16 (1996) 4952.
31. P.Friedlander et al., Molec. Cell. Bio., 16 (1996) 4961.

32. T.Girinsky et al., Cancer Res., 55 (1995) 3726.
33. A.J. D'Ercole, Endocrinol. Metab. Clinics N. Amer., 25 (1996) 573.
34. K.Tu et al., Cancer Lett., 93 (1995) 147.
35. H.Okazawa et al., J. Cell. Biol., 132 (1996) 955.
36. K.Tsukada et al., Biochem. Biophys. Res. Commun., 2105 (1995) 1076.
37. J.Yong-Keun et al., JBC, 271 (1996) 5112.
38. S.Rosenzweig et al., Adv. Exp. Med. Biol., 343 (1993) 159.
39. S.R.Plymate et al., J. Clin. Endocrinol. Metab., 81 (1996) 3709.
40. J.E.McNeil, J. Urol. Clin. N. Amer., 17 (1990) 477.
41. G.Dong et al., J. Clin. Endocrinol. Metab., 82 (1997) 2198.
42. S.V.Allander et al., J. Biol. Chem., 269 (1994) 10891.
43. S.E.Nunn et al., 10th Internat. Congress Endocrinol. Program & Abstr,(1996) 776.
44. J.A.Figueroa et al., J. Clin. Endocrinol. Metab., 80 (1995) 3476.
45. Z.Pietrzkowski et al., Cancer Res., 53 (1993) 1102.
46. P.Angelloz-Nicoud and M.Binoux, Endocrinol., 136 (1995) 5485.
47. M.K.Tennant et al., J. Clin. Endocrinol. Metab., 81 (1996) 3774.
48. M.K.Tennant et al., J. Clin. Endocrinol. Metab., 81 (1996) 411.
49. T.A.Stamey et al., J. Urol., 141 (1989) 1076.
50. P.Cohen et al., J. Clin. Endocrinol. Metab., 76 (1993) 1031.
51. H.Kanety et al., J. Clin. Endocrinol. Metab., 77 (1993) 229.
52. A.Christensson et al., Europ. J. Biochem., 194 (1990) 755.

Molecular Mechanisms to Regulate the
Activities of Insulin-like Growth Factors
K. Takano, N. Hizuka and S-I. Takahashi (Editors)

Evolutionary Aspects of the IGF System

Chris Collet, Judith Candy and Vicki Sara

Centre for Molecular Biotechnology, School of Life Science, Queensland University of Technology, GPO Box 2434, Brisbane, QLD 4001 Australia

1. INTRODUCTION

Insulin and the IGF molecules provide an excellent model system for examining the phenomenon of gene duplication and diversification from a hybrid progenitor to produce hormones of quite different function (1). Three aspects of the functional and structural diversification of the insulin and IGF hormone systems are of significant interest. Firstly, there are major changes in insulin and IGF peptide structure which include the acquisition and loss of coding regions and these are concomitant with changes in proteolytic processing of the pro-peptide. Secondly, insulin, IGF-I and IGF-II show significant differences in hormone responsiveness and temporal and spatial patterns of gene expression. Finally, there are significant differences between the two hormone systems in their mechanisms of action as the cell- and tissue-specific responses to circulating IGFs, in contrast to insulin, are modulated by a suite of specific binding proteins.

We have recently begun a project that examines the evolution of the IGF system in the radiation of the early chordates. The pioneering work of Drs Steiner, Chan and colleagues who have isolated the cDNAs encoding the insulin-like and proto-IGF molecules from lower vertebrates and protochordates (2-4) forms the basis of our approach. There are some anomalies in the currently accepted model of IGF ligand evolution, however, and the recent finding of insulin-like and IGF-like gene in a tunicate (5) suggests the two hormone systems diverged much earlier than thought. Our perspective is from the evolution of the genomic DNA (exons, introns, regulatory regions and intergenic DNA) and we hope to provide an insight into the diversification of these two important hormone systems and also on aspects of genome evolution spanning 1 billion years. In this paper, we present some of our ideas on evolution of the components of the IGF system. We also present the results of our initial analyses of a bony fish model system, the barramundi *Lates calcarifer*.

2. EVOLUTION OF THE IGF LIGANDS

2.1. The pathway and timing of IGF gene duplications

A pathway for divergence of the IGF genes from the evolutionary-more ancient insulin gene can be proposed based on human linkage relationships. In humans, the genes encoding tyrosine hydroxylase (TH), insulin (INS) and IGF-II form an extremely tight linkage group separated by 2.7 and 1.4 kb of flanking DNA, respectively (6,7). Their adjacent chromosomal location suggests the proto-IGF gene originated as the product of a tandem duplication of insulin. Similarly, mapping data suggests the divergence of IGF-I and IGF-II is the by-product of a partial chromosomal duplication involving chromosome 11 and its translocation to chromosome 12. The duplication of part of the short arm chromosome 11 includes TH, IGF-II, lactate dehydrogenase A (LDH-A), parathyroid hormone (PTH), the proto-oncogene Hras1, calcitonin and catalase. Paralogues of these loci occur on chromosome region 12q22-24, i.e., phenylalanine hydroxylase (PAH), IGF-I, LDH-B, PTH-related peptide, Kras2, calcitonin II and another catalase (8).

The currently accepted model of IGF evolution is derived from comparative analysis of cDNA sequences obtained from taxa which represent each of the major vertebrate lineages. A single insulin-like peptide (ILP) has been found in the cephalochordate amphioxus *Branchiostoma californiensis* with 48% sequence identity to the B- and A-domains of both vertebrate insulin and the IGFs (3). Separate insulin and proto-IGF genes have been characterised from the agnathan hagfish *Myxine glutinosa* (3,9). The agnathan proto-IGF has equal sequence identity to and characteristics of both mammalian IGF-I and IGF-II. These findings led to the proposal that amphioxus ILP represents the gene ancestral to both the vertebrate insulins and IGFs (2,3). These results suggest that the duplication event which gave rise to the separate insulin and proto-IGF genes occurred between the radiation of the cephalochordate and agnathan lineages. Distinct and characteristic IGF-I and IGF-II genes first appear in the Chondrichthyes (4) suggesting that all the gene duplication events had occurred prior to the rise of the major vertebrate lineages.

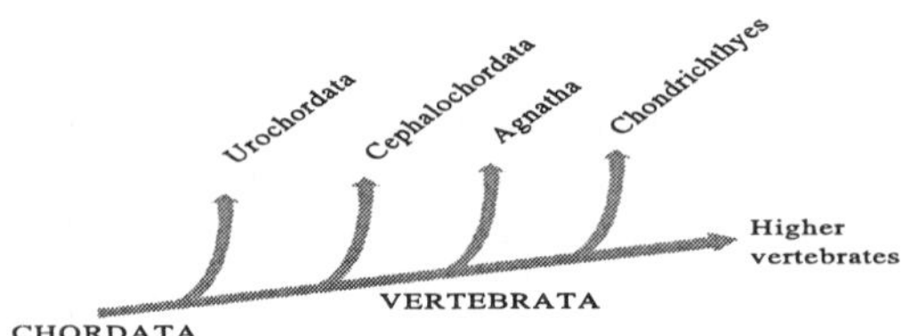

Figure 1. Schematic representation of the evolutionary relationships of the lower vertebrates (Agnatha or jawless fish and Chondricthyes or cartilaginous fish) and the proto-chordates (Urochordata or tunicates and Cephalochordata or amphioxus). The bony fish (Class Osteichthyes) diverged from the main tetrapod line after the divergence of the cartilaginous fish.

2.2. Anomalies in the proposed model

The direction of our project arose from the anomalies in the proposed model which raised doubts about the nature of the gene proposed as ancestral to both the vertebrate insulins and IGFs. The putative ILP peptide has paired basic residues which, theoretically, should permit cleavage of the C-domain and, in this respect, the gene is similar to insulin. The carboxy-terminal D- and E-domains of the amphioxus ILP gene are contiguous within the same exon of the A-domain and the E-domain might be subject to proteolytic cleavage since it contains a basic residue at its amino end. These latter characteristics are reminiscent of the proIGF genes. The occurrence of the D-domain in the IGFs and, hence in the amphioxus ILP, is significant as this domain is thought to block interaction between IGFs and the insulin receptor (10). The acquisition of the additional coding regions was most likely by a mutation which abolished the termination codon permitting read through by RNA polymerase. Since both the vertebrate insulin and invertebrate ILP genes lack the carboxy-terminal domains present in the IGFs, their presence in the amphioxus ILP presents an anomaly. For amphioxus ILP to represent the gene ancestral to the vertebrate insulin and IGFs, the additional coding regions must have been gained and subsequently lost by a reciprocal mutation at precisely the same location. Furthermore, several substitutions known to be critical for stability of a structure commensurate with high biological activity of insulin and the IGFs are not conserved in the amphioxus ILP A or B domains (2). The finding of insulin- and IGF-I- immunoreactive material in the gut mucosal cells and the cells peripheral to the nerve ganglion of the urochordate seasquirt *Ciona* (11) raised further doubt about the nature of amphioxus ILP. An alternative to invoking reciprocal mutations, is that another insulin-like gene exists in the cephalochordates that does not have the carboxy-terminal domains. Implicit is the suggestion that insulin and IGF diverged earlier than the model predicts and that the cephalochordate ILP is not the gene ancestral to vertebrate insulin and IGFs.

Clearly we have been uncertain of the nature of the gene ancestral to the vertebrate insulin and IGFs and the time of divergence of these two important peptide hormone systems. The very recent finding of cDNA sequences encoding both an insulin-like and an IGF-like gene in the urochordate *Chelysoma productum* suggests the two hormones diverged much earlier than previously thought and lineages which diverged from the main tetrapod line early in the radiation of the Deuterostomes need to be examined.

2.3 Studies on a teleost fish species: the barramundi *Lates calcarifer*

The barramundi or sea bass is a diploid perciform species that grows to about 1.5 metres and is found in northern Australia. In any choice of experimental organism, there are a number of important considerations; the foremost is taste, while the second is exotic field-trip destinations. The barramundi fulfils these criteria admirably as the meat is very sweet and the animal inhabits the rivers and creeks of the tropical rainforest areas of far north Queensland. Our lab has characterised the cDNA sequences of the barramundi IGFs and the chromosomal regions encompassing these genes.

Anne Kinhult in our group isolated and characterised the barramundi IGF-I cDNA sequence and examined the patterns of gene expression (12,13). The putative barramundi IGF-I has 88 and 93 amino acid sequence identity with homologous Ea4 prepropeptide sequences from salmonid and other perciform species, respectively. Expression of the IGF-I gene is GH-dependent in mammals (14). In the salmonids, four IGF-I transcripts (Ea-1, -2, -3 and -4) arise by alternate splicing of a cassette exon and two of these, Ea1 and Ea3, are GH-responsive (15-17). In contrast, Ea1 and Ea3 could not be detected in any juvenile barramundi tissue examined (liver, brain, heart, kidney, gill, spleen) whereas Ea2 could be found in liver and brain and Ea4 in liver, brain, spleen, kidney, heart and gill. Nor could Ea1 and Ea3 transcripts be induced by administration of recombinant bream GH (12,13) and there was no change in the Ea2:Ea4 ratio. Labelled rbGH did appear to bind liver membrane preparations and could be competed out by cold barramundi or bream GH suggesting an interaction between bream GH and barramundi GH-receptor. The absence of the transcripts in barramundi is unusual since alternate splicing does generate transcripts which either lack (Ea2) or include the cassette exon (Ea4). Experiments are underway to examine wild caught barramundi and to test the possibility that, in this species, IGF-II rather than IGF-I mediates the effect of GH.

We have also isolated the genomic region encompassing IGF-I from barramundi and are currently putting the sequence together for analysis. The structure of this gene in our species is identical to that of the salmonid IGF-I genes with a five exon/four intron organisation including a small cassette exon of 36 bases. The large size of intron 2 in salmonids is mirrored in barramundi where the intron appears to exceed 20 kb.

The nucleotide sequence of the gene encoding IGF-II in barramundi (18) can be compared directly with the chum salmon homologue which is in the databases (19). The barramundi IGF-II gene spans about 5.5 kb and comprises four exons, of 163, 151, 239 and >1079 bp, separated by three introns of 943, 1543 and 1450 bp respectively. The chum salmon gene is of similar size and has a structure identical to that of barramundi IGF-II including location of introns and putative promoter elements. Interestingly, there are ten extended blocks (32 - 146 bp) of high sequence identity (66 - 94%) located within the introns of the fish IGF-II genes, one in intron 1, two in intron 2 and seven in intron 3, surrounded by regions of low similarity (<30%). The only other genomic sequence for which a direct comparison can be made between these two species is for the growth hormone gene (20,21) where the five introns are neither conserved in sequence nor size. Two hypotheses may account for extensive conservation of sequence elements in the introns of the fish IGF-II genes. Firstly, some or all of the conserved sequence blocks may have some functional role in either expression of the IGF-II gene or in chromosomal house-keeping. Secondly, there may have been insufficient time for the sequences to diverge since divergence of the Perciformes and Salmoniformes 40 - 50 Mya (22). The latter is unlikely given that the five introns of the growth hormone genes of these two species share no similarity to each other in sequence or size. Although the rates of molecular evolution of non-coding DNA have been reported to vary substantially between genes (23,24), even a low rate of nucleotide substitution within the IGF-II region could not be expected to give rise to the contrasting patterns of sequence conservation and divergence evident.

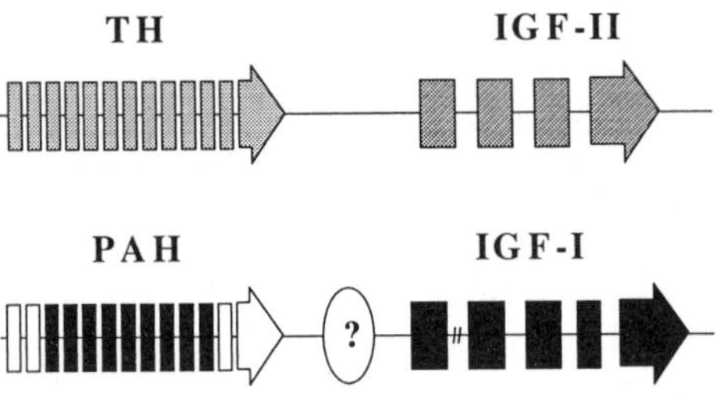

Figure 2. Tyrosine hydroxylase and IGF-II are sparated by 2.5 kb in the barramundi. Absent from this region is the insulin gene, situated between TH and IGF-II in mammals. We have characterised IGF-I and begun to sequence the gene encoding phenylalanine hydroxylase (PAH) in barramundi. Linkage between PAH and IGF-I remains to be demonstrated. Unshaded exons denote those exons that remain to be sequenced.

Unlike the mammalian IGF-II genes which span 30 kb of chromosomal DNA and have multiple mRNA transcripts arising through the use of stage-specific promoters and from differential polyadenylation, the basic structure of the fish IGF-II genes is relatively simple and compact. In mammals and chicken, a single exon encodes the 24 amino acid signal peptide and 28 residues of the B-domain of the respective IGF-II prepropeptides. In the fish IGF-II genes, the 47 residue signal peptide is encoded by two exons, 25 residues are encoded by exon 1 while exon 2 encodes 22 amino acids of the signal peptide and 28 residues of the B-domain. In this regard, the structure of the fish IGF-II genes is like that of the genes encoding IGF-I in salmon (25) and barramundi (unpublished data) suggesting that the ancestral IGF genes had a four exon/three intron structure. The acquisition of stage- and tissue-specific promoters in the IGF-II genes of higher vertebrates appears to have been accompanied by loss of the exon encoding the first half of the signal peptide.

We have also characterised the barramundi gene encoding TH which lies immediately 5' of IGF-II. Barramundi TH has a similar exon/intron structure to the human TH gene (26). The fish TH peptide shows about 87% identity to the TH peptides of mammals and quail (27). Interestingly, the insulin gene is absent from the intergenic region between TH and IGF-II. In the barramundi, approximately 2.5 kb of chromosomal DNA separates the coding regions of exon 13 of TH and exon 1 of IGF-II (Fig. 2). The insulin gene from the chum salmon spans 1.3 kb of genomic DNA (28) and the barramundi gene should be of similar size. No sequences related to insulin were identified in the barramundi intergenic region nor were any significant similarities noted between this region and any other sequences in the nucleic acid databases. In this species then, the insulin gene may be between PAH and IGF-I. If this is so, then the insulin gene of bony fish may be an ancient paralogue of the insulin gene of higher vertebrates arising as a consequence of differential silencing of the duplicates in the different vertebrate lineages. The insulin gene between TH and IGF-II may have been silenced in the bony fish after divergence of this lineage from the main tetrapod lineage while the INS gene between PAH and IGF-I may have been silenced in the lineage leading to mammals. We have begun a characterisation of the regions encompassing PAH and IGF-I in barramundi to investigate this possibility. Recently, we have isolated PAH and can now begin to search for the insulin gene.

3. EVOLUTION OF THE IGF-BINDING PROTEINS

3.1 IGF-BPs in lower vertebrates

Our knowledge of the number and nature of the IGF-binding proteins (IGF-BPs) in the lower vertebrates and protochordates is virtually non-existent. At least three IGF-BPs in bony fish have been detected (29). Their structural and functional relationships with the binding proteins of the higher vertebrates is unknown although there are some obvious similarities in regulation (29). Three IGF-binding proteins have also been detected in the serum of the agnathan lamprey (*Geotria australis*) with a major band at 50 kDa and minor protein species at 32 and 28 kDa (30). These latter findings suggests this group of proteins are an ancient multigene family that may have arisen at a point in evolutionary time concordant with or, at least, soon after the divergence of IGF from insulin. In this regard, the IGF-specific binding proteins may have played an immediate role in inhibiting the insulin-like activities of the proto-IGF as well as providing some degree of tissue- and developmental-specificty in the mode of action of the newly evolved IGF ligand.

3.2 The pathway and timing of IGF-BP gene duplications

As the examination of human linkage groups points to a pathway for the evolution of the IGF ligands, a similar exercise also provides some information on the possible pathway and the timing of the duplication events for the evolution of the IGF-binding proteins. In humans, IGF-BPs 1 and 3 are separated by 20-40kb on chromosome region 7p14-p12 and IGF-BPs 2 and 5 are separated by 20kb of DNA on chromosome region 2q33-q34 (31,32). The gene pairs are also tightly linked in mouse, with the genes encoding IGF-BPs 2 and 5 located 5kb apart (33). The genes in each pair are arranged with opposite transcriptional orientations in a tail-to-tail fashion. IGF-BPs 1 and 2 are more closely related to each other as are IGF-BPs 3 and 5. IGF-BP4 has been localised to chromosomal region 17q12-q21.1 and is more closely related to IGF-BPs 1 and 2 (32,33). IGF-BP6 is found at region 12q13 and would appear to be the most divergent member of the IGF-BP family (34).

The existence of the gene pairs suggests that local tandem gene duplication and inversion has occurred once in the evolution of the IGF-BPs and there have been two or three chromosome duplication events. The order of duplication events is not apparent from phylogenetic analysis of the IGF-BP sequences as the length of the sequences does not provide for robust results. The partial chromosome duplications appear to be large as other gene families have been coduplicated including genes for epidermal growth factor receptors and their oncogene homologues, cytokeratin, fibrillar-type collagen, retinoic acid receptor, actin, zinc finger proteins, glucose transporter proteins, mysoin light chain and the entire HOX gene clusters (8). Sequences of the fibrillar-type collagens and some of the HOX genes have formed the basis of phylogenetic reconstruction to establish the number and order of duplication events (35). The divergence of the four syntenic groups is thought to have been a three step process whereby the linkage group containing IGF-BPs 2 and 5 is thought to have diverged earliest followed by the syntenies containing IGF-BPs 1 and 3, then IGF-BP6 and finally IGF-BP4.

The amphioxus genome contains a single HOX gene cluster with the ancestral archetypical organisation of individual homeobox-containing genes (36). The agnathan species of lamprey contains either two, or possibly three, HOX gene clusters (36). The latter agrees with the finding of three proteins which bind IGF in ligand blotting studies (30). Teleost fishes have four HOX gene clusters as do all the higher vertebrates suggesting that the major rounds of genome expansion in the main vertebrate lineage had been complete by the divergence of the bony fish (35,36).

3.3 More IGF-binding proteins?

The model implies that the ancestral syntenic group, at least immediately prior to the chromosomal duplication, contained an IGF-BP gene pair. Each subsequent partial chromosome duplication would be expected to give rise to another IGF-BP gene pair. This is certainly evident in the clustering of IGF-BPs 1 and 3 but not with IGF-BPs

4 or 6. The model implies that eight IGF-BPs should be found in vertebrate taxa above the cartilaginous fish and, thus, a closer examination of the number of binding proteins is warranted in vertebrate taxa. Notwithstanding the possible loss of duplicated genes, it is possible that two other IGF-BPs exist in vertebrates and these could be found 3' of IGF-BPs 4 and 6 (Fig. 3). Given that some of the IGF-BPs are characterised by low levels of expression and restricted tissue- and developmental-specificity, it remains a distinct possibility that not all IGF-BPs have been detected.

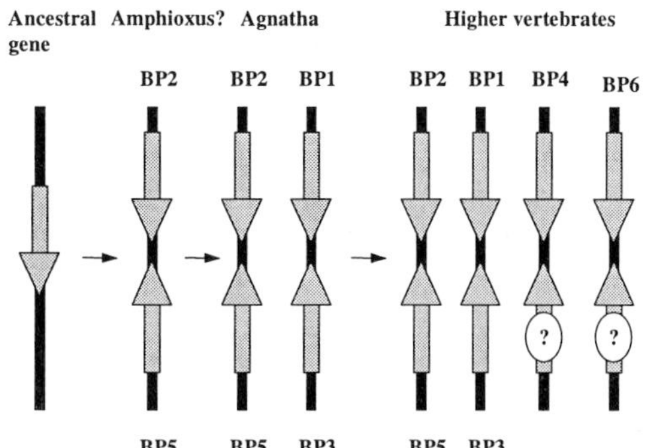

Figure 3. The IGF binding proteins are linked to the HOX gene clusters (8). The proposed model (35) of HOX gene cluster duplication implies that there may be two as yet unidentified binding proteins in vertebrates. These would be located 3' of IGF-BPs 4 and 6.

The model also implies that at least a single IGF-BP gene is present in amphioxus, although this remains to be demonstrated. The finding of an IGF-BP gene pair in amphioxus would, however, suggest the divergence of function for the insulin and IGF hormone systems before the radiation of the cephalochordate lineage and further support the notion that a functional IGF system is present in the Urochordata.

3.4 Other (types of) IGF-binding proteins?

Recently, claims have been made of the isolation and characterisation of the seventh and eigth members of the IGFBP multigene family. IGF-BP7 was isolated from human mammary epithelial cells (37). This protein, termed mac25, had also been previously isolated from human meningioma cells (38) and from the mouse (39). High affinity binding of the IGF ligands to non-glycosylated recombinant human mac25 has been demonstrated (40) and the level of affinity of mac25 to these ligands is 5 - 6 -fold and 20 - 25 -fold lower than IGFBP-3 binding to the IGF-I and IGF-II ligands, respectively. Sequence similarity between mac25 and the IGFBPs 1 to 6 was reported at 40% (37). The observed sequence similarity is restricted to the N-terminal domain, however, and no significant sequence similarity occurs between mac25 and the IGFBPs outside of this domain.

The conserved N-terminal cysteine-rich domain of the IGF-binding proteins is also found in other secreted extracellular proteins that form a recently recognised family of growth regulators, denoted the CCN family (41). This family of proteins includes the human connective growth factor (CTGF) and its mouse homologue fisp-12, mouse cyr61 and its chicken homologue cef-10, and the nov gene which is over-expressed in chicken nephroblastoma cells. The three CCN proteins (CTGF/fisp-12; cyr61/cef-10; nov) contain five separate structural domains. Four of the domains show similarity to other protein families, from the N- to C-terminals these domains show similarity with the IGFBPs (21 - 38%), Von Willebrand factor type C repeat (23 - 41%), a motif thought to bind both soluble and insoluble glycoconjugates and a dimerization domain found in Von Willebrand factor, the mucins and TGF-β (21 - 26%), respectively. The central, fifth domain is variable between CCN proteins. The N-terminal domain shared between the IGFBPs and the CCN family comprises 12 cysteine residues including the highly conserved GCGCCxxC motif and is clearly involved in some form of ligand binding. Zumbrunn and Trueb (42) recently reported the sequence of a serine protease which shares with the IGFBPs and the CCN proteins the same N-terminal domain. This serine protease has, because of the N-terminal domain, been classed as a protease of IGF-BP3 although there is no *a priori* reason to believe that proteases which cleave IGF-BP3 should also bind the ligand. The N-terminal domain of the IGFBPs is located within a single exon and it would appear that the origin of the ancestral IGF-BP gene presents another example of gene evolution by exon shuffling. No obvious homologies exist for the variable central domain of the IGF-BPs or the cysteine-rich C-terminal domain.

Undoubtedly mac25 does bind, with high affinity, IGF ligand, but it does not fall into the IGFBP family *per se* as it does not have the conserved cysteine-rich C-terminal domain shared by the latter. It does appear, however, to fall into that group of CCN proteins that have the growth-factor binding domain. Given that the function and mechanisms of action of the IGFBPs are not yet fully understood, nor for that matter are the same attributes of the CCN proteins, it may be premature to consider the CCN proteins as *bona fide* members of the IGFBP family.

A preliminary report has also proposed connective tissue growth factor as IGF-BP8 although it was found to bind IGF-I and IGF-II with only relatively low affinity (43). Unfortunately the proposed grounds for inclusion of CTGF as an IGF-BP, that is the presence of the GCGCCxxC motif, would exclude IGF-BP6, with a highly divergent motif (GCTEAGGC), from the family.

4. CONCLUDING REMARKS

The comparative evolutionary approach presents a valid scientific approach in unravelling IGF function and mechanisms of action in higher vertebrates. How far down the evolutionary tree we are required to go to find the minimal IGF system has yet to be determined although it is now apparent that insulin and the IGFs diverged earlier in the radiation of the chordates than previously thought. In this regard the origins of the IGFs may parallel more closely the protochordate (and early deutero-stomian?) Innovations in regulated development rather than the evolution of the hepatopancreatic system as a major endocrine organ in vertebrates.

REFERENCES

1. D.F. Steiner, S.J. Chan, J.M. Welsch and S.C.M. Kwok, Ann. Rev. Genet. 19 (1985) 463

2. S.J. Chan, Q. Cao and D.F. Steiner, Proc. Natl. Acad. Sci. (USA) 87 (1990) 9319
3. S. Nagamatsu, S.J. Chan, S. Falkmer and D.F. Steiner, J. Biol. Chem. 266 (1991) 2397
4. S.J. Duguay, S.J. Chan, T.P. Mommsen and D.F. Steiner, FEBS Lett. 371 (1995) 69
5. J.E. McRory and N.M. Sherwood, DNA and Cell Biol., 16 (1997) in press
6. K.L. O'Malley and P. Rotwein, Nucl. Acids Res. 16 (1988) 4437
7. P. de Patger-Holthuizen, M. Jansen, F. van Schaik, R. van der Kammen, C. Oosterwijk, J.L. Van den Brande and J.S. Sussenbach, FEBS Lett. 214 (1987) 259
8. L. Lundin, Genomics 16 (1993) 1
9. S.O. Emdin, D.F. Steiner, S.J. Chan and S. Falkmer, NATO ASI Ser A: Life Sci 103 (1985) 363
10. Y-C. Chu, S-Q. Hu, L. Zong, G.T. Burke, S. Gammeltoft, S.J. Chan, D.F. Steiner and P.G. Katsoyannis, Biochemistry 33 (1994) 11278
11. M. Reinecke, D. Betzler, S. Falkmer, K. Drakenberg and V.R. Sara, Histochemistry 99 (1993) 277
12. A. Kinhult, Ph.D. thesis, Queensland Unversity of Technology, Brisbane, (1996)
13. A. Kinhult, V. Sara and P. Hoeben, submitted to J. Mol. Endocrinol.
14. V.R. Sara and K. Hall, Physiol. Rev. 70 (1990) 591
15. S.J. Duguay, P. Swanson and W.W. Dickhoff, J. Mol. Endocrinol. 12 (1994) 25
16. S.J. Duguay, J. Lai-Shang, D.F. Steiner, B. Fukenstein and S.J. Chan, J. Mol. Endocrinol. 16 (1996) 123
17. M.J. Shamblott and T.T. Chen, Mol. Mar. Biol. Biotech. 2 (1993) 351
18. C. Collet, J. Candy, N. Richardson and V.R. Sara, Biochem. Genet. (1997) in press
19. A. Palamarchuk, A. Nektrutenko, O. Gritzenko and V. Kavsan, Genbank (1996) X97225.
20. X. Shen., Y. Wang, M. Welt, D. Liu and F.C. Leung, Genbank (1993) L04688
21. D.L Yowe and R.J. Epping, Gene 162 (1995) 255
22. J.S. Nelson, Fishes of the world, 2nd edn, Wiley, New York, (1984)
23. M. Bulmer, K.H. Wolfe and P.M. Sharp, Proc. Natl. Acad. Sci. USA 88 (1994) 5974
24. S. Easteal and C. Collet, Mol. Biol. Evol. 11 (1994) 643
25. V.M. Kavsan, V.A. Grebenjuk, A.P. Koval, A.S. Skorokhod, C.T. Roberts Jr. and D. LeRoith, DNA and Cell Biol. 13 (1994) 555
26. K.L. O'Malley, M.J. Anhalt, B.M. Martin, J.R. Kelsoe, S.L. Winfield and E.I Ginns, Biochemistry 26 (1987) 6910
27. M. Faquet, B. Grima, A. Lamouroux and J. Mallet, J. Neurochem. 50 (1988) 142
28. A.P. Koval, A.I. Petrenko, A.I. and V.M. Kavsan, Nucl. Acids Res. 17 (1989) 4916
29. M. Reinecke and C. Collet, Int. Rev. Cytol., in press (1997)
30. Z. Upton, S.J. Chan, D.F. Steiner, J.C. Wallace and F.J. Ballard, Growth Reg. 3 (1993) 29
31. S.V. Allander, PhD thesis, Karolinska Hospital, Stockholm (1994)
32. E. Ehrenborg, PhD thesis, Karolinska Institute, Stockholm (1994)
33. K. Kou, P.L. James, D.R. Clemmons, N.G. Copeland, D.J. Gilbert, N.A. Jenkins and P. Rotwein, Genomics 21 (1994) 653
34. S. Shimasaki, L. Gao, M. Shimonaka and N. Ling, Mol. Endocrinol. 5 (1991) 938
35. W.J. Bailey, J. Kim, G.P. Wagner and F.H. Ruddle, Mol. Biol. Evol. 14 (1997) 843
36. P.W.H. Holland, J. Garcia-Fernandez, N. Williams and A. Sidow, Dev. Suppl. (1994) 125
37. K. Swisshelm, K. Ryan, K. Tsuchiya and R. Sager, Proc. Natl. Acad. Sci. USA 92 (1995) 4472
38. M. Murphy, M. Pykett, P. Harnish, K. Zang and D. George, Cell Growth Diff. 4 (1993) 715
39. M. Kato, H. Sato, T. Tsukada, Y. Ikawa, S. Aizawa and M. Nagayoshi, Oncogene 12 (1996) 1361
40. Y. Oh, S.R. Nagalla, Y. Yamanaka, H.S. Kim, E. Wilson and R.G. Rosenfeld RG, J. Biol. Chem. 271 (1996) 30322
41. P. Bork, FEBS Lett. 327 (1993) 125
42. J. Zumbrunn and B. Trueb, FEBS Lett. 398 (1996) 187
43. K.S. Kim, S.R. Nagalla, Y. Oh, E. Wilson, C.T. Roberts Jr and R.G. Rosenfeld, Proc. Endo. Soc. (1997) 352

The IGF system in the brain - response to injury and therapeutic potential.

P.D. Gluckman, J. Guan, A. Scheepens and C.E. Williams.

Research Centre for Developmental Medicine and Biology, School of Medicine, University of Auckland, Private Bag 92019, Auckland, New Zealand.

1. SUMMARY

It is now well established that much cell death after acute brain injury is due to inappropriate activation of apoptosis. Similarly, apoptosis appears to be a central process in many forms of neurodegeneration. Our studies have demonstrated that following acute brain injury, induced by hypoxia-ischemia, specific components of the IGF system are induced in the brain: those induced within the phase of delayed cell death are IGF-1, IGFBP-2, IGFBP-3 and IGFBP-6 whereas other components such as IGF-2 and IGFBP-5 are induced only during the repair/scarring phase. Administration of IGF-1 to the injured brain 2-6 hours after injury dramatically reduced neuronal and glial cell loss, presumably by interfering with apoptosis. Our studies show that exogenous IGF-1 is specifically targeted to neurones in the region where cell death would have occurred by an IGFBP, presumably IGFBP-2. Our data also raises the possibility that in addition to IGF-1 itself having a neuroprotective action it may also act as a prohormone for the tripeptide GPE which has an independent neuroprotective action. The potential use of IGF-1 to treat both acute brain injury and neurodegenerative brain disease is under active consideration.

2. THE MECHANISMS OF ACUTE BRAIN INJURY

It is now clear that the mechanisms of injury and response are comparable across various forms of acute brain injury including those induced by ischemia (e.g. stroke) and asphyxia (e.g. birth asphyxia) provided that the insult is reversed by resuscitation or reperfusion.

Experimental observations have shown that in relationship to a time limited asphyxial insult, neuronal death occurs in two distinct phases. The first phase occurs during the insult and the immediate reperfusion/reoxygenation phase and is termed the primary phase. After a latent period of some hours there is a delayed or secondary phase in which cell loss occurs for at least 72 hours [1-4]. The mechanisms of cell death in these two phases have largely been elucidated using a variety of animal models. In the primary phase cell death occurs through a series of interrelated mechanisms including free radical formation, intracellular calcium accumulation, excitotoxicity, cytotoxic edema and necrotic lysis of the cell membrane. While these mechanisms may play some role in delayed cell death, the major mechanism in the secondary phase is the inappropriate induction of a wave of apoptosis [2]. In addition the activation of microglia may contribute via the release of cytotoxic cytokines [5]. Obviously the relative importance of these two phases depends on the nature and severity of the insult. However, recent evidence clearly shows that the magnitude of the delayed phase is a major determinant of neurological outcome [6].

The degree of brain injury is obviously not only determined by injuring mechanisms but also by potential protective responses. Studies in acute brain injury first raised the possibility that endogenous growth factors might be released in the brain presumably to limit cell death [7]. We took this observation as the starting point for our studies which were designed to test the hypothesis that the brain would induce endogenous neurotrophins to limit the apoptotic processes [7,8]. In this review we focus primarily on the IGF system: however other systems including the TGF-beta system [9] are also induced by injury.

3. THE RESPONSES OF THE CEREBRAL IGF SYSTEM TO ACUTE INJURY

Our initial studies were performed in 21 day old rats subject to a graded unilateral hypoxic-ischemic (HI) injury [10]. In situ hybridisation and other molecular biological techniques, including immunohistochemistry, were used to evaluate which neurotrophins were induced in response to the HI injury. We showed the induction of a number of transcriptional factors including *c-Jun* and *c-Fos* [11,12].

Initial studies on the neurotrophins focused on the nerve growth factor family. Induction of BDNF, and to a lesser extent NGF-beta and NT3, could be observed after injury but only on the non-injured side. That induction was blocked by anticonvulsant therapy suggesting that these neurotrophins are induced by seizures rather than brain injury *per se*.[13]

In contrast, we found very specific induction of components of the IGF system. IGF-1 was markedly induced, in proportion to the degree of injury, and in regions where neuronal death occurred. In the lesser insult induction was largely confined to the cortex, whereas with the more severe injury IGF-1 mRNA could be demonstrated throughout the injured hemisphere (Beilharz *et al* in preparation). There was no induction of IGF-1 mRNA until three days after injury and it was maximal at 5 days. It is tempting to speculate that this is casually related to the decline in apoptosis which commences at that time. The induced IGF-1 mRNA was largely of the Ea form with only a slight and delayed induction of the Eb form. This suggests that the induction is independent of the GH axis [14]. Induction was dependent on start site 3 in Exon 1 (Beilharz and Gluckman, unpublished). Immunohistochemistry demonstrated that initial expression was associated with GFAP positive astrocytes and with microglia [14]. In contrast, IGF-2 was not induced until 6 to 10 days following the injury and expression was limited to microglia within the region of the cortical infarct [15].

There is also a specific pattern of induction of the IGFBPs. IGFBP-1 expression is low and not affected [8]. IGFBP-2 is markedly induced in similar regions to IGF-1 with a similar time-course and it's expression is primarily from GFAP positive astrocytes.[16]. IGFBP-3 is also induced over a similar time course and is maximal at 3 days post injury. However IGFBP-3 is primarily induced by microglia and the expression is much less than for IGFBP-2 [Beilharz *et al* in preparation and 16]. IGFBP-4 is suppressed on the injured side. IGFBP-5 is induced within white matter tracks and thalamic neurones but not until 6 to 10 days after injury [17]. IGFBP-6 is also induced in the injured hemisphere. However, expression is restricted to blood borne neutrophils which enter the brain, ependymal cells lining the third and lateral ventricles and to the choroid plexus. There is minimal staining seen on microglia and some astrocytes surrounding the infarct (Scheepens and Gluckman, unpublished results).

The IGF-1 receptor mRNA showed an equivocal and transient increase in the first few hours after injury but there were no substantial changes over the subsequent period of study.

4. IGF-1 AS A NEURONAL RESCUE AGENT

Our primary hypothesis was that earlier administration of exogenous IGF-1, that is within a few hours of injury, would prove to be neuroprotective. In adult rats using a similar model of unilateral HI, we showed that a single dose of IGF-1 induced a dose dependent reduction in cortical infarction from 94% in controls to 27% at the 50 μg dose [8]. In contrast, IGF-2 and insulin at equimolar doses were not protective. There was a dose dependent reduction in neuronal loss in all areas of the brain, while the reduction in the incidence of cortical infarction showed that glial protection was also conferred [18]. Recently we have shown that IGF-1 is neuroprotective until 6 hours after the injury (Guan and Gluckman, unpublished results). However IGF-1 is not protective if given prior to injury [8]. We conclude that IGF-1 acts early in the post-asphyxial apoptotic cascade to inhibit delayed cell death.

We also demonstrated that IGF-1 was effective in the infant rat subject to HI injury (Zhang *et al*, unpublished). We used the chronically instrumented fetal sheep in late gestation subject to 30 min of total cerebral ischemia to further evaluate the therapeutic potential of IGF-1. A single dose of 0.1 to 1 μg IGF-1 administered into the lateral cerebral ventricle reduced secondary neuronal loss in all areas of the brain. Furthermore, we found a concomitant reduction in the incidence of post-asphyxial seizures and of secondary cytotoxic edema [19]. Surprisingly, infusion of IGF-1 into the lateral ventricle was not as effective as a bolus administration (Gunn, Bennet *et al*, unpublished). This further suggests that IGF-1 acts to block specific and early steps in the progression of programmed cell death rather than the subsequent cascade of events.

5. TRANSPORT OF IGF-1 IN CEREBRAL TISSUE - THE ROLE OF IGFBP'S

Tritiated IGF-1 was injected into the lateral cerebral ventricle of adult rats with unilateral brain injury. It rapidly accumulated in the injured hemisphere at higher concentrations than in the non-injured hemisphere [20]. Autoradiography showed that IGF-1 was initially present in the white matter tracts and the perivascular spaces where volume conduction of proteins can be rapid and is mediated by the transmitted vascular pulsation. High resolution autoradiography showed that the IGF-1 specifically accumulated on the cell bodies of neurones [21]. The transport of IGF-1 was saturable and blocked by both cold IGF-1 or IGF-2 suggesting the role of IGFBPs in transporting and targeting IGF-1 [20,21].

When des 1-3 IGF-1, which has low affinity for the IGFBPs, was administered in the adult rat model a 10 fold higher dose than of IGF-1 was needed to confer neuroprotection, suggesting that IGFBPs are required either to transport or to target exogenous IGF-1 to the injured cells [22].

6. MECHANISM OF THE NEURONAL RESCUE ACTION OF IGF-1

IGF-1 is known to inhibit neuronal apoptosis. For example, IGF-1 has been shown to block developmental apoptosis in chick spinal motorneurones [23]. Furthermore, in pathological paradigms, IGF-1 has been shown to block apoptosis in neurones subject to hypoglycaemia, hypoxia, osmotic stress or beta amyloid toxicity *in vitro* [24,25]. It has been suggested that IGF-1 has anti-apoptotic effects via several mechanisms. In the short term after the injury, IGF-I may delay the onset the apoptosis via a PI_3-kinase or MAP kinase pathway which in turn leads to actions on survival pathways, such as the Bcl system or inhibition with caspases [26-28].

As des 1-3 IGF-1 is also neuroprotective [22], it is reasonable to assume an action via the IGF-1 receptor. However, it has been suggested that in the brain there is a protease that cleaves the IGF-1 N terminus to des 1-3 IGF-1 and a tripeptide, GPE. While we have not been able to demonstrate the production of GPE from IGF-1 in neural cells *in vitro* (Cooper, Skinner *et al*, unpublished), we investigated whether this tripeptide could have a neuroprotective action. When administered 2 hours after injury in the adult rat, it was neuroprotective. The scope of protection was similarly broad to that of IGF-1 covering a range of neuronal phenotypes [29]. A neuroprotective action has also been demonstrated in an organotypic hippocampal culture using NMDA induced neurotoxicity [29]. The neuroprotective action of GPE could be demonstrated at timepoints when the NMDA antagonist MK801 was not neuroprotective, suggesting an action independent of the glutamate receptor. GPE has previously been suggested as a glutamate antagonist [30]. Thus the mode of action must remain speculative.

Therefore, we conclude that IGF-1 is highly likely to have two neuroprotective actions: one mediated via the IGF-1 receptor and one by acting as a prohormone for GPE.

7. CONCLUSIONS

Neuronal rescue is the strategy of administering an agent after acute brain injury to limit secondary neuronal death. This will be the most useful form of neuroprotective therapy. Our studies with IGF-1 suggest it is more potent than other neuronal rescue therapies and is active at times well removed from the insult. The clinical application of such therapies to stroke, perinatal asphyxia and other forms of acute brain injury is obvious. IGF-1 may also have a role in the treatment of neurodegenerative disease where apoptosis is an important component. Irrespective of whether GPE occurs naturally or not, our observations suggest that it may constitute a novel approach to neuroprotection.

8. ACKNOWLEDGMENTS

These studies are supported by the Health Research Council of New Zealand and NeuronZ Ltd. We thank our colleagues Drs G. Werther, V. Russo, B. Johnston, E. Beilharz, M. Klempt and N. Klempt for their contributions to the studies described. We also thank our many colleagues who have contributed to this work including Drs A. Gunn, L. Bennet, A. Butler, S. Skinner, G. Cooper, E. Beilharz, Mr E. Sirimanne and Mr R. Zhang.

REFERENCES

1. C.E. Williams, A.J. Gunn and P.D. Gluckman, Stroke 22 (1991) 516.
2. E.J. Beilharz, C.E. Williams, M. Dragunow, E.S. Sirimanne and P.D. Gluckman, Mol. Brain Res. 29 (1) (1995 Mar) 1-14.

3. C.E. Williams, A.J. Gunn, E.C. Mallard and P.D. Gluckman, P.D. Ann. Neurol. 31 (1992) 14-21.
4. A. Lorek, Y. Takei, E.B. Cady, J.S. Wyatt, J. Penrice, A.D. Edwards, D. Peebles, M. Wylezinska, H. Owenreece, V. Kirkbride, C.E. Cooper, R.F. Aldridge, S.C. Roth, G. Brown, D.T. Delpy and E.O.R. Reynolds, Pediatr. Res. 36 (1994) 699.
5 McNeill, H., Williams, C., Guan, J., Dragunow, M., Lawlor, P., Sirimanne, E., Nikolics, K., and Gluckman, P. (1994) Neuroreport 5, 901-904.
6. S.C. Roth, A.D. Edwards, E.B. Cady, D.T. Delpy, J.S. Wyatt, D. Azzopardi, J. Baudin, J. Townsend, A.L. Stewart, and E.O. Reynolds, Dev. Med. Child Neurol. 34 (1992) 285.
7. M. Nieto-Sampedro, E.R. Lewis, C.W. Cotman, M. Manthorpe, S.D. Skaper, G. Barbin, F.M. Longo and S. Varon, Science 217 (1982) 860.
8. P.D. Gluckman, N.D. Klempt, J. Guan, E.C. Mallard, E. Sirimanne, M. Dragunow, M. Klempt, K. Singh, C.E. Williams and K. Nikolics, Biochem. Biophys. Res. Commun. 182 (1992) 593.
9. N.D. Klempt, E. Sirimanne, A.J. Gunn, M. Klempt, K. Singh, C.E. Williams and P.D. Gluckman, Brain Res. Mol. Brain Res.(1-2) (1992 Mar) 93-101.
10. E.S. Sirimanne, J. Guan, C.E. Williams and P.D. Gluckman, J. Neurosci. Methods 55 (1994) 7-14.
11. A.J. Gunn, M. Dragunow, R.L. Faull and P.D. Gluckman, Brain Res. 531 (1990) 105-106.
12. M. Dragunow, D. Young, P. Hughes, G. MacGibbon, P. Lawlor, K. Singleton, E. Sirimanne, E. Beilharz and P. Gluckman, Mol. Brain Res. 4 (1993) 347-352.
13. Dragunow, M., Beilharz, E., Sirimanne, E., Lawlor, P., Williams, C., Bravo, R., and Gluckman, P. *Brain Res.* 25(1-2):19-33, 1994.
14. Rotwein, P., Bichell, D.P., and Kikuchi, K. [Review]. *Molecular Reproduction & Development* 35(4):358-63; discussion 363-4, 1993.
15. E.J. Beilharz, N.S. Bassett, E.S. Sirimanne, C.E. Williams and P.D. Gluckman, Mol. Brain Res.(1) (1995 Mar) 81-91.
16 Klempt, N.D., Klempt, M., Gunn, A.J., Singh, K. and Gluckman, P.D. (1992) Mol. Brain Res. 15, 55-61.
17. E.J. Beilharz, N.D. Klempt, M. Klempt, E. Sirimanne, M. Dragunow and P.D. Gluckman, Mol. Brain Res. 18 (1993) 209-215.
18. J. Guan, C.E. Williams, M. Gunning, E.C. Mallard and P.D. Gluckman, J. Cereb. Blood Flow Metab. 13 (1993) 609.
19. B.M. Johnston, E.C. Mallard, C.E. Williams and P.D. Gluckman, J. Clin. Invest. 97 (1996) 300-308.
20. Guan, J., Skinner, S.J.M., Beilharz, E.J., Hua, K.M., Hodgkinson, S.S., Gluckman, P.D. and Williams, C.E. (1996) Neuroreport 7, 632-636.
21. J. Guan, E.J. Beilharz, S.J.M. Skinner, P.D. Gluckman and C.E. Williams, J. Cereb. Blood Flow Metab. (1997) submitted.
22. GUAN, J., Williams, C.E., Skinner, S.J.M., Mallard, E.C. and Gluckman, P.D. (1996) Endocrinology 137, 893-898.
23. Neff NT. Prevette D. Houenou LJ. Lewis ME. Glicksman MA. Yin QW. Oppenheim RW. Journal of Neurobiology.24(12):1578-88, 1993 Dec.
24. I. Tamm and T. Kikuchi, J. Cell Physiol. 143 (1990) 494-500.
25. Dore, P., Kar, S., and Quirion, R. Proc Natl Acad Sci *USA* 94:4772-4777, 1997.

26. S.R. Dmello, K. Borodezt and S.P. Soltoff, J. Neurosci. 17 (5) (1997 Mar) 1548-1560.
27. G. Kulik, A. Klippel and M.J. Weber,Mol. & Cell. Biol. 17 (3) (1997 Mar) 1595-1606.
28. M. Parrizas and D. LeRoith, Endocrinology 138 (3) (1997 Mar) 1355-1358.
29. J.Guan, J.Saura, S.Gatti, SJM. Skinner, L.Curatalo, C.Peters, CE. Williams, M.Dragunow, C.Post and PD. Gluckman. Nature (submitted)
30. Bourguignon, JP., Alvarez Gonzalez ,ML., Gerard,A. and Franchimont, P. (1994) Endocrinology 134: 1589-1592.

Molecular Mechanisms to Regulate the
Activities of Insulin-like Growth Factors
K. Takano, N. Hizuka and S-I. Takahashi (Editors)

The IGF-1 paradox in cerebellar granule cells: prevention of death via apoptosis opens the route to death via glutamate-triggered necrosis.

P. Calissano[‡#], M.T. Ciotti[‡], C. Galli[‡], D. Mercanti[‡], N. Canu[#], L. Dus[‡], C. Barbato[‡], O.V. Vitolo[‡], C. Zona[#]

‡ Institute of Neurobiology - C.N.R., Via C.Marx 15, I - 00137 Roma Italy
# Dipartimento di Medicina Sperimentale - Università di Roma Tor Vergata

Introduction[1]

As amply documented, IGF-I acts as a pleiotropic agent for the survival and terminal differentiation of several types of nerve cells [1,2]. In cerebellum, this growth factor as well as its receptor and the specific class of IGF-I binding proteins are developmentally regulated so as to suggest a crucial role in circuit formation [3]. The site of synthesis of this somatomedine and its way of delivery to target cells are still obscure, althougth an orthograde supply from innervating neurones to target cells has been shown from the inferior olive to the cerebellum [4]. The paradox to which it is mentioned in the title and which will constitute the thread of this paper originates from the finding that, as it will be reported, IGF-I is the most potent trophic factor capable of supporting terminal differentiation of cerebellar granule cells and of keeping under control the program of cell death via apoptosis. Prevention from the suicide internal program, however, allows the production and release in culture of a polypeptide endowed with a powerful up modulating action on glutamate receptors. Under its action, these nerve cells express functional glutamate receptors, and become susceptible to the toxic action of this neurotransmitter with consequent death via necrosis.

Results

Identification of a neurite outgrowth adhesion comlex (NOAC).

In 1990 we started a systematic search for trophic factors that could be involved in terminal differentiation of cerebellar granule cells and in the epigenetic cues that may modulate death of these neurones via glutamate-triggered necrosis. During the course of these studies we identified in rabbit serum a protein complex endowed with the property of inducing neurite outgrowth and their

[1] Abbreviations: CGC=Cerebellar Granule Cells. IGF=Insulin-like Growth Factor. hrIGF=human recombinant IGF. BME=Basal Medium (Eagle). FGF=Fibroblast Growth Factor. TNF=Tumor Necrosis Factor.

fasciculation, hence the acronym of NOAC (neurite outgrowth adhesion complex) [5]. We found that when these neurones are explanted in culture and grown in the presence of different quantities of NOAC, 30-50% survive and grow an intricate network of neurites whose fasciculation to form large bundles is directly proportional to the concentration of this protein complex (fig.1).

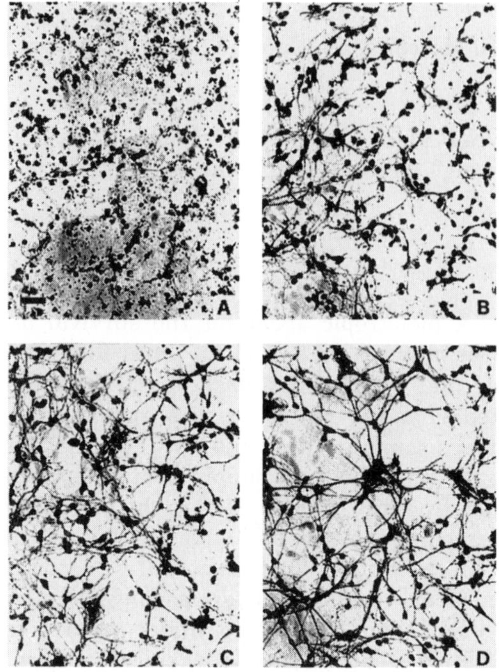

Figure 1. Concentration-dependent effect of NOAC on primary cultures of rat cerebellar granule cells. Neuronal cells were grown for 72 h in the growth medium (BME + KCl 25 mM) alone or supplemented with increasing concentrations of NOAC. A = no additions. B = 10 μg/ml. C = 25 μg/ml. D = 50 μg/ml. Cultures were fixed and Coomassie-blue stained.

A thorough analysis of the phenotypic properties of NOAC-cultured granules as compared to sister cultures grown in whole foetal calf serum (FCS), revealed that these neurones exhibit a marked resistance to the otherwise lethal action of glutamate, the excitatory aminoacid employed by these nerve cells to communicate and receive chemical inputs from other neurones. As it had been already elegantly shown during those years, glutamate toxicity is due to an excessive Ca^{2+} influx via activation of two ionotropic glutamate receptors known as NMDA and kainate [6, 7]. It was therefore logical to assess the extent of Ca^{2+} influx in NOAC-grown CGCs, an analysis that revealed a marked reduction of the inward entrance of this cation as compared to that detectable in FCS-cultured neurones [5]. This finding had two possible explanations: a) a decreased permeability of NMDA and kainate receptor-channels in NOAC-treated neurones or, b) a decreased number, or surface expression, of these same receptors. This second hypothesis revealed to be correct. Studies performed by measuring both the actual $^{45}Ca^{2+}$ influx during glutamate exposure [5] as well as the currents evoked by kainate [8, 9] revealed that NOAC-cultured neurones had a markedly reduced influx of this cation following glutamate exposure, a reduced kainate binding capacity, and markedly diminished kainate currents and voltage operated Na^+ channels [5, 8, 9, 10]. We referred to this specific phenotype as excitatory aminoacid minus (EAA-) as compared to the fully differentiated phenotype detectable when these neurones are grown in whole rabbit or foetal calf serum and referred to as EAA+. The (EAA-) phenotype was morphologically identical, as evaluated with different markers, to (EAA+) with the exception of a markedly reduced exposure of glutamate receptors and of voltage operated Na^+ channels.

In search of the glutamate sensitizing activity

Altogether these findings clearly indicated that during the purification of NOAC from rabbit serum or from other sera, (including that of human origin) a substance, endowed with the property of up-modulating the expression of glutamate receptors, was lost or discarded. In its presence, neurones bearing the (EAA-) phenotype observed when grown in the presence of purified NOAC, would acquire the (EAA+) phenotype characteristic of sister cultures grown in whole sera. Such a substance was defined as glutamate sensitizing activity or GSA and a systematic search for its identification and possible isolation was undertaken. A dual approach was carried out to this aim: a) a series of fractionation steps performed with classical chromatographic procedures of rabbit serum followed by tests on each fraction to assess the possible presence of GSA; b) tests on purified growth factors or hormones of various origin and properties to ascertain whether they were endowed, besides their well identified activities, of up-regulating glutamate receptors and converting the (EAA-) into a (EAA+) phenotype. This latter approach revealed to be faster since we found that, among several growth factors, IGF-I (but not IGF-II) or all other tested substances and factors, was fully capable of changing the phenotype resistant to glutamate into a phenotype sensitive to this excitatory and toxic aminoacid (table I) [10].

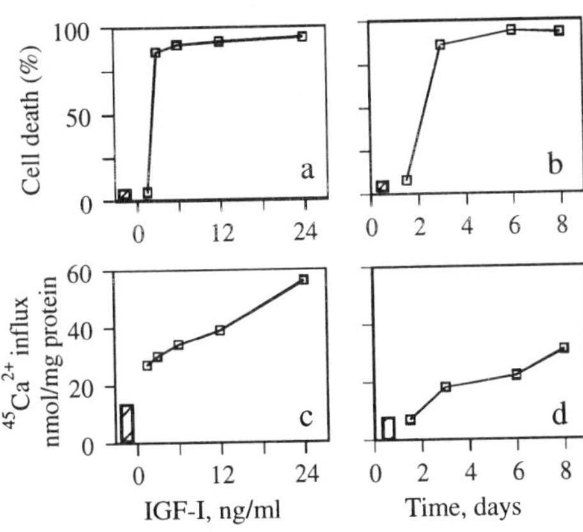

Figure 2. Concentration and time dependence of IGF-I effect. Cells were cultured for 8 days with a constant amount of NOAC (25 μg/ml) plus various concentrations of IGF-I (a,b) and tested for their response to a 100 μM glutamate pulse in terms of cell death (a) and $^{45}Ca^{2+}$ entry (b). Bars at lower left indicate the corresponding values in the presence of NOAC alone. (c and d) Cultures were grown in the presence of NOAC (25 μg/ml) with IGF-I (12 ng/ml) added at various times. In all cases, cell death or $^{45}Ca^{2+}$ influx measured in the absence of glutamate has been subtracted. Each experiment was performed in duplicate and repeated with three different cell preparation. Values reported are from a typical experiment and variations were within 10%. (from ref. 10).

A thorough analysis of the properties of CGCs grown in NOAC+IGF-I demonstrated that in the presence of this somatomedine CGCs acquire a phenotype very similar to FCS-grown sister cultures although the neuritic network is markedly reduced or absent (see fig. 6). Thus, glutamate-triggered Ca^{2+} influx markedly increases after IGF-I incubation (fig. 2), kainate evoked and voltage operated Na^{+} channels are more effective [9, 10] (fig. 3) and, as a result of this global up-modulation, the number of CGCs undergoing death after a 100 μM glutamate pulse lasting for 30 min. shifts from 10-20% to 80-90% (see also Table I).

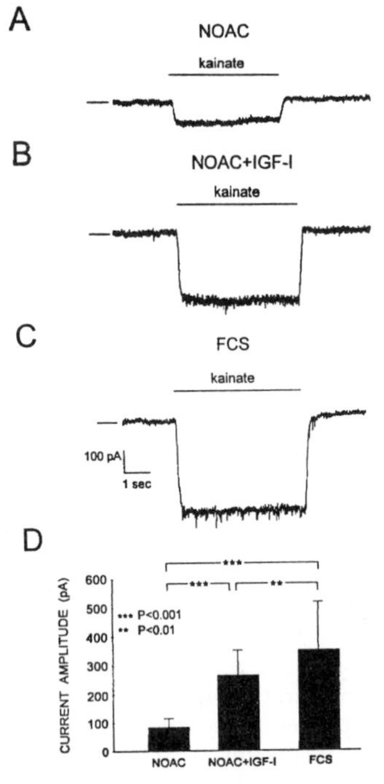

Figure 3 . Inward current induced by bath application of kainate in NOAC cell (A), in NOAC+ IGF-I cell (B) and FCS cell (C) at a holding potential of - 60 mV. Series resistance in the three types of cells during whole-cell recording was 10-18 MΩ. Compensation of series resistance was routinely used. The time of drug application is indicated by bars. The cells were 10 days old and were similar in diameter. (D) Histogram of theamplitude of the currents with standard deviation induced by 500 μM kainate in NOAC cells (n=39), NOAC + IGF-I cells (n=35) and FCS cells (n=45). (from ref. 8).

This finding seemed to indicate that the glutamate sensitizing activity present in whole serum could be IGF-I itself or a substance endowed with very similar properties and that during isolation steps of NOAC it was lost or discarded. However, a deeper analysis of these data revealed that, although this conclusion was correct in general terms, IGF-I was not identifiable with GSA itself.

IGF-I exerts a "permissive" role for the production and release of GSA.

In the presence of NOAC only a 30-50% of the whole neuronal population survived a 8 days incubation *in vitro* while with the addition of IGF-I such population increased to 80-90%, a value overimposable to that detectable when CGCs are grown in the presence of whole foetal calf serum (see Table I).

We interpreted this finding as an indication of an aspecific, trophic action exerted by this somatomedine on CGCs i.e. a pleiotropic effect distinguishable from the more specific action of up-modulating glutamate receptors [10]. As it will be shown, on the contrary, the trophic action is indirectly responsible for up-modulating the action exerted by IGF-I.

Table I. Effect of various growth factors on CGCs survival, glutamate-stimulated $^{45}Ca^{2+}$ influx and cell death after 100 μM glutamate pulse.

Culture conditions	Cells (x 10^4)	^{45}Ca influx (μmoles/mg prot.)	Cell death after glutamate (%)
F C S	212	57.0	90
I G F - I	170	56.4	89
NOAC	78	12.0	19
NOAC + IGF-I	198	59.0	93
" IGF-II	86	33.6	51
" NGF	75	13.8	21
" TNFa	67	5.8	11
" PDGF	75	6.0	10
" aFGF	65	12.0	19
" bFGF	89	8.6	14

Cerebellar granule cells were grown for 8 days in the presence of the substance(s) reported and tested for glutamate-stimulated $^{45}Ca^{2+}$ entry (100 μM glutamate, 10 minutes) and, in sister cultures, for the extent of cell death ensuing 24 hours later after a 30 minutes pulse of 100 μM glutamate. Growth Factors concentrations used: NOAC 25 μg/ml; IGF-I and IGF-II 25 ng/ml; NGF 100 ng/ml; bFGF 25 ng/ml; aFGF 100 ng/ml; TNFa 30 U/ml; PDGF 10 ng/ml.

Thus, while the studies on IGF-I interaction with CGCs were continuing and led to the discovery of a very effective antiapoptotic action (see below), we found by independent routes of studies that the response of CGCs to glutamate, NMDA or kainate is not an invariant property but it markedly depends upon the cell density of plating: the higher this parameter the higher the sensitivity to glutamate; the opposite was also true, namely the lower the plating, the lower the response to glutamate [11]. An analogous situation occurs if we change the volume of medium of culture and keep constant the density of plating: the lower the volume, the higher the sensitivity and response to glutamate (Fig. 4).

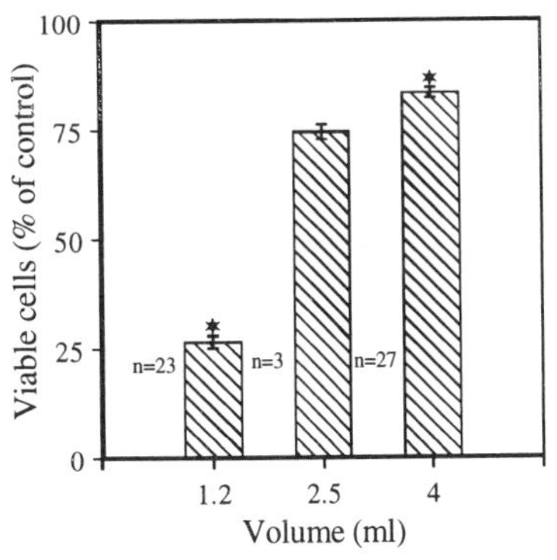

Figure 4. Effect of different volumes of medium on glutamate sensitivity. Cerebellar granule cells were plated at a density of $2.8x10^5$ cells/cm^2 in 12 well clusters and cultured for 8 days in BME+10% FCS at the indicated volumes. Notice that when cells are grown in 4.0 ml rather then in 1.2 ml of medium most granule cells are resistant to glutamate. (from ref. 11)

Moreover, if we add to low density cultures a conditioned medium (CM) obtained from high density cultures grown either in 10% FCS or in the presence of 25 ng/ml of IGF-I for 6 days, we induce - within a 24-48 hours period - a response to glutamate analogous to that detectable if cells are grown at high density or low volume (Fig. 5a). Moreover, conditioned media prepared both in FCS or in hrIGF-I accelerate by several days the onset of glutamate sensitivity (Fig. 5b)

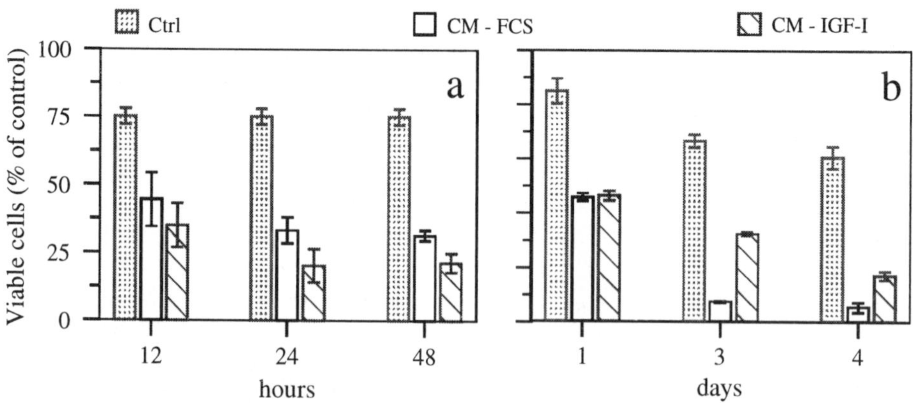

Figure 5. Time course of the action of conditioned media. (a) Cells were plated in the high-volume condition and incubated for 6 DIV, after which medium was replaced with conditioned medium derived from high-density cultures grown for 8 days in FCS (CM-FCS) or hrIGF-I (CM-IGF-I). Twelve, 24 and 48 h after the change of medium, cells were tested for their glutamate sensitivity. Parallel cultures without a change of medium served as controls (Ctr). (b) Cells were plated under the high-density condition, and after 1 day of culture the medium was replaced with 8-day conditioned medium (CM-FCS or CM-IGF-I). One, 3 and 4 days after the change of medium, cells were tested for glutamate sensitivity. Sister cultures were grown without any change of medium (control, Ctr). Notice the accelerating action of conditioned media on the onset of glutamate sensitivity. (from ref. 11).

This was not true if sister cultures were grown in FCS or in IGF-I at low density and their conditioned medium added, after 6 days of culture, to other cultures grown at low density or in high volume. In this case no substantial improvement in their sensitivity to glutamate was achieved. The action of the CM is abolished in the presence of actinomicin D and is specific for CGCs, since GABAergic neurones, present in culture in the amount of 1%-2% did not exhibit any change of their sensitivity to glutamate after CM incubation [11]. These series of studies led to hypothesise, and subsequently to prove, that CGCs synthesise and release in culture a substance identical in properties to IGF-I in conferring sensitivity to previously resistant neurones and therefore defined with the same acronym of GSA. The next question was therefore if IGF-I and GSA were the same substance (produced by CGCs and released in culture) or, alternatively, they were different but endowed with the same property. To some extent neither the first nor the second hypothesis revealed to be fully correct but a third explanation became apparent.

Thus, if GSA was in fact a pool of IGF-I synthesised and released from CGCs - an hypothesis agreeable with previous findings of the presence of this somatomedine in cerebellum - then antibodies to this factor added to high density or low volume culture in excess of three order of magnitude over its presumed concentrations in culture should be capable of blocking its glutamate sensitizing activity. However, addition of a specific monoclonal antibody, that previous studies had shown to fully block IGF-I activity, did not alter the extent of glutamate sensitivity ensuing in culture as a function of cell density or volume of plating (not shown). Furthermore, addition of 25 ng/ml IGF-I to low density or to high volume cultures did not cause any change in the low sensitivity, exhibited by this type of cultures [11]. These studies clearly showed that IGF-I was not endowed with GSA activity and that this substance was structurally and functionally distinct from the somatomedine.
On the other hand, our previous findings had shown that in the presence of IGF-I, NOAC-cultured neurones developed a glutamate sensitive phenotype[10]. How to reconcile such apparent discrepancy? The answer was soon found in the light of the findings about the glutamate sensitivity as a function of CGCs density in culture. Thus, due to its trophic action, IGF-I allows a survival of a much larger mass of granule cells cultured in NOAC and this, in turn, allows the production and release in culture of an amount of GSA sufficient to induce a shift from an (EAA-) into an (EAA+) phenotype of a large population of neurones (see also Table I).

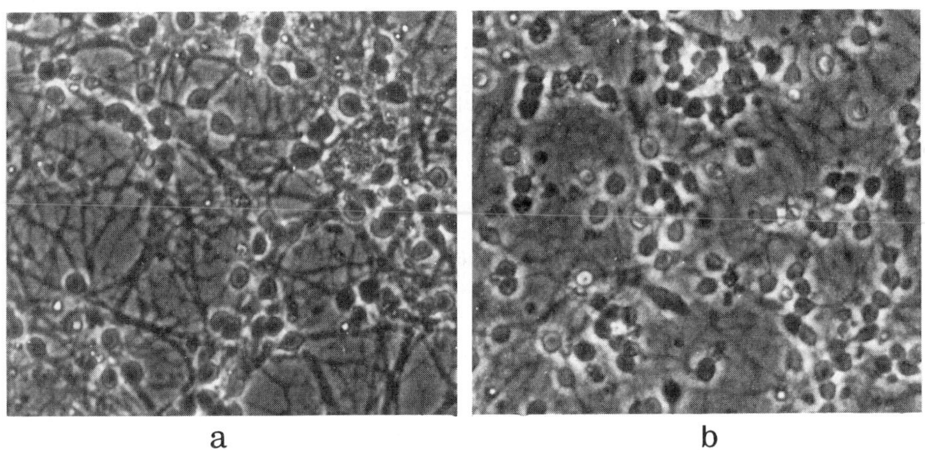

Figure 6. Phase contrast photomicrograph of cerebellar granule cells grown for 6 days in BME (K25) plus 10% FCS (a) or 25 ng/ml IGF-I (b). Notice the well preserved morphology of neurones and the reduced neuritic network in IGF-I cultures.

These considerations led us to conclude that IGF-I exerts a very effective pleiotropic action allowing the survival of a large population of neurones. Since the response to glutamate is a direct function of such survival, (i.e. of the total cell mass/volume of medium) in the presence of this somatomedine neurones become fully responsive to glutamate because they are healthy and very abundant. Altogether, these findings indicate that IGF-I is not endowed, per se, of a glutamate sensitizing activity but that it exerts a "permissive" action of its production via its trophic, generalized effect on CGCs. It is noteworthy that, among all growth factors and hormones tested, this somatomedine is the most effective agent [11].

IGF-I exerts an antiapoptotic action downstream the Ca_i drop.

While the studies on IGF-I effects on CGCs were in progress, our research group unequivocally demonstrated that the shift of KCl from 25 mM, previously reported to be optimal for CGCs survival, to 5.0 mM, triggers a membrane signal leading to activation of a program of cell death via apoptosis [12]. DNA fragmentation, sensitivity to cycloheximide and actinomicin D, formation of apoptotic bodies, were all detectable signs of this type of death. Indeed, years before Gallo *et al.* [13] had already shown that CGCs deprived of depolarising concentrations of KCl massively die, but at that time no attention was paid to the type of death, since the existence of a suicide program in all cells, including those of neural origin, often accompanied by the formation of apoptotic bodies, was still a knowledge confined to cell biologists analysing cell death in other cell types.

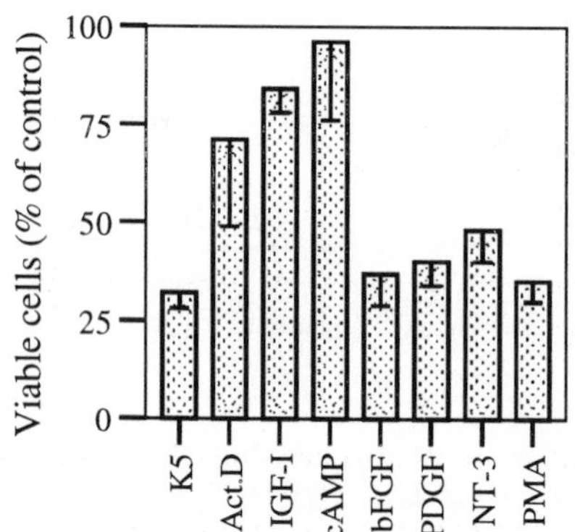

Figure 7. Survival of neurones after treatment with various agents in low K^+. Neurones were switched from culture medium containing 10% FCS and 25 mM KCl to serum free medium containing 5 mM K^+ and no additives (K5) or with actinomycin D (Act.D, 1 μg/ml), IGF-I (25 ng/ml), forskolin (10 μM; to increase cAMP), bFGF (100 ng/ml), PDGF-A (20 ng/ml), NT-3 (50 ng/ml) or phorbol-12-myristate-13-acetate (PMA, 100 nM). Control represents survival in serum-free medium containing 25 mM KCl. Survival was quantified by fluorescein diacetate staining 48 hours after treatment. Each bar represents mean (-SD) of five randomly chosen microscopic fields taken from three culture dishes. (from ref. 12).

Subsequent studies aimed at revealing the early, triggering steps of this process demonstrated, as partially expected also on the basis of similar studies conducted on NGF-deprived sympathetic cells, that the KCl shift from 25 to 5 mM caused a partial, selective closure of L-type, voltage-dependent, calcium channels [14]. A systematic search for possible antiapoptotic factors or substances, carried out once the general aspects of this suicide program were established, demonstrated once again and as partially expected on the basis of previous findings, that IGF-I is the most powerful antiapoptotic agent among all factors, hormones and other trophic substances tested [13,14] (Fig. 7).

The only, substance having similar, antiapoptotic action are cAMP and purine derivative such as adenosine, ADP or NAD [15]. As shown in fig. 7, although the Ca_i drop following the KCl shift is the triggering event, IGF-I exerts its antiapoptotic action without affecting the Ca_i concentration.

A series of experiments that would be too long to enumerate have shown that IGF-I acts downstream the Ca_i drop allowing survival of CGCs also at concentrations of this cation that would otherwise activate the internal program of cell death [14].

How does IGF-I operate in CGCs to keep such suicide program under control? Although we have no precise hints about this problem, recent studies performed in our laboratory indicate that this somatomedine very effectively helps to keep the oxidative phosphorylation of mitochondria in a normal state while in the absence of IGF-I they demonstrate early signs of functional unbalancement. This progressive functional impairment is causally accompanied by production of reactive oxygen species that probably play a role in early events of the apoptotic program [16].

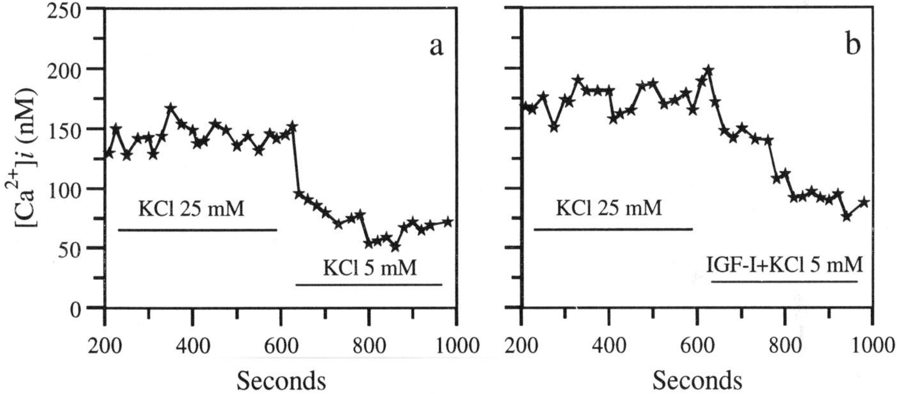

Figure 8. Rapid change in intracellular $[Ca^{2+}]i$ induced by lowering extracellular KCl. Neurones that had been grown for 6-7 days in standard culture conditions were kept in 25 mM KCl during the entire procedure of washing, loading with fura-2 and equilibrating. Neurones were shifted to 5 mM KCl with no addition (a), 25 ng/ml IGF-I (b). Values represent means with SEM of 20 cells. (redrown from ref. 14).

Discussion

The studies that we have carried out on IGF-I action in CGCs clearly indicate that the major, actually unique role exerted by this somatomedine, is that of an essential survival factor. Such a role is directly documented by its most effective action, among all hormones and growth factor tested, in allowing survival of CGCs in the absence of any other potential trophic substance. No other hormone, growth factor, or vitamin is capable of supporting the survival of a large population of cultured CGCs in pure Basal Eagle Medium in the absence of other substances generally constitutive of chemically defined media as it is IGF-I at the physiological concentrations of 10-25 ng/ml. Such trophic action is best evidenced by the the antiapoptotic effect revealed when CGCs are deprived of depolarising concentrations of KCl generally employed to sustain their survival in vitro. Such high, depolarising concentrations of potassium, are postulated to mimick an in vivo situation of electrical activity with continuous depolarising stimuli and consequent inward of Ca^{2+}. Such *in vitro* potassium stimulated or *in vivo* evoked Ca^{2+} influx, would keep under control the suicide program always ready to start its destructive project. If we consider that also adenosine analogs are endowed with such antiapoptotic activity we may draw the conclusion that at least 3 distinct mechanisms may keep silent the suicide program of CGCs: 1) the electrical stimuli which *in vivo* normally impinge upon these neurones from afferent nerve fibers and *in vitro* are mimicked by high, depolarising, concentrations of KCl; 2) a physiological release of purine derivatives such as adenosine, ADP or NAD from adjacent nerve or glial cells surrounding CGCs; 3) a possible supply from the inferior olive, from Purkinje cells or other as yet unidentified sources of IGF-I. Such supply, by allowing survival of CGCs could *in vivo* play an indirect permissive role for the production of the glutamate sensitizing activity (GSA) that up modulates the functional expression of glutamate receptors and the acquisition of the mature, adult phenotype characteristic of these neurones.

Acknowledgements

This work has been carried out under a research contract with NE.FA.C. within the National Research Plan Neurobiological Systems of the Ministero della Ricerca Scientifica e Tecnologica.

References

1. Hepler J.E. and Lund P.K., Molecular biology of the insulin-like growth factors: relevance to nervous system function. Mol. Neurobiol. 4 (1990) 93-127
2. Carson M.J., Beheringer R.R., Brinster R.L. and McMorris F.A., Insulin-like growth factor I increases brain growth and central nervous system myelination in transgenic mice. Neuron 10 (1993) 729-740.
3. Torres-Aleman I., Pons S. and Arevalo M.A., The insulin-like growth factor I system in the rat cerebellum; developmental regulation and role in neuronal survival and differentiation. J. Neurosci. Res. 39 (1994) 117-126.
4. Nieto-Bona M.P., Garcia-Segura L.M. and Torres-Aleman I., Orthograde transport of Insulin-like growth factor I from the inferior olive to the cerebellum. J. Neurosci. Res. 3 (1993) 520-527.
5. Mercanti D., Galli C., Liguori M., Ciotti M.T., Gullà P. and Calissano P., Identification of the serum complex which induces cerebellar granule cell in

vitro differentiation and resistance to excitatory amino acids. Eur. J. Neurosci. 4 (1992) 733-744.
6. Choi D.W. The role of glutamate neurotoxicity in hypoxic-ischemic neuronal death. Ann. Rev. Neurosci. 13 (1990) 171-182.
7. Choi D.W. Glutamate neurotoxicity and diseases of the nervous system. Neuron, 1 (1988) 623-634.
8. Zona C., Ragozzino D., Ciotti M.T., Mercanti D., Avoli M., Brancati A. and Calissano P., Sodium and late potassium currents are reduced in cerebellar granule cells cultured in the presence of a protein complex conferring resistance to excitatory amino acids. Eur. J. Neurosci. 5 (1993) 1479-1484.
9. Zona C., Ciotti M.T., Calissano P., Human recombinant IGF-I induces the functional expression of AMPA/Kainate receptors in cerebellar granule cells. Neurosci. Lett. 186 (1995) 75-78.
10. Calissano P., Ciotti M.T., Battistini L., Zona C., Angelini A., Merlo D. and Mercanti D., Recombinant human insulin-like growth factor I exerts a trophic action and confers glutamate sensitivity on glutamate-resistant cerebellar granule cells. Proc. Natl. Acad. Sci. USA 90 (1993) 8752-8756.
11. Ciotti M.T., Giannetti S., Mercanti D. and Calissano P., A glutamate-sensitizing activity in conditioned media derived from rat cerebellar granule cells. European J. Neurosci. 8 (1996) 1591-1600.
12. D'Mello S.R., Galli C., Ciotti T. and Calissano P., Induction of apoptosis in cerebellar granule neurones by low potassium: inhibition of death by insulin-like growth factor 1 and cAMP. Proc. Natl. Acad. Sci. USA , 90 (1993) 10989.
13. Gallo V., Kingsbury A., Balasz R., Jorgensen O.S., The role of depolarisation in the survival and differentiation of cerebellar granule cells in vitro. J. Neurosci. 7 (1987) 2203-2213.
14. Galli C., Meucci O., Scorziello A., Werge T. W., Calissano P. and Schettini G., Apoptosis in cerebellar granule cells is blocked by high KCl, forskolin and IGF-1 through distinct mechanisms of action: the involvement of intracellular calcium and RNA synthesis. J. Neurosci. 15 (1995) 1172-1179.
15. Vitolo O.V., Ciotti M.T., Galli C., Borsello T. and Calissano P., Adenosine blocks apoptosis in cultured rat cerebellar granule cells. NeuroReport (1997) submitted.
16. Atlante A., Gagliardi S., Marra E. and Calissano P., Neuronal apoptosis is accompanied by rapid impairment of cellular respiration and is prevented by scavengers of reactive oxygen species. Neurosci. Lett. (1997) submitted.

Molecular Mechanisms to Regulate the
Activities of Insulin-like Growth Factors
K. Takano, N. Hizuka and S-I. Takahashi (Editors)

IGFs and IGFBPs in Organogenesis: Development of Normal and Abnormal Kidneys

Victor K. M. Han[1,2,3,4] and Douglas G. Matsell[1,2]

Departments of Paediatrics[1], Anatomy and Cell Biology[2], Biochemistry[3] and Obstetrics and Gynecology[4]
MRC Group in Fetal and Neonatal Health and Development
University of Western Ontario
The Lawson Research Institute and the Child Health Research Institite
London, Ontario, Canada

ABSTRACT

Insulin-like growth factors (IGFs) and IGF binding proteins (IGFBPs) are important paracrine factors that regulate growth and differentiation of developing organs including the kidney. IGF-II and IGFBPs are expressed during development of the kidney in a specific spatial and temporal manner. In the kidney, IGF-II mRNA is expressed abundantly in the metanephric blastema, the S-shaped glomerulus and the renal mesenchyme, whereas the IGFBP-2 to -5 mRNAs are expressed in the specific cell types of the developing nephron, e.g. IGFBP-2 mRNA is expressed in the ampulla of the ureteric duct, induced metanephric blastema and the S-shaped nephron. The pattern of mRNA localization suggests that the focal expression of IGFBP gene at a specific developmental stage participates in the differentiation of cells, and that a single or a combination of IGFBPs is utilized for this purpose. The expression pattern of IGF-II and IGFBP-2 and -3 genes is altered in multicystic dysplastic kidney disease identified in early gestation human fetal kidneys and in cystic dysplastic obstruction at 0.5 gestation in the ovine fetus. These observations suggest that IGF-II and IGFBPs are involved in the pathogenesis of obstructive renal dysplasia. The paracrine IGF-IGFBP interactions occurring in specific spatio-temporal manner are important mechanisms by which the IGF system regulates organogenesis.

1. INTRODUCTION

Insulin-like growth factors (IGFs) and IGF binding proteins (IGFBPs) have been shown to be important in fetal growth (1) as well as the development of many different organs and tissues (2), including the kidneys (3). Organogenesis involves multiple processes including cellular proliferation, differentiation, migration, aggregation, cell-cell interaction, apoptosis and maturation, and the IGFs and IGFBPs play significant roles in most, if not all, steps (4). Nephrogenesis represents an interesting paradigm of organogenesis which involves not only all of the different developmental processes of epithelial and mesenchymal cells arising from the primitive germ layer, the mesoderm, but also crucial cell-cell and cell-matrix interactions which involve several growth factors including IGFs and IGFBPs (3). Failure or altered expression of these mediators may be important in the pathophysiology of abnormal renal development such as congenital polycystic kidney disease (5) or the multicystic dysplastic kidney disease (MCDK).

2. NEPHROGENESIS

The human kidney develops as metanephros in the lateral part of the dorsal mesenchyme at about fifth week of gestation. It consists of the mesenchymal cord called the "blastema", and the epithelial tissue derived from the Wolffian duct called the ureteric duct. The ureteric duct emerges from the most caudal portion of the Wolffian duct as it opens into the cloaca, and grows dorsally towards the caudal portion of the blastema. While the ureteric duct invades the mesenchyme and branches there, epithelial renal vesicles appear around the tips (ampullae). Consequently, when new vesicles are added while the older ones continue to develop, a progressive centripetal series of maturing nephrons may be detected in histological sections. This appositional development continues over an extended period of intrauterine period and proceeds in many species during the early postnatal period. In human embryos, the terminal nephrons appear in the thirty second week (6).

2.1. Normal Nephrogenesis

Although the existence of the primitive kidneys, the pronephros and mesonephros, is still being questioned, the mammalian kidney develops from the metanephros, undifferentiated mesenchyme derived from the lateral mesoderm of the embryo. Formation of the differentiated segments of the mature nephron and its supporting interstitium results from the reciprocal induction of the ureteric duct and the metanephric blastema. In the human fetus the process of nephrogenesis begins at 6-8 weeks gestation and is complete by 36 weeks gestation. This alone is an important concept particularly when in utero fetal surgery is contemplated to relieve urinary obstruction.

The ureteric duct is derived from the mesonephric duct (Wolffian duct), a vestige of the mesonephric kidney. Branching from the mesonephric duct as it joins with the cloaca, the ureteric duct migrates dorsally to the metanephric blastema. Under the inductive influence of the undifferentiated metanephric blastema, the ureteric duct migrates into the blastema and undergoes successive generations of branching, like a growing tree, each with an ampullary tip capable of exerting its own inductive influence on the undifferentiated metanephric blastema. Induction of the metanephric blastema is a well studied paradigm of mesenchymal-epithelial transformation and proceeds through a well described spectrum of events.

After successful approximation of the ureteric duct/ampulla and the metanephros, epithelial induction appears histologically as an aggregation of cells which progressively become polarized and begin to elaborate an extracellular matrix/basement membrane. With further proliferation of this cell mass, and likely through selective cell death, and segmental cellular differentiation, an S-shape nephron forms in which tubular and glomerular elements become defined. At this stage and in the human fetus at about 8-10 weeks gestation the primitive glomerular structure becomes vascularized. A distinctive glomerular visceral and parietal epithelial cell layer develop which with further development fold in on themselves to form the fetal glomerulus and an identifiable glomerular filtration barrier. Within each developing nephron, tubular segment elongation and differentiation occur, with the distal tubule establishing continuity with the collecting duct, a derivative of the ureteric duct. At 36 weeks gestation further induction of nephrons ceases, however final morphogenesis continues, with enlargement of the glomerulus (which at this stage is 1/3 the size of the adult glomerulus), elongation of the tubules, and terminal functional differentiation of the specialized epithelia along the length of the nephron.

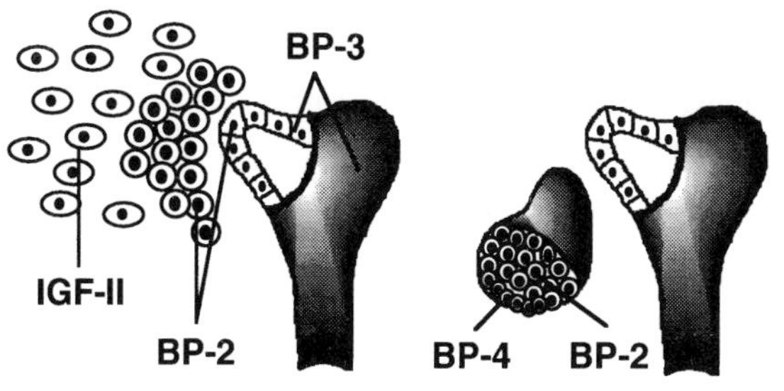

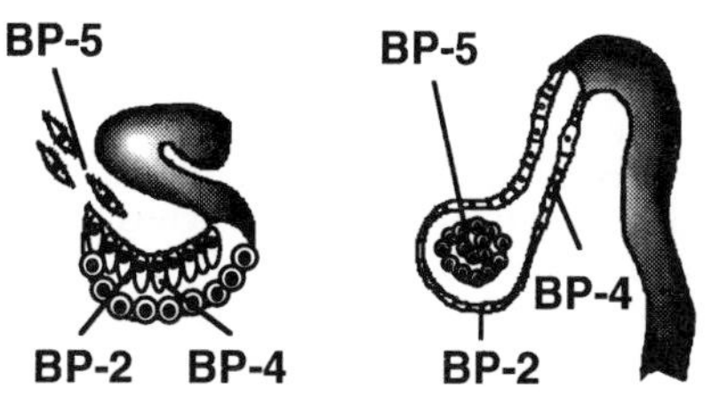

2.2. Species differences

The developmental process above describes the nephrogenesis in most mammalian kidneys. However, there are significant differences among various species particularly in their temporal pattern (7). The most significant being the difference between the rodents and other species. In the former, nephrogenesis occurs late in gestation (at 0.55 of gestation) and continues in the postnatal life for about 2 weeks. In humans, primates and sheep, nephrogenesis occurs early (0.12 of gestation) and is more or less completed before term. Although it has been described that nephrogenesis can continue to occur in many species during the early postnatal period (8), most of the described postnatal development occurs most prominently in the rodent species. It is therefore important to have such species differences in mind, when studying the temporal expression pattern of putative growth factors such as IGFs in renal development.

2.3. IGF system in nephrogenesis

Since nephrogenesis is an excellent example of epithelial-mesenchymal interaction during development, our laboratory has conducted extensive studies into the expression and the role of IGFs and IGFBPs during the process. We have reported the expression of IGF and IGFBP genes in the human (3, 9), and the sheep (10). Others have reported the expression pattern in rodents. Human fetal kidney from the early second trimester fetuses (13-14 weeks), is an excellent model because it consists of all different stages of nephrogenesis which is developing in a centripetal fashion, and allows us to follow the expression of IGF and IGFBP genes both spatially and temporally.

2.3.1. Human:

In general, IGF mRNAs (IGF-II in particular), were expressed predominantly in the mesenchymal cell types (undifferentiated metanephric blastema and differentiated renal mesenchyme), whereas IGFBP mRNAs were expressed in both the epithelial and mesenchymal cell types, with each IGFBP demonstrating a specific spatial and temporal pattern. IGF-II mRNA was expressed abundantly in the cells of the undifferentiated metanephric blastema and supporting mesenchyme and to a lesser extent in the cells of the developing S-shape nephron, including those destined to become visceral and parietal glomerular epithelium. In the maturing nephron, IGF-II mRNA was expressed abundantly in the mesenchymal cells, and less abundantly in the mesangium and visceral epithelium of the maturing glomerulus.

IGFBP-2 mRNA was expressed most dramatically in the early developing stages of the nephron. At these stages, the mRNA was identified in the cells of the induced aggregated blastema, the differentiated visceral and parietal glomerular epithelial cells of the S-shape nephron, the differentiating proximal tubular epithelial cells and to the actively proliferating cells of the ampulla of the ureteric duct. In the more mature glomerulus, the IGFBP-2 mRNA was identified in the differentiated visceral and parietal glomerular epithelial cells, and the proximal tubular epithelial cells. IGFBP-2 mRNA was the only IGFBP mRNA expressed in the ampulla.

IGFBP-3 mRNA was expressed specifically in the ureteric duct of the early developing stages of the nephron with the exception of the ampulla. In addition it was observed in the small nests of cells within the undifferentiated mesenchyme. In the mature kidney, epithelial cells of those structures derived from the ureteric duct, including the stratified epithelium of the collecting system, exclusively expressed IGFBP-3 mRNA.

In the early developing stages of the nephron, IGFBP-4 mRNA was expressed discretely in the differentiating glomerular visceral and parietal epithelial cells of the aggregated blastema and the S-shape nephron. With further maturation of the glomerulus, IGFBP-4 mRNA was expressed in the visceral glomerular epithelial cells and the proximal tubular epithelium.

Although IGFBP-5 mRNA was expressed predominantly in the more developed kidney, it was also observed in the epithelial cell layer of the renal capsule and outer zone of the metanephric blastema. IGFBP-5 mRNA was identified specifically in the rays of undifferentiated mesenchymal cells extending from the cortex to medulla of the kidney, and in the mesenchymal cells extending into the glomerular vascular cleft of the S-shape nephron. IGFBP-5 mRNA was expressed abundantly in the mesangium of the glomeruli of the more mature kidney. In both the early developing and maturing nephron, only low levels of IGF-I, IGFBP-1, and IGFBP-6 mRNA expression were observed.

A summary of cellular sites and relative abundance of expression of each IGF and IGFBP mRNA in the early developing nephron and the maturing nephron in the human fetal kidney are shown in Tables 1 and 2 respectively.

Table 1. IGF /IGFBP mRNA expression in the early developing nephron.

	IGF I	IGF II	BP1	BP2	BP3	BP4	BP5	BP6
Uninduced metanephric blastema	+	++++	-	-	-	+	-	+
Induced/Aggregated blastema	+	++++	-	+++	-	++	-	+
S-shape/Comma:								
Committed tubular	+	-	-	+++	-	-	-	+
Committed epithelial	+	++++	-	++++	-	++++	-	+
Mesenchymal derived	+	-	-	-	-	-	+++	-
Ureteric duct	+	-	-	-	+++	-	-	+
Ampulla	+	-	-	+++	-	-	-	+

Table 2. IGF/IGFBP mRNA expression in the maturing nephron.

	IGF I	IGF II	BP1	BP2	BP3	BP4	BP5	BP6
Maturing glomerulus:								
Endothelial	+	-	-	-	-	-	-	-
Epithelial	+	+++	-	++++	-	++++	-	+
Mesangial	+	-	-	-	-	-	++++	-
Proximal tubule +	-	-	++	-	+++	-	+	
Distal tubule	+	-	-	-	++	-	-	+
Collecting system	+	-	-	-	++++	-	-	+
Supporting mesenchyme	+	++	+	-/+	+	++	++++	+

2.3.2. Other species

The significant differences between the expression patterns of IGF and IGFBP mRNAs in the rodent and human nephrogenesis are: (1) the temporal pattern of IGF-II and IGFBP gene expression, (2) IGFBP-2 mRNA is more prominently expressed in the proximal tubules as the differentiation progresses, and (3) no significant IGFBP-3 mRNA being expressed in the ureteric duct. In the sheep, the most prominent difference from the human is the significant expression of IGFBP-2 mRNA in the proximal convoluted tubules. It is notable that in general, the expression pattern of IGF-II and IGFBP genes in the developing kidneys is similar among species, particularly in the cells that are involved in epithelial-mesenchymal interactions such as the metanephric blastema and the ampullae.

3. ABNORMAL NEPHROGENESIS

Multicystic dysplastic kidney disease (MCDKD) is the most common form of renal dysplasia and is characterized by architectural disorganization of the kidney and by varying degrees of histopathological changes including the presence of immature glomeruli, primitive tubules surrounded by fibromuscular collars, nests of metaplastic cartilage, and cysts derived from tubular and glomerular structures. The pathogenesis of MCDKD is unknown. Our studies on two MCDKD kidneys that were obtained at an early stage of development (14 and 19 weeks gestation) (11, 12) showed identified islands of spatially dislocated metanephric blastema adjacent to zones containing all the normal structural elements of nephrogenesis, including aggregates of induced mesenchyme, S-shaped bodies and maturing glomeruli, proximal and distal tubules. Renal cysts were lined with epithelia varying from a flattened squamous to a cuboidal morphology and immunochemical, lectin binding and molecular markers suggest their origin from all portions of the nephron including Bowman's space, proximal tubule and collecting duct. These findings indicate that in the early stages of MCDKD normal nephrogenesis occurs in what appears to be a normal metanephric blastema. An intrinsic abnormality in the branching morphogenesis of the ureteric duct may be responsible for the development of the histopathological changes described (11).

3.1. IGF system in abnormal morphogenesis

Using *in situ* hybridization and immunohistochemistry, we demonstrated that IGF-II, IGFBP-2, and IGFBP-3 expression are altered in the early fetal multicystic dysplastic kidneys (12). In comparison, kidneys obtained from the postnatal patients with MCDKD were studied. Since the fetal kidney are at the stage when cystic dysplastic changes are still occurring, we deduced that the alterations in the expression of IGF-II, IGFBP-2 and IGFBP-3 may play at least in part in the pathophysiology of dysplasia. In the fetal dysplastic kidneys, IGF-II mRNA is not only abundantly expressed in the metanephric blastema that is displaced, but also in the periductular fibromuscular collars of the cysts. In addition to the normally located and displaced blastema, IGF-II immunoreactivity is evident in the S-shape nephron, the mesangial area of the developing glomeruli and in the tubular epithelial cells. The epithelia of most cysts show IGF-II immunoreactivity in both fetal and postnatal dysplastic kidneys.

In the dysplastic fetal kidneys, IGFBP-2 mRNA is expressed in the early induced metanephric blastema aggregates, and in the differentiating epithelial cells of the S-shape nephron indicating that the normal pattern of expression exists but spatially displaced. Additionally, IGFBP-2 mRNA is present in the epithelia of the various cysts, although predominantly in the cuboidal epithelia in the multicystic dysplastic kidneys of all ages. In the fetal kidneys IGFBP-2 peptide immunoreactivity is localized to the cells of the metanephric blastema and to the normal proximal and distal tubular epithelia. In addition, IGFBP-2 immunoreactivity is recognized in the epithelia of cysts at all stages of development including the postnatal kidneys, mirroring IGF-II peptide localization. IGFBP-3 mRNA and peptide are low or absent from these epithelia of the cysts. The pattern of expression of other IGFBP genes are similar to that of the normal kidney. The altered expression of IGF-II, IGFBP-2 and IGFBP-3 genes in the developing multicystic renal dysplasia in the fetal kidney suggests a role for the IGF system in the progressive histopathological changes of this disorder.

One of the etiologies of cyst formation in MCDKD may be the obstruction of tubular urine flow at critical stages of early nephrogenesis. Thereafter renal tubular epithelial cysts may develop from increased epithelial cell proliferation or hyperplasia, tubular basement membrane abnormalities, or translocation of the epithelial basolateral membrane Na/K ATPase pump. Several factors may be involved in cyst epithelial cell hyperplasia including intrinsic or acquired defects in the properties of the tubular epithelial cells. The epithelium may fail to differentiate or it may dedifferentiate, maintaining a proliferative phenotype. Altered expression of local growth factors or their receptors, with or without a normal epithelium, may also contribute to the development of cysts. Increased abundance of IGF-II mRNA and peptide, together with IGFBP-2 immunoreactivity in the epithelia of cysts, suggest that the increased IGF-II and IGFBP-2 cause epithelial hyperplasia and/or alter differentiative functions of the epithelia, which may be the pathophysiologic mechanism of cyst formation in MCDKD.

4. CONCLUSIONS

Insulin-like growth factor-II is abundantly expressed in the metanephric blastema of the developing mammalian kidney. Aggregation that follows the induction by the ureteric duct is accompanied by the expression of the specific IGFBP, the IGFBP-2. We believe that IGFBP-2 plays a crucial role in targeting IGF-II to the developing S shaped glomerulus and the nephron. An increase in the expression of IGF-II and IGFBP-2 in the displaced metanephric blastema and the epithelia of the cysts of the fetal multicystic dysplastic kidney as the disease process is in its progressive phase, indicate that these peptides play an important role in the pathophysiology of this congenital disease.

REFERENCES

1. Liu J-P, Baker J, Perkins AS, Robertson EJ, Efstratiadis A. Cell (1993)75:73-82.
2. Han VKM, AJ D'Ercole, PK Lund. Science (1987) 236:193-197.
3. Matsell DG, Delhanty PJD, Stenpaniuk O, Goodyer C, Han VKM. Kidney Int (1994) 46:1031-1042.
4. Han VKM, Hill DJ. (1992) In: Schofield PN (ed) The Insulin-like Growth Factors. Structure and Biological Functions. Oxford University Press, Oxford, New York, pp 178-220.
5. Horikoshi S, Kubota S, Martin GR, Yamada Y, Klotman PE. Kidney Int (1991) 39: 57-62.
6. Osathanondh V and Potter E. Arch Pathol (1963) 76: 290-302.
7. Hoar RM, and Monie IW. In, Developmental Toxicology, Kimmel CA and Buelke-Sam J, eds, (1981) pp 13-33, Raven Press, New York.
8. Jokalainen P. Acta Anat Suppl (1963) 47: 1-73.
9. Han VKM, PK Lund, DC Lee, AJ D'Ercole. J. Clin. Endocrinol. Metabol (1988) 66:422-429.
10. Delhanty PJD and Han VKM. Endocrinology (1993) 132: 41-52.
11. Matsell DG, Bennett T, Goodyer P, Goodyer C, Han VKM. J Lab Invest (in press)
12. Matsell DG, Bennett T, Armstrong RA, Goodyer P, Goodyer C, Han VKM. J Lab Invest (in press)

Molecular Mechanisms to Regulate the
Activities of Insulin-like Growth Factors
K. Takano, N. Hizuka and S-I. Takahashi (Editors)

Expression of IGF-I and IGFBPs in rat kidney and the effect of GH and nutrition

S. Kobayashi[a] and H. Nogami[b]

[a]Second Department of Internal Medicine, National Defense Medical College, 3-2, Namiki, Tokorozawa, 359, Japan

[b]Department of Anatomy, School of Medicine, Keio University 35, Shinanomachi, Shinjyukuku, Tokyo, 160, Japan

Glomerular sclerosis is a common feature of many forms of human renal disease and frequently contributes towards a relentless and inevitable decline in renal function (1). When renal mass is reduced, whether by disease or surgical ablation, the remaining kidney tissue increases in size and function. Since this compansatory renal growth (CRG) plays an important role for glomerular sclerosis, we decided to advance our understanding of progressive renal disease by investigating the pathogenesis of CRG. It has been known that dietary manipulation can influence the outcome of progressive renal disease associated with a decrease in CRG (2-5). Numerous studies suggest that dietary protein restriction can reserve renal function and structure in various models of experimental renal disease. Frequently overlooked, however, is the fact that low protein diet can also lead to decreased food intake, and growth retardation (6), making it difficult to know whether the beneficial effect is due to a protein restriction per se or overall food restriction (7). In this regard, our previous study has clearly shown that 40% calorie restriction prevents end-stage renal pathology in remnant kidney of 5/6 nephrectomized rat model regardless of the extent protein intake (4,5,7). Although the mechanisms still remains unknown, we suggested that growth retardation is a necessary prerequisite for such protective effects in the rat.

The kidney is a well known production site of growth hormone (GH)-stimulated insulin-like growth factor-I (IGF-I) (8-9), which has been shown to play an important role in CRG (5,10-11). Recently, much attention has been focused on GH-IGF-I-IGFBPs axis in kidney. It has become more obvious that IGF-I increases glomerular filtration rate and renal plasma flow in rat (12) and human (13). Since IGF-I is bound to IGF-binding proteins (IGFBPs), the actions of IGF-Ishould be considered together with those of IGFBPs. It has been shown that IGFBPs are important modulators of the biological actions of IGF-I in kidney (14). Recently, we showed that GH and fasting affect expression of IGFBPs mRNAas well as that of IGF-I mRNA in liver (15). Despite the fact that GH contributes to glomerulosclerosis probably through IGF-I (16), the manner in which GH alters IGFBPs mRNA in the kidney has not been explained. Furthermore, much of our knowledge of the roles of GH and IGF-I has been obtained using hypophysectomized rats as a GH deficient model (17-19). This model, however, lacks several other trophic hormones which could also be capable of modulating IGF-I or IGFBP gene expression. In contrast, the spontaneous dwarf rat (SDR) lacks GH only due to a mutation in the GH gene (20). It is therefore believed to be a better model for studying the effect of changes in GH levels on renal IGF-I and IGFBPs expression.

We examined IGF-I and IGFBP-1, 2, 3, 4, 5 mRNA expression in kidneys of SDR rats before and after GH administration in addition to changes in expression in this organ as a result of a 48 hours fasting. Then, we also compared renal expressions of IGF-I and these IGFBPs mRNAs following nephrectomy in SDR rats with those of control Sprague-Dawley rats.

1. Materials and Methods

SDR (dr) rats were supplied by the Roussel Morishita Co. Ltd., Shiga, Japan, and housed in a light (14 h lighting per day) and temperature (23℃) controlled room with free access to tap water and food. These rats were obtained by mating affected parents (dr/dr). In addition to the 19 adult SDRs (8 weeks-old, mean initial body weight 64 g), 12 age-matched male Sprague-Dawley rats (SD) were also used which were purchased from CLEA Japan Inc. (Tokyo, Japan). The animals were divided into three groups, one for studying the effect of GH (11 female SDRs) and the other, for studying the effect of fasting (8 male SDRs), and the last for studying the effect of uninephrectomy (16 male SDRs).

1.1. Effect of GH

SDRs received twice daily injections for 5 days of 0.15 mg ovine GH (NIDDK oGH, kindly supplied by the National Hormone and Pituitary Program, Baltimore, MD). Ovine GH supplied as a lyophilized powder was dissolved in 0.03 M $NaHCO_3$/0.9% NaCl (pH 9.2) containing 1mg/ml of bovine serum albumin (BSA), and stored at -80 ℃. The GH was prepared daily by diluting a stock solution with saline. Control SDR animals received vehicle injections for 5 days. The animals were sacrificed 6 h after the last injection. This experiment was carried out using three groups, normal (SD rats), SDR, and SDR+GH. In addition, in order to examine the acute effect of GH, a single injection of oGH (0.75 mg) was given and SDRs were sacrificed 6 h later. Kidneys were removed from these animals under light ether anesthesia and used for RNA extraction. Experiments were carried out twice on a different day, and each was performed in duplicate or triplicate. The mRNA data were obtained for each group of 4-6 SDR or SD rats.

1.2. Effect of fasting

SDRs were subjected to fasting conditions or fed normally for 2 days with free access to tap water. The control SD rats were either fasted or fed normally for

2 days. The SDRs and normal SD rats of the groups which were fed were allowed access to water only for 2 hrs before sacrifice. The four groups studied were as follows: normal control, normal fasted, SDR control, and SDR fasted. Experiments were carried out twice on a different day and each experiment was performed in duplicate.

1.3. Effect of uninephrectomy

Uninephrectomy was performed on all animals under pentobarbital anesthesia between 1:00 and 3:00 pm. Kidneys removed from each rat served as controls. Eight nephrectomized SDRs received twice daily injections for 2 days of 2.5mg/kg body weight ovine GH. The first GH injection was given immediately after coming out of anesthesia on the day of operation, followed by two injections at 9:00 am and 6:00 pm on the next day. The final injection was administered at 9:00 am on the third day and the animals were sacrificed 6 h after the last injection. Control SDR animals received vehicle injections for 2 days. Kidneys were removed from these animals under light ether anesthesia and stored at -80 °C for RNA extraction. Thus, the RNA analyses were carried out using five groups; normal SD rats (normal), nephrectomized SD rats (normal:Nx), SDR rats (SDR), nephrectomized SDR rats (SDR:Nx), and GH-injected nephrectomized SDR rats (SDR:Nx+GH). Experiments were carried out three times on a different day, and each was performed in duplicate or triplicate.

1.4. Northern blot analysis

The animals were sacrificed under light ether anesthesia and kidneys were removed rapidly and stored at -80 ℃ prior to RNA extraction. The extraction was carried out as described previously (16). Poly A+ RNA was isolated using oligo-dT cellulose (Sigma, St. Louis, MO). The RNA was separated on a 1% agarose gel containing formalin and transferred onto a nylon membrane (Hybond N, Amersham, Arlington Heights, IL) by capillary action. The blot was prehybridized in a mixture of 50% formamide, 6xSSC (1xSSC=150 mM NaCl/15 mM sodium citrate, pH7.0), 0.5% sodium dodecyl sulfate (SDS), 0.1% each of polyvinylpyrollidone, Ficoll and BSA, and 0.1 mg/ml of sheared denatured herring sperm DNA, at 42℃ for 2h. The blot was then incubated in a fresh mixture of the same composition containing 2-5x10^6 cpm/ml of ^{32}P-labeled probe. After 20 h incubation, the blots were washed 3 times for 10 min with 2xSSC/0.1% SDS at room temperature and then with 0.1xSSC/0.1% SDS at 65℃ for 1h. They were then exposed to Kodak XAR-2 film at -80℃ for 1-14 days. When the chicken actin cDNA probe was used for hybridization, the final wash was carried out at 45℃ in the same buffer as described above.

<u>cDNA probes:</u> Rat IGF-I cDNA (prigf1-1, 0.5 kb insert) was kindly supplied by Dr. Graeme I. Bell, Howard Hughes Medical Institute, The University of Chicago, Chicago, IL. The cDNA of rat IGFBPs were the gifts from Dr.D'Ercole (University of North Carolina, Chapel Hill, NC.), and had been obtained by amplification from rat liver cDNA by the polymerase chain reaction and cloned into plasmid Bluescript.

The cDNA inserts were excised from the vector by appropriate enzyme digestion and purified by agrose gel electrophoreses. The IGFBP-1 (355bp), IGFBP-2 (360bp), IGFBP-3 (430 bp), IGFBP-4 (321bp) and IGFBP-5 (651 bp) cDNA probes contained the sequence corresponding to the coding region of bp436-791, bp613-959, bp 328-755, bp 503-703 and bp558-1201 of each respective cDNA. The random priming method was used for ^{32}P-labeling using the reagents obtained from Amersham (Arlington Heights, IL). The specific activity of the labeled probes was between 0.5 - 1x10^9 cpm/mg DNA.
Quantification of the relative abundance of the mRNA: Blots hybridized to ^{32}P-labeled probes were analyzed using a bioimaging analyzer (BAS-2000, Fuji Film, Tokyo, Japan) for quantification of radioactivity of each mRNA. Blots were then exposed to X-ray film (XAR-5, Kodak, Rochester, NY) at -80℃ for 2 to 14 days.

1.5. Statistical analysis:

All data were expressed means $\pm$ SE. Tests performed consisted of Student's t test and, for multiple compaisons, one way analysis of variance followed by Student-Newman-Keuls. P value of less than 0.05 was considered statistically significant.

2. Results

2.1. Effect of GH treatment on renal IGF-I and IGFBP gene expression in SDR

Northern blot analysis of renal poly(A^+) RNA revealed that IGF-I mRNA is heterogenous with respect to the size, which ranged between 0.7-7.5 kb. All these IGF-I mRNA subclasses were consistently seen both in SDRs and in normal rats. Quantitative analyses showed a decreased IGF-I mRNA level in the SDR (~30% of the normal rat). GH treatment of SDRs for 6 hr increased IGF-I mRNA up to 49% of normals (data not shown) and for 5 days induced a marked increase in the IGF-I mRNA level which became fully normalized. In addition, the IGFBP-1 mRNA level in SDR rats was found to have increased to 180%, compared to that of normal rats. GH treatment of SDRs for 6 hr decreased 147% of normals (data not shown) and GH treatment for 5 days lowered IGFBP-1 mRNA level to a fully normalized level. Although IGFBP-4 mRNA in SDRs increased to a level comparable to that of IGFBP-1 mRNA, GH treatment had no effect on the IGFBP-4 mRNA level in SDR rats. The IGFBP-2 and IGFBP-3 mRNA levels in SDRs were found to be almost normal and repeated chronic GH administrations failed to produce any change. IGFBP-5 mRNA showed a tendency to increase, but did not reach a statistically significant level (p=0.14), and furthermore it was unaffected by GH treatment.

2.2. Effect of fasting on IGF-I and IGFBP mRNA levels

Fasting for 48 h resulted in a reduction in renal IGF-I mRNA levels in SDRs as well as in normal rats. Although the baseline IGF-I mRNA level was about 3 times less in the SDRs than in normal rats, no difference was seen in the degree of reduction induced by fasting between the normal and SDR group (42% and 45% reduction, respectively). In contrast, fasting induced a 3-fold increase in IGFBP-1 mRNA level of normal rats, compared to that of normally fed rats. In the SDR, IGFBP-1 mRNA also increased by 50% with fasting, however IGFBP-2, 3, 4, 5 mRNA levels did not, as was also the case with normal rats.

2.3. Effect of uninephrectomy

In normal SD rats, Nx resulted in significant decreases in IGFBP-1 mRNA and IGFBP-5 mRNA levels to 64.7±4.9% and 53.1±5.0% those of the normal kidneys, respectively. Nephrectomy did not reduce the IGF-I mRNA level in normal SD rats. The levels of IGFBP-2, IGFBP-3, and IGFBP-4 mRNA were likewise unchanged following nephrectomy.

In SDR rats, Nx significantly lowered the levels of IGFBP-1, IGFBP-4, and IGFBP-5 mRNA to 59±2.8, 45±12, and 60±7.9 % of the control values of kidneys before nephrectomy, respectively. However, Nx did not alter the levels of IGF-I, IGFBP-2, and IGFBP-3 mRNA. GH injection of nephrectomized SDR rats fully normalized the decreased IGFBP-4 mRNA level, while levels of IGFBP-1 and IGFBP-5 mRNA were not reversed.

3. Discussion

We have demonstrated that the expression of mRNA encoding IGF-I, IGFBP-1, and IGFBP-4 is altered in the SDR kidney. The relevance of GH status and renal IGFBP-4 and IGFBP-5 mRNA levels was examined in the present study for the first time. IGF-I mRNA was shown to be reduced in the SDR kidney and was normalized by GH treatment, similar to results obtained with hypophysectomized rats (19, 21), confirming that GH, but not other pituitary hormones, is responsible for its regulation. IGFBP-1 and IGFBP-4 mRNA levels were higher in SDRs than in normal rats, suggesting negative regulation by GH in the former. GH was effective in reducing the IGFBP-1 mRNA level although it was unable to lower the increased level of IGFBP-4 mRNA. The reason why increased IGFBP-4 mRNA was not reversed by GH remains unknown.

It was of interest to note that changes in IGFBP mRNA levels brought about by GH-modulation are different in liver and kidney. We have recently shown (15) that hepatic expression of mRNA encoding IGF-I and IGFBP-3 were reduced and IGFBP-1 mRNA was elevated in SDR. GH administration normalized these levels, indicating the association of GH but not other pituitary hormones with hepatic expressions of these genes. Although IGFBP-2 mRNA was elevated in the SDR liver, GH administration failed to effect any change. In addition, there was no change in the expression of IGFBP-4 mRNA in liver (15). These different results demonstrate altered response of IGFBP mRNA by GH-modulation between liver and kidney. There were similarities, however, and these were that IGF-I mRNA decreased and IGFBP-1 mRNA increased with a full return to normal values with GH, suggesting that GH up-regulates IGF-I and down-regulates IGFBP-1 in both liver and kidney. These results are in agreement with those reported by Chin et al. who demonstrated by in in situ hybridization that IGF-I and IGFBP-1 mRNA are both localized in renal medullary thick ascending limbs of Henle's loops in normal rat kidney where their expression appears to be inversely regulated, directly or indirectly, by GH (22). The demonstration of reciprocal changes in IGF-1 versus IGFBP-1 gene expression in the kidney suggests a local interaction between IGF-1 and IGFBP-1 in the regulation of their respective mRNA levels. We have recently shown that tubular hypertrophy induced by furosemide is associated with an increase in IGFBP-1 mRNA, but not with an increase in IGF-I mRNA (23). IGFBP-2, IGFBP-3, and IGFBP-5 levels in SDR kidney were found to be similar to those in the normal rats although there was a tendency for the IGFBP-5 mRNA level to increase. Although we cannot exclude the possibility that IGFBP-2, -3, and -5 mRNAs are regulated at least in part by GH, the major regulatory mechanism(s) of mRNA expression of these proteins involves factors which are different from those responsible for the regulation of other IGFBPs examined in this study. IGFBP-3 was formerly known as the GH-dependent IGF binding protein, because its presence in serum could be correlated with the secretory status of GH. Recent evidence, however, suggests that IGF-I, rather than GH, is a major regulator of IGFBP-3 (24). It has been shown that in hypophysectomized rats IGFBP-3 mRNA is elevated in kidney whereas it is reduced in liver with an accompanying decrease in the serum level of IGFBP-3, suggesting that a different pituitary-derived or -dependent factor may be involved (21).

With SDRs, we were able to examine the effect of fasting on kidney IGF-I and IGFBP mRNA levels in a GH-deficient model without having to consider changes in the serum GH level. In normal rats, fasting resulted in a decrease in IGF-I mRNA in association with an increase in IGFBP-1 mRNA. Despite the reduced level of IGF-I mRNA in the SDR, fasting for 48 h resulted in further reduction of IGFBP-1 mRNA. This finding suggests the presence of other nutrition-sensitive factors involved in the regulation of IGF-I gene expression both in kidney and in liver, as previously reported by us (15). We can conclude that the presence of other pituitary hormones does not alter this nutritional effect. The reduced level of circulating insulin that occurs with fasting may be responsible for this since reduction of hepatic IGF-I mRNA in the diabetic animal can be restored by insulin injection (25). The IGFBP-1 mRNA level was elevated in kidneys of both the SDR and normal rat as a result of fasting. Although it is known that reduction of GH secretion during fasting contributes to an increase in hepatic IGFBP-1 mRNA (26), the finding that the IGFBP-1 mRNA level was likewise elevated in the SDR which has a total lack of GH suggests the presence of other nutrition-sensitive factor(s) involved in the regulation of IGFBP-1 gene expression in the kidney. Insulin may be responsible since the serum IGFBP-1 level is regulated at least in part by this hormone (27). Unfortunately, the serum insulin response to fasting was not measured in the present study. With respect to mRNA levels of other binding proteins, those of IGFBP-2, -3, -4, -5 remained unchanged in both normal rats and SDRs. We have, however, recently reported (15) that, in the SDR liver, the IGFBP-2 mRNA level increased as a result of fasting as was the case in normal rats, suggesting that the response of IGFBP mRNA to fasting differs according to tissue type.

Finally, we examined the effect of nephrectomy on renal expression of the GH-IGF-I-IGFBP axis. Uninephrectomy did not change IGF-I mRNA level at day 2 in either normal SD rats or in spontaneous GH-deficient dwarf rats (SDR). Our observation that IGF-I mRNA was not changed 2 days after uninephrectomy is compatible with previous findings (8, 19). It has already been noted (8,19) that IGF-I is not involved in the early (1-2 days) hypertrophic response after unilateral nephrectomy, but that GH and/or IGF-I participate at later phases of compensatory renal hypertrophy (3rd-5th days or thereafter). Thus, in the initial stages of hypertrophy, it may be possible that locally produced IGFBPs are involved in modulating the action of IGF-I. In this regard, the results of changes in the levels of IGFBP-1 to 5 in contralateral kidney after uninephrectomy are informative. Although we did not observe morphologically in the present study, It has been already known that in GH-deficient rats, compensatory renal hypertrophy and glomerulosclerosis after renal ablation do not fully develop (28). Therefore, the lack of a difference in IGFBP-1 and IGFBP-5 expression between GH normal rats (SD) and GH-deficient rats (SDR) suggests a lower likelihood that these IGFBPs may participate in compensatory renal hypertrophy leading to glomerulosclerosis. On the other hand, it is of interest that changes in IGFBP-4 mRNA levels brought about by nephrectomy in GH-deficient SDR rats were different from those in normal SD rats. In normal SD rats, Nx did not affect the IGFBP-4 mRNA level, while in GH-deficient rats (SDR) Nx lowered this level to 45 % of the control values before nephrectomy. The lowered IGFBP-4 mRNA level was completely reversed by GH injection. These findings suggest that GH has an inhibitory effect on Nx-induced reduction of the IGFBP-4 mRNA level. The previous report showed that the renal IGFBP-4 mRNA level increased in SDR rats and that GH was not directly related to the regulation of expression of IGFBP-4 gene (11). However, as demonstrated by the present study, GH injection of nephrectomized SDR rats increased the IGFBP-4 mRNA level. We speculate that some unknown factors altered by nephrectomy and GH interact to increase expression of kidney IGFBP-4 mRNA. The difference in the IGFBP-4 mRNA response to nephrectomy between normal SD and SDR rats should be considered to be highly significant since again it has been shown that unlike SD rats, glomerulosclerosis does not fully develop in SDR rats following renal ablation (28). Furthermore, taking into consideration that the proximal nephron grows out of proportion to the rest of the nephron in compensatory renal hypertrophy (29) and that IGFBP-4 mRNA alone among the six IGFBPs was localized to proximal tubules (30), a decreased level of IGFBP-4 mRNA following nephrectomy in GH-deficient rats may play a role in reducing the development of renal hypertrophy and glomerulosclerosis.

4. Conclusions

We demonstrated in kidney that GH upregulated IGF-I and downregulated IGFBP-1 mRNA and that fasting for 48 h downregulated IGF-I and upregulated

IGFBP-1 mRNA. We also demonstrated that the IGFBP-4 mRNA level was elevated in the SDR kidney, although it was unaltered in the SDR liver. Compared with the previous study, the present data suggest that IGFBP mRNA, particularly that of IGFBPs-2, -3, -4, 5 are regulated differently in different tissues. Since GH and nutritional status are associated in the regulation of IGFBPs in the kidney, nutrition-dependent growth appears to be regulated by the GH-IGF-IGFBP axsis. The difference in response to nephrectomy of normal SD and SDR rats with respect to IGFBP-4 mRNA remains open to speculation, but appears to have an important implication for glomerulosclerosis. Further investigation of the physiological role of IGFBPs in kidney should be undertaken in the future.

REFERENCES

1. H.G. Rennke and P.S. Klein, Am. J. Kidney. Dis., 13 (1989) 443-456.
2. B.M. Brennner, T.W. Meyer and T.H. Hostetter, New. Engl. J. Med., 307 (1982) 652-659.
3. J.R. Diamond, Am. J. Physiol., 258 (Renal Fluid Electrol Physiol 27) (1990) F1-F8.
4. D.C. Tapp, W.G. Wortham, J.F. Addison, D.N. Hammonds, J.L. Barnes and M.A. Venkatachalam, Lab. Invest., 60 (1989) 184-195.
5. S. Kobayashi and M.A. Venkatachalam, Kidney Int., 42 (1992) 710-717.
6. J.Neugarten, H.D. Feiner, R.G. Schacht and D.S. Baldwin, Kidney Int., 24 (1983) 595-560.
7. D.C. Tapp, S. Kobayashi and M.A. Venkatachalam, Semin. Nephrol., 9 (1989) 343-353
8. M.R. Hammerman, Am. J. Physiol., 257 (Renal Fluid Electrol Physiol 26) (1989) F503-F514.
9. S.B. Miller, P. Rotwein, J.D. Bortz, P.J. Bechtel, V.A. Hansen, S.A. Rogers and M.R. Hammerman, Am. J. Physiol., 259 (Renal Fluid Electrol Physiol 27) (1990) F251-F257.
10. A.H. El Nahas, J.E. LE Carpentier, A.H. Bassett and D.J. Hill, Kidney Int., 36 (1989) S15-S19.
11. H.P. Guler, J. Zapf, E. Scheiwiller and E.R. Froesch, Proc. Natl. Acad. Sci. USA., 85 (1988) 4889-4893.
12. R. Hirschberg and J.D. Kopple, J. Clin. invest., 83 (1989) 326-330.
13. R. Hirschberg, G. Brunori, J.D. Kopple and H.P. Guler, Kidney Int., 43 (1993) 387-397.
14. S. Kobayashi, T. Arai, A. Hishida and D.R. Clemmons. J. Am. Soc. Nephrol., 3 (1992) 471.
15. H. Nogami, T. Watanabe and S. Kobayashi, Am. J. Phsiol., 267 (Endocrinol Metab 30) (1994) E396-E401.
16. T. Doi, L.J. Striker, C. Quaife, F. Conti, R. Palmiter, R. Behringer, R. Brinster and G.E. Striker, Am. J. Pathol., 131 (1988) 398-403.
17. J.D. D'Ercole, A.D. Stiles and L.E. Underwood, Proc. Natl. Acad. Sci. USA., 81

(1984) 935-939.
18. G.T. Ooi, C.C. Orlowski, A.L. Brown, R.E. Becker, T.G. Unterman and M.M. Rechler, Mol. Endocrinol., 4 (1990) 321-328.
19. R. Lajara, P. Rotwein, J.D. Bortz, V.A. Hansen, J.D. Sadow, C.R. Betts, S.A. Rogers and M.R. Hammerman, Am. J. Physiol., 257 (Renal Fluid Electrol Phsiol 26) (1989) F252-F261.
20. H. Nogami, T. Takeuchi, S. Suzuki, S. Okuma and H. Ishikawa, Endocrinology., 125 (1989) 964-970.
21. A.L. Albiston and A.C. Herington, Endocrinology., 130 (1992) 497-502.
22. E. Chin, J. Zhou and C. Bondy, Endocrinology., 130 (1992) 3237-3245.
23. S. Kobayashi, D.R. Clemmons, H. Nogami, A.K. Roy and M.A. Venkatachalam, Kidney Int., 47 (1995) 818-828.
24. C. Camacho-Hubner, D.R. Clemmons and A.J. D'Ercole, Endocrinology., 129 (1991) 1201-1206.
25. C.I. Pao, P.K. Farmer, S. Begovic, S. Goldstein, G.J. Wu and L.S. Phillips, Mol. Endocrinol 6 (1992) 969-977.
26. L.J. Murphy, C. Seneviratne, P. Moreira and R.E. Reid, Endocrinolgy., 128 (1991) 689-696.
27. D.K. Snyder and D.R. Clemmons, J. Clin. Endocrinol. Metab., 71 (1990) 1632-1636.
28. H. Yoshida, T. Mitarai, M. Kitamura, T. Suzuki, H. Ishikawa, A. Fogo and O. Sakai, Am. J. Kidney. Dis., 23 (1994) 302-312.
29. L.G. Fine, Kidney Int., 29 (1986) 619-634.
30. R. Rabkin, M. Brody, L.H. Lu, C. Chan, A.M. Shaheen and N. Gllet, J. Am. Soc. Nephrol., 6 (1995) 1511-1518

Molecular Mechanisms to Regulate the
Activities of Insulin-like Growth Factors
K. Takano, N. Hizuka and S-I. Takahashi (Editors)

Regulatory role of interleukin1β and nitric oxide on vascular smooth muscle cell proliferation in primary culture

T. Bourcier* and A. Hassid

Department of Physiology and Biophysics, University of Tennessee, Memphis, TN 38163 USA

1. INTRODUCTION

Atherosclerosis, restenosis, and organ transplantation are conditions or events associated with vascular pathology, manifested in part by the excessive proliferation of vascular smooth muscle cells (SMC) in the subendothelial space of injured blood vessels, contributing significantly to neointimal formation and vessel occlusion[1,2]. This is the result, in part, of both migration and replication of medial and resident intimal smooth muscle cells[3]. There is significant *in vivo* and *in vitro* evidence to implicate several growth factors and inflammatory cytokines in regulating the intimal expansion of SMC. These factors have both direct and indirect effects on SMC growth, and also function as a network to potentiate and/or inhibit each others' activities. However, most of the cellular and biochemical effects of growth factors and cytokines are derived from *in vitro* studies of secondary cultures of SMC that express a cytoskeletal phenotype similar to that described for modulated SMC of the neointima[3]. In contrast, relatively few studies have focused on the influence of these factors on SMC replication in primary culture that more closely resembles the cytoskeletal phenotype of contractile, differentiated smooth muscle cells *in vivo*. Recently we have observed that the mitogenic responsiveness of rat aortic smooth muscle cells in primary culture to nitric oxide and cGMP is diametrically opposite to that observed in secondary cultures[4]. Inasmuch as inflammatory cytokines, notably IL-1β, induce SMC to produce nitric oxide[5], one would predict divergent effects of IL-1β on smooth muscle cell replication in primary vs. secondary culture. Demonstration of such divergent effects of growth factors and cytokines associated with a change in cytoskeletal phenotype would suggest a more complex regulation of smooth muscle cell growth by these factors than heretofore appreciated. Thus, the current study was undertaken to investigate the interaction between growth factors, interleukin-1β and nitric oxide relative to mitogenesis in primary cultures of rat aortic smooth muscle cells.

---

* Present address: Vascular Medicine & Atherosclerosis Unit, Brigham & Womens Hospital, Boston, MA 02115 USA. This work supported by NIH grant HL44671

2. METHODS

2.1. Cell Culture

Smooth muscle cells were obtained from the thoracic aortae of male Sprague-Dawley rats, as described[6]. Briefly, medial tissue cleaned of adventitia were enzymatically dispersed in minimal essential medium containing 0.2 mM $CaCl_2$, 15 U/ml elastase, 200 U/ml collagenase, 0.4 mg/ml trypsin inhibitor, and 1.7 mg/ml bovine serum albumin. Dissociated cells were washed free of enzyme and triturated to yield $0.8\text{-}1.5 \times 10^6$ cells/aorta, mostly as single cells. Cells were then seeded into Primaria culture dishes at a density of $1.8\text{-}2.3 \times 10^4$ cells/cm^2 and cultured for the first two days in serum-free DMEM/Ham's F12 (1:1) medium supplemented with insulin (5 µg/ml), transferrin (5 µg/ml), and selenous acid (5 ng/ml) in a humidified atmosphere of 5% CO_2/95% air. Most cells (~95%) attached to the culture surface within the first few hours after seeding and were spreading over the surface after overnight culture. Fetal bovine serum was then added to 10% final concentration and cultured for an additional 3-5 days. Each individual experiment presented here represents results from one such cell isolate, from a pool of cells obtained from 4-6 rat aortae.

2.2. DNA synthesis and cell proliferation

DNA synthesis was measured via [^{3}H]thymidine incorporation in mitogenically quiescent aortic smooth muscle cells. Quiescent cells were cultured 22h in the absence or presence of experimental agents, and [^{3}H]thymidine was added during the last 2 hrs of incubation. For cell cycle analysis, ethanol-fixed cells were stained with propidium iodide in the presence of type III RNAse using a DNA-prep device (Coulter, Hialeah, FL). Samples were analyzed for DNA content within one hour of staining on an Epics Profile Flow Cytometer (Coulter) at an excitation wavelength of 488 nm. The multiple-option cell cycle analysis software (Phoenix Flow Systems, San Diego, CA) was used to determine the percentage of cells in various phases of the cell cycle. Cell proliferation was determined after 2d incubation of cells with experimental agents by counting cells in a calibrated Coulter counter (Model ZM, Hialeah, FL).

2.3. Determination of cGMP

Following experimental incubations, culture media were collected into tubes containing 1mM isobutylmethylxanthine for determination of extracellular cGMP by radioimmunoassay. Samples were stored at -20°C until analysis.

2.4. Statistical Analysis

Statistical differences for paired samples were evaluated by the non-parametric Mann-Whitney U-test. For analysis of variance we used the non-parametric Kruskal-Wallis test, followed by the Dunn procedure. $P<0.05$ was considered statistically significant.

3. RESULTS

3.1 Nitric oxide and FGF-2 are co-mitogens for primary vascular SMC

Agents that stimulate the cGMP signal transduction pathway, such as nitric oxide and cGMP analogs, inhibit serum and growth factor-induced mitogenesis in secondary cultures of vascular smooth muscle cells[7]. In this study, the influence of cGMP agonists on growth factor-induced mitogenesis in primary cultures of vascular smooth muscle cells was investigated. Primary SMC were incubated with increasing concentrations of the NO-donor S-nitroso-N-acetylpenicillamine (SNAP) in the absence or presence of a maximally mitogenic concentration of FGF-2 (10 ng/ml), and [^{3}H]thymidine incorporation was measured 24h later. SNAP had little to no effect on its own, but amplified FGF-2-induced thymidine incorporation by ~3 fold over FGF-2 alone in concentration-dependent fashion (Fig 1). A structurally dissimilar NO-donor, SIN-1, was also effective in amplifying the mitogenic response to FGF-2. Moreover, N-acetylpenicillamine, identical to SNAP except for the NO moiety, was ineffective in amplifying FGF-2-induced thymidine incorporation (data not shown). Thus, nitric oxide synergistically amplifies the mitogenic response to FGF-2 in primary vascular smooth muscle cells.

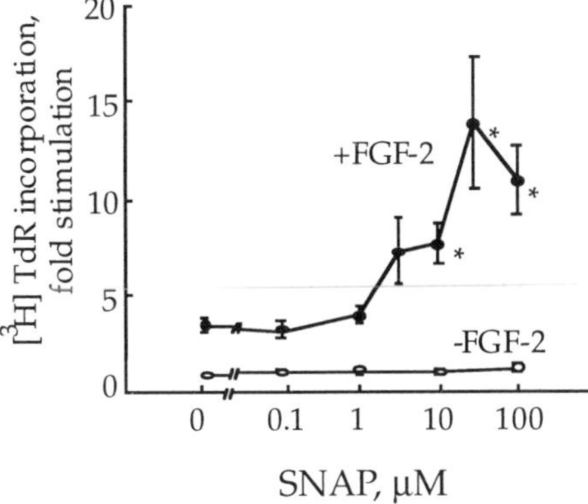

Figure 1. NO-donor S-nitroso-N-acetylpenicillamine (SNAP) amplifies FGF-2-induced thymidine incorporation in primary cultures of rat aortic SMC. Results are expressed as fold stimulation over serum-free medium, (range 60.5 to 328 cpm/µg protein) and are the means ±SEM of 5-12 separate experiments each performed in quadruplicate. *Asterisks indicate significance at $p<0.05$ relative to the response to FGF-2 alone.
Data from: Hassid et al. 1994. Am J Physiol. 267:H1043-H1048

Experiments were next performed to determine if the co-mitogenic effect of NO was mediated by cGMP as the second messenger. Figure 2a shows that a hydrolysis-resistant analog of cGMP, 8-bromo-cGMP, dose-dependently amplified thymidine incorporation induced by FGF-2 while having negligible effects of its own. On the other hand, 8-bromo-cAMP was ~10-fold less potent than 8-bromo-cGMP in amplifying the mitogenic response to FGF-2 (Fig 2b), suggesting that cAMP was unlikely to mediate the co-mitogenic effect of 8-

bromo-cGMP or NO. Moreover, the phosphodiesterase inhibitor, zaprinast (M&B 22,948) enhanced the potency of SNAP by ~1.5 orders of magnitude, while increasing 100 μM SNAP-induced cGMP levels from 7.15±0.34 to 41.9±1.62 (SE) *f*mol/μg protein (n=4 wells; similar results were obtained in 2 additional experiments). These results provide evidence to support the notion that the effect of NO is, at least partially, mediated via cGMP as the second messenger. Thus, in striking contrast to its growth-inhibitory effect in secondary cultures of SMC, NO enhances FGF-2-induced mitogenesis in primary smooth muscle cells via a mechanism involving the cGMP pathway.

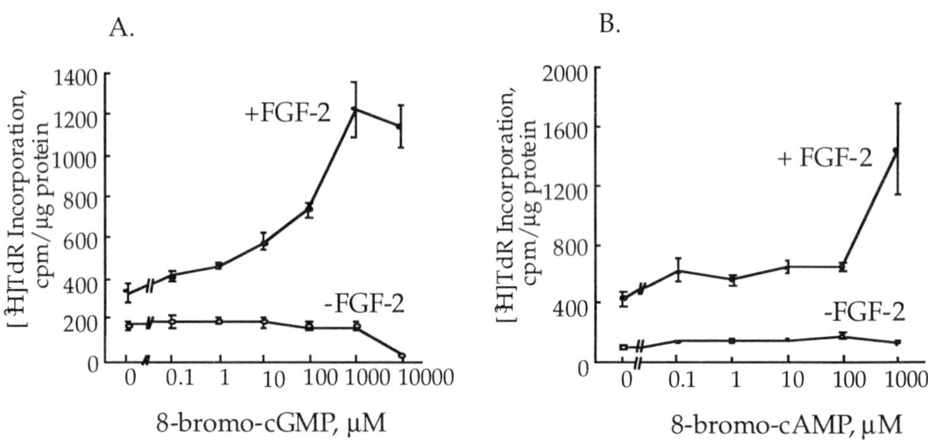

Figure 2. (a) 8-bromo-cGMP amplifies FGF-2-induced thymidine incorporation in primary SMC more effectively than 8-bromo-cAMP (b). Values are means±SEM from one of 3 experiments done in quadruplicate.
Data from: Hassid et al. 1994. Am J Physiol 267: H1040-H1048.

3.2 IL1β and FGF-2 are comitogens for primary vascular SMC

Because of the comitogenic effect of exogenous sources of NO in primary cells, it would be expected that endogenous production of NO by smooth muscle cells stimulated with cytokines, such as IL-1β, would elicit a similar response. Therefore, experiments were performed to investigate potential synergistic interactions between IL-1β and FGF-2 on DNA synthesis in primary vascular SMC. Figure 3 shows that femtomolar levels (1-10 pg/ml) of IL-1β alone had no independent mitogenic effect, whereas picomolar levels of IL-1β (≥100 pg/ml) increased thymidine incorporation by 2-4 fold. FGF-2 alone (50 ng/ml) increased thymidine incorporation by 4.36±0.82-fold (mean±SEM; n=4) relative to control. When combined with 10 pg/ml IL-1β, that alone was virtually non-mitogenic, thymidine incorporation was increased by ~2-fold over that induced by FGF-2 alone, and ~10-fold relative to control incubations in serum-free medium. As the concentration of IL-1β was increased to mitogenic levels (>100 pg/ml), the interaction with FGF-2 became additive rather than synergistic. The co-mitogenic effect of IL1β was abolished by an antagonist of the Type I IL1 receptor, IL1RA, without altering the response to FGF-2 alone, indicating the requirement for activation of the IL1 receptor (data not shown).

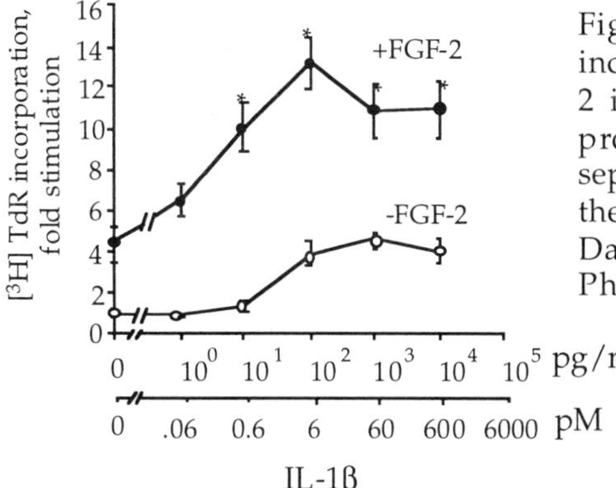

Figure 3. Synergistic increase of thymidine incorporation in response to IL-1β and FGF-2 in primary SMC. Values, normalized to protein, represent the mean±SEM of 4 separate experiments. *$p<0.05$ relative to the response to FGF-2 alone.
Data from: Bourcier et al. 1995. J Cell Physiol. 164:644-657.

To verify that the increases in thymidine incorporation accurately reflected DNA synthesis, the progression of cells through the cell cycle was followed over a 4 day period. As shown in figure 4a, a non-mitogenic concentration of IL-1β alone had no effect on the percentage of cells found in S-phase, whereas FGF-2 increased the S-phase population to 12.3% after 1 day of incubation. The combined use of IL-1β and FGF-2 further increased the fraction of cells found in S-phase to 18% after 1 day of incubation and this value decreased to control levels by day 2 and remained there for the duration of the experiment. The observed increases in the percentage of S-phase cells after 1 day of incubation were associated with a concomitant decrease in the percentage of cells in the G0/G1 phase. By day 2, the percentage of cells in the G0/G1 phase returned to control values, indicating that the cells had progressed through the G2/M-phase and completed the cell cycle. In separate cell proliferation experiments, FGF-2 increased the cell number by 25% over control after a 2 day incubation, and the combined use of IL-1β and FGF-2 increased cell number by 40.7%, as shown in Figure 4b. IL-1β alone had no significant effect on cell number. Taken together, these results support the view that the interaction of IL-1β with FGF-2 significantly enhances DNA synthesis and cell proliferation in vascular SMC.

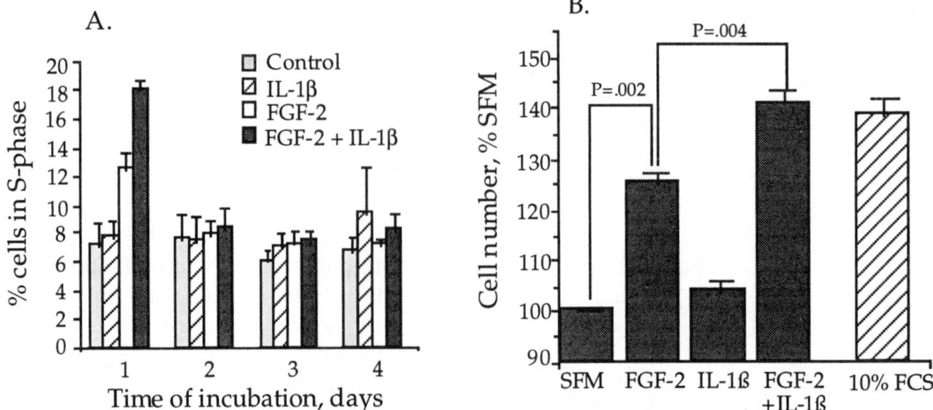

Figure 4. (a) Influence of treatment of primary aortic smooth muscle cells with IL-1β (10 pg/ml), FGF-2 (50 ng/ml), or a combination of the two, on the percentage of cells in S phase of the cell cycle. The percentage of cells in S-phase was determined every 24 h via flow cytometry. Values are the mean±SEM of 3-4 separate experiments. (b) Cell number after incubation with serum-free medium (SFM), FGF-2 (10 ng/ml), IL-1β (50 pg/ml), FGF-2 + IL-1β, or 10% FBS for 2 days. Results are the mean±SEM of 5 experiments, each in quadruplicate, and are normalized to the number of cells in SFM which ranged from 1.26-2.02 x 10^5 cells/well. Statistical significance was evaluated by the Mann-Whitney U test.
Data from: Bourcier et al. 1995. J Cell Physiol. 164:644-657.

3.3 IL-1ß enhances thymidine incorporation induced by platelet-derived growth factor-AB and epidermal growth factor, but not by insulin-like growth factor-1.
Growth factors other than FGF-2, including PDGF, EGF and IGF-1, are likely to play a role in vascular SMC proliferation *in vivo*[1]. We therefore determined the effect of IL-1β on PDGF-AB, EGF, and IGF-1-induced thymidine incorporation in primary aortic smooth muscle cell cultures. Similar to the synergistic interaction of IL-1β and FGF-2, PDGF-AB and EGF-induced thymidine incorporation in vascular smooth muscle cells was synergistically increased by IL-1ß by ~2 fold and 3.5 fold, respectively (data not shown). On the other hand, IL-1β and IGF-1 induced an additive rather than synergistic increase of thymidine incorporation. Thus, IL-1β appears to interact in synergistic fashion with several, but not all, growth factors to amplify mitogenesis.

3.4 The synergistic interaction of IL-1ß and FGF-2 is independent of the nitric oxide/cGMP pathway.
Several investigators have reported that IL-1β elevates cGMP levels in cultured SMC via induction of NO synthesis and subsequent activation of soluble guanylate cyclase[5,8]. Experiments were therefore performed to identify a

potential role for the NO/cGMP pathway in the synergistic interaction between IL-1β and FGF-2. Cells were stimulated with FGF-2, IL-1β, or the two together, in medium containing reduced arginine (30 μM), in the absence or presence of the NO-synthase inhibitor N^5-iminoethyl L-ornithine (L-NIO). Both thymidine incorporation and the extracellular accumulation of cGMP were measured 24 h later. As shown in Figure 5, FGF-2 alone had no effect on cGMP levels, whereas IL-1β alone increased cGMP by 2-fold relative to the levels found in serum-free medium. Interestingly, the modest increase in response to IL-1β was markedly enhanced by FGF-2, to levels ~8 fold over that found in serum-free medium. These increases in cGMP were totally blocked by 100 μM L-NIO. However, despite the complete blockade of IL-1β-induced increases in cGMP levels by L-NIO, potentiation of FGF-2-induced thymidine incorporation by IL-1β was decreased by only ~13% (Figure 5a). Thus, the mechanism for the co-mitogenic effect of IL-1β is largely independent of increases in NO/cGMP. This was further supported by experiments that showed a nearly additive effect of SNAP and IL-1β on FGF-2-induced thymidine incorporation, consistent with the view that the co-mitogenic effect of IL-1β is largely independent of the NO/cGMP pathway.

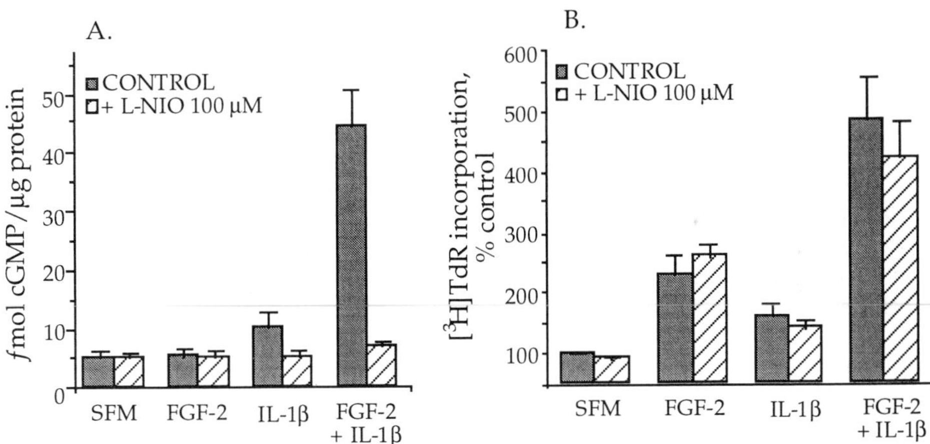

Figure 5. (a) Effect of the NO-synthase inhibitor L-NIO on cGMP accumulation in response to FGF-2 (10 ng/ml) and/or IL-1β(50 pg/ml). (b) Effect of L-NIO on thymidine incorporation in response to FGF-2 and IL-1β from the same experiment described in (a). Values for cGMP and thymidine were normalized to protein and are the mean±SEM from 3 separate experiments performed in triplicate. Data from: Bourcier et al. 1995. J Cell Physiol. 164:644-657.

4. DISCUSSION

The goal of this study was to investigate the influence of exogenous donors of NO and a stimulus for endogenous production of NO, IL-1β, on growth factor-induced mitogenesis in primary cultures of aortic SMC that exhibit a relatively differentiated phenotype. One principal finding of this study is that two structurally dissimilar NO donors, SNAP and SIN-1, enhance the mitogenic response to FGF-2 in primary SMC, without having an effect on their own. That 8-bromo-cGMP mimicked the comitogenic effect of the NO donors, whereas 8-bromo-cAMP was ~10-fold less effective, supports the view that cGMP serves as second messenger. In addition, the finding that NO amplifies FGF-2-induced mitogenesis in primary cells but inhibits mitogenesis in repetitively passaged cells[4,7] supports a variable effect of NO on smooth muscle cell proliferation, possibly dependent on SMC phenotype. Because exogenous sources of NO enhance the mitogenic response to FGF-2 in primary cells, it seemed reasonable to propose that IL-1β would also enhance the mitogenic response to growth factors by increasing the endogenous production of NO by SMC. Accordingly, a second principal finding of this study is that femtomolar concentrations of IL-1β enhance the mitogenic response to several mitogens of importance in vascular pathology, including FGF-2, EGF and PDGF, in primary SMC. However, the mechanism that underlies this comitogenic influence appears independent of the NO/cGMP pathway. Further study into the mechanism of comitogenesis between IL1β and FGF-2 excluded enhanced FGF-2 signaling events, including early or delayed tyrosine phosphorylation events, MAP kinase activity, and c-fos protooncogene expression. Taken together, these data support the view that SMC phenotype may markedly influence the response to and interaction between growth factors, IL1β, and NO/cGMP. We speculate that IL-1β and nitric oxide, and possibly the paracrine activity of IL-1β-induced increases of nitric oxide, could play a role in amplifying the mitogenic activity of growth factors *in vivo*, through separate mechanisms, at least in the early stages following vascular injury.

REFERENCES

1. R. Ross, Nature, 362 (1993) 801.
2. M. W. Liu, G. S. Roubin and S. B. r. King, Circulation, 79 (1989) 1374.
3. S. M. Schwartz, D. deBlois and E. R. O'Brien, Circulation Research, 77 (1995) 445.
4. A. Hassid, H. Arabshahi, T. Bourcier, G. S. Dhaunsi and C. Matthews, Am. J. Physiol., 267 (1994) H1040.
5. D. Beasley, J. H. Schwartz and B. M. Brenner, J Clin Invest, 87 (1991) 602.
6. T. Bourcier, M. Dockter and A. Hassid, J. Cell Physiol., 164 (1995) 644.
7. U. C. Garg and A. Hassid, J Clin Invest, 83 (1989) 1774.
8. V. B. Schini and P. M. Vanhoutte, J Cardiovasc Pharmacol, 11 (1989)

Molecular Mechanisms to Regulate the
Activities of Insulin-like Growth Factors
K. Takano, N. Hizuka and S-I. Takahashi (Editors)

THE YEAST TWO-HYBRID SYSTEM TO INVESTIGATE IGF-I RECEPTOR SIGNAL TRANSDUCTION.

R. W. Furlanetto[a], K. Frick[a], B. R. Dey[b], W. Lopaczynski[b], C. Terry[b] and S. P. Nissley[b]

[a]Department of Pediatrics, University of Rochester, Rochester, NY, USA

[b]Metabolism Branch, National Cancer Institute, National Institutes of Health, Bethesda, MD, USA

The biological effects of the insulin-like growth factors (IGFs) are mediated through the type I IGF receptor (IGFIR). This receptor is a transmembrane tyrosine-kinase and is closely related to the insulin receptor (IR). Binding of IGF activates the IGFIR and results in its autophosphorylation. This creates motifs which are recognized by cytosolic proteins involved in IGF signal transduction, some of which are also phosphorylated by the receptor. Several proteins involved in IGF signal transduction have been characterized, including IRS-1 and Shc, but it is clear that important proteins remain unidentified.

Understanding the mechanism by which the IGFs elicit their biological effects depends on identifying and characterizing the cellular proteins which interact with the receptor *in vivo*. A number of methods have been used to identify such proteins, including classic biochemical approaches (*i.e.* purification and sequencing) and expression library screening. Recently, we (1,2) and others (3,4) have used two-hybrid systems to investigate IGFIR signal transduction. Two-hybrid systems are yeast-based genetic assays for detecting protein-protein interactions (5,6). These systems offer a number of advantages over other methods for detecting interacting proteins. Because the interaction occurs in an *in vivo* environment, the proteins are more likely to maintain their native conformations than in *in vitro* assays. Possibly because of this, yeast two-hybrid assays are highly sensitive and will detect weak and transient interactions. These systems are also very versatile and can be used to determine if two known proteins interact, to define the motifs mediating the interaction and to identify novel proteins interacting with a protein of interest. Finally, the assays are rapid and easy to perform and do not require the use of radioactivity.

In this chapter we will discuss the use of the two-hybrid system to investigate IGF receptor signal transduction. Emphasis will be placed on describing the specific features of the system which we employed - the interaction trap - and the use of this system to map sites of interaction and to identify novel proteins which interact with the receptor. The application of this system to investigating the role of 14-3-3 proteins in IGFIR function is discussed elsewhere in this volume.

1. THE INTERACTION TRAP

Two-hybrid assays are yeast-based genetic systems for detecting protein-protein interactions. These systems take advantage of the fact that many eukaryotic transcription factors have a modular domain structure consisting of two distinct elements, a DNA binding domain and an activating domain. The DNA binding domain directs binding of the transcription factor to DNA sequences present in the regulatory region of specific genes; the activation domain activates transcription by contacting components of the transcriptional machinery. Although these two domains are normally part of the same protein, a covalent attachment is not necessary and activation can occur when these domains are on separate proteins, provided the proteins interact. In yeast two-hybrid systems, two hybrid proteins are employed. The first is a fusion protein containing the DNA-binding domain of a specific transcription factor, such as Gal4 or LexA, linked to the protein of interest; this hybrid protein is termed the bait. The second component is a hybrid protein composed of a known transcriptional activating domain fused either to a protein suspected to interact with the first protein (termed the prey), or, alternatively, to a cDNA library. Both hybrid proteins are expressed in a yeast strain that contains one or more reporter genes which are responsive to the transcription factor supplying the DNA-binding domain. If the activation domain hybrid associates with the DNA binding hybrid, an intact transcriptional activator is reconstituted and the reporter genes are activated.

A number of two-hybrid assays have been developed (5,6,7). They differ in the specific transcription factor which is employed (i.e. Gal4 or LexA) and certain other features, such as the nature of the activating domain and the reporter genes. In general, the various components are not interchangeable. Moreover, the different systems do not always behave similarly. For example, the interaction between the IGFIR and IRS-1 was readily detected using the interaction trap, a LexA-based two-hybrid system developed by Brent and coworkers (5), but no interaction of these proteins could be detected using the Gal4-based system of Fields and Song (6).

The interaction trap is a two-hybrid system which utilizes the bacterial LexA promoter and DNA-binding elements. A detailed description of this system can be found in reference 5. Briefly, this system has three components. The first is a shuttle plasmid (pEG202) used for constructing the LexA-bait hybrid. This plasmid encodes amino acids 1 to 202 of LexA, which includes the DNA binding and dimerization domains, and includes a multiple cloning site for inserting the cDNA of interest. The LexA-bait protein is constitutively expressed from this plasmid. This plasmid also contains a HIS3 selectable marker and the 2um origin of replication to allow selection and propagation in yeast. The second element is a shuttle plasmid (pJG4-5) used for the construction of the activation domain-prey/library hybrid. It encodes a 105 amino acid sequence composed of the SV40 large T nuclear localization signal, the bacterial B42 activation domain (AD) and the hemagglutinin (HA) epitope tag and includes an insertion site for the target or library cDNA. Expression of the activation domain-hybrid is under control of the Gal1 promoter and it is expressed when yeast are grown on galactose and repressed when they are grown on glucose. This plasmid also contains a TRP1 selectable marker and the 2um origin of replication. The third component is a yeast strain (EGY/pSH18-34) which contains two LexA-responsive reporter genes: a chromosomally integrated LEU2 gene and a plasmid (pSH18-34) encoded lacZ gene. In this system the leucine (leu) reporter is more sensitive than the lacZ reporter but lacZ (*i.e.* β-galactosidase) activity can be better quantitated using a solution assay (5). This yeast strain is auxotrophic for histidine, leucine, tryptophan and uracil. In this system, a positive interactor has a galactose inducible leu+/lacZ+ phenotype.

1.1. LexA-IGFIRß - The Bait

The properties of the bait hybrid are critical for the two-hybrid system. To be useful, the bait must be stably expressed, enter the nucleus, bind to the appropriate DNA element and it should not have intrinsic activating activity. In addition, if the parent protein has enzymatic activity (i.e., tyrosine-kinase activity), the bait should retain this activity.

The LexA-IGFIRß fusion hybrid which we employed encodes LexA (202) fused directly to residue 936-1337 of the IGFIR using the EcoR1 cloning site of pEG202. Amino acid 936 is six amino acids distal to the transmembrane domain of the IGFIR; the transmembrane domain was deleted because it could interfere with transport of the hybrid into the nucleus. To determine if the LexA-IGFIRß hybrid was stably expressed in yeast, extracts were prepared from yeast transfected with this bait and examined by Western blotting using a rabbit polyclonal antibody to the carboxy-terminal portion of the human IGFIR (1). With this bait construct, a band of the expected MW (approximately 68 kDa) was observed; this band was absent in yeast containing a control (LexA-bicoid) plasmid (1). To determine if the Lex-A-IGFIRß fusion hybrid entered the nucleus and bound to the LexA operators, a repression assay was performed using the reporter plasmid pJK101 (5). This assay is based on the observation that a non-activating LexA fusion represses transcription of a reporter gene that has LexA operators positioned between the TATA box and the upstream activating sequence, as found in pJK101. We found that the Lex-A-IGFIRß bait hybrid repressed transcription from pJK101, indicating that this bait is transported to the nucleus and binds to LexA operators (data not shown).

For a bait hybrid to be useful in the interaction trap, it must not have intrinsic activating activity. To determine if the LexA-IGFIRß bait plasmid fulfilled this criterion, yeast containing this bait were plated onto medium either lacking leucine or containing X-gal. Yeast expressing the LexA-IGFIRß hybrid described above did not activate either reporter (*i.e.* they did not grow in the absence of leucine or metabolize X-gal), indicating that this construct did not possess intrinsic activating activity and that it was suitable for use in the interaction trap. However, an observation made during these studies is noteworthy. In our initial experiments, we utilized a bait plasmid in which LexA (1-202) was joined to IGFIRß (936-1337) using the BamH1 site in the polylinker of pEG202; this introduced a linker sequence (QFPGINP) derived from the polylinker site of the vector. This construct had moderate intrinsic activating activity: *i.e.* it activated the leucine reporter but not the less sensitive lacZ reporter. Because the parent plasmid (i.e. pEG202) itself has weak activating activity which maps to the polylinker site, we reasoned that this sequence could be contributing to the activating activity of this bait construct. This led us to re-engineer the construct by deleting the polylinker sequences and this resulted in a markedly improved bait construct. This observation underscores the need for careful bait plasmid design. It is also noteworthy that a LexA-IRβ hybrid constructed using the same principles retained weak to moderate activating activity, indicating that the IRβ sequence itself is responsible for the activating properties of this construct.

To determine if the LexA-IGFIRß hybrid has tyrosine-kinase activity, extracts were prepared from yeast transfected with this bait or with a mutant LexA-IGFIRß hybrid in which lysine 1003 in the ATP binding site was changed to arginine (IGFIRß(KR)); this mutant has been shown to lack tyrosine-kinase activity in mammalian cells (8). The extracts were examined by Western blotting using an antiphosphotyrosine antibody. This analysis revealed that the fusion protein derived from the active receptor (i.e., the wild type) contained phosphotyrosine while that from the KR mutant did not, indicating that the ß subunit of the receptor is a constitutively active tyrosine-kinase when expressed as a LexA fusion protein in yeast (1). Since LexA binds to its operator as a dimer, it is possible that the phosphorylation occurs by a *trans* mechanism, as is believed to occur in mammalian cells.

1.2. Studies with IRS-1 and Shc

IRS-1 and Shc have been shown to interact with the IGFIR in mammalian cells (9,10). To test the utility of the interaction trap for studying IGFIR signal transduction, we used these proteins as model prey (1). Plasmids encoding either the AD-HAtag-IRS-1 (pJG-IRS-1) or AD-HAtag-Shc (pJG-Shc) hybrids or the parent plasmid (pJG4-5) were introduced into EGY/pSH18-4 containing either the LexA-IGFIRß(WT) or the LexA-IGFIRß(KR) bait plasmids. Multiple colonies were picked from the selection plates and tested for the galactose dependent leu+/lacZ+ phenotype expected for true interactors. As shown in Table 1, yeast containing the LexA-IGFIRß(WT) hybrid and either the AD-HAtag-IRS-1 or the AD-HAtag-Shc hybrid showed galactose dependent activation of the leucine and lacZ reporters; in contrast, yeast coexpressing the LexA-IGFIRß(KR) mutant bait hybrid and either the AD-HAtag-IRS-1 or the AD-HAtag-Shc hybrid did not activate the reporters. Thus, we concluded that under the conditions employed in the yeast interaction trap, the IGFIRß subunit interacts directly with IRS-1 and Shc and that an active tyrosine-kinase domain is essential for this interaction. These findings support the use of the interaction trap to investigate IGFIR signal transduction.

Table 1
IRS-1 and Shc interact with the IGFIR in the yeast interaction trap[1]

Plasmids		Colony Growth		β- Galactosidase Activity	
LexA Hybrid	AD-HAtag Hybrid	Glu	Gal	Glu	Gal
IGFIRβ	--	0	0	0	0
IGFIRβ	IRS-1	0	+2	0	+4
IGFIRβ	Shc	0	+4	0	+4
IGFIRβ(KR)	--	0	0	0	0
IGFIRβ(KR)	IRS-1	0	0	0	0
IGFIRβ(KR)	Shc	0	0	0	0

[1]The LexA-hybrids were introduced into yeast containing either pJG4-5, pJG-IRS-1, or pJG-Shc and the transformants isolated on selective plates. Transformants were assayed for their ability to grow in the absence of leucine (colony growth) or to activate lacZ (β-galactosidase activity) with either glucose (glu) or galactose (gal) as the carbon source. Plates were scored after 72 hours. The colony growth was scored a follows: 0-no growth; +1-microscope colonies; +2-visible colonies; +3-subconfluent colonies; and +4-confluent. The β-galactosidase activity ranged from: 0-white; 1-trace blue; +2-light blue; +3-blue; and +4-dark blue.

2. MAPPING SITES OF INTERACTION ON THE IGFIR AND ITS TARGETS

In addition to confirming an interaction between known proteins, yeast two-hybrid systems can also be used to map the interacting sites on proteins. For example, we have used the interaction trap to delineate the regions of IRS-1 and Shc which interact with the IGFIR and to identify the region of the receptor which interacts with these proteins. To delineate the regions of IRS-1 and Shc which interact with the IGFIR, plasmids encoding hybrids containing portions of IRS-1 or Shc fused to the activation domain-HA tagged moiety were introduced into EGY/pSH18-34 containing the LexA-IGFIRß(WT) hybrid and the phenotypes of the transformants were determined. The results are summarized in Table 2. As noted above, coexpression of LexA-IGFIRß and the AD-HA tagged hybrids containing the entire IRS-1(2-1242) coding sequence activated both reporter genes. Constructs containing IRS-1 residues 2-516 or 160-516 activated both reporter genes to levels similar to or greater than those observed with full length IRS-1. In contrast, constructs containing IRS-1 residues 2-160 or 203-516 did not activate the reporters. Constructs containing full length Shc (*i.e.* 1-473) or Shc residues 1–232 also activated the reporters to similar levels, suggesting that the carboxy-terminus of Shc is not necessary for its interaction with the IGFIR. In agreement with this, a construct containing Shc residues 233-473 and one containing only the SH2 domain of Shc (i.e. 378-473) did not activate the lacZ reporter and only weakly activated the leucine reporter. Western blotting with an αHA antibody indicated that the various hybrid proteins were expressed at similar levels (not shown). These observations suggest that an element necessary for the interaction of IRS-1 with the IGFIR is located between residues 160 and 203 and that the amino-terminal region of Shc, and not the SH2 domain, mediates its interaction with the IGFIR. These findings are similar to those reported for the interactions of IRS-1 and Shc with the insulin receptor (using the interaction trap) (11) and of Shc with the EGF receptor (using expression library screening) (12).

Table 2
Deletion Analysis of IRS-1 and Shc[1]

AD-HAtag Hybrid	Colony Growth		ß-Galactosidase Activity	
	Glu	Gal	Glu	Gal
IRS-1:2-1242	0	+2	0	+4
IRS-1:2-516	0	+4	0	+4
IRS-1:2-160	0	0	0	0
IRS-1:160-516[2]	0	+4	0	+4
IRS-1:203 -516[2]	0	0	0	0
Shc:1-473	0	+4	0	+4
Shc:1-232	0	+4	0	+4
Shc:233-473	0	0	0	0
Shc:378-473	0	+1	0	0

[1]Experiments were performed as described in the legend to Table I using the LexA-IGFIRβ(WT) hybrid. Colony growth and color were scored after 72 hours.
[2]These plasmids were obtained from Dr. T. Gustafson (11).

We have also used the interaction trap to identify the region of the receptor which interacts with IRS-1 and Shc. For these experiments, a series of mutant LexA-IGFIRβ bait hybrids were generated. These included constructs in which the carboxy-terminal 27 residues of the receptor (*i.e.* 1310Y) and 93 residues (*i.e.* 1244Y) were deleted and a construct in which tyrosine 950 in the juxtamembrane region was changed to phenylalanine (*i.e.* 1337F). These receptor baits were coexpressed with either the AD-HAtag-IRS(2-516) or the AD-HAtag-Shc(1-473) hybrid and the transformants were assayed for reporter gene expression. The results, using a quantitative solution assay for β-galactosidase activity, are shown in Figure 1. Coexpression of AD-HAtag-IRS-1 or Shc with the wild type LexA-IGFIR hybrid resulted in strong, galactose-dependent activation of the lacZ reporter gene, while their coexpression with the kinase-inactive mutant (*i.e.* KR) did not activate the reporter. Deletion of the carboxy-terminal 27 residues of the receptor (*i.e.* 1310Y) or deletion of the carboxy-terminal 93 residues (*i.e.* 1244Y) had essentially no effect on β-galactosidase expression, indicating that the carboxy-terminal region of the IGFIR is not involved in IRS-1 or Shc binding. (Differences of less than two-fold may be due to differences in expression levels of the bait constructs and are not considered significant.) In contrast, mutation of tyrosine 950 to phenylalanine (*i.e.* 1337F) decreased expression of the reporter to negligible levels, indicating that this tyrosine is essential for the interaction of the receptor with these proteins. Tyrosine 950 forms part of a

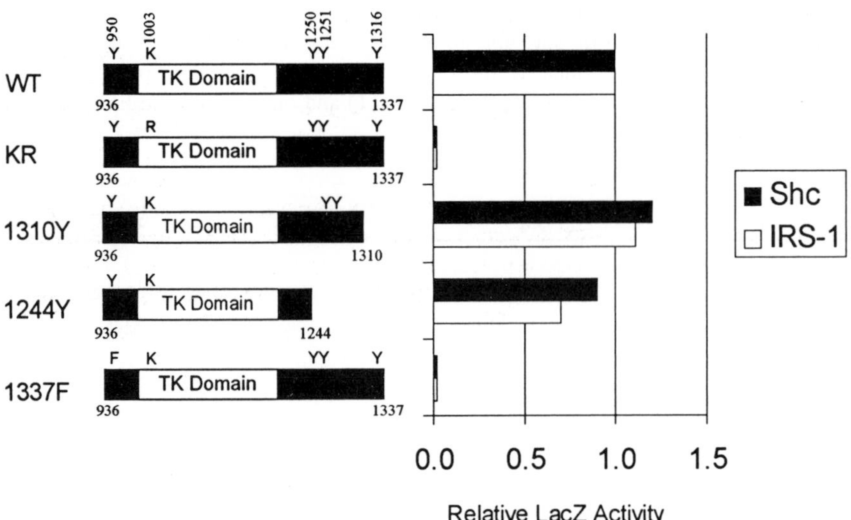

Figure 1. Mapping the binding site for IRS-1 and Shc on the IGFIR using the yeast interaction trap. The left panel shows schematic representations of the wild type IGFIR cytoplasmic domain and informative mutants. In the right panel is shown the β-galactosidase (lacZ) activity of the constructs when coexpressed with the AD-HAtag- Shc (closed bars) or IRS-1 (open bars) hybrids. β-galactosidase activity was measured using a quantitative solution assay (5) and activity is expressed relative to that of the wild type IGFIR bait construct with each prey. Four clones from each transfection were assayed and the mean values are shown. TK, tyrosine-kinase.

NPXY motif; these motifs have been shown to mediate the binding of IRS-1 and Shc to the insulin and other growth factor receptors (11,13,14).

3. IDENTIFYING NOVEL PROTEINS WHICH INTERACT WITH THE IGFIRβ

Yeast two-hybrid systems can also be used to identify novel proteins which interact with a protein of interest. To identify novel proteins which interact with the IGFIRβ we used the interaction trap to screen a human fetal brain library. This library was chosen because fetal brain contains many IGF responsive cell types and expresses a large number of transcripts. The library screen was done using the two step procedures recommended by Brent (5); this procedure helps prevent the loss of plasmids encoding toxic proteins. In the first step, the fetal brain library was used to transform EGY188/pSH18-34 containing the LexA-IGFIRß(WT) construct and transformants were selected on glucose containing medium. To select for interactors, these primary transformants were then plated onto galactose containing medium to induce expression of the AD-HA tagged library hybrids. We chose to do a double selection (i.e. simultaneously select for leucine and lacZ reporter expression) and isolate the fastest growing, darkest colonies, as these generally represent the strongest interactors. This selection yielded approximately 100 colonies after 3-4 days. However, colonies continued to appear for 7 days, many with weak or no β-galactosidase activity. These leu+/lacZ+ colonies were subcultured on selection medium and yielded 80 pure clones with a galactose dependent leu+/lacZ+ phenotype. These clones were further characterized by PCR analysis and restriction digestion. This analysis indicated that the 80 clones represented eight different cDNA inserts encoding all or parts of six different proteins.

When using the interaction trap, it is necessary to verify that the activation of the reporter genes is due to a specific interaction of the AD-HA tagged protein with the bait and not with the LexA DNA binding domain or with other portions of the transcriptional machinery. In practice, the selectivity of the interaction (i.e. reactivity with the IGFIR vs. IR) and the requirement for tyrosine-kinase activity can be conveniently determined at the same time. Experimentally, the unique plasmid clones were rescued from the yeast by growth in KC8, a strain of *E. coli* which contains an inactive trpC gene which is complemented by the TRP1 gene on the library plasmid. The plasmids were purified from the *E. coli* and used to transfect fresh EGY188/pSH18-34 expressing either LexA-IGFIRß, LexA-IGFIRß(KR) or LexA-IRβ and their phenotype determined. All eight clones activated both reporter genes when coexpressed with the wild type IGFIR bait but not when coexpressed with a kinase-negative receptor bait (i.e. KR). Five of these clones also activated reporter gene expression when coexpressed with the human IR cytoplasmic domain while three were specific for the IGFIR.

To identify the proteins encoded by these plasmids, the inserts were sequenced using primers which hybridize to the regions flanking the insertion site in pJG4-5. Four of these cDNAs encoded SH2 domains which, when isolated, had not been previously reported. Two of these clones encoded portions of a protein subsequently identified as Grb-10/Grb-IR (1,15,16). Grb-10 was initially identified by screening a bacterial expression library with the tyrosine phosphorylated carboxy-terminus of the EGF receptor (15). The authors noted that Grb-10 bound poorly to the EGF receptor and hypothesized that its endogenous ligand was another tyrosine-kinase. Subsequently, this protein was identified as binding to the IR using the yeast two-hybrid system and was called Grb-IR (16). The third clone encoded an SH2 domain containing protein most closely related to the amino-terminal SH2 domain of the p85 subunit of PI-3 kinase. This protein was subsequently identified by screening an adipocyte-derived, bacterial expression library with tyrosine-phosphorylated IRS-1; it was found to be a 55 kDa protein which is a new regulatory subunit of PI-3 kinase (termed $p55^{PIK}$) (17). Our findings suggest that, like Shc, $p55^{PIK}$ can also bind directly to the IGFIR and the IR. The fourth clone encoded a second SH2 domain containing protein most closely related to the amino-terminal SH2 domain of p85; this protein has not yet been identified. The fifth clone encoded a zinc finger containing protein of unknown function (18).

The remaining three clones also activated expression of the reporter genes when coexpressed with the wild type IGFIR but not the KR mutant. However, these clones did not activate the reporters when coexpressed with bait constructs containing the IR cytoplasmic domain. These clones encoded members of the 14-3-3 family of proteins (2). One clone encoded a hybrid protein containing the entire coding sequence of the ζ isoform of 14-3-3 coupled to the activation domain by a 36 amino acid linker peptide derived from the (normally) 5' untranslated region of the cDNA. The other two clones encoded 14-3-3β; they differed in the length of the (normally) 5' untranslated region of the cDNA and encoded linker peptides of 46 and 23 amino acids. These linker peptides differed from that encoded by the 14-3-3ζ construct, suggesting that the binding was due to the 14-3-3 proteins and not the linker peptides. This was confirmed using a 14-3-3β hybrid from which the entire linker sequence was deleted. 14-3-3 proteins and their potential role in IGFIR signal transduction are discussed elsewhere in this volume.

3.1 Confirming Interactions Observed in the Two-Hybrid System

Like other systems used to detect protein-protein interactions, interactions detected using the two-hybrid system must be verified using one or more independent approaches. These approaches include *in vitro* binding studies and/or co-immunoprecipitation from mammalian cells or insect cells expressing the proteins and, ultimately, functional analysis. It is possible, however, that some interactions identified using the yeast system may not be detectable using *in vitro* or co-immunoprecipitation techniques. For example, Shc-IGFIR or IR complexes have not been identified by immunoprecipitation analysis. In these circumstances, approaches aimed at defining the functional significance of the interaction, such as altering the expression level of the native protein or a mutated analog unable to interact with its normal partner, may be helpful.

4. SUMMARY

A yeast two-hybrid assay - the interaction trap - is a useful system for investigating IGFIR signal transduction. This system provides a rapid and sensitive means for confirming interactions between the IGFIR and proteins identified using other methods, for mapping sites of interaction on the receptor and its targets and for identifying new interactors. Observations made using the interaction trap must, however, be confirmed using other *in vitro* or *in vivo* procedures. It is likely that information gained from the interaction trap will offer new insights into the mechanisms by which the IGFs regulate their biological responses and how this differs from insulin receptor signal transduction.

REFERENCES

1. Dey B., Frick K., Lopaczynski W., Nissley S., Furlanetto R. Mol. Endo. 10 (1996) 631.
2. Furlanetto RW, Dey BR, Lopaczynski W, Nissley SP. Biochem. J. 323 (1997) (In Press).
3. Craparo A, O'Neill TJ, Gustafson TA. J. Biol. Chem. 270 (1995) 15639.
4. Rocchi S, Tartare-Deckert S, Sawaka-Verhelle D, Gamha A, Van Oberghen E. Endocrinology 137 (1996) 4944.
5. Gyuris J., Brent R. in *Current protocols in molecular biology*, eds. Ausubel F.M., Brent R., Kingston R.E., Moore D.D., Seidman J.G., Smith J.A., Struhl K. (John Wiley & Sons, NY) (1987) pp.13.14.1-13.14.17.
6. Fields S, Song O. Nature 340 (1989) 245.
7. Vojtek AB, Hollenberg S, Cooper J. Cell 74 (1993) 205.
8. Kato H, Faria TN, Stannard B, Roberts CT, LeRoith D. J. Biol. Chem. 268 (1993) 2655.
9. Meyers MG, Sun XJ, Cheatham B *et al.* Endocrinology 132 (1993) 1421.
10. Sasaoka T, Rose DW, Jhun BH, Saltiel AR, Draznin B, Olefsky JM. J. Biol. Chem. 269 (1994) 13689.
11. O'Neill TJ, Craparo A, Gustafson TA. Mol. Cell. Biol. 14 (1994) 6433.
12. Blaikie P, Immanuel D, Wu J, Li N, Yajnik V, Margolis B. J. Biol. Chem. 269 (1994) 32031.
13. Kavanaugh WM, Christoph WT, Williams LT. Science 268 (1995) 1177.
14. Obermeier A, Lammers R, Weismuller K-H, Jung G, Schlessinger J, Ulrich A. J. Biol. Chem. 268 (1993) 22963.
15. Ooi J, Yajnik V, Immanuel D, Gordon M, Moskow JJ, Buchberg AM, Margolis B. Oncogene 10(1995) 1621.
16. Liu F, Roth RA. Proc. Natl. Acad. Sci. USA 92 (1995) 10287.
17. Pons S, Assano T, Glasheen E *et al.* Mol. Cell. Biol. 15(1995) 4453.
18. Tommerup N, Vissing H. Genomics 27(1995) 259.

Molecular Mechanisms to Regulate the
Activities of Insulin-like Growth Factors
K. Takano, N. Hizuka and S-I. Takahashi (Editors)

Signal transduction mechanism of insulin and growth hormone

Kadowaki T., Yamauchi T., Tobe K., Ueki K., Tamemoto H., Kaburagi Y., Yamamoto-Honda R., Tsushima T.*, Yazaki Y.

Third Department of Internal Medicine, Faculty of Medicine, University of Tokyo, Bunkyo, Tokyo 113, Japan, and *Second Department of Internal Medicine, Tokyo Women's Medical College, Tokyo 162, Japan

1.Introduction

Insulin receptor substrate-1 (IRS-1) is the major substrate of insulin receptor and insulin-like growth factor-1 (IGF-1) receptor tyrosine kinases; it has an apparent relative molecular mass of 160-190,000 (Mr, 160-190-kDa) on SDS polyacrylamide gel[1-3]. Tyrosine-phosphorylated IRS-1 binds the 85-kDa subunit of phosphatidylinositol 3-kinase[4, 5] which is involved in the translocation of glucose transporters[6,7] and the abundant src homology protein (ASH)/Grb2[8,9] which is involved in activation of p21ras and MAP kinase cascade[10]. IRS-1 also has binding sites for Syp and Nck and other src homology 2 (SH2) signalling molecules[10].

To clarify the physiological roles of insulin receptor substrate-1 (IRS-1) in vivo, we made mice with a targeted disruption of the IRS-1 gene locus. Mice homozygous for targeted disruption of the IRS-1 gene were born alive but were retarded in embryonal and postnatal growth. They also had resistance to the glucose-lowering effects of insulin, insulin-like growth factor (IGF-1) and IGF-2. These data suggest the existence of both IRS-1-dependent and IRS-1-independent pathways for signal transduction of insulin and IGFs[11].

2. Identification of a 190-kDa protein as a novel substrate for the insulin receptor kinase functionally similar to insulin receptor substrate-1 (IRS-1)

In order to identify IRS-1-independent pathways, we examined the insulin-stimulated tyrosine-phosphorylated proteins in livers of wild type and IRS-1-deficient mice. Tyrosine phosphorylation of an 190-kDa protein (pp190) by insulin was significantly stimulated in livers of IRS-1-deficient mice, which was weakly observed in wild type mice in addition to IRS-1, We also demonstrated that pp190 was immunologically distinct from IRS-1 and was associated with both the 85-kDa subunit of phosphatidylinositol 3-kinase and the Grb2 /Ash molecule as IRS-1. We identified pp190 as a novel substrate for insulin receptor kinase (IRS-2), which can bind both PI3-kinase and Ash /Grb2, and whose tyrosine phosphorylation is specifically induced in IRS-1-dependent mice. These data suggested that pp190 may play some physiological roles in insulin's signal transduction; furthermore, induction of tyrosine phosphorylation of pp190 may be one of the compensatory mechanisms that substitute for IRS-1 in IRS-1-dependent mice[12)].

3. Insulin signalling and insulin actions in the muscles and livers of insulin-resistant, insulin receptor substrate 1-deficient mice

We next investigated the roles of IRS-1 and IRS-2 in the biological actions in the physiological target organs of insulin by comparing the effects of insulin in wild-type and IRS-1-deficient mice. In muscles from IRS-1-deficient mice, the responses to insulin-induced PI3-kinase activation, glucose transport, p70 S6 kinase and MAP kinase activation, mRNA translation, and protein synthesis were significantly impaired compared with those in wild-type mice. Insulin-induced protein synthesis was both wortmannin sensitive and insensitive in wild-type and IRS-1-deficient mice. However, in another target organ, the liver, the responses to insulin-induced PI3-kinase and MAP kinase activation were not significantly reduced. The amount of tyrosine-phosphorylated IRS-2 (in IRS-1-deficient mice) was roughly equal to that of IRS-1 (in wild-type mice) in the liver, whereas it was only 20 to 30% of that of IRS-1 in the muscles. These data suggest, (i) IRS-1 plays central roles in two major biological actions of insulin in muscles, glucose transport and protein synthesis; (ii) the insulin resistance of IRS-1-deficient mice is mainly due to resistance in the muscles; and (iii) the degree of compensation for IRS-1 deficiency appears to be correlated with the amount of tyrosine-phosphorylated IRS-2 (in IRS-1-deficient mice) relative to that of IRS-1 (in wild-type mice)[13)] (Figure 1).

4. Role of insulin receptor substrate-1 and pp60 (IRS-3) in the regulation of insulin-induced glucose transport and GLUT4 translocation in primary adipocytes

In fat, another target tissue of insulin glucose transport occurs through the translocation of GLUT4 from an intracellular pool to the cell surface. Phsphatidylinositol (PI) 3-kinase has been shown to be required in this process. Insulin is thought to activate this enzyme by stimulating its association with tyrosine-phosphorylated proteins such as insulin receptor substrate (IRS)-1,

IRS-2, Grb2-associated binder-1, and pp60. To study the role of these endogenous substrates in glucose transport, we analyzed adipocyted from IRS-1 null mice. In adipocytes from these mice, we showed that: 1) insulin-induced PI3-kinase activity in the antiphosphotyrosine immunoprecipotates was 54% of wild-type; 2)pp60 was the major tyrosine-phosphorylated protein that associated with PI3-kinase, whereas tyrosine phosphorylation of IRS-2 as well as its association with this enzyme was almost undetectable; and 3)glucose transport and GLUT4 translocation at maximal insulin stimulation were decreased to 52 and 68% of those from wild-type. These data suggest that both IRS-1 and pp60 play a major role in insulin-induced glucose transport in adipocytes, and that pp60 is predominantly involved in regulating this process in the absence of IRS-1[14] (Figure 1). This pp60 has recently been cloned and designated as IRS-3[15].

5. Growth hormone-induced tyrosine phosphorylation of EGF receptor as an essential element leading to MAP kinase activation and gene expression

Growth hormone (GH) binding to its receptor, which belongs to the cytokine receptor superfamily, activates Janus kinase (JAK) 2 tyrosine kinase, thereby activating a number of intracellular key proteins such as STAT (signal transducers and activators of transcription) proteins and mitogen-activated protein (MAP) kinases, which finally lead to GH's biological actions including gene expression. In contrast to receptor tyrosine kinases such as insulin and IGF-1 receptor tyrosine kinases, the signalling pathways leading to MAP kinase activation by GH are poorly understood but appear to involve Grb2 and Shc. We show that GH stimulated tyrosine phosphorylation of epidermal growth factor receptor (EGFR) and its association with Grb2, and concomitantly stimulated MAP kinase activity in liver, a major target tissue. Expression of EGFR and its mutants into CHO-GH receptor (GHR) cells revealed that GH-induced full activation of MAP kinase and c-fos expression required tyrosine phosphorylation sites of EGFR but not its intrinsic tyrosine kinase activity. Moreover, by also using dominant negative JAK2 and in vitro kinase assay, we demonstrated that tyrosine 1068 of EGFR was suggested to be one of the major phosphorylation and Grb2 binding sites stimulated with GH via JAK2. These data suggest that the role of EGFR in GH signalling is to be phosphorylated by JAK2, thereby providing docking[16] sites for Grb2 and activating MAP kinases and gene expression. This novel cross talk pathway may, together with the cross talk pathway between G-protein coupled receptors and EGFR[17,18] provide the first paradigm that the hormone and cytokine receptor superfamily transduces signals via associated nonreceptor tyrosine kinase by phosphorylating growth factor receptor and utilizing it as a docking protein independent of its receptor tyrosine kinase activity (Figure 2).

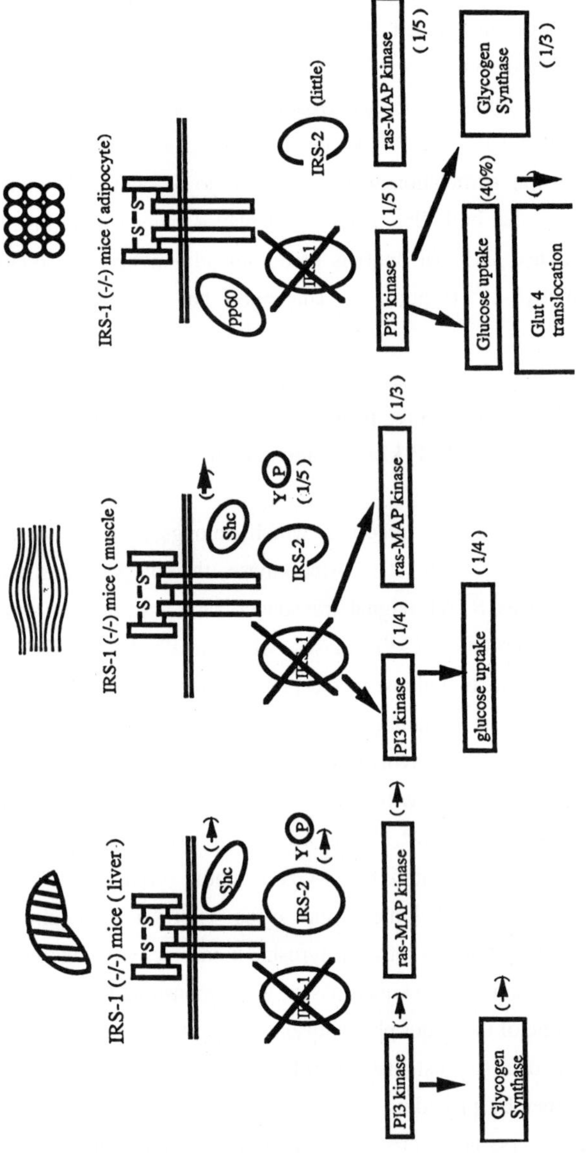

Figure 1. Insulin's signaling system in IRS-1 (-/-) mice

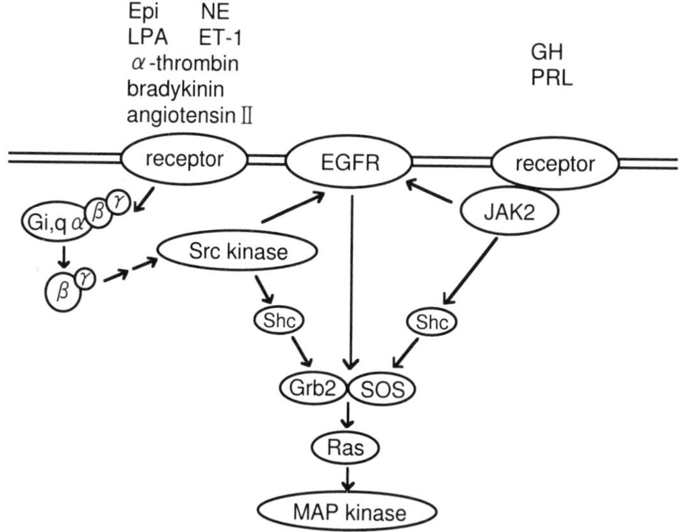

Figure 2. Cross-talk between G-protein coupled receptor and cytokine receptor superfamilies and EGF receptor

REFERENCES

1) White, M.F., Maron, R. and Kahnm C.R. : **Nature** 318, 183-186, 1985
2) Kadowaki, T. et al. : **J.Biol.Chem.** 262, 7342-7350, 1987
3) Sun, X.-J. et al. : **Nature** 352, 73-77, 1991
4) Backer, J.M. et al. : **EMBO J.** 11, 3469-3479, 1992
5) Yonezawa, K. et al. : **J.Biol.Chem.** 267, 25958-25966, 1992
6) Okada, T., Kawano, Y., Sakakibara, T., Hazeki, O. and Ui, M. : **J.Biol.Chem.** 269, 3568-3573, 1994
7) Hara, K. et al. : **Proc.Natn.Acad.Sci.USA**, in press.
8) Skolnik, E.Y. et al. : **Science** 260, 1953-1955, 1993
9) Tobe, K. et al. : **J. Biol.Chem.** 268, 11167-11171, 1993
10) White, M.F.and Kahn, C.R. : **J.Biol.Chem.** 268, 1-4, 1994
11) Tamemoto, H. et al. : **Nature** 372, 182-186, 1994
12) Tobe, K. et al. : **J.Biol.Chem.** 270, 5698-5671, 1995
13) Yamauchi, T. et al. : **Mol.Cell.Biol.** 16, 3074-3084, 1996
14) Kaburagi, Y. et al. : **J.Biol.Chem.** 272, 25839-25844, 1997
15) Lavan, B.E. et al. : **J.Biol.Chem.** 272, 11439-11443, 1997
16) Yamauchi, T.et al. : **Nature** 390, 91-96, 1997
17) Daub, H. et al. : **Nature** 379, 557-560, 1996
18) Luttrell, L.m. et al. : **J.Biol.Chem.** 242, 4637-4644, 1997

Molecular Mechanisms to Regulate the
Activities of Insulin-like Growth Factors
K. Takano, N. Hizuka and S-I. Takahashi (Editors)
1998 Elsevier Science B.V.

The insulin-like growth factor-I receptor and cellular signaling: Implications for cellular proliferation and tumorigenesis

D. Le Roith, A. P. Koval, A. A. Butler, S. Yakar, M. Karas, B. S. Stannard and V. A. Blakesley

Diabetes Branch, National Institutes of Health, 10 CENTER DR MSC 1770, BETHESDA MD, 20892-1770, USA.

1. INTRODUCTION

The insulin-like growth factor-I receptor (IGF-IR) belongs to a family of growth factor tyrosine kinase receptors that includes the insulin receptor (IR), the IGF-IR and the insulin receptor-related receptor (IRR). Each receptor is a oligomer of two αβ subunits bound by disulfide bridges and spanning the plasma membrane. In addition, many tissues and cells express hybrid receptors that are comprised of a hemireceptor each of the IR and IGF-IR. Only the IR and IGF-IR, however, have known functions [1-3].

One major quandry that has perplexed investigators for many years is the mechanism(s) whereby IR and IGF-IR, while highly conserved structurally, have different functions. The IR is primarily involved in metabolic functions whereas the IGF-IR is involved in pre-and postnatal development, growth and specialized postnatal tissue-specific functions. A number of physiological important systems exist that contribute to and maintain the separate functions of these receptors (Table 1). These include specific circulating IGF-binding proteins, tissue-specific expression of each receptor and the high affinity interaction of each ligand with its cognate receptor[4, 5].

Additionally, numerous investigators have suggested that the separate functions of the IR and IGF-IR may also be subserved by the use of specific intracellular substrates. Initial studies have identified a number of potential candidates (Table 1) [6-10].

Table 1. Separation of the Insulin and IGF Systems

Structural differences between --
- The ligands insulin and IGF-I and IGF-II.
- The insulin and IGF-I receptors.

Differential binding of --
- Ligands to their cognate receptors.
- IGFs to the IGF-binding proteins.

Divergent signaling
- Tissue distribution of insulin and IGF receptors differs
- Differential interactions with substrates
  - Insulin receptor interacts specifically with Stat 5, pp120, MAD5.
  - IGF-I receptor interacts specifically with isoforms of 14-3-3 proteins.

We chose to study the Crk family of adapter proteins as candidate substrates for the IGF-IR. We have demonstrated their role in IGF-IR signaling and identified some interesting differences between Crk-II and CrkL, the most ubiquitous members of this family of protooncogenes. These findings form the subject of this review and on-going studies in our laboratory.

2. IGF-IR Signaling

Two major pathways of signaling have been identified for the IGF-IR (Fig. 1). The insulin receptor substrate (IRS) family consists of four related proteins that have conserved pleckstrin homology (PH) domains at the N-terminal region, followed by a protein tyrosine binding (PTB) domain and, in the case of IRS-I, a SAIN (Shc and IRS-1 NPXY-binding) domain [11]. Both the PTB and SAIN domains are important for IRS binding to both insulin and IGF-I receptors[12, 13]. The remainder of the IRS molecule diverges, except for some very highly conserved motifs that surround specific tyrosine residues. Examples of these motifs include YXXM, YMXM and YXNI, which have been shown to be important for the binding of SH2 domains of other proteins to the tyrosine-phosphorylated IRS molecules. Ligand binding stimulates autophosphorylation of the receptor and activation of the tyrosine kinase. Activated receptor tyrosine kinase then phosphorylates the IRS docking proteins. Substrates including p85, the regulatory subunit of phosphoinositide 3' kinase (PI3K), protein tyrosine phosphatase-2/Syp, Grb2, and Nck bind to phosphorylated IRS molecules, thereby activating signaling pathways mediated by these proteins [14-20].

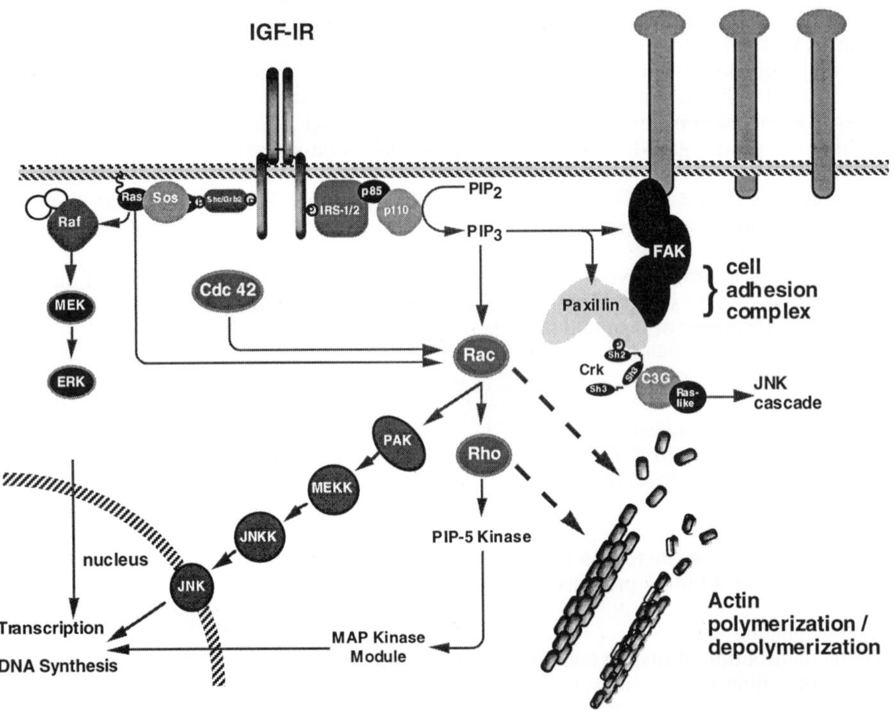

Figure 1. Insulin-Like Growth Factor-I Signal Transduction. The most well-characterized signaling pathways of the activated IGF-IR include the pathways involving the IRS family of proteins and Shc. Other cascades include the Crk family of proto-oncogenes and the small Ras-like G proteins Rac and Rho.

A second major pathway emanating from the IGF-IR occurs via Shc, which binds the receptor in the juxtamembrane region, becomes phosphorylated upon receptor activation and interacts with Grb2. Grb2 itself interacts with mSos resulting in activation of the Ras/Raf/MAP kinase cascade, a pathway that modulates the mitogenic response of the cell to growth factor stimulation. PI3K, on the other hand, is important for insulin-induced metabolic functions and critical for the anti-apoptotic effects of IGF-I. IGF-I-stimulation of the PI3K pathway activates the kinase Akt, thereby preventing apoptosis by an unknown pathway.

Other signaling pathways that may be activated by the IGF-IR include the Ras-related small G proteins Rac and Rho. These small G-proteins are involved in maintenance of cell morphology and cell motility. It has not been shown that Rac is directly activated by the IGF-IR; however, studies suggest that Rac mediates the effects of PI3K on cell motility [21].

3. Crk Family of Adapter Proteins

Crk proteins are the cellular homologs of the viral oncogene v-crk, an SH2 and SH3 domain-containing protein encoded by the avian sarcoma virus CT10. The proto-oncogene family of Crk proteins include Crk-I, a 21 kDa protein with one SH2 and one SH3 domain that is a splicing variant of the more abundant 40 kDa protein Crk-II. Crk-II has an additional C-terminal SH3 domain. As adapter proteins they lack any catalytic activity but contain an important tyrosine residue at amino acid 221 that is phosphorylated. Crk-like (CrkL) protein is a separate gene product of 36 kDa and shares ~60% overall homolgy with Crk-II, although the homology within the SH2 and SH3 domains is closer to 85%. Crk-II and CrkL are phosphorylated by c-Abl. CrkL is also a substrate for Bcr/Abl. The downstream substrates that associate with the Crk proteins include two guanine nucleotide exchange proteins, C3G and mSos. Both C3G and mSOS can activate Rap1 as well as the Ras/Raf /MAP kinase cascade. The Crk proteins are also involved in focal adhesion complexes and signaling from integrins following interaction with the extracellular matrix. Both Crk-II and CrkL are ubiquitousely expressed [22-26].

IGF-IR activation by ligand binding rapidly induces phosphorylation of both Crk-II and CrkL on tyrosine residues, suggesting that the Crk family of proteins is involved in IGF-IR signaling [27, 28]. To examine this possibility further, Crk-II and CrkL were overexpressed in 293 embryonic kidney cells and NIH-3T3 fibroblasts. A number of cell lines were created overexpressing different levels of the Crk proteins. Cell lines transfected with empty vectors or vectors with the cDNAs in the antisense orientation were used as controls. Using both thymidine incorporation and MTT assays, we demonstrated that cells overexpressing Crk-II and CrkL responded in an exaggerated manner to IGF-I stimulation compared with control cell lines. Furthermore, the exaggerated response was proportional to the level of Crk protein overexpression [28]. To determine the molecular mechanisms whereby the activated IGF-IR enhanced the tyrosine phosphorylation of the Crk proteins we examined the potential interactions of the Crk proteins with the IGF-IR and the IRS family of proteins. Using GST-fusion proteins containing various domains of the Crk proteins we demonstrated that the activated IGF-IR efficiently phosphorylated the GST-Crk-II protein. This phosphorylation was dependent on the presence of the SH2 domain and tyrosine 221 in the spacer region between the two SH3 domains. When tyrosine 221 was mutated, the interaction of GST-Crk-II with the IGF-IR was markedly enhanced, leading us to conclude that phosphorylation of this specific tyrosine allowed the SH2 domain to interact at this site and cause the dissociation from the receptor. This proposed mechanism is reminiscent of the protein folding and intramolecular association of the SH2 domain of Src with its own phosphorylated tyrosine residue. Additionally, Crk-II has been shown to fold and an intramolecular association has been demonstrated by NMR analysis [29]. Following IGF-IR tyrosine phosphorylation, the SH2 domain of Crk-II binds to the juxtamembrane and C-terminal regions at specific tyrosines, e.g. Tyr 943/950 and Tyr 1250/1251, respectively. In contrast, CrkL apparently does not interact with the IGF-IR either in intact cells or *in vitro*.

Initial studies demonstrated that Crk-II associated with both IRS-1 and IRS-2 in lysates of 293 cells. This association was readily detected in unstimulated cells. Upon stimulation with

IGF-I, IRS-1 and IRS-2 became phosphorylated on tyrosine residues. Concurrently, Grb2 association with the IRS molecules increased, but Crk-II dissociated from the IRS proteins. While the mechanism for this dissociation is unknown it may be similar to the mechanism described above. When GST-fusion proteins of various domains of Crk-II and CrkL were used in *in vitro* precipitation assays, we failed to demonstrate a direct association of the Crk proteins with IRS-1 or IRS-2, suggesting that the association seen in whole cell lysates may involve another molecule(s). Crk-II and CrkL did, however, interact directly with another 180kDa tyrosine-phosphorylated protein that likely represents IRS-4 (unpublished observation). Crk-II interacted with IRS-4 in whole cell lysates, but CrkL did not. In addition, the SH2 domain of Crk-II showed the same affinity for IRS-4 as did full-length Crk-II. However, the N-terminal SH3 plus SH2 domains of CrkL were required for an interaction similar to that of the full-length CrkL molecule. These results suggest that Crk-II and Crkl differ in their mechanisms of binding substrates. These differences may explain some of interesting biological differences observed (see below).

4. Biological Differences Between Crk-II and CrkL

Crk-II and CrkL were readily phosphorylated on tyrosine residues following IGF-I stimulation of cells, suggesting that both adapter proteins mediated the mitogenic and oncogenic effects of IGF-I. To study the role of the Crk proteins in mediating the IGF-I signals, cells were stably transfected with Crk-II and CrkL cDNAs. Individual clones were tested for protein expression using antibodies specific for Crk-II and CrkL in western blot analyses. Of interest, cells overexpressing Crk-II showed increased apoptosis proportional to the levels of Crk-II protein being expressed. CrkL overexpressing cells, however, did not show increased apoptosis; instead they demonstrated continued cell growth even in serum-free medium. To further examine this difference, Crk-II and CrkL expressing cells were tested for the ability to form colonies in a soft agar, indicative of a transformed phenotype. Whereas cell lines overexpressing CrkL formed significant numbers of large colonies, those overexpressing Crk-II failed to form colonies and were indistinguishable from control cells (Table 2).

Table 1. Anchorage-Independent Colony Formation of Cells Expressing Crk Proteins

Clone	Colonies # ± sem	Clone	Colonies # ± sem
Crk II		CrkL	
3T3-7	27.3 ± 2.8	3T3-L-22	161.3 ± 3.8
3T3-9	19.2 ± 1.2	3T3-L-24	73.0 ± 12.3
Control OP-2	12.3 ± 0.3	Control OP-5	19.0 ± 1.5

Cell suspension were plated at 1000 cells/ml in a 0.2% agar overlay in the presense of 10% FBS. Colonies greater than 0.05mm in diameter were scored after three weeks of culture. The mean number of colonies ± standard error of the mean are presented for 3 experiments.

We then tested the various cell lines by FACS analysis to determine the progression of cells through the cell cycle. At low density with exponentially growing cells (0 h), the control cells and cells overexpressing either Crk-II or CrkL showed similar proportions of cells in the S phase of the cell cycle (Table 3). However, at high density (24 and 30 h) the proportions of control cells and cells expressing Crk-II in S phase fell markedly. CrkL-expressing cells, however, had a significantly higher percentage of cells in S phase. These data are consistent with the view that CrkL-expressing cells continued to progress through the cell cycle under conditions where control cells and cells expressing Crk-II were leaving the cell cycle. Similarly, in serum-free medium the proportion of control and Crk-II-expressing cells in S phase decreased but CrkL-expressing cells continued the move through the cell cycle. Furthermore, IGF-I treatment increased the proportion of control cells and cells expressing CrkL in S phase, but IGF-I alone did not change the percentage of Crk-II-expressing cells in S phase.

Table 2. Cell Cycle Progression of Cells Expressing Crk Proteins

Time	OP2		OP9		3T3-7		3T3-9		3T3-L-22		3T3-L-24	
IGF-I	-	+	-	+	-	+	-	+	-	+	-	+
-24h	37		30		31		31		46		36	
0h	7		12		6		4		28		22	
24h	9	29	9	19	8	12	3	6	26	40	20	32
30h	8	14	8	14	7	9	3	4	25	34	18	25

NIH-3T3 control cells (clones OP2 and OP9) and cells overexpressing Crk-II (clones 3T3-7 and 3T3-9) or CrkL (clones 3T3-L-22 and 3T3-L-24) were plated onto 100mm dishes and allowed to attach. At -24h the medium was replaced with serum-free medium. Beginning at 0h the cells were supplemented (+) or not supplemented (-) with 100nM IGF-I. At given time points the cells were trypsinized and studied by FACS analysis to determine the precentage of cells in each stage of the cell cycle. The percentages of cells in S phase are presented. Four separate experiments were performed with similar results.

We conclude from these studies that while both Crk-II and CrkL are endogenous substrates for the IGF-I receptor, they have different functions. These separate functions most likely are determined by differential substrate specificity. Studies are on-going in our laboratory to elucidate the substrate specificity of the Crk proteins and the mechanisms involved in their specific responses to IGF-IR activation.

5. Conclusions

It is becoming apparent that the signaling pathways emanating from cell surface receptors is extremely complex. While the IRS proteins and Shc are major functional substrates for the insulin and IGF-I receptors, other substrates may be just as important in conferreing the specificity of action of these two receptors. Elucidation of these substrates and their functions is clearly the next important step in unravelling the mysteries of insulin and IGF-I cell biology.

REFERENCES

1. Ullrich, A, Bell, JR, Chen, EY, Herrera, R, Petruzzelli, LM, Dull, TJ, Gray, A, Coussen, L, Liao, Y-C, Tsubokawa, M, Mason, A, Seeburg, PH, Grunfeld, C, Rosen, OM, and Ramachandran, J. 1985. Nature. 313: 756-761.
2. Ullrich, A, Gray, A, Ram, AW, Yang-Feng, T, Tsubokawa, M, Collins, C, Henzel, W, Le Bon, T, Kathuria, S, Chen, E, Jacobs, S, Francke, U, Ramachandran, J, and Fujita-Yamaguchi, Y. 1986. EMBO J. 5: 2503-2512.
3. LeRoith, D, Werner, H, Beitner-Johnson, D, and Roberts, CT, Jr. 1995. Endo Rev. 16: 143-163.
4. Blakesley, VA, Scrimgeour, A, Esposito, D, and LeRoith, D. 1996. Cytokine & Growth Factor Rev. in press: .
5. Clemmons, DR. 1997. Cytokine and Growth Factor Reviews. 8: 45-62.
6. Chen, J, Sadowski, HB, Kohanski, RA, and Wang, L-H. 1997. Proc Natl Acad Sci USA. 94: 2295-2300.
7. Craparo, A, Freund, R, and Gustafson, TA. 1997. JBC. 272: 11663-11669.
8. Laviola, L, Giorgino, F, Chow, JC, Baquero, JA, Hansen, H, Ooi, J, Zhu, J, Reidel, H, and Smith, RJ. 1997. J Clin Invest. 5: 830-837.
9. O'Niell, TJ, Zhu, Y, and Gustafson, TA. 1997. JBC. 272: 10035-10040.
10. Najjar, SM, Blakesley, VA, Calzi, SL, Kato, H, Le Roith, D, and Choice, CV. 1997. Biochemistry. 36: 6827-6834.
11. Gustafson, TA, He, W, Craparo, A, Schaub, CD, and O'Neill, TJ. 1995. MCB. 15: 2500-2508.
12. Craparo, A, O'Neill, TJ, and Gustafson, TA. 1995. JBC. 270: 15639-15643.

13. He, W, Craparo, A, Zhu, Y, O'Niell, TJ, Wang, L-M, Pierce, JH, and Gustafson, TA. 1996. JBC. 271: 11641-11645.
14. Backer, JM, Myers, MG, Jr, Shoelson, SE, Chin, DJ, Sun, X-J, Miralpeix, M, Hu, P, Margolis, B, Skolnik, EY, Schlessinger, J, and White, MF. 1992. EMBO J. 11: 3469-3479.
15. Lowenstein, EJ, Daly, RJ, Batzer, AG, Li, W, Margolis, B, Lammers, R, Ullrich, A, Skolnik, EY, Bar-Sagi, D, and Schlessinger, J. 1992. Cell. 70: 431-442.
16. Myers, MG, Jr, Sun, XJ, Cheatham, B, Jachna, BR, Glasheen, EM, Backer, JM, and White, MF. 1993. Endo. 132: 1421-1430.
17. Myers, MG, Jr, Wang, L-M, Sun, XJ, Zhang, Y, Yenush, L, Schlessinger, J, Pierce, JH, and White, MF. 1994. MCB. 14: 3577-3587.
18. Myers, MG, Sun, XJ, and White, MF. 1994. Trends Biochem Sci. 19: 289-293.
19. Sun, XJ, Crimmins, DL, Myers, MG, Jr, Miralpeiz, M, and White, MF. 1993. MCB. 13: 7418-7428.
20. Sun, XJ, Wang, L-M, Zhang, Y, Yenush, L, Myers, MG, Jr, Glasheen, E, Lane, WS, Pierce, JH, and White, MF. 1995. Nature. 377: 173-177.
21. Leventhal, PS, and Feldman, EL. 1997. TEM. 8: 1-6.
22. Matsuda, M, Hashimoto, Y, Moroya, K, Hasegawa, H, Kurata, T, Tanaka, S, Nakamura, S, and Hattori, S. 1994. MCB. 14: 5495-5500.
23. Salgia, R, Uemura, N, Okuda, K, Li, J-L, Pisick, E, Sattler, M, de Jong, R, Druker, B, Heisterkamp, N, Chen, LB, Groffen, J, and Griffin, JD. 1995. JBC. 270: 29145-29150.
24. Tanaka, S, Hattori, S, Kurata, T, Nagashima, K, Fukui, Y, Nakamura, S, and Matsuda, M. 1993. MCB. 13: 4409-4415.
25. ten Hoeve, J, Morris, C, Heisterkamp, H, and Groffen, J. 1993. Oncogene. 8: 2469-2474.
26. Teng, KK, Lander, H, Fajardo, JE, Hanafusa, H, Hempstead, BL, and Birge, RB. 1995. JBC. 270: 20677-20685.
27. Beitner-Johnson, D, and LeRoith, D. 1995. JBC. 270: 5187-5190.
28. Beitner-Johnson, D, Blakesley, VA, Shen-Orr, Z, Spiegel, S, and LeRoith, D. 1996. JBC. 271: 9287-9290.
29. Rosen, MK, Yamazaki, T, Gish, GD, Kay, CM, Pawson, T, and Kay, LE. 1995. Nature. 374: 477-479.

Molecular Mechanisms to Regulate the
Activities of Insulin-like Growth Factors
K. Takano, N. Hizuka and S-I. Takahashi (Editors)
1998 Elsevier Science B.V.

Differences between insulin and IGF-I signaling

P. Nissley[a], B.R. Dey[a], K. Frick[b], W. Lopaczynski[a], C. Terry[a], and R.W. Furlanetto[b]

[a]National Cancer Institute, National Institutes of Health, Bldg. 10, Rm 4N115, Bethesda, MD 20892

[b]Department of Pediatrics, University of Rochester Medical Center, 601 Elmwood Ave, Rochester, NY 14642

Recently a receptor (DAF-2) which is 35% identical to the human insulin receptor and 34% identical to the human IGF-I receptor has been identified in the nematode *C. elegans* (1). This receptor, found in a worm which dates to a time before nematodes and mammals diverged 700 million to 800 million years ago, probably represents a homolog of the ancestor of these two receptors which have diverged during evolution so that the modern insulin receptor regulates glucose and fuel metabolism and the modern IGF-I receptor regulates growth.

Part of the explanation for the differences in the *in vivo* action of insulin and IGFs has to do with their different physiology (site of production, control of secretion etc.). Part of the explanation is the relative distribution of the insulin receptor and IGF-I receptor in different tissues and cell types. Intracellularly, the signaling pathways for the insulin receptor and the IGF-I receptor that have been described to date are very similar (phosphorylation of IRS-1/IRS-2 and Shc, activation of Ras and PI 3-kinase signaling pathways) (2). However, Lammers et al. (3) reported that the cytoplasmic domain of the IGF-I receptor was 10 times more effective than the insulin receptor cytoplasmic domain in stimulating DNA synthesis. The cytoplasmic portions of the two receptors are different; the distribution of tyrosine residues is not identical and the amino acid sequence homology in the carboxy tail of the receptors is only 44% (Fig. 1) (4). There is evidence that there may be different signaling pathways emanating from the carboxy tail of the two receptors. For example, Baserga's laboratory has defined amino acid residues in the carboxy tail of the IGF-I receptor that are required for transformation as measured by the growth of cells in soft agar (5). These residues include tyrosine 1251 and a cluster of serines (1280-1283) (5, 6). Tyrosine 1251 and this serine cluster are not represented in the human insulin receptor. Indeed, a chimeric receptor in which the carboxy 99 amino acids of the insulin replaces the C-terminus of the IGF-I receptor, was not capable of causing transformation (5).

Thus, it seems likely that signaling pathways that are selective for the IGF-I receptor or the insulin receptor will be discovered. In this brief review we will describe recent work from our laboratories and from other investigators that provides evidence for selective interaction of three proteins (14-3-3, Grb10, and MAD2) with either the IGF-I receptor or the insulin receptor.

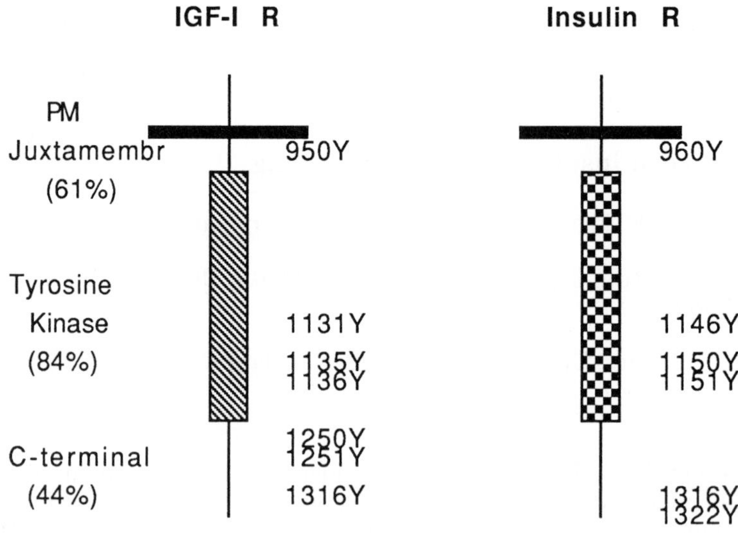

Figure 1. Domain structure of the cytoplasmic portions of the IGF-I receptor and insulin receptor. The cytoplasmic portions of single β chains in the heterotetrameric receptors are shown. Receptor domains are described on the left and the percent homology between the two receptors is given. Tyrosine residues known to undergo phosphorylation upon receptor activation are shown. PM: plasma membrane.

1. 14-3-3 PROTEINS

1.1. Yeast two-hybrid interaction trap

We have used the yeast two-hybrid system to identify proteins that bind to the IGF-I receptor (7, 8). This method is discussed in detail in the chapter by Furlanetto et al. in this volume. Briefly, yeast are cotransformed with a DNA binding domain plasmid containing the cDNA of the cytoplasmic domain of the IGF-I receptor and an activation domain plasmid containing a cDNA library. If a protein encoded by a cDNA in the library binds to the cytoplasmic domain of the IGF-I receptor, a functional transcription factor (DNA binding domain plus activation domain) is reconstituted and transcription of two reporter genes, lacZ (β-galactosidase activity) and LEU2 (colony growth on leucine dropout plates) is stimulated.

A screen of a human fetal brain library with the IGF-I receptor using the LexA based yeast two-hybrid system identified two members of the 14-3-3 family of proteins, 14-3-3β and 14-3-3ζ (8). When the cytoplasmic domain of the IGF-I receptor was coexpressed with 14-3-3β or 14-3-3ζ, the lacZ and LEU2 reporter genes were activated. Interaction of 14-3-3 proteins with the IGF-I receptor required autophosphorylation of the IGF-I receptor since a kinase negative receptor (K1003R) showed no interaction. Interestingly, the cytoplasmic domain of the wild type insulin receptor did not bind to 14-3-3 proteins, raising the possibility that binding of 14-3-3β and 14-3-3ζ to the IGF-I receptor *in vivo* might help to explain differences in signaling by the IGF-I receptor and the insulin receptor.

1.2. 14-3-3β binds to the IGF-I receptor *in vitro*

We were able to duplicate our yeast two-hybrid system findings that 14-3-3β interacts with the IGF-I receptor but not with the insulin receptor with *in vitro* experiments using receptor preparations from fibroblasts overexpressing the IGF-I receptor or the insulin receptor together with GST-14-3-3β beads (8). NIH-3T3 cells overexpressing the IGF-I receptor (NWTc48 cells from Derek LeRoith) were stimulated by addition of IGF-I, and cell lysates were tested for ability to bind to GST-14-3-3β Sepharose. Following incubation, material bound to the beads was solubilized and resolved by SDS-PAGE. Immunoblotting was performed with an IGF-I receptor antiserum or with a phosphotyrosine antibody. There was no binding of IGF-I receptor to the GST-14-3-3β Sepharose in lysates from serum-starved cells. As early as 5 min following addition of IGF-I to the cells, IGF-I receptor in the lysate bound to the GST-14-3-3β beads and binding reached maximum levels at 16-24 hours. When the same experiment was performed with NIH 3T3 cells overexpressing the insulin receptor (3006 cells from Simeon Taylor) there was no binding of the insulin receptor to GST-14-3-3β beads after addition of insulin to the cells. Immunoblotting with the phosphotyrosine antibody showed that the insulin receptor was autophosphorylated following addition of insulin to the resting cells. These results indicate that the IGF-I receptor isolated from mammalian cells also interacts selectively with 14-3-3β and that this interaction requires receptor activation.

In other experiments we were also able to show that the baculovirus-expressed cytoplasmic domain of the IGF receptor purified from Sf9 insect cells also bound to GST-14-3-3β beads but not to GST beads (8). Thus, both *in vitro* experiments using IGF-I receptor overexpressed in NIH-3T3 cells and *in vitro* experiments using IGF-I receptor overexpressed in insect cells demonstrate that interaction of the receptor with 14-3-3 proteins is not confined to the yeast two-hybrid system.

1.3. *In vivo* binding of 14-3-3 to baculovirus expressed IGF-I receptor

To examine binding of 14-3-3 proteins to the IGF-I receptor *in vivo* we analysed the cytoplasmic domain of the IGF-I receptor purified from the Sf9 insect cells for the presence of 14-3-3 proteins (8). The baculovirus-expressed IGF-I receptor cytoplasmic domain was expressed as a $(His)_6$ tagged fusion protein, allowing a one step, rapid purification on a metal affinity column. When the purified receptor was analysed by SDS-PAGE and immunoblotted with 14-3-3 antibody, 14-3-3 was seen to co-purify with the receptor. In contrast, 14-3-3 was not seen when a lysate from uninfected Sf9 cells was purified on the metal affinity column and analysed in parallel.

1.4. Phosphorylation of 14-3-3 by the IGF-I receptor tyrosine kinase

Mammalian 14-3-3 proteins contain multiple tyrosines which are conserved among isoforms, and Bcr-Abl phosphorylates the τ isoform of 14-3-3 on tyrosine (9). We showed that 14-3-3β was a substrate for the IGF-I receptor tyrosine kinase in the yeast two-hybrid system and *in vitro* (8). Yeast were cotransformed with the LexA IGF-I receptor DNA binding plasmid and the 14-3-3β activation domain plasmid and extracts were analysed by SDS-PAGE and immunoblotting with phosphotyrosine antibody. A band corresponding to the size of the activation domain 14-3-3β fusion protein was phosphorylated on tyrosine. This band was not phosphorylated on tyrosine in extracts from yeast cotransformed with the kinase negative IGF-I receptor and 14-3-3β. Similarly, the purified Sf9 expressed IGF-I receptor cytoplasmic domain phosphorylated recombinant 14-3-3β and ζ *in vitro*.

1.5. IGF-I receptor binding site for 14-3-3β

We mapped the binding site for 14-3-3β on the IGF-I receptor by coexpressing mutant constructs of the receptor with 14-3-3β in the yeast two-hybrid system (Fig. 2) (8). Deletion of the extreme carboxy tail (aa1310-1337) resulted in a 20% decrease in reporter gene expression. Further deletion to aa 1244 abolished interaction with 14-3-3β. There are two tyrosines (1250, 1251) in the important segment aa 1244-1310. However, mutation of these tyrosines to phenylalanine did not decrease reporter gene expression below the level observed with wild type receptor. 14-3-3 proteins have been reported to bind to a phosphoserine based motif in Raf, RSXpSXP (10). A similar motif, SSSSLP, is found in the critical segment aa 1244-1310 of the IGF-I receptor. When the serine and proline residues were changed to alanine (AAAALA), interaction with 14-3-3β was greatly diminished. Similarly, changing a single serine (1283) to alanine resulted in a substantial decrease in reporter gene expression. If, in addition, the carboxy-terminal 27 amino acids of the IGF-I receptor was deleted, binding to 14-3-3β was totally abolished.

Competitive binding experiments utilizing a receptor peptide (aa1269-1286) demonstrated a requirement for serine 1283 phosphorylation in the binding of 14-3-3β to the IGF-I receptor. NIH3T3 cells overexpressing the IGF-I receptor were stimulated with IGF-I and cell lysates were incubated with GST-14-3-3β Sepharose. Receptor peptides in which serines 1272 and 1283 were phosphorylated or unphosphorylated were included in the incubation. Binding of receptor to GST-14-3-3β beads was blocked when serine 1283 was phosphorylated; the peptide containing 1272 phosphoserine did not inhibit binding. We conclude that 14-3-3β binds mainly to the IGF-I receptor SSSpSLP motif around phosphoserine 1283 with perhaps an additional minor site in the carboxy-terminal 27 amino acids of the receptor. Craparo et al. (11) have recently reported that 14-3-3ε also interacts with a phosphoserine based motif on the IGF-I receptor and that 14-3-3ε does not bind to the insulin receptor.

Members of the family of 14-3-3 proteins have been reported to be binding partners for the products of oncogenes as well as proteins involved in the signaling pathways for proliferation, apoptosis, and cell cycle regulation. These proteins include Bcr, Bcr-Abl, middle-T antigen, Raf, protein kinase C, PI 3-kinase, BAD, A-20, and cdc25 phosphatase (12-18). The x-ray crystallographic structure of 14-3-3 proteins shows a dimer in which bundles of antiparallel helices form an amphipathic groove which is large enough to accomodate two alpha helices (19, 20). This structure suggests that 14-3-3 proteins, complexed as homodimers or heterodimers, could function as a bridge to hold signaling complexes together. For example, it has been shown that c-Raf and A-20 coimmunoprecipitated in a 14-3-3 dependent manner (17). In all cases thus far examined 14-3-3 proteins bind to phosphoserine residues present in an RSXpSXP motif (10). A similar motif

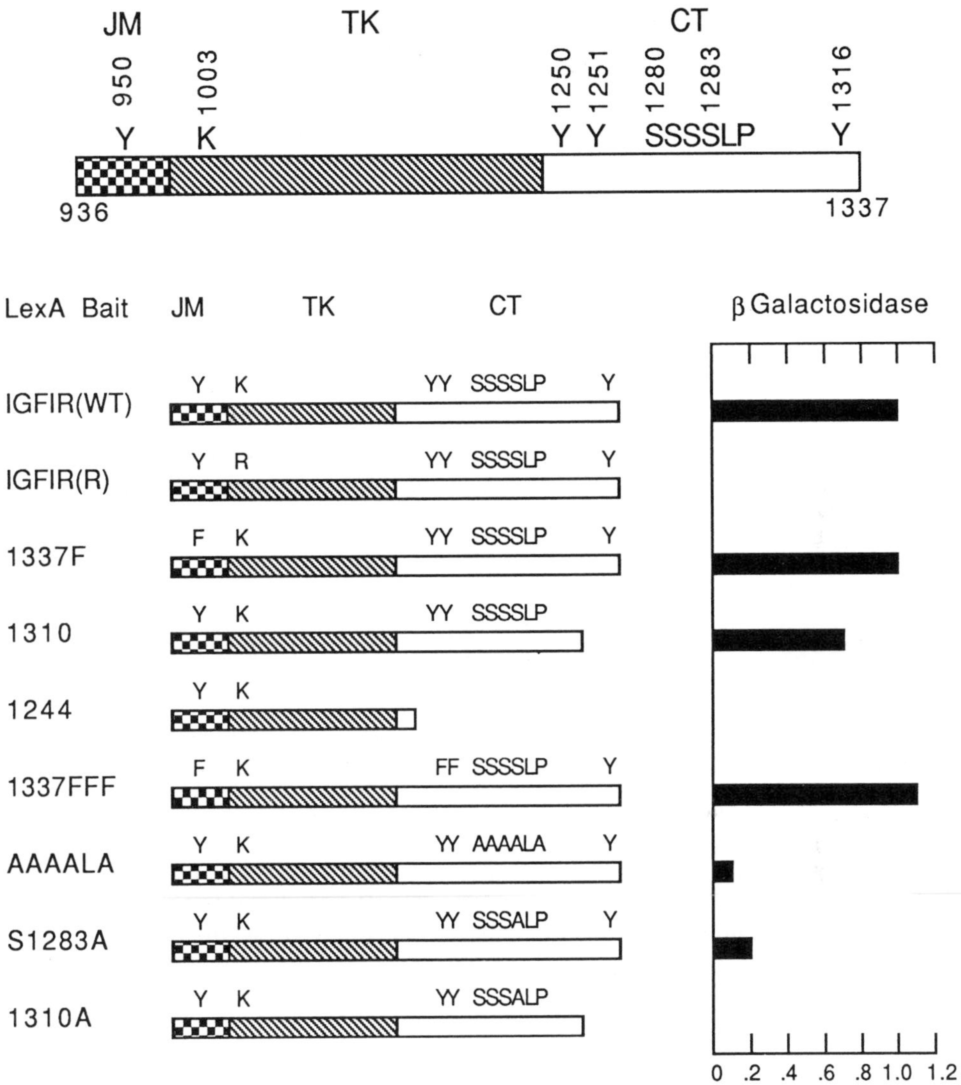

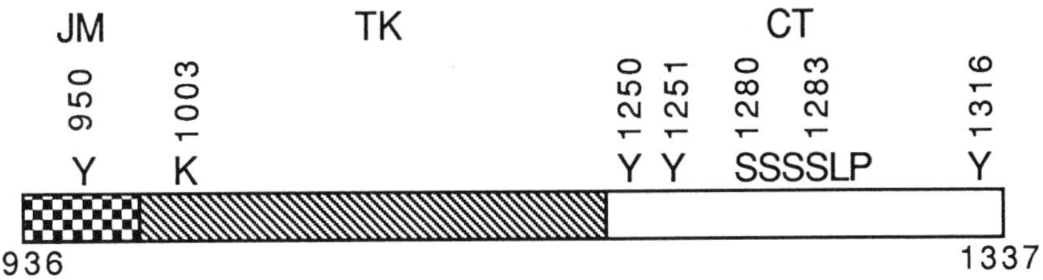

Figure 2. Mapping the IGF-I receptor binding site for 14-3-3β in the yeast two hybrid system. At the top is a representation of the domain structure of the cytoplasmic portion of the receptor showing the residues that were mutated. On the left are shown representations of the receptor constructs that were coexpressed with 14-3-3β. On the right are shown results for readout from the LacZ gene (relative β galactosidase activity). Adapted from Furlanetto et al. (8). JM: juxamembrane; TK: tyrosine kinase; CT: carboxy-terminus.

(SSSpSLP) was identified as the binding site in the IGF-I receptor. The absence of an arginine residue at position minus 3 in the IGF-I receptor is unique and suggests that this residue is not an absolute requirement for binding.

To date we have not identified 14-3-3 proteins in the IGF-I receptor immunoprecipitates from whole cell lysates. It is possible that the interaction of 14-3-3 proteins with the IGF-I receptor is transient, such as seen for an enzyme-substrate interaction. We have shown that 14-3-3β is a substrate for the receptor tyrosine kinase (8). Perhaps tyrosine-phosphorylated 14-3-3 is released rapidly from the receptor and then performs another function in the cell. In any case, the observation that 14-3-3 proteins interact with the IGF-I receptor but not with the insulin receptor may lead to an understanding of some of the differences in signaling pathways emanating from these two receptors.

2. GRB10

We and others have identified Grb10 as a binding partner of the insulin receptor and the IGF-I receptor by using yeast two-hybrid systems (7, 21-25). Recently Smith's laboratory has provided evidence for preferential binding of Grb10 to the insulin receptor in intact cells, suggesting that Grb10 may be selectively involved in signaling pathways emanating from the insulin receptor (25).

Figure 3. Domain structure of Grb10 showing the amino terminal proline rich domain, the central pleckstrin homology (PH) domain, and the carboxy terminal src homology 2 (SH2) domain.

2.1. Splice variants of Grb10

Grb10 cDNAs have been identified in mouse and human and splice variants have been noted in both species (21, 24-27). The Grb10 domain structure (Fig. 3) consists of a proline rich domain in the amino terminal segment, a pleckstrin homology (PH) domain in the central region, and a carboxy terminal SH2 domain. Liu and Roth (21) cloned a human Grb10 (Grb-IR) with a 46 aa deletion in the proximal portion of the PH domain and a 55 aa extension at the amino terminus. Gustafson's and Shoelson's laboratories identified the same human Grb10 cDNAs encoding a complete PH domain and lacking the 55 aa amino-terminal extension (24, 26). Two mouse splice variants have been identified that differ in the length of the segment between the proline rich domain and the PH domain (25, 27). Relative to the human Grb10 there is either a 80 aa or 55 aa insertion in the two mouse splice variants. Because the human splice variants exhibit differences in the PH domain which has been implicated in protein:lipid and protein:protein interactions, it is possible that the protein products of these alternatively spliced mRNAs could exhibit different signaling properties.

2.2. Evidence against Grb10 interacting with IRS-1/IRS-2

In most experiments in which various cells were stimulated with insulin and cell lysates were tested for binding to GST-Grb10 beads, the insulin receptor was detected in immunoblots with receptor antibody or phosphotyrosine antibody, but IRS-1 was not seen (22, 26). Gustafson's laboratory did identify a band the size of IRS-1 in these GST-Grb10 pull down experiments but speculated that IRS-1 could have been bound to the coprecipitating insulin receptor (24). Similarly, in cells overexpressing the insulin receptor and HA-tagged Grb10, IRS-1 was not detected in the HA antibody immunoprecipitates, but the insulin receptor was detected (21). Thus, most experiments were unable to detect an interaction of IRS-1 with Grb10. Unlike SH2 domain containing molecules such as Syp or p85 which bind IRS-1 and insulin receptor/IGF-I receptor, Grb10 appears to bind exclusively to the receptors. This should make experiments designed to define the signaling role of Grb10 easier to interpret.

2.3. Sites of interaction with the insulin and IGF-I receptors

To date, studies to determine the binding site for Grb10 on the insulin and IGF-I receptors have produced conflicting results. Hansen et al. (22) provided evidence that the Grb10 SH2 domain interacted with the carboxy tail of the insulin receptor. GST-Grb10 SH2 was tested for binding to phosphopeptide affinity columns representing sequences around phosphotyrosine residues in the activated insulin receptor. GST-Grb10 SH2 bound only to a peptide containing carboxy tail phosphotyrosine 1322. In addition, an activated mutant receptor which lacked the carboxy-terminal 43 amino acids did not bind to GST-Grb10 SH2 beads (22). In contrast to these results, Gustafson's laboratory reported that the Grb10 SH2 domain retained interaction with mutant insulin receptor constructs lacking the carboxy-terminal 30 residues (including tyrosines 1316 and 1322) in the yeast two-hybrid system (24). Moreover, they showed equal binding of wild type and mutant insulin receptors to Grb10 in GST-Grb10 SH2 domain pull down experiments using lysates from cells expressing a mutant insulin receptor in which tyrosines 1316 and 1322 were changed to phenylalanine. In a control experiment, this mutant receptor did not bind to GST-p85 SH2 beads. In support of this, Shoelson's laboratory also reported that a 43 aa carboxy tail deletion mutant of IR bound normally to GST-Grb10 and GST-Grb 10 SH2 (26).

Baserga's laboratory also examined binding of Grb10 SH2 to IGF-I receptor mutant constructs in a yeast two-hybrid system (23). Based on the finding of positive interaction of receptors deleted beyond residue 1245, 1293, and 1310, but no interaction of a receptor deleted beyond aa 1229 with Grb10 SH2, they concluded that the segment aa 1229-1245 contained the Grb10 SH2 binding site. In addition, they reported a positive interaction when the carboxy 120 amino acids of the IGF-I receptor alone was coexpressed with Grb10 in the yeast two hybrid system. There is general agreement that binding of Grb10 to the insulin receptor and to the IGF-I receptor in the yeast two-hybrid system requires autophosphorylation of the receptors. It should be pointed out that phosphotyrosine based motifs are unlikely to be directly involved in the interactions described by Baserga since the IGF-I receptor fragment 1229-1245 does not contain tyrosine and the carboxy tail of the receptor would not be expected to undergo tyrosine phosphorylation in the absence of the kinase domain.

Shoelson's laboratory found that full length GST-Grb10 was more effective than GST-Grb10 SH2 in pulling down the insulin receptor from 3T3-L1 adipocytes, suggesting that domain(s) other than the SH2 domain were engaged in binding to the receptor (26). An obvious candidate would be the PH domain. For example, in the case of binding of IRS-1 to the insulin receptor both the PH domain and the PTB domain are important (28). Further evidence that full length Grb10 binds to additional sites on the insulin receptor is that phosphopeptides representing receptor sequences in the juxtamembrane domain around tyrosine 960, tyrosines 1146, 1150, and 1151 in the kinase domain, or tyrosines 1316 and 1322 in the carboxy tail, did not compete for binding of insulin receptor to full length Grb10, but the peptide containing phosphotyrosine 960 and the kinase domain phosphopeptide competed for binding of insulin receptor to the Grb10 SH2 domain (26). These results also suggest that there are at least two binding sites on the receptor for the Grb10 SH2 domain.

We have attempted to map the binding site on the IGF-I receptor for Grb-10 SH2 domain using the yeast two hybrid system (BRD, unpublished data). Deletion of the carboxy-terminal 93 amino acids from the receptor or changing tyrosines 950, 1250, 1251, and 1316 to phenylalanine did not result in dramatic decreases in the interaction of the IGF-I receptor with Grb10 SH2 domain. These results are consistent with there being more than one binding site for Grb10 SH2 domain in the IGF-I receptor.

2.4. Function of Grb10 in receptor signaling

Liu and Roth (21) compared Grb-IR transfected CHO-IR cells with CHO-IR cells for tyrosine phosphorylation of endogenous proteins in response to insulin. Expression of Grb-IR (Grb10) resulted in decreased tyrosine phosphorylation of p60GAP-associated protein, in decreased tyrosine phosphorylation of IRS-1, and in decreased PI 3-kinase activity in anti-phosphotyrosine immunoprecipitates. These results could be interpreted as meaning that Grb-IR serves to inhibit signaling by the insulin receptor or, alternatively, that Grb-IR links the insulin receptor to an alternative signaling pathway. In the case of inhibition of phosphorylation of IRS-1 in the presence of Grb-IR overexpression, it could be argued that Grb-IR is simply inhibiting binding of IRS-1 to the juxtamembrane tyrosine 960 based motif.

O'Neill et al. (24) reported that microinjection of Grb10 SH2 domain into Rat 1 fibroblasts overexpressing the insulin receptor resulted in a 50% decrease in insulin and IGF-I stimulated DNA synthesis as measured by BrdU incorporation. There was no decrease in EGF or serum stimulated DNA synthesis. Presumably the microinjected Grb10 SH2 domain is occuping sites on the insulin and IGF-I receptors that were normally occupied by endogenous Grb10 and thereby blocked signaling. These experiments suggest that Grb10 plays a role in a signaling pathway leading to DNA synthesis.

2.5. Preferential association of Grb10 with the insulin receptor

To determine the relative amount of Grb10 associated with the insulin receptor and the IGF-I receptor, Smith's laboratory utilized R- fibroblasts from IGF-I receptor knockout mice (25). These R- cells had been transfected with cDNAs for the insulin receptor or the IGF-I receptor. Cells were stimulated with the appropriate ligand and the amount of Grb10 found in receptor immunoprecipitates was quantitated by immunoblotting. The amount of Grb10 found in the IGF-I receptor immunoprecipitate following IGF-I stimulation was considerably less than the amount of Grb10 associated with the insulin receptor following insulin stimulation. Antiphosphotyrosine immunoblotting of whole cell lysates showed that the level of phosphorylation of the IGF-I receptor was actually somewhat greater than the level of phosphorylation of the insulin receptor. These results suggest that *in vivo* Grb10 may be selectively involved in insulin receptor signaling pathways.

3. MAD2

Gustafson's laboratory identified the human homolog of a yeast cell cycle checkpoint regulatory protein called MAD2 in a yeast two-hybrid screen using the cytoplasmic domain of the insulin receptor as bait (29). MAD2 did not bind to the IGF-I receptor cytoplasmic domain. Binding of MAD2 to the insulin receptor cytoplasmic domain in the yeast two-hybrid system did not require autophosphorylation of the receptor. The binding site for MAD2 was localized to the carboxy-terminal 30 amino acids of the receptor. It was proposed that the unphosphorylated insulin receptor binds MAD2 better than the phosphorylated receptor and that upon binding of insulin to the receptor, MAD2 is released from the receptor to perform some function or, alternatively, allow binding of another signaling molecule to the receptor. This was supported by the results of an experiment in which baculovirus-expressed insulin receptor was purified and phosphorylated *in vitro*. The phosphorylated and unphosphorylated receptors were analyzed by SDS-PAGE and blotting with radiolabeled GST-MAD2. The GST-MAD2 bound to the unphosphorylated receptor but not to the phosphorylated receptor. In contrast to these results, GST-MAD2 pull down experiments showed that GST-MAD2 bound almost as well to receptor from insulin stimulated cells as to receptor from resting cells, and insulin receptor was found in MAD2 immunoprecipitates from both resting and insulin stimulated cells. Although the binding of MAD2 was selective for the insulin receptor the function of MAD2 in insulin receptor signaling remains to be defined.

4. CONCLUSION

14-3-3 proteins, Grb10, and MAD2 appear to interact selectively with either the insulin receptor or the IGF-I receptor, and therefore, are candidates for being signaling molecules on pathways that distinguish these two receptors.

REFERENCES

1. K.D. Kimura, H.A. Tissenbaum, Y.Liu and G. Ruvkun, Science, 277 (1997) 942.
2. D. LeRoith, H. Werner, D. Beitner-Johnson, and C.T. Roberts Jr., Endocr. Rev., 16 (1995) 143.
3. R. Lammers, A. Gray, J. Schlessinger, and A. Ullrich, EMBO J., 8 (1989) 1369.
4. A. Ullrich, A Gray, A. W. Tam, T. Yang-Feng, M. Tsubokawa, C. Collins, W. Henzel, T. Le Bon, S. Kathuria, E. Chen, S. Jacobs, U. Francke, J. Ramachandran and Y. Fujita-Yamaguchi, EMBO J., 5 (1986) 2503.
5. A. Hongo, C. D'Ambrosio, M. Miura, A. Morrione and R. Baserga, Oncogene, 12 (1996) 1231.
6. S.Li, M. Resnicoff and R. Baserga, J. Biol. Chem. 271 (1996) 12254.

7. B.R. Dey, K. Frick, W. Lopaczynski, S.P. Nissley and R.W. Furlanetto, Mol. Endocrinology, 10 (1996) 631.
8. R.W. Furnanetto, B.R. Dey, W. Lopaczynski and S.P. Nissley, Biochem. J., (1997) in press.
9. G.W. Reuther, H. Fu, L.D. Cripe, R.J. Collier and A.M. Pendergast, Science, 266 (1994) 129.
10. A.J. Muslin, J.W. Tanner, P.M. Allen and A.S. Shaw, Cell, 84 (1996) 889.
11. A. Craparo, R. Freund and T.A. Gustafson, J. Biol.Chem., 272 (1997) 11663.
12. D. Morrison, Science, 266 (1994) 56.
13. P. Acs, Z. Szallasi, M.G. Kazanietz and P.M. Blumberg, Biochem. Biophys. Res. Commun., 216 (1995) 103.
14. N. Bonnefoy-Berard, Y.-C. Liu, M. von Willebrand, A. Sung, C. Elly, T. Mustelin, H. Yoshida, K. Ishizaka and A. Altman, Proc. Natl. Acad. Sci. USA, 92 (1995) 10142.
16. D.S. Conklin, K. Galaktionov and D. Beach, Proc. Natl. Acad. Sci. USA, 92 (1995) 7892.
17. C. Vincenz and V.M.Dixit, J. Biol. Chem., 271 (1996) 20029.
18. J. Zha, H. Harada, E. Yang, J. Jockel and S.J. Korsmeyer, Cell, 87 (1996) 619.
19. B. Xiao, S.J. Smerdon, D.H. Jones, G.G. Dodson, Y. Soneji, A. Aitken and S.J. Gamblin, Nature, 376 (1995) 188.
20. D. Liu, J. Bienkowska, C. Petosa, R.J. Collier, H. Fu and R. Liddington, Nature, 376 (1995) 191.
21. F. Liu and R.A. Roth, Proc. Natl. Acad. Sci. USA, 92 (1995) 10287.
22. H. Hansen, U. Svensson, J. Zhu, L. Laviola, F. Giorgino, G. Wolf, R.J. Smith, and H. Riedel, J. Biol. Chem., 271 (1996) 8882.
23. A. Morrione, B. Valentinis, S. Li, J.Y.T. Ooi, B. Margolis, and R. Baserga, Cancer Res., 56 (1996) 3165.
24. T.J. O'Neill, D.W. Rose, T.S. Pillay, K. Hotta, J.M. Olefsky and T.A. Gustafson, J. Biol. Chem., 271 (1996) 22506.
25. L. Laviola, F. Giorgino, J.C. Chow, J. A. Baquero, H. Hansen, J. Ooi, J. Zhu, H. Riedel, and R.J. Smith., J. Clin. Invest. 99 (1997) 830.
26. J.D. Frantz, S. Giorgetti-Peraldi, E.A. Ottinger and S.E. Shoelson, J. Biol. Chem. 272 (1997) 2659.
27. J. Ooi, V. Yajnik, D. Immanuel, M. Gordon, J.J. Moskow, A.M. Buchberg and B. Margolis, Oncogene, 10 (1995) 1621.
28. L. Yenush, K.J. Makati, J. Smith-Hall, O. Ishibashi, M.G. Myers Jr. and M.F. White, J. Biol. Chem., 271 (1996) 24300.
29. T.J. O'Neill, Y. Zhu and T.A. Gustafson, J. Biol. Chem., 272 (1997) 10035.

Molecular Mechanisms to Regulate the
Activities of Insulin-like Growth Factors
K. Takano, N. Hizuka and S-I. Takahashi (Editors)

The Multiple Roles of the IGF-I Receptor in Cell Growth

Renato Baserga, Marco Prisco, and Mariana Resnicoff

Kimmel Cancer Center, Thomas Jefferson University, 233 S. 10th Street, 624 BLSB, Philadelphia, Pennsylvania 19107, USA

1. INTRODUCTION

It has been known for a long time that the growth of tumors depends not only on the rate of cell proliferation, but also on the rate of cell death (1). An increase in cell proliferation and a decrease in the rate of cell death (apoptosis) cumulatively result in an increase in cell number, which is the most relevant characteristic of tumors. From this perspective, the insulin-like growth factor I receptor (IGF-IR) plays a very important role in tumor growth, because it stimulates cell proliferation, confers anchorage-independence and protects cells from cell death in general, and apoptosis in particular. Indeed, a functional impairment of the IGF-IR actually achieves the desirable goals of inhibiting cell proliferation and of inducing massive apoptosis in vivo, leading to inhibition of tumorigenesis. In this presentation, I will briefly summarize the functions of the IGF-IR related to growth, and I will then focus on some recent findings from our laboratory that are of unusual interest.

2. THE IGF-IR RECEPTOR IN NORMAL AND ABNORMAL GROWTH

The role of the IGF-IR in mitogenesis, transformation and protection from apoptosis has been reviewed in detail in two recent publications by Baserga et al. (2,3). The following comments are based on those reviews, to which the reader is referred for extensive references.

One way by which the IGF-IR activated by its ligands (IGF-I and IGF-II, and supraphysiological concentrations of insulin) contributes to cell proliferation is by stimulating mitogenesis. This, of course, has been known for a long time, indeed since IGF-I and IGF-II still went under the names of somatomedins (4). Since then, a substantial literature has confirmed that the IGF-IR transmits a mitogenic signal in a great variety of cells, ranging from fibroblasts, to epithelial cells, hemopoietic cells, bone cells and chondrocytes, smooth muscle cells, and several others, including stem cells (3). The mitogenicity of the IGF-IR in vivo has been formally demonstrated by the elegant experiments of Efstratiadis and co-workers (5,6), in mice with a targeted disruption of the IGF-IR genes. Clinical observations in man and animals confirm an in vivo role of the IGFs in development and post-natal growth. Often, to be mitogenic, the IGF-IR needs the co-operation of other growth factors, such as PDGF or EGF (7), but if the receptor is even modestly overexpressed, then the cells are stimulated to proliferate by IGF-I only, without the need of other growth factors (8).

As mentioned above, a second way by which the activated IGF-IR increases cell number in a given cell population is by reducing the rate of cell death, more specifically by protecting cells from apoptosis, induced by either physiological or pathological agents. The protective effect of IGF-I against apoptosis is a more recent finding that the discovery of its mitogenicity, going back to the early '90s (9), but it is now a well established phenomenon, well documented by many laboratories. It rests on two separate but converging lines of evidence: on the one hand, overexpression of the IGF-IR protects cells from a variety of apoptotic injuries (2,3), on the other hand, when the function of the IGF-IR is impaired by antisense strategies or by dominant negatives, cells, especially tumor cells, undergo massive apoptosis (10,11).

A population of cells can also increase in number because the cells can proliferate even under unfavorable environmental conditions, a property that is characteristic of tumor cells or transformed cells. The growth advantage of transformed cells is especially apparent in certain conditions, like growth in anchorage-independence, where normal cells fail to grow. An overexpressed IGF-IR transforms cells, but this is not particularly remarkable, because a great number of overexpressed gene products (including other growth factor receptors) can transform cells. What makes the IGF-IR more interesting is the finding that R- cells (12), which are 3T3-like fibroblasts derived from mouse embryos with a targeted disruption of the IGF-IR genes (see above), are refractory to transformation by a variety of cellular and viral oncogenes that readily transform 3T3 cells expressing even low levels of IGF-IRs (3). In other words, a functional IGF-IR is quasi-obligatory in the establishment and maintenance of the transformed phenotype.

Thus, the IGF-IR contributes to the expansion of cell populations in different ways: by stimulating mitogenesis, by decreasing the rate of cell death, and by allowing cell transformation. A corollary is that targeting the IGF-IR for therapeutic interventions has the advantage of inhibiting abnormal cell proliferation by 3 different mechanisms.

The mitogenic, transforming and anti-apoptotic functions of the IGF-IR map on various domains of the β subunit. The results of the extensive mutational analysis for these 3 functions are summarized in Table 1.

Table 1
Summary of the Mitogenic, Transforming and Anti-apoptotic Activities of the IGF-I Receptor and its Mutants

Receptor	Mitogenicity	Transformation	Anti-apoptosis
wt	+	+	+
d 1229	+	-	+
d1245	+	-	+
d1270	ND	+	+
d1293	+	+?	+
d1310	+	+	+
K 1003	-	-	-
Y950F	-	-	+
Y3F	-	-	-
Y1250F	+	+	+
Y1251F	+	-	-
S1280/1283A	+	-	?
1293/1294(FL)	+	+	-
Y1316F	+	+	+

These data are summarized from ref. 3, 15, 17 and 20. Mitogenicity is the ability to make cells grow in SFM supplemented solely with IGF-I; transforming activity is the ability to make cells form colonies in soft agar; and protection from apoptosis as protection of FL5.12 cells from apoptosis induced by IL-3 withdrawal or of 3T3 cells from okadaic acid-induced apoptosis. ND = not done

3. THE IGF-IR IS A PHYSIOLOGICALLY RELEVANT TARGET OF P53 IN APOPTOSIS OF MURINE HEMOPOIETIC CELLS

Werner et al. (13) have established a relationship between the IGF-IR and the pro-apoptotic protein p53, by demonstrating that wild type p53 represses transcription from the IGF-IR promoter, thereby decreasing the level of its expression. We therefore decided to investigate the relationship between p53 and the IGF-IR in apoptosis induced in murine hemopoietic 32D cells by IL-3 withdrawal; these cells are suitable for such studies because they have a very low level of IGF-IRs. The question we asked, specifically, was whether the IGF-IR is a physiologically relevant target of p53 in the process of apoptosis in 32D cells.

For this purpose, after establishing in preliminary studies that the IGF-IR does indeed protect 32D cells from apoptosis caused by IL-3 withdrawal, we stably transfected into 32Dtsp53 cells (expressing a thermo-sensitive mouse p53 protein) two different plasmids, both expressing the wild type human IGF-IR cDNA: in one plasmid, the cDNA was under the control of the CMV promoter; in the other plasmid, it was under the control of a rat IGF-IR promoter. Our rationale was that, according to the findings of Werner et al. (13), at 32°C, the p53, being in wild type conformation, should have reduced the IGF-IR number in cells expressing the IGF-IR under the control of the IGF-IR promoter, but not in cells expressing the IGF-IR cDNA under the control of the viral promoter (unresponsive to p53). This turned out to be the case. The question then became whether or not the failed down-regulation of the IGF-IR number would inhibit the onset of apoptosis in 32Dtsp53 cells. The results by Prisco et al. (14) were very clear. At 39^0 (p53 in mutant conformation), all transfectants survive in the absence of IL-3 , and only the parental 32D cells died. However, at 32^0 (p53 in wild type conformation), the 32D cells with the IGF-IR under the control of the viral promoter (high receptor levels) survived, but the other transfectants (with down-regulated IGF-IRs) underwent apoptosis. These experiments demonstrate that, in 32D cells, p53 cannot mediate apoptosis induced by IL-3 withdrawal, unless the IGF-IR is down-regulated.

4. IGF-II AND THE INSULIN RECEPTOR

The importance of the activated IGF-IR in murine development has been rigorously demonstrated in vivo by Efstratiadis and co-workers who showed that mouse embryos with a targeted disruption of the IGF-IR and IGF-II genes (Igfr-/-/mice) have a size at birth that is only 30% the size of wild type littermates (5,6). Subsequent experiments by the same group have further elucidated the role of the IGF system in murine embryo growth. Mouse mutants inheriting maternally a targeted disruption of the imprinted Igf2r gene have increased serum and tissue levels of IGF-II and exhibit overgrowth of 135% in respect to wild type littermates. These mutants usually die perinatally, the explanation being that, in the absence of IGF-IIR-mediated turnover, the IGF-IR is overstimulated by the excess IGF-II, resulting in heart defects. Consistent with this hypothesis is the finding that Igf2r mutants are completely rescued, when they carry a second mutation, eliminating either IGF-II or the IGF-IR. Triple mutants lacking IGF-IR, IGF-IIR and IGF-II are non-viable dwarfs, 30% in size, like the double mutants lacking IGF-II and IGF-IR. These experiments indicated that IGF-II signals also through an unknown or unidentified receptor, designated as XR. More recently, genetic evidence from double mutants lacking both the IGF-IR and the IR led to the conclusion that XR is actually the IR (Louvi, Accili, Taylor and Efstratiadis, personal communication).

We have taken advantage of R- cells, the 3T3-like mouse embryo fibroblasts (12), that we developed from mouse embryos with a targeted disruption of the IGF-IR genes (see above), to demonstrate in tissue cultures that IGF-II (but not IGF-I) can stimulate cell proliferation through the IR. For this purpose, we generated from R- cells clones overexpressing the IR. Since these cells do not have IGF-IRs, any effect of IGF-II has to be attributed either to an unknown receptor, XR, or to the IR. The results (16) are summarized in Table 2, from which it is clear that IGF-II stimulates cell proliferation through the IR, since R-cells, the parental cell line, cannot be stimulated (no XR). Interestingly, IGF-I does not stimulate the proliferation of R- cells overexpressing the insulin receptor, suggesting that the effect of IGF-II is ligand-specific.

Table 2
IGF-II Stimulates Cell Proliferation through the Insulin Receptor

cell line	growth factor	percent increase (decrease)
R-	insulin	(12)
	IGF-II	(18)
	IGF-I	(12)
R-/IR	insulin	22
	IGF-II	55
	IGF-I	(5)
R-/IR cl.2	IGF-II	14

All growth factors at 50 ng/ml. Growth expressed as percent increase over serum-free medium. R- cells have no IGF-I receptors and $5x10^3$ insulin receptors per cell; R-/IR have $5x10^5$ insulin receptors per cells; R-/IR clone 2 cells have 10^5 insulin receptors per cell (16).

5. PROTECTIVE ROLE OF THE IGF-IR IN APOPTOSIS

There are reports that several apoptotic-inducing agents or modalities are inhibited by an appropriate activation of the IGF-IR. They include IL-3 withdrawal (9,17), c-myc overexpression (18), etoposide, high serum concentrations, ICE expression, TNF α, TGF-β1, osmotic shock, serum withdrawal (19), okadaic acid (20) and p53 (14). The variety of the procedures used to induce apoptosis suggests that the overexpressed wild type IGF-IR may really have a widespread anti-apoptotic effect. Indeed, it may be a generalized inhibitor of apoptosis, with a range wider than other inhibitors, as the following example will show.

Since we have shown that the activated IGF-IR can protect mouse embryo fibroblasts from apoptosis induced by okadaic acid (20), we asked whether the general inhibitor of ICE activation, the baculovirus p35 protein, could also inhibit OKA-induced apoptosis. p35 prevents cell death in many types of organisms, including mammalian cells, and inhibits the ICE family proteases, that have been suggested to be mediators of all apoptotic cell death (21). Indeed, p35 can protect from ionizing radiation-induced apoptosis, where another general inhibitor of ICE activation, CrmA, fails. Mouse embryo R508 cells (8) were stably transfected with a plasmid expressing the baculovirus p35 protein, and two clones were selected for study. The results (Fig. 1) shows that expression of p35 does not protect R508 cells from OKA-induced apoptosis. Indeed, apoptosis seem to be increased in these cells by the expression of p35. Even at 24 hrs., this trend was evident: at this time, and with a concentration of 25 nM OKA, cell recovery ($x10^4$) was 3.8 for R508 cells, and 1.2 and 1.0 for the two clones of R508 expressing the p35 protein.

It therefore seems that the IGF-IR can protect cells from apoptosis, in conditions in which even the p35 baculovirus protein fails to protect.

	Cell Line	p6	R503	R508	R508/p35-23	R508/p35-27
Okadaic Acid	20 nM	27.3	19.9	4.4	0.87	0.7
	25 nM	25.1	7.7	2.0	0.4	0.5

MW (kDa) 36.8, 27.2 ←p35

Fig. 1 p35 Baculovirus Protein Does not Protect R508 Cells from Okadaic Acid-induced Apoptosis.
Cells were seeded and tested as usual, the starting concentrations of cells being $5x10^4$ cells per dish. The number of cells was determined after 48 hrs. The gel beneath the table shows that the two clones of transfected R508 cells do express the p35 protein.

6. CONCLUSIONS

The IGF-IR is rapidly emerging as a crucial receptor in the control of cell proliferation. It is present in the great majority of cells in the animal body, it is important in development (5,6), in normal growth (2,3), in the establishment and maintenance of the transformed phenotype (12), and in protection from apoptosis (10,20). Targeting of the IGF-IR has already been shown to be an effective way to induce apoptosis of tumor cells in vivo, and to inhibit tumorigenesis and metastases (3) in experimental animals. The fact that the functions of the IGF-IR map to different domains of the β subunit make this receptor a gold mine for the study of old and new signaling pathways. On the more practical side, the results obtained by several laboratories with transplantable tumors of animals make the IGF-IR a suitable candidate for anti-cancer therapy.

REFERENCES

1. R. Baserga and W. E. Kisieleski. J. Nat. Cancer Inst. 28 (1962) 331-339.
2. R. Baserga, M, Resnicoff, C. D'Ambrosio and B. Valentinis. Vitamins and Hormones 53 (1997) 65-98.
3. R. Baserga, A. Hongo, M. Rubini, M. Prisco and B. Valentinis. Biochim. Biophys. Acta 1332 (1997) 105-126.
4. D. R. Clemmons and J. J. Van WykSomatomedin: physiological control and effects on cell proliferation. In: Tissue Growth Factors (R. Baserga ed.) Springer-Verlag, Berlin, pp. 161-208 (1981).
5. J-P. Liu, J. Baker, A. S. Perkins, Robertson, E. J. and Efstratiadis, A. Cell 75 (1993) 59-72.

6. J. Baker, J-P Liu, E. J. Robertson, and A. Efstratiadis. Cell 73 (1993) 73-82.
7. C. D. Stiles, G. T. Capone, C. D. Scher, N. H. Antoniades, J. J. Van Wyk and W. J. Pledger. Proc. Natl. Acad. Sci. USA 76 (1979) 1279-1283.
8. M. Rubini, A. Hongo, C. D'Ambrosio and R. Baserga. Exp. Cell Res. 230: (1997) 284-292.
9. J. A. McCubrey, L. S. Stillman, M. W. Mayhew, P. A. Algate, R. A. Dellow and M. Kaleko. Blood 78 (1991) 921-929.
10. M. Resnicoff, D. Abraham, W. Yutanawiboonchai, H. Rotman, J. Kajstura, R. Rubin, P. Zoltick and R. Baserga. Cancer Res. 5 (1995) 2463-2469.
11. C. D'Ambrosio, A. Ferber, M. Resnicoff and R. Baserga. Cancer Res. 56 (1996) 4013-4020.
12. C. Sell, M. Rubini, R. Rubin, J-P. Liu, A. Efstratiadis and R Baserga. Proc. Natl. Acad. Sci. USA, 90 (1993) 11217-11221.
13. H. Werner, E. Karnieli, F. J. Rauscher, III and D. LeRoith. Proc. Natl. Acad. Sci. 93 (1996) 8318-8323.
14. M. Prisco, A. Hongo, M. G. Rizzo, A. Sacchi and R. Baserga. Mol. Cell. Biol. 17 (1997) 1084-1092.
15. A. Hongo, C. D'Ambrosio, M. Miura, A. Morrione and R. Baserga. Oncogene 12 (1996) 1231-1238.
16. A. Morrione, B. Valentinis, S-Q Xu, G. Yumet, A. Louvi, A. Efstratiadis and R. Baserga. Proc. Natl. Acad. Sci. 94 (1997) 3777-3782.
17. R. O'Connor, A. Kauffmann-Zeh, Y. Liu, S. Lehar, G. I. Evan, R. Baserga and W. A. Blattler. Mol. Cell. Biol. 17 (1997) 427-435.
18. E. A. Harrington, M. R. Bennett, A. Fanidi and G. I Evan. EMBO J. 13 (1994) 3286-3295.
19. M. Parrizas, A. R. Saltiel and D. LeRoith. J. Biol. Chem. 272 (1997) 154-161.
20. C. D'Ambrosio, B. Valentinis, M. Prisco, K. Reiss, M. Rubini and R. Baserga. Cancer Res. (1997) in press.
21. P. A. Henkart. Immunity 4 (1996) 195-201.

Molecular Mechanisms to Regulate the
Activities of Insulin-like Growth Factors
K. Takano, N. Hizuka and S-I. Takahashi (Editors)

Role of IGF-II in Wilms Tumourigenesis and Overgrowth Disorders

Anthony E Reeve

Cancer Genetics Laboratory, Department of Biochemistry, University of Otago, P.O. Box 56, Dunedin, New Zealand.

ABSTRACT

Genomic imprinting plays an important role in a number of human diseases including cancer. Wilms tumour and other embryonal tumours have provided the most useful and practical experimental models for studying imprinting mechanisms in tumourigenesis. Insulin-like growth factor 2 (IGF2) is an embryonal growth factor which has been implicated in the onset of Wilms tumour. The IGF2 gene is imprinted, that is, only the paternal allele is transcribed into mRNA while the maternal allele is silent. We have shown that the normal imprinting of the IGF2 gene is lost in some Wilms tumours. The consequence of this event is that IGF2 is expressed from both alleles, thereby leading to increased transcription in Wilms tumours. Loss of the IGF2 imprint (LOI) occurs very early in the tumourigenesis pathway and may lead to increased mitogenesis or decreased apoptosis in a precursor population of cells. We have recently defined an entity termed the 'IGF2 overgrowth disorder' in which loss of IGF2 imprinting occurs constitutionally and in a mosaic fashion. If loss of IGF2 imprinting is widespread, excessive somatic overgrowth and tumour predisposition may occur. Conversely, if the loss of IGF2 imprinting is confined to a few tissues (e.g. kidney) the only manifestations of this disorder may be sporadic Wilms tumour.

WILMS TUMOUR AND CHROMOSOME 11

Genes on chromosome 11 have been the subject of much Wilms tumour research. Starting with the observation by Uta Francke and colleagues that patients with the Wilms tumour-aniridia-genitourinary association (WAGR) had deletions of band 11p13[1], this work ultimately led to the cloning of the WT1 gene[2]. Subsequently it was found that more than one third of sporadic Wilms had loss of heterozygosity exclusively involving the 11p15 region, suggesting that another gene was also involved in the tumourigenesis pathway[3-5]. Although there are many genes in the 11p15 region, as will be discussed below, IGF2 is the strongest candidate.

WILMS TUMOUR AND GENOMIC IMPRINTING

Early investigations of chromosome 11p LOH in Wilms tumour were based around the assumption that a tumour suppressor gene was involved according to the Knudson 'two hit' model. One of the assumptions of this model is that the first mutation can occur on either of a diploid pair of chromosomes, and therefore the second event, namely chromosome loss of heterozygosity (LOH), should occur randomly. In Wilms tumour, however, 11p LOH always involves the maternal allele with retention of the paternal allele. Wilkins first proposed that this event must have involved a critical imprinted gene(s) located within the distal region of the short arm of chromosome 11[6], and subsequently other models were proposed[7]. These imprinting models were described before the discovery of imprinted genes within chromosome 11p15. As discussed below, there is now strong evidence that imprinted genes play a critical role in Wilms tumourigenesis.

WILMS TUMOUR AND 'DOUBLE DOSE' OF INSULIN-LIKE GROWTH FACTOR II

Wilms tumour was the first human tumour in which IGF2 was shown to be transcribed at high levels[8, 9]. IGF2 has since been shown to be transcribed in a wide variety of tumours[10-12]. From the early studies on Wilms tumour it was not possible to determine whether IGF2 was causal in tumourigenesis or simply a consequence of the embryonal nature of this tumour. IGF2 was nevertheless an attractive candidate because of its mitogenic properties and its gene location within the chromosome 11p15 region which frequently undergoes LOH and simultaneous reduplication in Wilms tumour.

An important clue for the role of IGF2 came with the discovery that the IGF2 gene was imprinted in mice i.e. only the paternal allele was transcribed in expressing tissues[13]. Following this finding, it was hypothesised by Little and colleagues that in Wilms tumour, maternal 11p15 LOH and reduplication of the paternal 11p15 region could lead to a 'double dose' of IGF2[14]. The assumption was made that IGF2 could lead to the proliferation of a precursor cell population, although IGF2 is now also known to play a role in delaying apoptosis[15]. Regardless of the role played by IGF2, the mechanism by which it could promote the early steps of tumourigenesis was in contradiction to the original two-hit tumour suppressor gene model.

LOSS OF IGF2 IMPRINTING AND WILMS TUMOUR

Although the IGF2 'double dose model' provides an attractive explanation for at least some of the steps in Wilms tumourigenesis, 11p15 LOH occurs in only 30-40% of Wilms tumours[16-19]. This apparent paradox was resolved when two groups investigated the imprinting status of the silent IGF2 maternal allele in a series of Wilms tumours which retained 11p15 heterozygosity. In 19 tumours analysed by

RT-PCR using a polymorphism in exon 9 of IGF2, 74% expressed IGF2 from the maternal allele, consistent with the 'double dose' model[20, 21]. Nevertheless, not all tumours in this series had IGF2 loss of imprinting (LOI) or 11p15 LOH, indicating that if the model is correct, it cannot provide the complete answer. It is possible in tumours with neither IGF2 LOI nor 11p15 LOH that up-regulating mutations might be present within the IGF2 gene or the IGF2 signal transduction pathway. The IGF2 type 2 receptor which normally down regulates IGF2 is one strong candidate given that mutations in this gene have been found in other tumour types[22].

Several reasons indicate that 11p15 LOH and IGF2 LOI may occur early in tumourigenesis. RFLP analysis of tumours with 11p15 LOH indicates that that maternal allele loss is essentially complete[16-19]. If LOH occurred at a late stage of tumourigenesis then incomplete loss would be seen in keeping with the evolution of a late-stage clone of proliferating cells. A similar reasoning can be applied to IGF2 LOI, where the ratio of expressed maternal/paternal alleles is approximately 1:1. It is therefore likely that the majority of Wilms tumours arise from an early clone of cells within the kidney that have either 11p15 LOH or IGF2 LOI. Clonal expansion due to enhanced mitosis or delayed apoptosis, followed by additional genetic events such as 1p or 16q LOH[23-25] could then lead to the formation of the rapidly dividing tumour. It is interesting that approximately 1% of newborn infants have localised areas of immature cells (nephrogenic rests)[26] which are known to express IGF2 at high levels. Possibly these rests provide a source of target cells within which additional genetic mutations may occur. Rests have been shown to contain mutations within the WT1 gene, but it is not yet clear whether this is a frequent event[27].

There are two potential difficulties with the IGF2 dosage model. First, this model has been derived by DNA or RNA and not protein analysis. Verification of the model should require a demonstration of increased IGF2 protein expression in Wilms tumours. However, to show that protein levels are increased will be considerably more difficult. Comparisons, using western blotting with the tissue of origin, namely, the nephrogenic mesenchyme in the developing kidney will be needed. Wilms tumours are, however, frequently heterogeneous consisting of three or more cellular types. Furthermore, the cellular milieu in the developing kidney is vastly different to a Wilms tumour therefore making expression comparisions essentially impossible with existing technologies. Second, and similarly, the levels of IGF2 transcripts in 11p15 LOH and IGF2 LOI tumours should be two fold higher compared to tumours with normal monoallelic IGF2 expression. As with making comparisons of IGF2 protein expression levels, this analysis is also extremely difficult because of the histologic heterogeneity of Wilms tumours. For example, we have previously reported that variations in IGF2 mRNA levels of up to 30 fold were observed by northern blot analysis in different regions of the same tumour[28].

CHROMOSOME 11 CONTAINS AN IMPRINTED DOMAIN ENCOMPASSING IGF2 AND H19

The H19 gene located approximately 90kb from IGF2 is also imprinted, however unlike IGF2, expression is from the maternal allele[29, 30]. H19 is an unusual gene in that its product consists of an RNA which is not translated. The expression patterns of IGF2 and H19 have been examined in Wilms tumours resulting in their classification into three groups[31-33]: (i) tumours with biallelic IGF2 expression and absent H19 expression (ii) tumours with 11p15 LOH and very low expression of H19 and (iii) tumours with normal reciprocal expression patterns of H19 and IGF2 expression similar to that seen in embryonal kidney .

Tilghman and colleagues have proposed that the reciprocal allelic expression pattern of H19 and IGF2 is due to the competition of their respective promoters with a shared H19 enhancer element[34]. Silencing of the paternal H19 allele is associated with DNA methylation. Several groups have shown in Wilms tumours with IGF2 LOI the silenced maternal H19 allele is also methylated[31-33]. Conversely, IGF2 LOI has been shown to result in extensive demethylation of the maternal IGF2 allele[33]. Tumour specifc changes in the allele specificity of expression and methylation have led to the concept that IGF2 LOI results in the maternal H19/IGF2 locus adopting a paternal epigenotype. These changes in DNA methylation do not appear to be a random process occurring during tumourigenesis because in tumours with normal the normal imprinted expression patterns of IGF2 and H19, the characteristic differential methylation pattern was maintained. Furthermore, the methylation pattern of other 11p15 genes showed no correlation with H19/IGF2 imprinting status indicating that the epigenetic switch was not a consequence of global changes in DNA methylation[33].

Other imprinted genes in close vicinity to H19/IGF2 might also play a role in tumourigenesis. The gene for the cyclin dependent kinase inhibitor $p57^{KIP2}$ (CDKN1C) is located approximately 700kb centromeric of IGF2 and has been shown to be imprinted in the mouse with expression from the maternal allele[35]. It is tempting to consider that $p57^{KIP2}$ could act as a tumour suppressor gene following maternal 11p15 LOH, however its role in Wilms tumourigenesis is controversial. In transformation assays $p57^{KIP2}$ expression did not correlate with tumour suppression[36], and in Wilms tumours there was only a general association between $p57^{KIP2}$ expression levels and 11p15 LOH[37]. Surprisingly, $p57^{KIP2}$ expression was maintained but with expression from the paternal allele following 11p15 LOH[38, 39]. Furthermore, the maintenance of $p57^{KIP2}$expression following maternal 11p15 LOH is not consistent with $p57^{KIP2}$ being a tumour suppressor gene in Wilms tumour[39]. Unlike the specific pattern of imprinted expression in the mouse, in human tissues $p57^{KIP2}$ expression is 'leaky' with significant expresssion from the paternal allele[38, 39]. It is possible that this 'leaky' expression could signal the existence of a mosaic cell population. Consistent with this scenario are the observations that $p57^{KIP2}$ is expressed only from the paternal allele in tumours with

11p15 LOH and that there is significant variation in the ratio of maternal : paternal allelic expression amongst different tissues.

The notion of a coordinately regulated imprinted domain including p57^{KIP2}, H19 and IGF2 seems unlikely given that the pattern of allelic expression and expression levels of p57^{KIP2} are unchanged in Wilms tumour which have IGF2 LOI[39]. This is in contrast to the Prader-Willi syndrome which has several coordinately regulated imprinted genes spread over 2MB. Recently the KCNA9 (KVLQT1 potassium channel) gene which is located between p57^{KIP2} and IGF2 was shown to be imprinted with expression predominantly from the maternal allele[40]. It is interesting that the exons for this gene span a large interval within which maternal translocations occur on rare occasions in the Beckwith -Wiedemann syndrome (BWS) (see below). The KCNA9 gene has several alternatively spliced RNAs of which two do not appear to be translated in a similar manner to H19[40]. This raises the possibility that KCNA9 could act as an imprint control centre which could have a long-distance effect on imprinted genes in this region.

IGF2 OVERGROWTH DISORDER, BECKWITH-WIEDEMANN SYNDROME, AND EMBRYONAL TUMOURIGENESIS

The Beckwith Wiedemann syndrome is intriguing because in addition to the most common feature, namely tissue overgrowth, approximately 10% of cases develop a range of embryonal malignancies. Wilms tumour is the most common neoplasm with associated congenital anomalies including macroglossia, visceromegaly, omphalocoele, hepatomegaly and nephomegaly. The BWS is also associated with high birth weight and advanced bone age. The role of IGF2 in the BWS was first proposed by Olshan[41] and there are now several types of evidence for its role in this syndrome. First, IGF2 is expressed at high levels in the embryonal tissues which display overgrowth and malignancy in the BWS[42]. Second, the inheritance pattern of familial forms of BWS show genetic linkage to the INS-IGF2 locus[43, 44]. Third, a substantial proportion (20-30%) of sporadic BWS demonstrate paternal uniparental disomy (UPD) of chromosome 11p, suggesting that an imprinted gene(s) is involved[45, 46]. Paternal UPD involves the loss of maternal 11p alleles suggesting that a maternally expressed 11p gene could be involved. However, rare cases of BWS involve trisomy 11p with two paternal copies and one maternal copy of the 11p chromosomal region[47]. Thus, it is unlikely that BWS arises by the loss of an 11p encoded maternally expressed growth suppressor. Fourth, paternal duplication of mouse chromosome 7 (which contains the Igf2 gene) results in fetal overgrowth[48]. Fifth, rare translocations in the BWS occur within the maternal 11p15 region and have been documented to lead to biallelic IGF2 expression[49]. Finally, the strongest evidence for the role of IGF2 in BWS was provided by Weksberg and colleagues where IGF2 imprinting was found to be relaxed in 4/6 BWS cases[50]. More recent reports indicate that biallelic IGF2 expression may occur in up to 80% of BWS cases[51]. Presumably, as a result of IGF2 transcription from both alleles, increased levels of the IGF2 protein resulted in enhanced growth.

Recently the maternally expressed p57^{KIP2} gene was shown to be mutated in 2/9 cases of BWS[52]. Forty additional cases of BWS analysed in our laboratory were mutation negative suggesting that this event may be uncommon[53]. Interestingly, a mouse p57^{KIP2} knockout was shown to have some features characteristic of BWS, for example adrenal hyperplasia, renal medullary dysplasia and omphalocoele[54]. However, the most striking BWS feature, namely tissue overgrowth, was not observed. It is possible that p57^{KIP2} point mutations in the BWS and the loss of function mutations in the p57^{KIP2} knockout mice have different phenotypic consequences. It is also possible that mutations in p57^{KIP2} could interact with components of the IGF2 pathway at some point. Whether this is the case, perhaps resulting in biallelic IGF2 expression or down regulation of the IGF2 type 2 receptor, remains to be determined.

We have recently shown that biallelic IGF2 expression occurs in the developing kidney of patients with sporadic Wilms tumour[53]. This epigenetic event occurs very early in the tumourigenesis pathway and may lead to increased mitogenesis or decreased apoptosis in a precursor population of cells. At the other extreme, we have shown that biallelic IGF2 expression is present constitutionally in four children with generalised overgrowth and no other manifestations of the BWS except Wilms tumour and/or nephromegaly[55]. In all cases of BWS, the various manifestations are variable i.e. mosaic. These observations have led to the suggestion of an IGF2 overgrowth syndrome/disorder[55]. Thus where loss of IGF2 imprinting is widespread, extensive somatic overgrowth may occur, while intermediate levels of mosaicism may lead to the variable phenotypic features of the BWS. At the extreme point, if loss of IGF2 imprinting is confined to one or only a few tissues (e.g. kidney, liver or adrenal) then the only consequence of this limited mosaicism may be a predisposition to embryonal tumour development.

ACKNOWLEDGEMENTS

The author wishes to thank the following people who have co-authored several references cited herein and have contributed to the development of concepts in this paper: David Becroft, Michael Eccles, Ian Morison, Osamu Ogawa, Kesei Okamoto Michael Sullivan, Takanobu Taniguchi and Kankatsu Yun. This work was supported by the Health Research Council of New Zealand, NZ Lottery Grants Board and the Cancer society of New Zealand.

REFERENCES

1. Francke, U., Holmes, L. B., Atkins, L. Riccardi, V. M., Cytogenet. Cell Genet, 24, (1979) 185-192.
2. Gessler, M., Poustka, A., Cavenee, W., Neve, R. L., Orkin, S. H. Bruns, G. A. P., Nature, 343, (1990) 774-777.

3. Mannens, M., Slater, R. M., Heyting, C., Bliek, J., de Kraker, J., Coad, N., de Pagter-Holthuizen, P. Pearson, P. L., Hum. Genet., 81, (1988) 41-48.
4. Reeve, A. E., Sih, S. A., Raizis, A. M. Feinberg, A. P., Mol. Cell Biol., 9, (1989) 1799-1803.
5. Baird, P., Wadey, R. Cowell, J., Oncogene, 6, (1991) 1147-1149.
6. Wilkins, R. J., The Lancet, i, (1988) 329-331.
7. Scrable, H., Cavenee, W., Ghavimi, F., Lovell, M., Morgan, K. Sapienza, C., Proc. Nat. Acad. Sci., 86, (1989) 7480-7484.
8. Reeve, A. E., Eccles, M. R., Wilkins, R. J. W., Bell, G. I. Millow, L. J., Nature, 317, (1985) 258-260.
9. Scott, J., Cowell, J., Robertson, M. E., Priestley, L. M., Wadey, R., Hopkins, B., Pritchard, J., Bell, G. I., Rall, L. B., Graham, C. F. Knott, T. J., Nature, 317, (1985) 261-262.
10. Shapiro, D. N.&Helman, L. J., Oncogene, 11, (1995) 2503-2507.
11. Suzuki, H., Ueda, R., Takahashi, T., Takahashi, T., Nature Genetics, 6, (1994) 332-333.
12. Zhan, S., Shapiro, D. N. Helman, L. J., J. Clin. Invest., 94, (1994) 445-448.
13. DeChiara, T. M., Robertson, E. J. Efstratiadis, A., Cell, 64, (1991) 849-859.
14. Little, M., van Heyningen, V. Hastie, N., Nature, 351, (1991) 609-610.
15. Evan, G. I., Wyllie, A. H., Gilbert, C. S., Littlewood, T. D., Land, H., Brooks, M., Waters, C. M., Penn, L. Z. Hancock, D. C., Cell, (1992)
16. Koufos, A., Hansen, M. F., Lampkin, B. C., Workman, M. L., Copeland, N.G., Jenkins, N. A. Cavenee, W. K., Nature, 309, (1984) 170-174.
17. Orkin, S. T., Goldman, D. S. Sallan, S. E., Nature, 309, (1984) 172-174.
18. Reeve, A. E., Housiaux, P. J., Gardner, R. J. M., Chewings, W. E., Grindley, R. M. Millow, L. J., Nature, 309, (1984) 174-176.
19. Fearon, E. R., Vogelstein, B. Feinberg, A. P., Nature, 309, (1984) 176-178.
20. Rainier, S., Johnson, L. A., Dobry, C. J., Ping, A. J., Grundy, P. E. Feinberg, A. P., Nature, 362, (1993) 747-749.
21. Ogawa, O., Eccles, M. R., Szeto, J., McNoe, L. A., Yun, K., Maw, M. A., Smith, P. J. Reeve, A. E., Nature, 362, (1993) 749-751.
22. Souza, A. T. D., Hankins, G. R., Washington, M. K., Orton, T. C. Jirtle, R. l., Nature Genet., 11, (1995) 447-449.
23. Maw, M. A., Grundy, P. E., Millow, L. J., Eccles, M. R., Dunn, R. S., Smith, P. J., Feinberg, A. P., Law, D. J., Paterson, M. C., Telzerow, P. E., Callen, D. F., Thompson, A. D., Richards, R. I. Reeve, A. E., Cancer Res., 52, (1992) 3094-3098.
24. Coppes, M. J., Bonetta, L., Huang, A., Hoban, P., Chilton-MacNeill, S., Campbell, C. E., Weksberg, R., Yeger, H., Reeve, A. E. Williams, B. R. G., Genes, Chrom. & Cancer, 5, (1992) 326-334.
25. Grundy, P. E. Telzerow, P. E., Breslow, N., Moksness, J., Huff, V. Paterson, M. C., Cancer Research, 54, (1994) 2331-2333.
26. Bennington, J. L.&Beckwith, J. B., Tumors of the kidney,renal pelvis and ureter, in Washington DC: Armed Forces of Pathology. Atlas of tumor Pathology. Second Series, Fascile, 12. 1974,

27. Park, S., Bernard, A., Bove, K. E., Sens, D. A., Hazen-Martin, D. J., Garvin, A. J. Haber, D. A., Nature Genet., 5, (1993) 363-367.
28. Reeve, A. E., in Genomic Causes and Consequences, ed.R. Ohlsson, K. Hall and M. Ritzen. 1995, Cambridge University Press. 209-223.
29. Bartolomei, M. S., Zemel, S. Tilghman, S. M., Nature, 351, (1991) 153-155.
30. Zhang, Y.&Tycko, B., Nature Genet., 1, (1992) 40-44.
31. Moulton, T., Crenshaw, T., Hao, Y., Moosikasuwan, J., Lin, N., Dembitzer, F., Hensle, T., Weiss, L., McMorrow, L., Loew, T., Kraus, W., Gerald, W. Tycko, B., Nature Genet., 7, (1994) 440-447.
32. Steenman, M. J. C., Rainier, S., Dobry, C. J., Grundy, P., Horon, I. L. Feinberg, A. P., Nature Genet., 7, (1994) 433-439.
33. Taniguchi, T., Sullivan, M. J., Ogawa, O. Reeve, A. E., Proc. Nat. Acad. Sci. (USA), 92, (1995) 2159-2163.
34. Tilghman, S. M., Harvey Lecture Series, 87, (1993) 69-84.
35. Hatada, I.&Mukai, T., Nature Genet., 11, (1995) 204-206.
36. Reid, L. H., Crider-Miller, S. J., West, A., Lee, M.-H., Massague, J. Weissman, B. E., Cancer Research, 56, (1996) 1214-1218.
37. Chung, W.-Y., Yuan, L., Feng, L., Hensle, T. Tycko, B., Hum. Mol. Genet., 5, (1996) 1101-1108.
38. Matsuoka, S., Thompson, J. S., Edwards, M. C., Barletta, J. M., Grundy, P., Kalikin, L. M., Harper, J. W., Elledge, S. J. Feinberg, A. P., Proc. Nat. Acad. Sci. (USA) 93, (1996) 3026-3030.
39. Taniguchi, T., Okamoto, K., Reeve, A. E., Oncogene, 14, (1997) 1201-1206.
40. Lee, M. P., Hu, J.-J., Johnson, L. Feinberg, A. P., Nature Genet, 15, (1997) 181-185.
41. Olshan, A. F., Cancer Genet. Cytogenet., 21, (1986) 303-307.
42. Han, V. K. M., D'ercole, J. Lund, P. K., Science, 236, (1987) 193-197.
43. Koufos, A., Grundy, P., Morgan, K., Aleck, K. A., Hadro, T., Lampkin, B. C., Kalbakji, A. Cavenee, W. K., Am. J. Hum. Genet., 44, (1989) 711-719.
44. Ping, A. J., Reeve, A. E., Law, D. J., Young, M. R., Boehnke, M. Feinberg, A. P., Am. J. Hum. Genet., 44, (1989) 720-723.
45. Henry, I., Bonaiti-Pellie, C., Chehensse, V., Beldjord, C., Schwartz, C., Utermann, G. Junien, C., Nature, 351, (1991) 665-667.
46. Grundy, P., Telzerow, P., Paterson, M. C., Haber, D., Berman, B., Li, F. Garber, J., The Lancet, 338, (1991) 1079-1080.
47. Okano, Y. et al., 1986, Japan J. Hum. Genet.31, (1986) 365-372.
48. Ferguson-Smith, A. C., Cattanach, B. M., Barton, S. C., Beechey, C. V. Surani, M. A., Nature, 351, (1991) 667-670.
49. Mannens, M., Hoovers, J. M. N., Redeker, E., Verjaal, M., Feinberg, A. P., Little, P., Boavida, M., Coad, N., Steenman, M., Bliek, J., Niikawa, N., Tonoki, H., Nakamura, Y., de Boer, E. G., Slater, R. M., John, R., Cowell, J. K., Junien, C., Henry, I., Tommerup, N., Weksberg, R.,Pueschel, S. M., Leschot, N. J. Westerveld, A., Eur. J. Hum. Genet., 2, (1994) 3-23. Weksberg, R.,
50. Weksberg, R., Shem, D. R., Song, Q. L. Squire, J., Nature Genet., 5, (1993) 143-149.
51. Reik, W. and Maher, E. R., TIG, 13, (1997) 330-334.

52. Hatada, I., Ohashi, H., Fukushma, Y., Kaneko, Y., Inoue, M., Komoto, Y., Okada, A., Ohishi, S., Nabetani, A., Morisaki, H., Nakayama, M., Niikawa, N. Mukai, T., Nature Genet., 14, (1996) 171-173.
53. Okamoto, K., Morison, I. M., Reeve, A. E., Tommerup, N., Wiedemann, H. R. Friedrich, U., J. Med. Genet., in press, (1997)
54. Zhang, P., Liégeois, N. J., Wong, C., Finegold, M., Hou, H., Thompson, J. C., Silverman, A., Harper, J. W., DePinho, R. A. Elledge, S. J., Nature, 387, (1997) 151-158.
55. Morison, I. M., Becroft, D. M., Taniguchi, T., Woods, C. G. Reeve, A. E., Nature Medicine, 2, (1996) 311-316.

Molecular Mechanisms to Regulate the
Activities of Insulin-like Growth Factors
K. Takano, N. Hizuka and S-I. Takahashi (Editors)

The IGF System in Breast Cancer

D. Yee, J.G. Jackson, C-N. Weng, J.L. Gooch, and A.V. Lee

Department of Medicine, Division of Medical Oncology, University of Texas Health Science Center, San Antonio, Texas 78284-7884

1. INTRODUCTION

A seminal advance in the treatment of breast cancer was the recognition that the estrogen receptor (ER) had a function in regulating malignant cellular proliferation. The relevance of ER was demonstrated several ways. First, ovarian estrogen has long been known to regulate breast cancer growth [1]. Second, after ER was isolated and reagents were created to identify its expression, it was found to be expressed in the majority of breast cancer specimens [2]. Third, ER expression in primary breast cancer was associated with prognosis suggesting that it played an important biological function [3]. Fourth, a cell line derived from an ER-positive (ER+) tumor was stimulated by estradiol and inhibited by an anti-estrogen (tamoxifen) *in vitro* [4]. Fifth, tamoxifen was an effective anti-breast cancer agent *in vitro* and in animal model systems. Lastly, tamoxifen also has proven to be effective treatment for patients bearing ER+ tumors [5]. Thus, identification of a key regulatory pathway and synthesis of an agent to block this pathway has proven to be one of the most successful examples of targeted treatment for cancer. Furthermore, since tamoxifen is relatively free of side effects, it is currently being tested as a preventative agent in breast cancer.

Thus, ER and tamoxifen have proven to be the "holy grail" for breast cancer researchers. Surely if this key growth regulatory pathway exists, then there must be other pathways that could also be inhibited to successfully treat breast cancer. Numerous studies have suggested that the insulin-like growth factors could have important growth regulatory properties for breast cancer cells. In fact, much of the evidence initially gathered to support ER's function in breast cancer have also been gathered to support that the IGFs have a role in regulating breast cancer growth.

2. EXPRESSION OF IGF SYSTEM COMPONENTS IN BREAST CANCER

Virtually every breast cancer cell line studied expresses type I and type II IGF receptors and IGF binding proteins. When IGF system components are examined in primary breast cancer specimens, similar patterns of expression are found [6-11]. Thus, most breast cancer cells express the necessary effectors to respond to the IGFs. Furthermore, expression of some of the IGF system components are associated with early recurrence in patients with breast cancer. For example, we have found that high levels of IGFBP-4 are associated with early recurrence in patients with small tumors [12]. Similarly, high levels of the insulin receptor substrate-1 (IRS-1), a key signaling molecule activated by the type I IGF receptor (IGFR1), also identifies patients with early recurrence (13 - see below).

However, high IGFR1 levels are associated with a favorable prognosis [14]. While this seems to be contradictory to the suggestion that the IGFs stimulate breast cancer growth there are several possible explanations. First, high IGFR1 levels could represent tumors that are not stimulated by the IGFs. That is, low levels of IGFR1 are caused by downregulation of the receptor after IGF stimulation. Alternatively, IGFR1 may be similar to ER in breast cancer. While high ER levels are associated with more differentiated tumors and favorable outcome, the ER pathway still represents a key growth regulatory pathway for many breast cancer cells. Thus, the pattern of expression of the IGF system components in breast cancer suggest that they may play a role in regulating tumor cell proliferation.

3. STIMULATION OF BREAST CANCER GROWTH BY THE IGFs

Both insulin and the IGFs are potent mitogens for breast cancer cells. In serum-free cell culture systems, IGF-I can substitute for estradiol in stimulating the growth of ER-positive breast cancer cell line MCF-7 [6,15,16]. However, the expression of IGFR1 does not necessarily mean that the cells will respond to IGF-I. For example, the ER-negative breast cancer cell line MDA-MB-231 does not respond to insulin or IGF-I, yet expresses abundant levels of receptor. In these cells, expression of the PC-1 glycoprotein may make them refractory to insulin stimulation [17]. The MDA-MB-468 cell line also does not appropriately respond to insulin or IGF-I. This cell line expresses little IRS-1 which may account for its defects in post-receptor signal transduction [18]. Thus, while cells are responsive to the IGFs, the mere presence of IGFR1 does not necessarily result in signal transduction and mitogenesis.

Despite this observation, inhibition of IGFR1 function has been shown to block breast cancer cell growth. Arteaga et al. has demonstrated that antibody directed against IGFR1 inhibits tumor growth of the MDA-MB-231 cell line grown as a xenograft tumor in athymic mice [19]. We have shown similar results using a polyethylene glycol conjugated form of IGFBP-1 to neutralize IGF action. In an athymic mouse model, we have shown that IGFBP-1 blocks xenograft growth of MDA-MB-231 cells and ascites tumor growth of MDA-MB-435 cells [20]. However, neither of these cell lines is particularly sensitive to IGF-I *in vitro*. In fact, as noted above, the MDA-MB-231 is not appropriately stimulated by either insulin or IGF-I. Thus, these anti-IGF strategies may have resulted in inhibition of host IGF function required for tumorigenesis rather than inhibition of tumor cell IGFR1. It has also been shown that athymic mice with low levels of endocrine IGF-I do not support the growth of the IGF-responsive MCF-7 cells [21]. Even though these animals were supplemented with estradiol, growth was suboptimal in the absence of IGF-I. While it is still possible that IGF-I host effects are required for tumorigenesis, these data support the idea that direct stimulation of the breast cancer cell by IGF-I is required for optimal tumor growth. Thus, one can speculate from these data that the IGFs have important roles in regulating both host and cancer cell functions required for tumor growth.

4. SIGNALING PATHWAYS ACTIVATED BY THE IGFs IN BREAST CANCER CELLS

It has been shown that a variety of signaling pathways can be activated by insulin and IGF-I [22]. Several proteins can directly bind activated IGFR1 and stimulate these downstream pathways. For example, the insulin receptor substrate proteins, IRS-1 and IRS-2, have been identified as key components of both the insulin receptor and IGFR1 signaling cascade. In addition, the SHC adaptor protein may also be directly activated by IGFR1. Several other adapter proteins (crk, grb10, nck, p85 subunit of phosphatidyl inositol 3 kinase) may also directly engage activated IGFR1. Many of these studies were performed in cell lines of varying origin or in model transfection systems. To better understand the activation of post-receptor signaling molecules in breast cancer cells, we utilized the MCF-7 breast cancer cell line. As mentioned, this cell line is exquisitely responsive to the IGFs, thus identification of the endogenous activated signaling molecules in this cell line could identify additional targets for inhibiting IGF action.

We have shown that a 185kDa protein is the predominant species phosphorylated after IGF-I treatment of MCF-7 breast cancer cells [23]. Because this protein migrates at the appropriate position for IRS-1, we used immunoprecipitation to show that a major component of this phosphorylated band was indeed IRS-1. The findings in the MCF-7 cell line are consistent with our studies of primary tumors. As mentioned, levels of IRS-1 expression were associated with patient outcome [13]. Women with small tumors and high IRS-1 levels had inferior survival compared to women with small tumors and low IRS-1. In fact, women with small tumors (<2cm) and high IRS-1 levels had disease-free survivals comparable to women with large tumors. Since a key component of the IGF signaling pathway identifies patients with poor outcome, these data support the idea that IGF-I stimulation could lead to a more aggressive tumor.

The related adaptor protein, IRS-2, migrates with a molecular mass similar to that of IRS-1. We used IRS-2 immunoblots to demonstrate its expression in most breast cancer cell lines and tissues. In most breast cancer cell lines and tissues, IRS-1 and IRS-2 are co-expressed [24]. However, in IGF responsive cell lines (MCF-7, T47-D, ZR75), IRS-1 is the predominant protein species phosphorylated after ligand activation of IRS-1. Thus, we conclude that IRS-1, and not IRS-2, is the predominant signaling molecule activated by IGF-I in IGF-responsive breast cancer cells. In our cells, we do not detect phosphorylation of SHC by IGF-I, however, other investigators have identified this protein to be activated in breast cancer cell lines [25]. While differences in MCF-7 strains could account for these discrepancies in the identified signaling pathways, it is agreed that IGF action in this breast cancer cell line initiates a cascade of intracellular signaling events involved in cell proliferation.

5. IGF AND ESTROGEN RECEPTOR CROSSTALK

The IGFs are mitogens for many normal and cancer cell types. Are there any features of the IGFs that make them particularly important for the growth of human breast cancer cells? Since ER is a key growth regulatory protein for breast cancer cells, and IGF-I has been documented to result in ligand independent ER activation [26], we examined the ability of IGF-I to activate ER in MCF-7 breast cancer cells. We found that IGF-I can activate an estrogen receptor element (ERE) reporter construct in MCF-7 breast cancer cells [27]. While the level of activation was approximately half that induced by estradiol, we saw consistent ligand independent activation that was blocked by anti-estrogens. However, when IGF-I and estradiol were given together, we saw levels of activation higher than those achieved with estradiol alone. These data suggest that IGF-I may function to enhance ER's transactivating functions and thereby increase its growth stimulatory effects. Since IGF-I and estradiol synergistically enhance growth of MCF-7 breast cancer cells, these data provide a potential explanation for this synergism.

It has also been suggested that estradiol can enhance IGF action by increasing levels of IGFR1 [28]. While this could also be a potential mechanism to explain the synergistic effects of estradiol and IGF-I on growth, increasing levels of IGFR1 by transfection does not significantly enhance the growth properties of MCF-7 cells [29,30]. Specifically, they do not become estrogen independent or tumorigenic. However, transfection of MCF-7 cells with an IRS-1 expression construct enhanced their growth properties [31]. These data are consistent with our observation that

IRS-1 appears to be the predominant signaling molecule activated by IGFR1. Taken together, these data suggest that IRS-1 could be the key determinant of IGF action in breast cancer cells by serving as the rate limiting molecule in the IGF signal transduction pathway.

Our initial results showed that estradiol and tamoxifen did not regulate IRS-1 expression or activation over 8 hours in MCF-7 cells [27]. Our experiments demonstrating that IGF-I enhanced liganded and unliganded ER transactivation while tamoxifen inhibited this transactivation were performed over an 8 hour time period. Thus, we concluded that the influence of IGF-I on ER function was not likely to be due to changes in the IGF signal transduction pathway. However, when MCF-7 cells were incubated for longer periods of time with estradiol, we saw that IRS-1 levels were increased over control cells. Furthermore, the levels of tyrosine phosphorylation of IRS-1 were correlated with the increased levels of expression. Thus, one effect of estradiol is to enhance and maintain levels of a central component of the IGF signaling pathway, IRS-1, over time, perhaps allowing the breast cancer cells to remain responsive to IGF-I.

Thus, the data shows that crosstalk between ER and IGFR1 exists. In fact, both pathways serve to reinforce the other. Figure 1 shows how these pathways may potentially interact. IGF treatment of breast cancer cells results in ligand independent activation of ER, but more importantly, enhances ER's transcriptional response to estradiol. Since mitogen activated protein kinase [32] and cyclin D1 [33] may both enhance ER function, these signaling molecules downstream of IGFR1 could be potential links between the two systems. While the genes regulated by ER necessary for breast cancer cell growth have not been precisely defined, key elements of cell cycle control, including cyclin D1, are regulated by estradiol [34]. Our data show that estradiol treatment of MCF-7 cells results in the enhanced expression of IRS-1, a key component of the IGF signaling pathway. We hypothesize that the IGF signaling pathway, like the basic machinery of cell cycle control, may be specifically augmented by estradiol in breast cancer cells. Therefore, IGF's ability to enhance ER function, and vice versa, suggest that the IGFs may have a special role in regulating breast cancer growth.

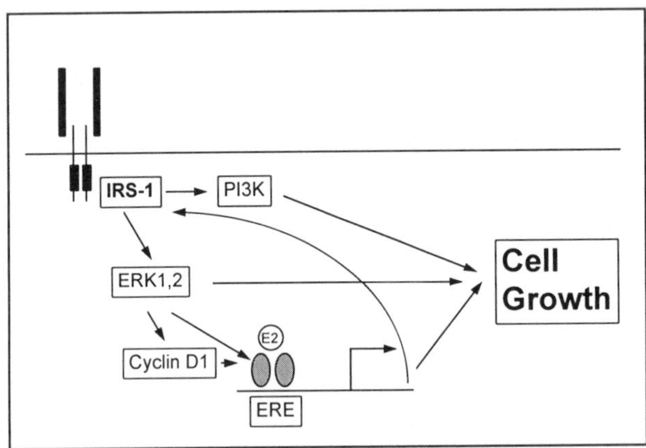

Figure 1 - Crosstalk between estrogen receptor and IGF signaling pathways.

6. SUMMARY

It is clear that the IGFs are potent stimulators of breast cancer growth *in vitro*. Data taken from animal model systems and primary human tumors also suggest that IGFs are growth regulatory in these more complex systems. New data show that IGF signaling pathways and ER function are intertwined in breast cancer cells suggesting a critical role for the IGFs in the growth regulation of breast cancer. Inhibitors of ER action have proven to be of enormous therapeutic benefit in the treatment of breast cancer. While the data have not yet shown that direct inhibition of IGF action on breast cancer cells results in tumor growth inhibition, antagonism of the IGF pathway could also have important therapeutic applications.

REFERENCES

1. Beatson GT: On the treatment of inoperable cases of carcinoma of the mamma. Suggestions for a new method of treatment with illustrative cases. Lancet 2:104-107, 1896
2. McGuire WL, Carbone PP, Vollmer EP: Estrogen Receptors in Human Breast Cancer. Raven Press, New York, 1975,
3. McGuire WL, Clark GM: Prognostic factors and treatment decisions in axillary node-negative breast cancer. N Engl J Med 326:1756-1761, 1992
4. Levenson AS, Jordan VC: MCF-7: The first hormone-responsive breast cancer cell line. Cancer Res 57:3071-3078, 1997
5. Jordan VC: Tamoxifen, A Guide for Clinicans and Patients. PRR, Huntington, NY, 1996,
6. Cullen KJ, Yee D, Sly WS, Perdue J, Hampton B, Lippman ME, Rosen N: Insulin-like growth factor receptor expression and function in human breast cancer. Cancer Res 50:48-53, 1990
7. Foekens JA, Portengen H, vanPutten WLJ, Trapman AM, Reubi J-C, Alexieva-Figusch J, Klijn JGM: Prognostic value of receptor for insulin-like growth factor I, somatostatin, and epidermal growth factor in human breast cancer. Cancer Res 49:7002-7009, 1989
8. Bonneterre J, Peyrat JP, Beuscart R, Demaille A: Prognostic significance of insulin-like growth factor I receptors in human breast cancer. Cancer Res 50:6931-6935, 1990
9. Yee D, Favoni RE, Lippman ME, Powell DR: Identification of insulin-like growth factor binding proteins in breast cancer cells. Breast Cancer Res Treat 18:3-10, 1991
10. Clemmons DR, Camacho-Hubner C, Coronado E, Osborne CK: Insulin-like growth factor binding protein secretion by breast carcinoma cell lines: correlation with estrogen receptor status. Endocrinology 127:2679-2686, 1990
11. McGuire SE, Hilsenbeck SG, Figueroa JA, Jackson JG, Yee D: Detection of insulin-like growth factor binding proteins (IGFBPs) by ligand blotting in breast cancer tissues. Cancer Lett 77:25-32, 1994
12. Yee D, Sharma J, Hilsenbeck SG: Prognostic significance of insulin-like growth factor-binding protein expression in axillary lymph node-negative breast cancer. J Natl Cancer Inst 86:1785-1789, 1994
13. Rocha RL, Hilsenbeck SG, Jackson JG, Van Den Berg CL, Weng C-N, Lee AV, Yee D: Insulin-like growth factor binding protein-3 and insulin receptor substrate-1 in breast cancer: correlation with clinical parameters and disease-free survival. Clin Cancer Res 3:103-109, 1997
14. Papa V, Gliozzo B, Clark GM, McGuire WL, Moore D, Fujita-Yamaguchi Y, Vigneri R, Goldfine ID, Pezzino V: Insulin-like growth factor-I receptors are overexpressed and predict a low risk in human breast cancer. Cancer Res 53:3736-3740, 1993
15. Osborne CK, Bolan G, Monaco ME, Lippman ME: Hormone responsive human breast cancer in long-term tissue culture: effect of insulin. Proc Natl Acad Sci USA 73:4536-4540, 1976
16. Myal Y, Shiu RPC, Bhaumick B, Bala M: Receptor binding and growth-promoting activity of insulin-like growth factors in human breast cancer cells (T-47D) in culture. Cancer Res 44:5486-5490, 1984

17. Belfiore A, Costantino A, Frasca F, Pandini G, Mineo R, Vigneri P, Maddux B, Goldfine ID, Vigneri R: Overexpression of membrane glycoprotein PC-1 in MDA-MB231 breast cancer cells is associated with inhibition of insulin receptor tyrosine kinase activity. Mol Endocrinol 10:1318-1326, 1996
18. Sepp-Lorenzino L, Rosen N, Lebwohl DE: Insulin and insulin-like growth factor signaling are defective in the MDA MB-468 human breast cancer cell line. Cell Growth Differ 5:1077-1083, 1994
19. Arteaga CL, Kitten LJ, Coronado EB, Jacobs S, Kull Jr. FC, Allred DC, Osborne CK: Blockade of the type I somatomedin receptor inhibits growth of human breast cancer cells in athymic mice. J Clin Invest 84:1418-1423, 1989
20. Van Den Berg CL, Stroh C, Hilsenbeck SG, Weng C-N, McDermott MJ, Cox GN, Yee D: Polyetheylene glycol conjugated insulin-like growth factor binding protein inhibits growth of breast cancer in athymic mice. Eur J Cancer 33:1108-1113, 1997
21. Yang XF, Beamer WG, Huynh H, Pollak M: Reduced growth of human breast cancer xenografts in hosts homozygous for the lit mutation. Cancer Res 56:1509-1511, 1996
22. Yenush L, White MF: The IRS-signalling system during insulin and cytokine action. Bioessays 19:491-500, 1997
23. Yee D, Jackson JG, Kozelsky TW, Figueroa JA: Insulin-like growth factor binding protein-1 (IGFBP-1) expression inhibits IGF-I action in MCF-7 breast cancer cells. Cell Growth Differ 5:73-77, 1994
24. Yee D, Gooch JL, Jackson JG: IGF-I, insulin, and IL4 activate IRS1 in human breast cancer cells: differential IRS1 tyrosine phosphorylation by IGF-I is associated with increased MAPK and PI3K activitation. Proc AACR 38:#2910, 1997
25. Nolan MK, Jankowska L, Prisco M, Xu S-Q, Guvakova MA, Surmacz E: Differential roles of IRS-1 and SHC signaling pathways in breast cancer cells. Int J Cancer, in press
26. Aronica SM, Katzenellenbogen BS: Stimulation of estrogen receptor-mediated transcription and alteration in the phosphorylation state of the rat uterine estrogen receptor by estrogen, cyclic adenosine monophosphate, and insulin-like growth factor-I. Mol Endocrinol 7:743-752, 1993
27. Lee AV, Weng CN, Jackson JG, Yee D: Activation of estrogen receptor-mediated gene transcription by IGF-I in human breast cancer cells. J Endocrinol 152:39-47, 1997
28. Stewart AJ, Johnson MD, May FEB, Westley BR: Role of insulin-like growth factors and the type I insulin-like growth factor receptor in the estrogen-stimulated proliferation of human breast cancer cells. J Biol Chem 265:21172-21178, 1990
29. Guvakova MA, Surmacz E: Overexpressed IGF-I receptors reduce estrogen growth requirements, enhance survival, and promote E-cadherin-mediated cell-cell adhesion in human breast cancer cells. Exp Cell Res 231:149-162, 1997
30. Daws MR, Westley BR, May FEB: Paradoxical effects of overexpression of the type I insulin-like growth factor (IGF) receptor on the responsiveness of human breast cancer cells to IGFs and estradiol. Endocrinology 137:1177-1186, 1996
31. Surmacz E, Burgaud J-L: Overexpression of insulin receptor substrate 1 (IRS-1) in the human breast cancer cell line MCF-7 induces loss of estrogen requirements for growth and transformation. Clin Cancer Res 1:1429-1436, 1995
32. Kato S, Endoh H, Masuhiro Y, Kitamoto T, Uchiyama S, Sasaki H, Masushige S, Gotoh Y, Nishida E, Kawashima H, Metzger D, Chambon P: Activation of the estrogen receptor through phosphorylation by mitogen-activated protein kinase. Science 270:1491-1494, 1995

33. Zwijsen RM, Wientjens E, Kompmaker R, van der Sman J, Bernards R, Michalides RJ: CDK-independent activation of estrogen receptor by cyclin D1. Cell 88:405-415, 1997
34. Planas-Silva MD, Weinberg RA: Estrogen-dependent cyclin E-cdk2 activation through p21 redistribution. Mol Cell Biol 17:4059-4069, 1997

Molecular Mechanisms to Regulate the
Activities of Insulin-like Growth Factors
K. Takano, N. Hizuka and S-I. Takahashi (Editors)

rhIGF-I/IGFBP-3 (SomatoKine) Therapy for the Treatment of Osteoporosis

S. Adams, D. Rosen, A. Sommer
Celtrix Pharmaceuticals, 3055 Patrick Henry Drive, Santa Clara, CA 95054

1. INTRODUCTION

Advancing age is associated with significant loss of bone and muscle, which often progress to osteoporosis, and loss of function. Although osteoporosis is defined as a loss of bone mass that increases risk of fracture, there exists compelling evidence to support the assertion that the poor outcomes associated with this disease are also the result of significant losses in muscle mass, or sarcopenia [1]. Therefore, osteoporosis therapies directed at reversing or preventing bone loss alone may not address the full scope of the pharmacological intervention that is indicated. Since rhIGF-I/IGFBP-3 has pharmacological activities in muscle and bone, it represents an intriguing alternative to osteoporosis therapies limited to bone-specific activities. This paper summarizes the preclinical as well as the recent clinical investigations that provide a strong rationale for the use of rhIGF-I/IGFBP-3 for the treatment of osteoporosis.

2. PRECLINICAL EXPERIENCE WITH rhIGF-I/IGFBP-3

Animal models of osteoporosis have provided striking evidence of the therapeutic potential of rhIGF-I/IGFBP-3. The effects on bone and muscle in ovariectomized rats are consistently greater than could be achieved with similar doses of free IGF-I without the binding protein. The most impressive effects on bone and muscle were observed at doses of rhIGF-I/IGFBP-3 that can not be safely administered as free IGF-I.

Although administration of 0.9 mg/kg IGF-I administered to rats for 8 weeks significantly suppressed the bone resorption as a result of ovariectomy, this dose did little to result in an actual improvement in bone formation or effect bone volume. When the dose was increased to 2.6 mg/kg of IGF-I, there is some improvement in bone formation parameters, but bone resorptive surfaces are comparable to the ovariectomized controls. These results were consistently less impressive when compared to rats that received equimolar doses of IGF-I in the form of binary complex rhIGF-I/IGFBP-3 [2-3].

These same rat experiments have clearly demonstrated that in contrast to free IGF-I, the effects of IGF-I/IGFBP-3 are dose dependent, with the best effects seen at the highest dose administered (7.5 mg/kg IGF-I administered bound to IGFBP-3). The equimolar dose of IGF-I represents an intolerable dose. Even a dose of 2.6 mg/kg of free IGF-I administered as a subcutaneous bolus resulted in significant hypoglycemia, whereas doses of rhIGF-I administered as a complex with IGFBP-3, as high as 7.5 mg/kg had no effects on serum glucose.

It has been reported that rhIGF-I/IGFBP-3 administered less frequently (three times/week or as infrequent as daily) also resulted in significant stimulation of bone formation [4]. Even at this early stage of preclinical evaluation, the pharmacokinetic and pharmacological properties of rhIGF-I/IGFBP-3 provides insight that intermittent dosing may be a viable approach to its use as a therapeutic agent for osteoporosis.

3. PHASE I CLINICAL STUDIES WITH rhIGF-I/IGFBP-3

The administration of rhIGF-I/IGFBP-3 in normal volunteers has conclusively demonstrated what had been predicted by preclinical studies in rats and monkeys. rhIGF-I/IGFBP-3 allows the safe administration of IGF-I at doses that would have been lethal with free IGF-I. Clearance of rhIGF-I/IGFBP-3 is also dramatically delayed when compared to free IGF-I, allowing significant improvements in systemic exposure for any given dose of IGF-I.

The first Phase I study was performed where patients received 0.3, 1.0, 3.0 or 6.0 mg/kg of rhIGF-I/IGFBP-3 (equivalent to 60, 200, 600 and 1200 μg IGF-I). Previous FDA recommendations had stated that the maximum safe dose of IGF-I that could be safely administered as an intravenous bolus of as an infusion was 24μg/kg/hour [5]. The safe and well-tolerated administration of 3.0 mg/kg (equivalent to 600 μg/kg IGF-I) represents a safety margin of 25X. Dose-limiting symptomatic hypoglycemia and hypophosphotemia was observed in this study at the 6.0 mg/kg dose (equivalent to 1200 μg/kg IGF-I). There were no other significant observations noted at this dose [6].

This study therefore demonstrates that administration of IGF-I with the binding protein IGFBP-3, allow for the safe administration of IGF-I at doses many-fold greater than those that can be safely administered with free IGF-I. This work provides us with human clinical experience that substantiates the claims suggested in earlier preclinical studies that rhIGF-I/IGFBP-3 provides a far greater margin of safety than free IGF-I.

An additional Phase I study was conducted in healthy women 55-70 years of age, who received rhIGF-I/IGFBP-3 by continuous subcutaneous infusion for 7 days. Subjects received 0, 0.5, 1.0 or 2.0 mg/kg/day (equivalent to 100, 200 and 400 μg/kg IGF-I respectively). Mean circulating levels of IGF-I after administration of 1.0 mg/kg/day was 700 ng/ml at steady state, which was reached at approximately days 2 to 3 after initiation of the infusion. These serum concentrations were sustained throughout the 7 days of administration In this study, rhIGF-I/IGFBP-3 was safely administered at doses up to 2.0 mg/kg/day. Mild headache, fatigue and nausea were seen at the 2.0 mg/kg dose. No evidence of edema was observed. The superior safety profile of 1.0 mg/kg/day in this volunteer study led us to select a dose of 1.0 mg/kg/day for future development efforts [7].

In addition to demonstration of safety, this study provided evidence that rhIGF-I/IGFBP-3 administered by continuous infusion resulted in rapid increases in markers of bone and connective tissue metabolism. Interestingly, it was noted that markers of bone metabolism and collagen synthesis were increased within several days after initiation of administration of rhIGF-I/IGFBP-3, but makers of bone resportion did not rise as quickly. In addition, resorption markers decreased rapidly after withdrawal of rhIGF-I/IGFBP-3, whereas markers of bone formation remained elevated for as much as one week after withdrawal of rhIGF-I/IGFBP-3 [7]. This sustained bone formation activity also suggests that less frequent or intermittent therapy may provide optimal bone formation.

4. DISCUSSION

The role of IGF-I in both bone and muscle metabolism is well established. Various preclinical studies have clearly demonstrated the anabolic activities of IGF-I and have led to its subsequent clinical evaluation in a variety of catabolic states [8-11]. Unfortunately, the use of free IGF-I has met serious limitations in practical application. Not only have side effects been significant and often difficult to control in clinical studies, evidence for significant therapeutic efficacy has often been lacking. The lack of impressive therapeutic efficacy of IGF-I and growth hormone in various catabolic and degenerative states are likely related to the inability to administer sufficient doses of IGF-I. The improved safety profile that has been demonstrated with rhIGF-I/IGFBP-3 allows safe administration of up to 25 fold higher than is considered safe by intravenous bolus injection [6]. Animal studies with rhIGF-I/IGFBP-3 have demonstrated increased anabolic activities in muscle and bone when doses of rhIGF-I/IGFBP-3 are administered which are not possible with free IGF-I [2-3]. Therefore, rhIGF-I/IGFBP-3 represents an opportunity to administer doses that allow entry into therapeutic ranges that have not been possible with free IGF-I.

The pharmacological activity of rhIGF-I/IGFBP-3 on bone and muscle metabolism provides the basis of a strong rationale for its use as a therapeutic agent to treat osteoporosis. Although it is widely recognized that osteoporosis is a disease of bone, a major risk of fracture appears to correlate to the risk of falling. Agents that might enhance muscle strength should result in a decreased risk of falls and thereby effect fracture risk.

Human clinical studies have substantiated a prolonged serum half-life of rhIGF-I/IGFBP-3, and the elevation of bone formation markers for at least one week after withdrawal of treatment. These qualities of rhIGF-I/IGFBP-3 indicate that intermittent dosing regimens may represent a viable therapeutic approach to the treatment for osteoporosis.

6. REFERENCES

1. Marcus, R. Relationship of Age-Related Decreases in Muscle Mass and Strength to Skeletal Status. *J Geront*, Series A, Vol. 50A (Special Issue), 86-87, 1995.

2. Bagi CM, Brommage B, DeLeon L, Adams S, Rosen D, Sommer A. Benefit of Systemically Administered rhIGF-I and rhIGF-I/IGFBP-3 on Cancellous Bone in Ovariectomized Rats. *J Bone Miner Res* 9(8):1301-12, 1994.

3. Bagi CM, DeLeon E, Brommage R, Rosen D, Sommer A. Treatment of Ovariectomized Rats with the Complex of rhIGF-I/IGFBP-3 on Cancellous Bone Mass and Improves Structure in the Femoral Neck. *Calcified Tissue Internat* 57:40-46, 1995.

4. Nurusawa K, Nakamura T, Suzuki K, Matsioka Y, Lee L-J, Tanaka H, Seino Y. The Effects of Recombinant Human Insulin-Like Growth factor (rhIGF)-1 and rhIGF-I/IGF Binding Protein-3 Administration on Rat Osteopenia Induced by Ovariectomy with Concommittant Bilateral Sciatic Neurectomy. J Bone Min Res 10:1853-1864, 1995.

5. Malozowski S, Stadel B. Risks and Benefits of Insulin-Like Growth Factor. *Ann Int Med.* 121:549, 1994.

6. Sanders M, Moore J, Clemmons D, Sommer A, Adams S. Safety, Pharmacokinetics and Biological Effects of Intravenous Administration of rhIGF-I/IGFBP-3 (SomatoKine) to Healthy Subjects. *Endocrine Society 79th Annual Meeting*, Minneapolis OR12-5, 1997.

7. Adams S, Sanders M, Ste. Marie A, Rosen D, Sommer A. Administration of rhIGF-I/IGFBP-3 (SomatoKine) Stimulates Bone Formation in Elderly Women. J Bone Min Res Vol 12, S1, 1997.

8. Jacob R, Barrett E, Plewe G, Fagin KD, Sherwin RS. Acute Effects of Insulin-Like Growth Factor I on Glucose and Amino Acid Metabolism in the Awake Fasted Rat Compared with Insulin. *J Clin Invest* 83(5): 1717-23, 1989

9. Tomas FM, Knowles SE, Owens PC, et. al. Effects of Full-Length and Truncated Insulin-Like Growth Factor-1 on Nitrogen Balance and Muscle Protein Metabolism in Nitrogen-Restricted Rats. *Biochem. J.* 128(1):97-105, 1991.

10. Clemmons DR, Smith BA, Underwood LE. Reversal of Diet-Induced Catabolism by Infusion of Recombinant Insulin-Like Growth Factor-I in humans. *J. Clin. Endocrinol. Metab.* 75(1):234-238, 1992.

11. Hatton J, Rapp R, Kudsk K, et.al. Intravenous Insulin-Like Growth Factor (IGF-I) in Moderate-to-Severe Head Injury: A Phase II Safety & Efficacy Trial. *Neurol Focus* 2 (5) 1997.

Molecular Mechanisms to Regulate the
Activities of Insulin-like Growth Factors
K. Takano, N. Hizuka and S-I. Takahashi (Editors)

Role of insulin-like growth factor-I in gastrointestinal growth and repair

Corinna-B. Steeb[a], Cheryl A. Shoubridge[a], Jasmine Lamb[b], Gordon S. Howarth[a] and Leanna C. Read[a]

[a]*Child Health Research Institute and Cooperative Research Centre for Tissue Growth and Repair, 72 King William Rd. North Adelaide, South Australia 5006, Australia.*

[b]*Queensland Pharmaceutical Institute; Griffith University, Nathan, Queensland 4111, Australia.*

Abstract. The growth promoting action of insulin-like growth factor-I (IGF-I) following systemic infusion of IGF-I and comparable analogues of IGF-I on tissues of the gastrointestinal tract in adult and newborn rats have been well established. More recently, evidence is emerging that orogastrically administered IGF-I stimulates proliferation of the intestinal epithelium and the maturation of digestive enzymes, however species differences may exist. In neonatal rats, IGF-I responsiveness appears to increase with increasing GH-dependency and, as such the time just before weaning seems to be the most IGF-I responsive period. Under gut compromised conditions, such as seen following administration of the chemotherapeutic drug, Methotraxate, the return of normal villus and crypt architecture is accelerated by systemically administered IGF-I at doses between 2-5μg/g/day. Similarly, administration of IGF-I stimulates growth of the intestinal mucosa and returns histopathological features of colonic inflammation towards normal values, in a rat model of colonic damage induced by 2% Dextran Sulphate Sodium. However, potentially harmful effects of IGF-I have been implicated in conditions of chronic bowel inflammation. While most studies provide evidence to encourage the use of IGF-I in growth and mucosal repair, beneficial effects must be clearly dissected from potentially harmful effects of IGF-I if IGF-I is to be considered as a therapeutic agent in the treatment of IBD.

1. Introduction

Insulin-like growth factor-I (IGF-I) is a single polypeptide of about 7.5kDa, structurally homologous to proinsulin. IGF-I induces cell division and differentiation in a variety of tissues and cell types and is secreted from multiple tissues in addition to the liver. It is now accepted that IGF-I acts in an endocrine, paracrine and autocrine manner on a variety of target tissues. IGF-I activity is regulated by IGF binding proteins (IGFBPs) which bind IGFs with higher affinity than the IGF receptors. The recent discovery of several low affinity IGFBPs (IGFBP-7-10), and evidence of IGFBP action independent of IGFs, adds significantly to the complexity of the IGF system and its affects on target tissue [1]. For a comprehensive review of the IGF system see Jones and Clemmons [2].

Over the past years, several research groups have demonstrated that *the gastrointestinal tract is a major target organ* for IGF-I, stimulating growth, maturation and repair of gastrointestinal tissues [3-7]. The expression of IGF-I mRNA in the rat small intestine was first detected by Lund et al. in 1986 [8] and later confirmed in gastrointestinal tissues in a number of other mammalian species [9-10]. Expression of the type 1 IGF receptor has been demonstrated throughout the length of the gastrointestinal tract in several species [11-14]. IGF receptors are present in greater density in the crypt compared to the villus [11], indicating that IGF-I may play a direct role in regulating proliferation of crypt cells. A role for IGFBPs in the gut is demonstrated by the fact that LR^3IGF-I, an IGF-I analogue with a 13 amino acid N-terminal extension and greatly reduced affinities to IGFBPs [15], initiates massive stimulation of gut growth and maturation especially during the suckling period in rats, when infused systemically [16,17]. This leads to speculations about the therapeutic use of IGF-I peptides under conditions of gut compromised conditions, particularly during early postnatal life.

The purpose of this paper is to review the role and mechanism of IGF-I in the regulation of gut growth and maturation in the immature intestine. Furthermore, we examine the role of milk-derived IGF-I on gut growth. The role of IGF-I and their IGFBPs in conditions of gut repair will be discussed.

2. Endocrine role of IGF-I on gut growth and maturation

An endocrine role for IGF-I on the gastrointestinal tract stems from observations that systemically administered IGF-I or LR^3IGF-I stimulates growth and function of the gastrointestinal tract in normal adult rats, rats made catabolic by the treatment of dexamethasone or streptozotocin induced diabetes as well as in gut resected rodents [6,7,16-18]. These studies have shown that continuous infusion of IGF-I or LR^3IGF-I at doses ranging from 0.3 to 12.5μg/g/d improves body weight gain and increases the length and cross-sectional mass of the small bowel, particularly in the proximal intestinal region. We identified that IGF-I peptides increased epithelial proliferation and linear gut growth and improved absorptive function was demonstrated by reduced faecal nitrogen and fat excretion in rats with 70-80% jejuno-ileal resection, treated with IGF-I or LR^3IGF-I [6].

2.1. Systemically administered IGF-I organ growth

We now have extended our studies and examined the effects of systemically administered IGF-I or LR^3IGF-I on growth and maturation of the immature gut. Accordingly, we infused IGF-I peptides to normal suckling rats during two distinct lactation periods. The first infusion period, between days 6.5 and 13 *post partum* (*pp*), represents a time where tissue and organ growth is very rapid whereas the second infusion period, between days 12.5 and 19 *pp*, investigates the growth response closely to weaning, when the suckling rat pups become increasingly growth hormone (GH) dependent and independent from maternal nutrition.

On either day 6 or day 12 *pp*, suckling rat pups were divided into treatment groups receiving 0, 2, 5 or 12.5μg/g/d of IGF-I, or 2 or 5μg/g/d of the more potent analogue, LR³IGF-I. Peptides were delivered via Alzet mini-osmotic pumps (model 1007D), implanted on day 6 or 12 *pp* in the scapular region under 3-3.5% fluothane inhalation anaesthesia. Peptides were delivered for 6 and a half days.

Pump implantation in both age groups was well tolerated and body weight at the start of the infusion period was similar across all treatment groups compared to a groups of age matched untreated pups (data not shown). Plasma IGF-I levels increased dose-dependently following infusion of IGF-I but not LR³IGF-I, suggesting rapid clearance of the analogue from serum. However, despite the marked elevation of plasma IGF-I in the IGF-I treated rat pups, an increase in body weight gain during the early infusion period was only observed in rat pups treated with 5μg/g/d of LR³IGF-I [17]. In contrast, a marked increases in body weight was observed in the older pups treated with 12.5μg/g/d of IGF-I and both doses of LR³IGF-I (Figure 1). As infusion of IGF-I augmented circulating IGF-I levels to a similar extend in both age groups, the increased IGF-I responsiveness observed in the older rat pups does not correlate to plasma IGF-I levels but with the increasing GH dependence that occurs at this time in rodents [19].

Figure 1: Body weights of rat pups treated for 6.5 days with IGF-I or LR³IGF-I starting on day 6 or day 12 *pp*.

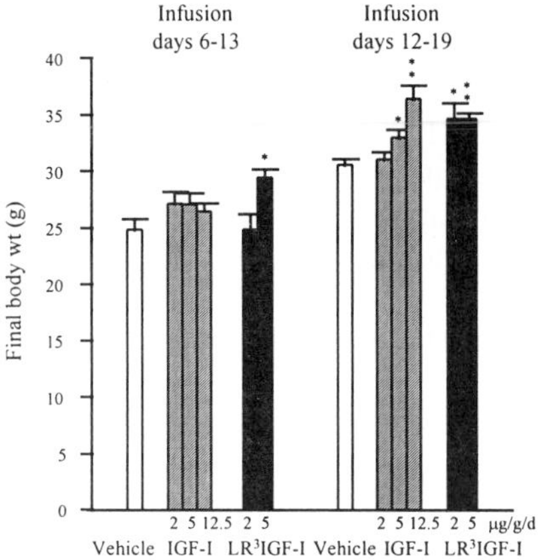

*Values are means ± SE, N=6-9. At each age group, IGF-I or LR³IGF-I groups were compared to the vehicle treated group. * p<0.05, ** p<0.01, determined by ANOVA.*

Figure 2: Weights of small (SI) and large bowel (LI) following treatment with IGF-I, LR³IGF-I or vehicle for 6.5 days.

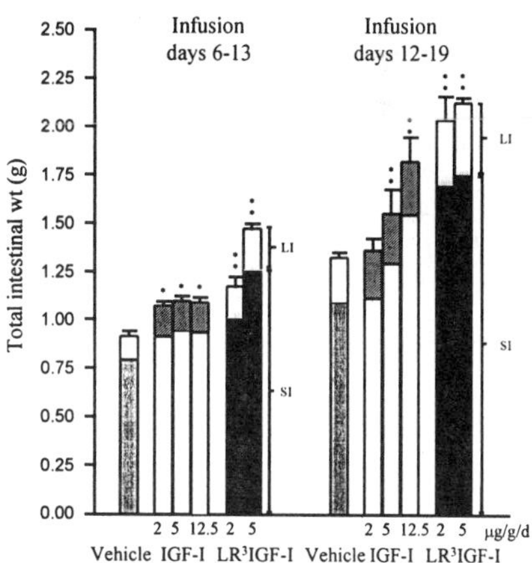

*Values are means ± SE, N=6-9, IGF-I or LR³IGF-I treated rats were compared to vehicle treated rats. *p<0.05, **p<0.01, ANOVA.*

Infusion of IGF-I or LR³IGF-I into suckling rats also selectively increased the wet tissue weight of several visceral organs, most notably the kidney (up to 85%) and the spleen (up to 76%). End-organ effects of IGF-I were observed for both infusion periods. This is in agreement with other studies [20,21].

2.2. IGF-I stimulates mucosal growth

Although the wet tissue weight of all gastrointestinal components increased by 10-18% following infusion of IGF-I, a dose dependent response was not demonstrated during the early infusion period. This is in stark contrast to dose-dependent increases in small bowel weight in the older rats, by 17% and 40% following infusion of 5 or 12.5μg/g/d of IGF-I, respectively. LR³IGF-I was more potent for all growth parameters in both age groups, increasing total gut weight by 60% above control values following administration of the highest dose of LR³IGF-I. Histological observations included lengthening of villi, increased crypt depth and thickening of the muscularis externa. Most importantly however, infusion of IGF-I and LR³IGF-I increased intestinal length by up to 27% and 36% after infusion of the highest dose of LR³IGF-I in the younger and older rat pups, respectively (Table 1). It may therefore be speculated that IGF-I has considerable potential in resection because it increases both intestinal length and cross-sectional mass and hence absorptive capacity of the intestine.

Table 1: Total intestinal length in rat pups treated with or without IGF-I or LR³IGF-I for 6.5 days

Treatments	Infusion days 6-13	Infusion days 12-19
	Length (cm)	Length (cm)
Vehicle	37.8 ± 0.8	40.9 ± 0.8
IGF-I		
2μg/g/d	39.4 ± 1.3	43.4 ± 2.2
5μg/g/d	39.7 ± 1.5	47.4 ± 0.9**
12.5μg/g/d	41.1 ± 0.6**	51.2 ± 1.8**
LR³IGF-I		
2μg/g/d	42.3 ± 1.4**	49.6 ± 1.2**
5μg/g/d	47.1 ± 0.7**	54.3 ± 1.3**

*Values are means ± SE, N=6-9 pups. *p<0.05, **p<0.01, ANOVA.*

The thymidine labelling index in duodenal sections from rat pups in both age groups indicated that the mechanism by which an increased mucosal mass was accomplished could be attributed to a significant increase in epithelial proliferation in IGF-I or LR³IGF-I treated rat pups. Similarly, Zhang et al. [22] have shown that stimulation of villus height, crypt depth and villus surface area by IGF-I treatment can be attributed to enhanced epithelial proliferation.

2.3. IGF-I accelerates the maturation of digestive enzymes

As a measure of intestinal maturation, we also examined the effects of systemic IGF-I administration on the pattern of expression of two major disaccharidases, LPH (lactase-phlorizin-hydrolase) and sucrase-isomaltase (SI). These two enzymes are reciprocally expressed during the early developmental period in rodents, with high activities of LPH and low activities of sucrase during the first 3 weeks of postnatal live. At the time of weaning (3 weeks *pp*) a steady decline of LPH occurs in parallel with increasing expression of SI. The developmental decline of lactase activity and appearance of sucrase expression is well documented and enzyme activities are modulated by a multitude of factors, including nutrients, glucocorticoids, thyroxine, and several growth factors, including EGF and gastrin (reviewed by Henning, Rubin and Shulman [23]).

To investigate if IGF-I induced gut growth is associated with gut maturation, we measured lactase and sucrase activity in jejunal tissue homogenates following the general method by Dahlqvist [24]. To further elucidate the effect on IGF-I on the pattern of disaccharidase expression in jejunal enterocytes we also characterised enzyme activity along the crypt-villus axis by histocytochemical detection of lactase or sucrase in cryostat-sectioned jejunum. We found that disaccharidases activity was modified in the jejunum of the older rat pups infused with the highest dose of LR^3IGF-I. Orogastric LR^3IGF-I appeared to accelerate the natural decline of lactase and increased the expression of sucrase, to levels usually measured at mid-weaning (days 21-24 *pp*). Quantitative expression of disaccharidases by jejunal enterocytes indicated a marked reduction in lactase expression along the entire length of jejunal villi following treatment with the highest dose of LR^3IGF-I. At the same time, sucrase activity, which was relatively low in the rats infused with either vehicle or IGF-I between 12 and 19 days *pp*, was switched on in the LR^3IGF-I treated rats, with increased enzyme activities detected in enterocytes in the mid-villus region. This provides evidence that infusion of LR^3IGF-I stimulates the precocious decline of lactase activity and at the same time initiates the reciprocal switch to sucrase expression.

3. Does orally delivered IGF-I stimulate gut growth and maturation?

It is well established that colostrum and milk contain a variety of hormones and bioactive peptides including IGF-I, which are known to stimulate both cellular growth and differentiation *in vivo* and *in vitro* [25]. In recent times, research activities have focused on aspects of gut growth and maturation in response to orally administered IGF-I to investigate the potential therapeutic use of oral IGF-I preparations. While the effects of oral IGF-I on mucosal growth are less clear, it appears that orogastric IGF-I stimulates digestive enzyme function. There may be, however, some species specific differences that account for the differences reported in the literature.

One of the first studies to investigate orogastric IGF-I effects on the gastrointestinal tract was carried out by Young et al. [26]. Their study showed that orogastric installation of 1μg/d of IGF-I delivered for 6 days to rat pups between days 10 and 16 *pp* significantly increased specific enzyme activity of maltase, lactase, alkaline phosphatase and amino-peptidase in the jejunum. No effects on intestinal growth were reported. In agreement, Xu and colleagues have shown that IGF-I stimulates several brush border enzymes but fails to induce intestinal growth in suckling rats [27]. In contrast, stimulation of intestinal mucosal growth in formulae fed, unsuckled, neonatal piglets have been reported by Burrin et al. [5], following orogastric administration of 3.5μg/g/d for a 4 day period. In their study, orogastric IGF-I increased small intestinal weight, protein and DNA content but not intestinal length. Effects on disaccharidase activity were not examined by these authors.

Recently, Houle et al. [28] have shown that orogastric supplementation of a much lower dose of IGF-I (0.2μg/g/d) significantly increased lactase and sucrase activities in the jejunum and proximal ileum, when added to commercially available piglet formula and administered to colostrum deprived piglets, born by caesarean section. Increased disaccharidase activity in their study was associated with histological parameters (villus height) but no detectable effects on whole body weight gain, organ weight or weight and length of the intestine in response to IGF-I were reported. Similarly, mucosal protein, DNA and RNA content, serum IGF-I/II, IGFBP profiles and intestinal IGFBP mRNA levels were comparable between piglets with or without supplementation of IGF-I in their diet.

3.1 Orogastric IGF-I stimulates disaccharidase expression

Studies were carried out in our own laboratories, to investigate the effects of orogastically administered IGF-I in naturally suckling rat pups. In these studies, supraphysiological doses of IGF-I (12.5μg/g/d) and LR^{3}IGF-I (5μg/g/d) were administered orogastrically by twice daily gavage to 6 or 12 day old rat pups for a period of 6 days. Control rats received saline vehicle, also by gavage. The study design paralleled the design for the systemic infusion study described above for direct comparison. Doses of IGF-I and LR^{3}IGF-I were chosen on the basis of their ability to evoke maximal growth and maturation responses when delivered systemically. We measured whole body weight, weight and length of gastrointestinal components, mucosal DNA and protein content as well as disaccharidase activity. Stability of IGF-I from intestinal or stomach flushings was also determined.

All suckling rat pups tolerated the treatment well. Growth rates between rats receiving twice daily gavage were comparable to the growth rates of normal untreated rats with zero mortality rates for both age groups. We found that orogastric IGF-I or LR^{3}IGF-I, administered to suckling rat pups, failed to induce whole body weight gain or growth of gastrointestinal components. No differences were observed between treatment groups for any of the growth parameters under investigation. Similarly, disaccharidase activity did not differ between treatment groups. We attributed these findings to the rapid degradation of IGF-I in the intestinal lumen. For instance, in flushings taken from the small intestine of 12 day old rat pups, the recovery of immunoprecipitable IGF-I was 35% at 2.5 minutes and no intact IGF-I recoverable at 20 minutes. Although we did not observe any effects of orogastric IGF-I on intestinal growth or maturation, it is feasible that IGF-I may stimulate the maturation of gastric mucosa, because degradation of IGF-I was greatly reduced with up to 86% of ^{125}-I-IGF-I being recovered after 2 hours, in flushings taken from the stomach of the suckling rats.

Taken together, these studies indicate that orogastric IGF-I exerts local effects on the tissues of the intestinal and perhaps the gastric mucosa. Trophic effects of orogastric IGF-I appears to be associated with increased digestive enzyme activity rather then growth. It is not clear at present if the mucosal growth response observed in newborn piglets relates to the markedly greater dosage of IGF-I or to differences in the experimental design.

3.2 Does orogastric IGF-I stimulate systemic growth effects?

There is indirect and direct evidence that milk-born IGF-I may be absorbed intact across the intestinal epithelium in newborn calves, piglets and rats, [14, 29, 30]. For instance, administration of ^{125}I-IGF-I to newborn calves and piglets results in 5-12% of immunoreactive IGF-I in plasma [14,30]. This indicates that in addition to the trophic effects of IGF-I on the intestinal mucosa, transport of IGF-I across the mucosal barrier occurs in newborn animals *in vivo*. However, orally absorbed IGF-I is unlikely to significantly contribute to the circulating IGF-I pool. We have found that unlike systemically administered IGF-I, which increases circulating levels of plasma IGF-I by up to 5 fold, orogastric administration of IGF-I (up to 12.5μg/g/d of IGF-I for 7 days) does not increase the total plasma IGF-I levels. Furthermore, growth effects of peripheral organs are not evident at this supraphysiological dose, suggesting that orogastically derived IGF-I has limited systemic effects [17].

These studies indicate that orally derived IGF-I stimulates digestive enzyme maturation and perhaps mucosal growth. The mechanism by which this occurs need to be elucidated further but it is likely that IGF-I initiates growth and differentiation by direct interaction with the type 1 receptor present on crypt enterocytes. Effects of orogastric IGF-I on mucosal growth are somewhat inconclusive, although it appears that IGF-I effects are limited to tissues of gastrointestinal tract. Possible species specific differences can however not be discounted.

4. IGF-I in gut repair

Further to the action of IGF-I on normal gut growth development, several studies have described IGF-I in conditions of mucosal repair. In a recent review by Lund and Zimmermann, the involvement of IGF-I on intestinal inflammation in IBD has been discussed in detail [31].

Under gut compromised conditions, such as are seen following the administration of the chemotherapeutic drug, Methotraxate, administration of 2-5μg/g/d of IGF-I promotes epithelial repair and returns histopathological features of inflammation to normal values. Similarly administration of IGF-I may be beneficial and act as a protective agent in conditions of ulcerative colitis [4]. In Crohn's disease, an inflammatory condition of the distal intestine and colon, marked by massive mucosal ulceration, mucosal and transmural damage and immune infiltration, surgical removal of bowel tissue is commonly used as treatment strategy following complications. Although resection by itself promotes adaptive mucosal growth, resection studies in animals have clearly shown that administration of IGF-I further stimulates mucosal re-growth and increases the absorptive capacity of the remnant small intestine [6, 7]. Adaptive growth of the small bowel is associated with altered expression of IGF-I and reduced expression of IGFBP-3 mRNA, indicating that IGF-I action in adaptive mucosal growth may be limited by the level of locally expressed IGFBP-3 [32].

Studies from our own laboratories have shown that sucutaneous administration of 296 or 742μg/d of IGF-I for 7 days results in accelerated re-growth of damaged colonic epithelium in an animal model of ulcerative colitis, induced by the oral consumption of dextran sulphate sodium (DSS). Although the effect of IGF-I on epithelial re-growth were minor, IGF-I administration reduced the thickness of the submucosal layer, indicating perhaps, a reduction in immune infiltration initiated by IGF-I.

Potentially harmful effects of IGF-I in IBD need consideration as IGF-I is a potent stimulator of immune cells and cytokines and recent evidence suggests that pro-inflammatory cytokines induce expression of IGF-I [31]. Furthermore, involvement of IGF-I in conditions of chronic inflammation stems from observations that IGF-I expression is markedly increased in animal models representative of chronic bowel inflammation. For instance, in chronic granulomatous inflammation, induced by peptidoglycan polysaccharides, increased expression of IGF-I mRNA is observed in mesenchymal cells of submucosal origin [33]. Localisation of cells showing increased expression of IGF-I demonstrates a fibroblast/smooth muscle-like morphology present at sites associated with increased collagen deposition and fibrosis [33]. Clearly, the beneficial effects of IGF-I in the treatment of chronic inflammation need to be dissected from the possibility of fibrogenic complications associated with IBD and particularly Crohn's, hence the use of IGF-I as a therapeutic agent in IBD needs to be assessed cautiously.

Acknowledgments

This work was supported by a Cooperative Research Centre Grant from the Australian Government. Technical assistance of Kerry Penning, Leanne Srpek, Anna Mercorella and Callum Gillespie is gratefully acknowledged.

References

1. Oh Y, Wilson E, Kim HS, Yang DH, Rutten MJ, Graham DL, Deveney CW, Hwa V and Rosenfeld RG. Proceedings of the Endocrine Society 1997;A267.
2. Jones JI and Clemmons DR. Endocrine Reviews 1995;16:3-34.
3. Lund PK. In: Dockray G and Walsh JH (eds) Gut Peptides: Biochemistry and Physiology: New York: Raven Press, 1994;587-613.

4. Read LC, Howarth GS, Steeb C-B and Lemmey AB. In: Baxter RC, Gluckman PD and Rosenfeld RG (eds) The insulin-like growth factors and their binding proteins. Amsterdam: Elsevier Publishing Co, 1994;409-416.
5. Burrin DG, Wester TJ, Davis TA, Amick S and Heath JP. Am J Physiol 1996;270:R1085-1091.
6. Lemmey AB, Martin AA, Read LC, Tomas FM, Owens PC and Ballard FJ. Am J Physiol 1991;260:E213-E219.
7. Vanderhoof JA, McCusker RH, Clark R, Mohammadpour H, Blackwood DJ, Harty RF and Park JHY. Gastroenterology 1992;102:1949-1956.
8. Lund PK, Moats-Staats BM, Hynes MA, Simmons JG, Jansen M, D'Ercole AJ and Van Wyk JJ. J Biol Chem 1986;261:14539-14544.
9. Lowe WL, Adamo M, Werner H, Roberts CT and LeRoith D. J Clin Invest 1989;84:619-626.
10. Han VK, Hill DJ, Strain AJ, Towle AC, Lauder JM, Underwood LE and D'Ercole AJ. Pediatr Res 1987;22:245-249.
11. Laburthe M, Rouyer-Fessard C and Gammeltoft S. Am J Physiol 1988;254:G457-G462.
12. Schober DA, Simmen FA, Handsell DL and Baumrucker CR. Endocrinology 1990;126:1125-1132.
13. Termanini B, Nardi RV, Finan TM, Parikh I and Korman LY. Gastroenterology 1990;99:51-60.
14. Baumrucker CR, Handsell DL and Blum JW. J Anim Sci 1994;72:428-433.
15. Francis GL, Ross M, Ballard FJ, Milner SJ, Senn C, McNeil KA, Wallace JC, King R and Wells JRE. J Mol Endocrinol 1992;8:213-223.
16. Steeb C-B, Trahair JF, Tomas FT and Read LC. Am J Physiol 1994;266:G1090-1098.
17. Steeb C-B, Shoubridge CA, Tivey DR and Read LC. Am J Physiol 1997;272:G522-G533.
18. Read LC, Tomas FM, Howarth GS, Martin AA, Edson KJ, Gillespie CM, Owens PC and Ballard FJ. J Endocrinol 1992;133:421-431.
19. Glasscock GF and Nicoll CS. Endocrinology 1981;109:176-184.
20. Skottner A, Clark RG, Fyklund L and Robinson ICAF. Endocrinology 1989;124 2519-2526.
21. Philipps AF, Persson B, Hall K, Lake M, Skottner A, Sanengen T and Sara VR. Pediatr Res 1988;23:298-305.
22. Zhang W, Frankel WL, Adamson WT, Roth JA, Mantell MP, Bain A, Ziegler TR, Smith RJ and Rombeau JL. Transplantation 1995;59:755-761.
23. Henning SJ, Rubin D and Shulman RJ. In Johnson LR (ed). Physiology of the gastrointestinal tract. New York: Raven Press, 1994;571-610.
24. Dahlqvist A. Analytical Biochem 1986;22:99-107.
25. Koldovsky O. In: Lebenthal E. (ed). Textbook of Gastroenterology and Nutrition in Infancy. Hormones in milk: Their possible physiological significance for the neonate. New York, Raven Press, 1989;87-119.
26. Young GP, Taranto TM, Jonas HA, Cox AJ, Hoog A and Werther GA. Digestion 1990;46:240-252.
27. Ma L and Xu RJ. Life Siences 1997;61:51-58.
28. Houle VM, Schroeder EA, Olde J and Donovan S. Pediatr Res 1997;42:78-86.
29. Philipps AF, Rao R, Anderson GG, McCracken DM, Lake M and Koldovsky O. Pediatr Res 1995;37:586-592.
30. Donovan SM, Chao JC-J, Zijlstra RT and Olde J. J Pediatr Gastroenterol Nutr 1997;24:174-182.
31. Lund PK and Zimmermann EM. In: Goodlad RA and Wright NA (eds) Clinical Gastroenterology. Cytokines and growth factors in Gastroenterology. Bailliere's, 1996;10:83-96.
32. Albiston AL, Taylor RG, Herington AC, Beveridge DJ and Fuller PJ. Mol Cell Endocrinol 1992:83:17-20.
33. Zimmermann EM, Sartor RB, McCall RD, Pardo M, Bender D and Lund PK. Gastroenterology 1993;105:399-409.

Molecular Mechanisms to Regulate the
Activities of Insulin-like Growth Factors
K. Takano, N. Hizuka and S-I. Takahashi (Editors)

Metabolic actions of insulin-like growth factors

J. Zapf, C. Schmid and E.R. Froesch

Division of Endocrinology and Diabetes, Department of Medicine, University Hospital, CH-8091 Zürich, Switzerland

1. INTRODUCTION

It is now well established that insulin-like growth factor (IGF) I mediates effects of growth hormone (GH) on growth. Evidence comes from both animal experiments and human trials. Thus, exogenously administered IGF I promotes growth and can qualitatively replace GH in hypophysectomized (hypox) rats (1), and it stimulates growth in growth-arrested diabetic rats (2) which lack GH and are unresponsive to GH action, and in patients with the GH insensitivity syndrome (Laron dwarfs) who do not respond to GH due to a GH receptor defect (3).

Because of its structural homology with insulin, IGF I can also mimic insulin actions. This ability, however, is limited by circulating IGF binding proteins (IGFBPs). Therefore, insulin-like effects are observed when the binding capacity of these BPs is exceeded resulting in increased free IGF levels (4), or when the circulating 40 kD IGFBP complex becomes saturated with IGF (5,6,7) resulting in increased IGF bioavailability to insulin-sensitive tissues.

We have used recombinant human (rh) IGF I to explore its insulin-like potential in rats and in humans and to compare the metabolic actions of administered IGF I with those of endogenous IGF I induced by GH. This comparison should answer the question whether IGF I, in addition to mediating GH actions on growth, also mediates those on metabolism. The studies show that exogenous IGF I exerts insulin-like actions depending on the mode and dose of administration, and suggest that not all GH actions on intermediary metabolism are mediated via GH-induced endogenous IGF I.

1.1. Animal studies

1.1.1. Acute metabolic effects of intravenously (iv) injected IGF in normal and hypox rats

Iv injection of 20 μg of IGF I or II into hypox rats causes a decrease of the blood sugar level similar to that achieved with a 6 mU of insulin (8). At the same time, glycogen synthesis in the diaphragm is enhanced. In contrast to insulin, however, IGF does not lower serum FFA levels. In normal rats rendered acutely diabetic by simultaneous injection of anti-insulin serum, IGF I prevents the rise of the blood sugar level and causes a transient suppression of serum FFA. Glycogen synthesis in diaphragm and lipid synthesis from glucose in adipose tissue are less stimulated by IGF than by insulin (fig. 1).

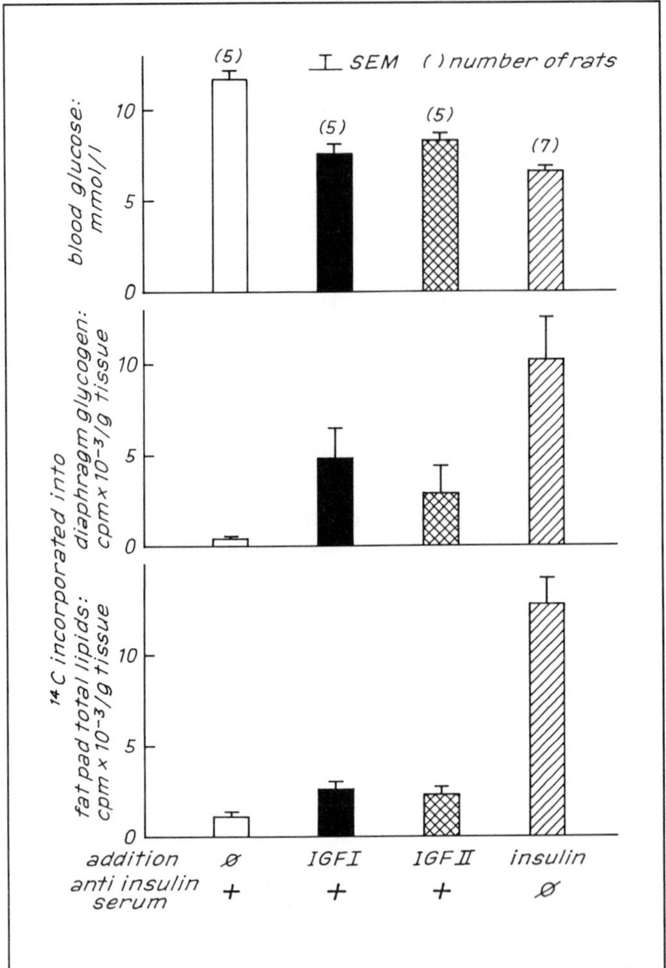

Figure 1. Acute effects of iv injected IGF I, IGF II (20 μg together with U^{14}-glucose and anti-insulin serum) or insulin (6 mU without antiserum) in normal rats 30 min after injection (modified from ref. 8)

Calculating the molar potency ratios between IGF I and insulin from the doses required to achieve similar effects on glucose homeostasis, IGF I is around 50-times less potent than insulin. This agrees with in vitro data obtained in mouse soleus muscle (9) where IGF I stimulates glucose uptake, glycolysis and glycogen synthesis with a 10- to 30-fold lower potency than insulin. It is likely that these effects of IGF I on skeletal muscle are mediated via both the insulin and the type 1 IGF receptor. In adipose tissue, functional type 1 IGF receptors have not been identified (10), and the action of IGF I must therefore be mediated via the insulin receptor (11, 12).

1.1.2. Metabolic effects of IGF I in hypox rats during constant subcutaneous (sc) infusion

When infused sc at a dose of 300 μg/day over 6 days into hypox rats, rhIGF I does not alter serum glucose (although transient hypoglycemia may occur a few hours after starting the infusion; 13) or serum FFA levels (Table 1). The most conspicuous changes are a dramatic fall of the serum urea and creatinine levels and an increase of the inorganic serum phosphorus (P_i). The same effects are observed during sc infusion with 200 mU/day of rhGH. They reflect positive nitrogen balance and growth as well as increased glomerular filtration and phosphate reabsorption by the kidney. IGF I appears to mediate these effects.

Table 1. Effects on serum hormones, ALS and IGFBPs, and on serum metabolites of rhIGF I (300 μg/d) or rhGH (200 mU/d) during a 6 day sc infusion into hypox rats

hormones, ALS and IGFBPs	rhIGF I	GH
IGF I	↑	↑
GH	Ø	↑
acid-labile subunit (ALS)	↓	↑
IGFBP-3	↑	↑
150 kD-complexed IGF I	↓	↑
40 kD-complexed IGF I	↑	↑
free IGF I	Ø	↑
insulin	(↓)	↑
leptin	↓	→
erythropoietin	↑	↑

serum metabolic indices	rhIGF I	GH
glucose	→	→
FFA	→	→
triglycerides	→ (↓)	→
cholesterol	→	→
urea	↓	↓
creatinine	↓	↓
inorg. phosphorus	↑	↑
alk. phosphatase	→ (↓)	→ (↓)
Ca^{2+}	(↓)	(↓)

The results presented in this table are a summary from 3 different infusion experiments.

The distribution of infused rhIGF I and endogenous GH-induced IGF I between the 150 and 40 kD IGFBP serum complexes shows considerable differences between the two treatment groups (Fig. 2): most of the infused IGF I is recovered in the 40 kD complex, and free IGF I is barely detectable. In contrast, endogenous IGF I in the GH-infused animals is recovered in both the 150 and the 40 kD complexes, and a significant amount of IGF I appears in the free form. Therefore, the above metabolic changes reflecting growth during rhIGF I infusion in hypox animals appear to be due to IGF I in the 40 kD complex while the latter as well as free and locally produced IGF I may account for these effects during GH infusion.

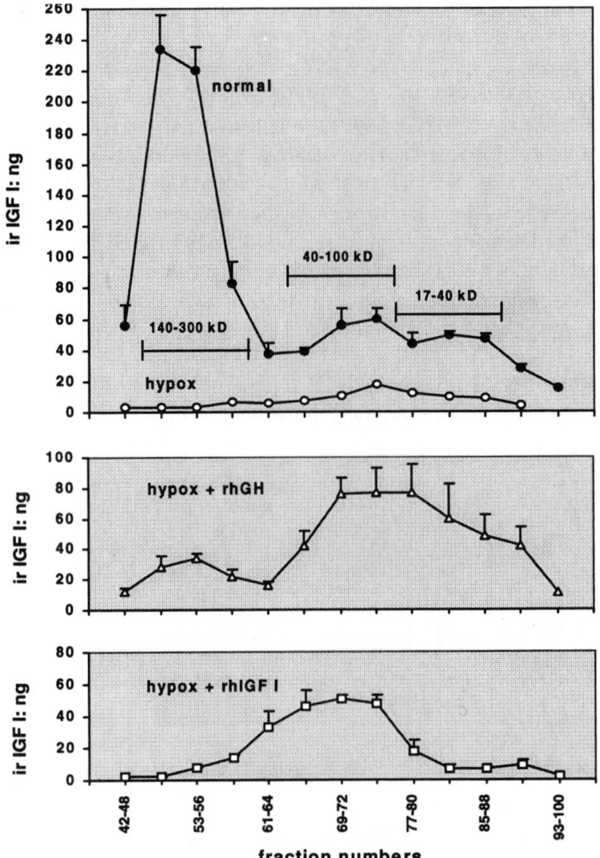

Figure 2. Molecular mass distribution of immunoreactive (ir) IGF I in serum (1 ml) of hypox rats infused for 6 days with saline, 200 mU/d of rhGH or 300 µg/d of rhIGF I, and in serum of untreated age-matched normal rats. Sera were gel-filtered over Sephadex-G200 and IGF I was measured in pooled fractions as described in ref. 13. All points are the means of 3 gel-filtration runs (2 runs for hypox animals). Bars indicate SEMs. ^{125}I-IGF I tracer eluted between fractions 89 and 100 (from Zapf and Froesch, IGF I actions on somatic growth. In: Handbook of Physiology, vol. 5; Hormonal Control of Growth [J.L. Kostyo, editor], in press).

1.1.3. Metabolic effects of high IGF I and IGF II doses (2.5 mg/rat x day) in streptozotocin-diabetic rats

Six days of treatment of streptozotocin-diabetic rats with high IGF I or II doses does not only result in dramatic growth, but also in significant metabolic changes (Zapf et al., in preparation). Surprisingly, however, these high IGF doses have only a weak or no (IGF II)

effect on the blood sugar level as compared to insulin, and the animals still exhibit significant glucosuria. In contrast, serum FFA and triglycerides are reduced to a similar extent as with 2.5 U/rat x day of insulin. It appears that the weak or lacking blood sugar-lowering effect is due to the fact that IGF-stimulated glucose uptake by skeletal muscle is offset by lacking suppression of hepatic glucose production. On the other hand, the infused IGFs appear to inhibit lipolysis and to enhance triglyceride synthesis in adipose tissue whose mass increases during the infusion. The decrease in serum triglyceride levels is probably due to increased lipoprotein lipase and enhanced clearing of triglycerides from the circulation and subsequent resynthesis of the released FFA in adipose tissue.

In contrast to hypox animals, the above IGF infusions caused a much smaller decrease of the serum urea level despite the dramatic effect on weight gain and skeletal growth. Decreased urea excretion due to impaired renal function in the diabetic animals may partly account for this difference.

In the diabetic rat, the infusion of insulin may be compared to GH infusion in hypox animals: insulin raises decreased serum IGF I levels because it restores impaired GH secretion (14) and GH responsiveness of the liver (15) and thereby promotes growth indirectly. Insulin itself has only weak growth promoting activity (16). In contrast, the actions of insulin on intermediary metabolism are direct and probably largely independent of the concomitant rise of IGF I. The observed metabolic effects of the infused IGF I, on the other hand, which occur in the absence of insulin, are due to its insulin-like actions. They usually appear when the binding capacity of IGFBPs in the circulation is overridden, resulting in increased free IGF. However during the IGF II infusion, which causes a similar reduction in serum FFA and triglycerides as IGF I and insulin, free IGF is not elevated. In this situation, all of the infused IGF II is bound to the 150 and 40 kD IGFBP complexes. Since the capillary passage of the 150 kD IGF/IGFBP complex is largely restricted, IGF II in the 40 kD complex must thus be biologically active.

1.2. Metabolic effects of administered IGF I in humans

In 1987, Guler et al. demonstrated that an IGF I bolus given iv to normal subjects rapidly lowered blood glucose and serum FFA levels and caused symptoms of hypoglycemia as observed with an insulin bolus (4). Subsequent studies showed that sc infusions of IGF I over several days in normal or GH-deficient adults caused several metabolic effects which were no longer insulin-like, but opposite to those of insulin, and which partly mimicked and partly opposed GH actions.

1.2.1. Healthy adults

A continuous sc infusion of rhIGF I (20 μg/kg x h) for 6 days in 2 healthy adults suppressed GH, insulin and C-peptide levels without significantly changing blood glucose. The suppression of insulin secretion in the absence of elevated serum glucose levels suggested that rhIGF I infusion enhanced insulin sensitivity and was interpreted as an "insulin sparing effect" of IGF I (17). In subsequent trials, using the euglycemic, hyperinsulinemic clamp Hussain et al. reported that sc infused rhIGF I in normal subjects, indeed enhanced, whereas GH decreased insulin sensitivity (18). The study also showed that exogenous IGF I, like GH, increased energy expenditure and lipid oxidation, and reduced protein oxidation. Energy required for anabolic processes is obviously supplied by enhanced oxidation of FFA to ketone bodies which rose during both IGF I and GH treatment (18).

Beside its effects on intermediary metabolism summarized in table 2, administered IGF I has effects on indices of bone turnover and renal function (table 3). Thus, it increases serum osteocalcin and procollagen type I carboxyterminal propeptide (PICP) levels and enhances the urinary deoxypyridinoline/creatinine ratio (19). These effects are shared with those of GH and indicate that exogenous IGF I rapidly activates bone turnover. At the kidney level, both IGF I and GH increase creatinine clearance, renal plasma flow (20) and calcitriol production (19). In contrast to GH, IGF I does not increase the serum phosphate level nor the maximal tubular phosphate reabsorption divided by the glomerular filtration rate (TmP/GFR) (19). Thus, exogenous IGF I does not appear to mimic the effect of GH on renal phosphate reabsorption in adult humans, which differs from findings in growing rats (21).

Table 2. Comparison between the effects of sc administered rhIGF I and GH in humans

I. Hormones, ALS, IGFBPs		
serum levels	rhIGF I	GH
- GH	↓	↑↑
- IGF I	↑↑	↑↑
- IGF II	↓↓	→
- acid-labile subunit (ALS)	→	↑
- IGFBP-2	↑	↓→
- insulin, C-peptide	↓	↑

II. Carbohydrate, lipid and protein metabolism		
serum levels	rhIGF I	GH
- glucose	→(↓)	↑
- FFA, ketone bodies	↑	↑
- triglycerides (VLDL)	↓	↑
- LDL-cholesterol	→ (↓)	→ (↑)
- Lp (a)	↓	↑
- urea	↓	↓
insulin sensitivity	↑	↓
lipid oxidation	↑	↑
energy expenditure	↑	↑
protein oxidation	↓	↓

1.2.2. Patients with type II diabetes

Sc injections of rhIGF I (2 x 120 μg/kg x day) for 5 days in 8 patients with type II diabetes improved fasting glucose levels, glucose tolerance (22) and lipid profiles (23) at reduced insulin and C-peptide levels compatible with increased insulin sensitivity during treatment. Similarly, Moses et al. observed a 3.4-fold increase in insulin sensitivity and improved glycemic control in patients with type II diabetes during a 6 week treatment with rhIGF I (2 x 100 μg/kg sc bid) (24).

A recent study in adults with insulin-dependent (type I) diabetes mellitus showed that rhIGF I treatment decreased insulin requirements and significantly lowered total cholesterol and triglycerides (25). It was suggested that the reduction of mean overnight GH concentrations by the administered rhIGF I was mainly responsible for the insulin sparing effect in the latter patients and for the improvement of the lipid profile.

Table 3. Comparison between the effects of sc administered rhIGF I and GH in humans

III. Bone turnover and calcium regulating hormones		
index	rhIGF I	GH
serum osteocalcin	↑	↑
serum procollagen I carboxyterminal peptide	↑	↑
urinary deoxypyridinoline/ creatinine ratio	↑	(↑)
serum calcium	→	(↑)
serum PTH	→	→
serum $1,25(OH)_2D_3$	↑	↑

IV. Renal effects/phosphate reabsorption		
index	rhIGF I	GH
- serum urea	↓	↓
- serum creatinine	↓	↓
- creatinine clearance	↑	↑
- renal plasma flow	↑	↑
serum Pi	→	↑
TmP/GFR	→	↑

1.2.3. GH deficient patients

In order to differentiate between the actions of rhIGF I in the absence and presence of GH and to investigate the combined effects of the two hormones, GH-deficient adults were treated for 7 days with rhIGF I (infused sc at a rate of 10 µg/kg x h), with rhGH (daily sc injections of 2 IU/m^2) or with the combination of the two hormones (26). RhIGF I caused the same metabolic effects as in normal subjects (Table 2), suggesting that these effects do not depend on suppressed GH secretion during IGF I treatment. RhIGF I infused on top of GH reduced elevated serum insulin and C-peptide levels and restored insulin sensitivity. The effects on energy expenditure, lipid oxidation, FFA and ketone bodies and on protein oxidation were additive during the combined treatment (26). In summary, infused rhIGF I has similar net effects as GH on lipid and protein metabolism, whereas it opposes GH effects on glucose metabolism.

Despite the fact that infused IGF I mimics the effect of GH on lipid metabolism, their mechanism of action differs: GH activates hormone-sensitive triglyceride lipase by enhancing the responsiveness of this enzyme to lipolytic hormones (27) and by decreasing the sensitivity of adipose tissue to the antilipolytic effect of insulin. In contrast, infused IGF I indirectly enhances lipolysis by inhibiting insulin secretion and thereby releasing the brakes on lipolysis. Exogenous and GH-induced endogenous IGF I therefore do not appear to be equivalent, but the action of endogenous IGF I may be masked by direct GH effects.

The same may be true for the opposite effects of GH (and GH-induced endogenous IGF I) and exogenous IGF I on carbohydrate metabolism and on insulin sensitivity. GH may act directly via GH receptors of skeletal muscle and liver, exogenous IGF I preferentially via type 1 IGF receptors on muscle (but not on liver, due to the absence of functional type 1 IGF receptors). Furthermore, GH administration raises, whereas IGF I suppresses serum insulin levels. Elevated serum insulin and GH levels may downregulate insulin receptors, alter their affinity and impair postreceptor signalling (28). Decreased insulin levels during IGF I infusion may cause the opposite and thus explain increased insulin sensitivity. Decreased insulin levels during IGF I treatment and increased insulin levels during GH treatment also explain the opposite effects of the two hormones on serum triglycerides. VLDL production by the liver is regulated by insulin (29). VLDL production falls when insulin levels decrease and rises when insulin levels increase.

Whether rhIGF I and GH reduce protein oxidation by the same or by different mechanisms is unclear. It is likely that both hormones act directly on skeletal muscle in this respect. GH inhibits proteolysis and stimulates protein synthesis in skeletal muscle. Inhibition of proteolysis and stimulation of protein synthesis has also been reported during IGF I infusion (30,31). GH may act on muscle protein metabolism directly and via IGF I.

2. SUMMARY AND CONCLUSIONS

1. The effects of administered IGF depend on the mode and dose of administration.
2. Iv bolus injections of IGF I in humans and in the rat cause insulin-like metabolic effects (hypoglycemia, decrease of FFA, enhanced glucose utilization). These effects are due to the transient rise of circulation free IGF I.
3. Metabolic effects of constantly sc infused IGF I in hypox rats (decreased urea and creatinine and increased P_i serum levels) essentially reflect positive nitrogen balance and growth as well as increased glomerular filtration and phosphate reabsorption by the kidney and are shared with GH. Indices of carbohydrate and lipid metabolism appear unaffected.
4. Sc infusion of high IGF doses in streptozotocin-diabetic rats relieves growth arrest, causes pronounced effects on lipid metabolism (decrease of serum FFA and triglycerides similar to insulin), but barely affects hyperglycemia (in contrast to insulin).
5. Sc infusion of IGF I in healthy and GH-deficient adults causes metabolic effects which are similar and additive (increased energy expenditure and lipid oxidation, decreased protein oxidation) or opposite to each other (insulin sensitivity). Differences between the mode of action of exogenous and GH-induced endogenous IGF I may be due to direct GH effects which may mask and offset metabolic effects of endogenous IGF I.
6. The improvement of fasting glucose levels, glucose tolerance and lipid profiles in IGF I-treated type I and type II diabetic patients continues to be an attractive supplement to insulin therapy.

REFERENCES

1. E.B. Hunziker, J. Wagner, J. Zapf. J. Clin. Invest., 93 (1994) 1078-1086
2. E. Scheiwiller, H.-P. Guler, J. Merryweather, C. Scandella, W. Maerki, J. Zapf, E.R. Froesch.. Nature 323 (1986) 169-171
3. R.G. Rosenfeld, A.L. Rosenbloom, J. Guevara-Aguirre. Endocr. Rev. 15 (1994) 369-390
4. H.P. Guler, J. Zapf, E.R. Froesch. N. Engl. J. Med. 317 (1987) 137-140
5. W.H. Daughaday, M. Kapadia. Proc. Natl. Acad. Sci. U.S.A. 86 (1989) 6778-6782
6. J. Zapf. T.E.M. 6 (1995) 37-42
7. J. Zapf, Ch. Hauri, E. Futo, M. Hussain, J. Rutishauser, Chr.A. Maack, E.R. Froesch. J. Clin. Invest. 95 (1995) 179-186
8. J. Zapf, C. Hauri, M. Waldvogel, E.R. Froesch. J. Clin. Invest.; 77 (1986) 1768-1775

9. C. Poggi, Y. Le Marchand-Brustel, J. Zapf, E.R. Froesch, P. Freychet. Endocrinology 105 (1979) 723-730
10. J. Massague, M.P. Czech. J. Biol. chem. 257 (1982) 5038-5045
11. GL. King, R. Kahn, M.M. Rechler, S.P. Nissley. J. Clin. Invest. 66 (1980) 130-140
12. J. Zapf, E. Schoenle, M. Waldvogel, I. Sand, E.R. Froesch. Eur. J. Biochem. 113 (1981) 605-609
13. J. Zapf, C. Hauri, M. Waldvogel, E. Futo, H. Häsler, K. Binz, H.P. Guler, C. Schmid , E.R. Froesch. Proc. Natl. Acad. Sci. U.S.A. 86 (1989) 3813-3817
14. L.M.S. Carlsson, R.G. Clark, A. Skottner, I.C.A.F. Robinson. J. Endocr. 122 (1989) 661-670
15. M. Maes, J.M. Ketelslegers, L.E. Underwood. Diabetes 32 (1983) 1060-1069
16. J. Zapf. J. Pediatr. Endocrinol. Metab. 10 (1997) 87-95
17. H.-P. Guler, Chr. Schmid, J. Zapf, E.R. Froesch. Proc. Natl. Acad. Sci. U.S.A.; 86 (1989) 2868-2872
18. M.A. Hussain, O. Schmitz, A. Mengel, A. Keller, J.S. Christiansen, J. Zapf, E.R. Froesch . J. Clin. Invest. 92 (1993) 2249-2256
19. T. Bianda, M.A. Hussain, Y. Glatz, R. Bopuillon, E.R. Froesch, C. Schmid. J. Int. Med. 241 (1997) 143-50
20. H.-P. Guler, K.-U. Eckardt, J. Zapf, C. Bauer, E.R. Froesch. Acta Endocr. (Copenh.) 121 (1989) 101-106
21. J. Caverzasio, C. Montessuit, J.P. Bonjour. Endocrinology 127 (1990) 453-59
22. P.D. Zenobi, S.E. Jaeggi-Groisman, W. Riesen, M. Roder, E.R. Froesch. J. Clin. Invest. 90 (1992) 2234-2241
23. P.D. Zenobi, P. Holzmann, Y. Glatz, W.F. Riesen, E.R. Froesch. Diabetologia 36 (1993) 465-469
24. A.C. Moses, C.J. Simon, L.A. Morrow, M. O'Brian, D.R. Clemmons. Diabetes 45 (1996) 91-100
25. P.V. Carroll, M. Umpleby, G.S. Ward, S. Imuere, E. Alexander, D. Dunger, P.H. Sönksen, D.L. Russell-Jones. Diabetes 46 (1997) 1453-1458.
26. M.A. Hussain, O. Schmitz, A. Mengel, Y. Glatz, J.S. Christiansen, J. Zapf, E.R. Froesch . J. Clin. Invest. 94 (1994) 1126-1133
27. J. Dietz, J. Schwartz. Metab. Clin. Exp. 40 (1991) 800-806
28. M. Muggeo, R.S. Bar, J. Roth, C.R. Kahn, P. Gorden. J. Clin. Endocrinol. Metab. 48 (1979) 17-25
29. G.M. Reavan. Diabetologia 38 (1995) 3-13
30. D.A. Fryburg. Am. J. Physiol. 267 (Endocrinol Metab 30) (1994) E331-E336
31. R.G. Douglas, P.D. Gluckman, K. Ball, B. Breier, J.H. Shaw. J. Clin. Invest. 88 (1991) 614-622

Molecular Mechanisms to Regulate the
Activities of Insulin-like Growth Factors
K. Takano, N. Hizuka and S-I. Takahashi (Editors)

IGF-1 Treatment During Early HIV Infection: Immunological and Hormonal Effects

F. Sattler[a], J. LoPresti[b], M. Dube[a], A.B. Montgomery[c], P. Jardieu[c], C. Spencer[b], M. Saad[b], J.T. Nicoloff[b] and R.G. Clark[d]. [1]

[a]Department of Medicine, Division of Infectious Disease and
[b]Division of Endocrine/Diabetes
University of Southern California School of Medicine, 1300 N. Mission Road, Rand Schrader Clinic, Room 351, Los Angeles, CA 90033, U.S.A.

[c]Genentech Inc., 1 DNA Way, San Francisco, CA 94080, U.S.A.

[d]Research Centre for Developmental Medicine and Biology, School of Medicine, University of Auckland, Private Bag 92019, Auckland, New Zealand.

INTRODUCTION

The immune system and the endocrine system are now recognized to interact at multiple levels. This chapter will focus on the anabolic hormones growth hormone (GH) and insulin-like growth factor-1 (IGF-1) which are now seen as modulators of immune function [1-3]. For example, animal studies show that IGF-1 is important to lymphocyte maturation [4] and function [5]. It has been proposed [3] that endocrine IGF may help restore a damaged immune system, suggesting that IGF-1 may be a useful therapeutic in immunodeficient states.

Infection with the human immunodeficiency virus (HIV) is associated with progressive and profound impairment of immune function and an associated tissue wasting. Ultimately, these derangements result in serious opportunistic infections, neoplastic complications, profound wasting, and finally in death [6]. Therapies are needed to improve immune function through immunomodulation while endocrine-cytokine-metabolic approaches are needed to correct abnormalities that predispose patients to excess energy expenditure, loss of lean body mass (LBM) and wasting. IGF-1 warrants study, because it possesses immune modulator properties, and because it is a second messenger for GH it shares some of its effects on body composition and LBM [7]. The latter is of critical importance since death is inevitable in cachectic cancer and AIDS patients when LBM declines by 30% [6] .

To evaluate the potential of IGF-1 as a therapy for HIV infected patients, we conducted a Phase I/II safety-pharmacokinetic-efficacy study of rhIGF-1 administration to immunodeficient patients with HIV. This investigation also allowed us to evaluate several endocrine systems important to normal anabolic-catabolic processes, in particular the thyroid

[1] This study was supported in part by grants from the United States Public Health Service, National Institutes of Health (DK-49308-03), NCCR GCRC (MO1 RR-43) and Genentech Inc.

hormone axis which, like the GH/IGF-1 axis, has also been implicated in the regulation of lymphoid cell development [8] and which has been shown to be affected by HIV infection [9-11]. Our results confirm that there is a severe hormonal dysregulation early in the course of HIV infection which could predispose patients to a loss of muscle mass and wasting. Treatment with near maximal doses of rhIGF-1, although safe in asymptomatic, immune deficient HIV patients, did not increase peripheral lymphocyte cell counts.

METHODS

Patients

The study population consisted of 11 HIV positive patients (9 men and 2 women) without prior history of AIDS defining conditions and who had CD_4 counts in the range of 250-500/dl, as determined by flow cytometry, over a 6 week period prior to the initiation of the study. The stability of the CD_4 counts was assessed by averaging duplicate test results on 3 occasions over this run-in period and was gauged to be satisfactory if the total counts did not vary by more than 80 cells/dl. No change in antiviral therapy was allowed during the preceding 6 weeks (mono-therapy or combination therapy with any nucleoside reverse transcriptase inhibitor) and subjects had to be willing to continue this treatment regimen for the subsequent 24 consecutive weeks of the study protocol.

A separate healthy non-infected control group was employed for assessing the influence of rhIGF-1 on the thyroid/IGF-1 response to a three week course of rhIGF-1 (0.1 mg/kg/day, subcutaneously). An additional study group of 2 additional HIV positive subjects underwent a voluntary fast for a 5 day period to determine the impact of this catabolic stress on altering thyroid hormone indices.

Study Procedures

The study protocol was approved by the Institutional Review Board (IRB) of the Los Angeles County-University of Southern California Medical Center. A written informed consent was obtained on all study subjects prior to enrolment. The HIV infected subjects were admitted to the General Clinical Research Center (GCRC) to initiate rhIGF-1 therapy (Genentech, Inc.). A 5mg dose was administered twice daily subcutaneously. This specific dose was selected as it represented the maximal tolerable daily total dose of 10 mg, as reflected by the incidence of side effects, which was previously determined in a study of HIV negative control subjects. All subjects remained on the GCRC for the initial 3 days of the study to allow close monitoring for hypoglycemia as well as for collection of blood samples for rhIGF-1 pharmacokinetic analysis. During this period, the study subjects were also taught how to self-administer rhIGF-1. Additionally, patients were instructed regarding the necessity of consuming a nutritionally balanced meal 1 hour prior to rhIGF-1 administration to minimize the risk of hypoglycemia. Study subjects were then followed as outpatients being evaluated every two weeks for new symptoms and signs, changes in body weight, blood urea nitrogen levels, presence of P24 antigenemia and CD_4 lymphocyte counts (lymphocyte phenotyping was done by averaging duplicate values from the same sample). Additional blood samples for routine chemistry (23 channel panels) and complete blood counts were also collected at the same intervals to monitor for other possible toxicities. An electrocardiogram, chest radiograph and urinanalysis were obtained at baseline and at the end of the twelve week treatment period. At the end of the 12 week treatment period, patients were re-admitted to the GCRC prior to the last dose of rhIGF-1 to undergo an additional set of serial blood collections.

Serial timed blood samples (5 min before and 0.5, 1, 1.5, 2, 3, 4, 5, 6, and 10 hr after sc injection) were collected after the first, second, and 168th dose of rhIGF-l. Blood was also collected at 7:55 AM and 10 AM on the morning of the first treatment day 1 (the baseline assessment since the only dose of rhIGF-1 on day 1 was at 6:00 PM), the second treatment day (dose 2), and day 8 for tri-iodothyronine (T3), thyroxine (T4), and thyroid stimulating hormone (TSH) which were measured as previously described [12]. Total IGF-l and IGFBP-3 were measured by methods that have been described previously [13]. Data are means ± SD's, unless otherwise stated.

RESULTS

Compliance and Safety

All 11 HIV patients completed the 12 weeks of therapy with rhIGF-1 and the 12 week follow-up period. The compliance rate was 96% based on evaluation of rhIGF-1 usage. Two subjects had adverse reactions which could be attributed to IGF-1 therapy. One patient complained of sweating on study day 29; later that day his random serum glucose was 54 mg/dL. This patient continued rhIGF-1 therapy without further hypoglycemic episodes. Another patient developed bilateral temperomandibular pain and morning join stiffness of the hands during the second week of therapy. Symptoms persisted during therapy but resolved two weeks after the last rhIGF-1 dose.

Pharmacokinetics of IGF-1 and Concentrations of IGFBP-3

Total serum IGF-1 concentrations in the 11 HIV positive subjects rose briskly after the initial treatment with subcutaneously administered rhIGF-1. The pre-dose and post-dose maximal concentrations were 78±42 and 290±61 ng/ml, respectively ($p<0.05$). By contrast, baseline and peak serum IGF-1 concentrations in 6 HIV negative control subjects given a single 0.07 mg/kg subcutaneous dose of rhIGF-1 were 141±65 ng/ml prior to therapy and 282±66 ng/ml after the dose of IGF-1 (Genentech, Inc., file data). Thus, endogenous basal IGF-1 concentrations in our HIV positive patients were about 50% lower than non-matched HIV negative control subjects. However, total serum IGF-1 concentrations were comparable following rhIGF-1 administration.

In the HIV positive study subjects, the pre-dose and post-dose maximal concentrations after the second dose (210±57 and 422±68 ng/ml, respectively, $p<0.01$), were significantly higher than after the first dose of rhIGF-1. By contrast, the area under the serum concentration time curve (AUC) was significantly ($p<0.05$) lower after the last (2717±2229) dose compared with that after the first dose (5866±2289). Concentrations of IGFBP-3 were higher ($p=0.001$) in 10 of the 11 patients (3.17±0.89 μg/ml) on the second day but were lower ($p=0.01$) at day 84 in nine of the 11 (2.66±0.96 μg/ml) compared to treatment day 2.

Thyroid Function

Prior to treatment with rhIGF-1, peripheral thyroid hormone metabolism as assessed by serum T_3/T_4 ratio values was significantly different in the 11 HIV study subjects as compared to the 5 healthy HIV-negative controls (22.2±1.8 vs. 14.9±0.9, $p<0.01$, respectively). Further, the T_3/T_4

ratio values were unaltered by the 12 weeks of rhIGF-1 therapy (21.9±4.5, p=NS). However, a similar 3 week course of rhIGF-1 therapy in the group of healthy, HIV-negative control subjects showed a significant increase in the T_3/T_4 ratio value (14.9±0.9 to 20.2±1.9, ($p<0.05$), respectively). Prior to IGF-1 administration, the average serum TSH concentration was 1.79±1.4mu/l at 7.55 AM and 1.66±1.22 mu/l at 11:00 AM which were within the normal range. In contrast, on study day 2, serum TSH values were slightly reduced from 1.56±1.10 to 1.23±0.68 (p=0.048), as well as on day 84 when they declined from 1.50±0.71 to 1.17±0.57 mu/l (p=0.003) at a similar clock time.

In the 5 HIV-negative control subjects who underwent a voluntary 5 day fast, an approximate 50% decline in the mean serum T_3/T_4 ratio value was noted (18.7±1.4 to 9.5±1.1, $p<0.01$). In contrast, the serum T_3/T_4 ratio minimally declined in the 2 HIV-positive volunteer subjects (21.9 to 20.4). Strict monitoring of the fast and the presence of urinary ketones confirmed compliance with the fasting regime.

Anabolic, Virologic and Immunologic Measurements

Total body weights and BUN did not change over the course of therapy with rhIGF-l in the 11 HIV positive subjects. No patient developed detectable P24 antigenemia by standard testing and acid dissociation. Monthly CD_4 lymphocyte counts for the 11 HIV subjects are shown in Figure 1. During the 12 weeks of therapy there was a downward trend in CD_4 counts from a mean value of 339 (269-407) at baseline to 287 (179-466) at week 4 and 271 (179-437) at week 8 ($p < 0.02$ by paired t-test). In fact, nine of the 11 patients had decreases in their CD_4 counts by week 8, but by week 12 the mean CD_4 count was 351 (236-486). During the 12 week follow-up period after the patients discontinued rhGF-I, but continued their original antiretroviral therapy, mean CD_4 values remained stable and similar to baseline

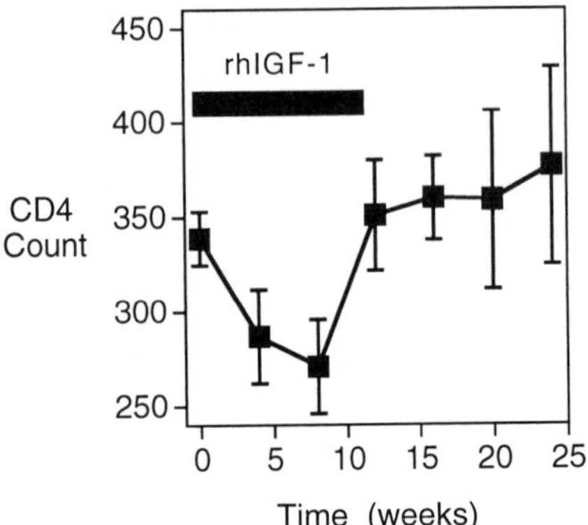

Fig. 1 Eleven patients with HIV (CDC stage A2) were treated with rhIGF-1 (5 mg, sc., bid) for 12 weeks (bar). Blood was drawn and CD_4 positive lymphocytes counted. Means ± SEM.

and week 12 values (Figure 1). Three patients had appreciable increases in their absolute CD_4 counts (increases of 248, 127, 90) compared to baseline. In each of these three cases peak values occurred between weeks 12 and 24 and each had experienced declines in their CD_4 counts during the prior treatment weeks. A fourth patient demonstrated a sharp decline in CD_4 cells at week 8 (280 cells compared to 401 at baseline) but values rose briskly to 525 at week 12 and remained around 400 for the 12 week follow-up period. A fifth patient had a slow but steady increase in CD_4 cells from 269 at baseline to 363 at the end of the 24th week (increase of 94 cells). Despite increases in these five patients, the mean/median values for the 11 study subjects remained very similar during the 12 week follow-up period compared to baseline since five other patients had declines in cells during the 12 week follow-up period (counts were stable in the sixth subject).

DISCUSSION

This study clearly establishes that rhIGF-1 can be administered subcutaneously twice daily for 12 weeks to asymptomatic HIV patients without providing any apparent deleterious effects on CD_4 immunity or HIV activity as assessed by serum P24 antigen. The only adverse effects involved hypoglycemia in one patient and jaw pain with arthralgias in another patient. These are known side effects of treatment with rhIGF-l but were not sufficiently severe to be treatment terminating.

On the other hand, rhIGF-1 therapy did not consistently increase T-lymphocytes and CD_4 counts during therapy. In fact, for the group as a whole, there was a modest decline in CD_4 cells of 52 and 55 cells at weeks 4 and 8, respectively, of rhIGF-1 therapy, although from weeks 12 through 24, counts for the group as a whole returned to baseline. However, five patients had increases of 90 to 248 cells from their nadir values during treatment and that these increases occurred during the post-treatment observation phase in four of the five. Whether this pattern represents the natural history of HIV in these five patients or reflects an increased turnover and production of CD_4 cells induced by IGF-1 is unclear. In bone marrow transplant patients rendered aplastic before transplantation, new lymphocytes do not completely repopulate lymphoid tissue for 8-10 weeks after the infusion of new bone marrow [14]. In an analogous manner, rhIGF-1 may cause a homing effect of mononuclear cells to lymphoid tissue early in the course of therapy resulting in a decrease in the blood pool of lymphocytes before such cells can be differentiated and expanded in the thymus and reticuloendothelial organs.

In a recent study [15], aged rhesus monkeys were infused with recombinant human growth hormone (rhGH) and rhIGF-1. Very different effects were seen in lymphocyte subsets in blood measured by flow cytometry compared to peripheral lymphoid organs. In peripheral blood, the percentage of CD_4 cells and the CD_4/CD_8 ratio fell during rhGH but were normalized by infusing both rhGH and rhIGF-1. In fact, this combination therapy tripled the percent of CD_4 cells and doubled the CD_4/CD_8 ratio in the spleen [15]. These differential effects in various body compartments may be due to the anabolic hormones altering lymphocyte trafficking as rhGH and rhIGF-1 appear to cause the accumulation of lymphocytes in lymphoid organs at the expense of lymphocytes in the circulation [5, 3]. Although lymphocyte numbers in lymphoid organs could not be evaluated in the current study, the decline of CD_4 cells in the peripheral blood of our patients during the 12 weeks of rhIGF-1 therapy is consistent with the more detailed studies performed in monkeys.

This primate data suggests that combined rhGH and rhIGF-1 might prove to be useful in humans in improving immune function, especially following damage to the immune system or in the immune senescence of the elderly. Although administration of growth factors are a standard treatment in humans for restoring hematopoietic cells following radiation, chemotherapy, or bone marrow transplantation, no growth factor therapy is currently available for speeding the recovery of lymphopoiesis. This undoubtedly explains why infections remain a major long-term problem after even using the most successful bone marrow transplantation regimens [16]. Clearly, further *in vivo* and clinical studies will be necessary to clarify these issues and to determine whether rhIGF-1 and/or rhGH may play a useful role in this process.

There was no clinically apparent anabolic effect of rhIGF-1 in the HIV-infected patients as reflected by the absence of weight gain and absence of decreases in blood urea nitrogen. However, this study was not designed to assess changes in anabolism or changes in lean tissue which may have taken place. However, there was some dysregulation in endogenous IGF-1 metabolism as IGF-1 concentrations in our HIV infected subjects were about 50% lower than those in non-matched HIV negative individuals. After the second dose of rhIGF-1 in the HIV study subjects, C_{min} and C_{max} concentrations were significantly higher compared with the first dose. This was likely the result of saturation of the vascular pool of IGFBP-3. In contrast, the AUC after the last dose was significantly less than after the first dose. The lower IGFBP-3 levels on study day 84 compared with baseline could explain the lower AUC of IGF-1 on day 84. The lower IGF- 1 concentrations at the end of study therapy are likely due to IGF-1 suppressing endogenous GH secretion [17] leading to lowered IGFBP-3 levels and ultimately lowered total serum IGF-1 later in the course of therapy. The lower IGFBP-3 concentrations at the end of rhIGF-1 supports this hypothesis. This feedback relationship needs to be considered in studies employing rhIGF-1 as hormonal monotherapy, and has led to the advocacy of treating with the combination of rhGH and rhIGF-1 [7, 18].

Endogenous growth hormone (GH) concentrations were measured and found to be quite low in our HIV study subjects. In fact, concentrations were non-detectable over 18 hours in five of the 11 HIV study subjects prior to study therapy despite use of a highly sensitive assay for GH. Although GH concentrations did not appear to be affected by rhIGF-1 treatment, any interaction may have been missed since specimen collections were not designed to optimally demonstrate normal pulsatile GH physiology. Even with these limitations in study design, it was apparent that this IGF-1 therapy was producing some expected biological actions on the GH/IGF-1 axis. The question then arose whether expected endocrine actions could be detected in some other system? For this purpose, the TSH/thyroid axis was tested.

Abundant evidence currently exists supporting the concept that activation of the GH/IGF-1 axis is associated with a parallel stimulation in the TSH/thyroid axis [19, 20]. In the current study, the 50% increase in the serum T_3/T_4 ratio value after rhIGF-1 therapy seen in the non-HIV control population documents this normal interaction. However, the absence of such a response in our HIV population clearly indicates that such linkage between the GH/IGF-1 and TSH/thyroid system is absent in many HIV infected patients. Further evidence supporting this notion of underlying derangement in the TSH/thyroid axis in HIV infected patients was the observed increase in basal serum T_3/T_4 value [21]. This increase was inappropriate in the setting of chronic infection with HIV, since peripheral conversion of T_4 to T_3, as reflected by the T_3/T_4 ratio, is normally depressed in chronic disease [9]. Moreover, the failure of the serum T_3/T_4 values to fall in 2 HIV study subjects who voluntarily underwent a fast further underscored this finding of a deranged system. Although a slight decrease in serum TSH values was noted, the magnitude of this change was probably of marginal significance.

The unexpected elevation in the T_3/T_4 ratio and depressed concentrations of endogenous IGF-1 prior to treatment with rhIGF-1 in asymptomatic patients with chronic HIV infection and lack of the normal responsiveness to provocative stimuli (rhIGF-1 treatment and starvation) suggests that a major dysregulation exists in the normal IGF-l/thyroid signalling pathway. These derangements also suggest that the signals responsible for the normal genesis of the low T_3 state to preserve energy during chronic disease are either not activated or muted by HIV infection. Moreover, this dysregulation in peripheral thyroid hormone metabolism results in excessive T_3 production from T_4 [9]. It is tempting to speculate that this increased peripheral conversion of T_4 to T_3 is related to the increases in resting energy expenditure that have been demonstrated early in the course of HIV disease predisposing these patients to severe loss of muscle mass and wasting later in their illness [6]. Because wasting contributes appreciably to morbidity and mortality in patients with advanced HIV infections [6], the severe derangements in the anabolic hormones already present during earlier stages of HIV warrants further investigation to define mechanisms which predispose these patients to excessive energy expenditure and weight loss.

These changes in the thyroid hormone axis may impact the activity of the endogenous GH system in patient with HIV, since it is well known that thyroid hormones influence GH sensitivity [22]. The changed thyroid hormone status during HIV infection may contribute to the reduced levels of rhIGF-1. In addition, the effects of treatment with rhIGF-1 and or rhGH may also be affected by a compromised thyroid hormone axis. Indeed, in HIV patients very large doses of rhGH have been found necessary to achieve an anabolic effect [23, 24]. It is possible that the multi-hormonal abnormalities in HIV [11] are associated with a resistance to GH, making such large doses of rhGH necessary. In a recent study [25, 26] low dose rhGH, even when combined with high dose rhIGF-1, lacked anabolic efficacy.

In animals thyroid hormones have been shown to stimulate lymphopoiesis [8] so it is possible that the thyroid status of HIV patients may effect their immunodeficiency. Therefore both the anabolic and the immunologic responses to rhIGF-1 and rhGH may be reduced if thyroid status is impaired. In fact, thyroid function and nutritional status are closely related in HIV-infected patients [27] and a small clinical trial using thyroid hormone replacement produced promising results [10]. Thus, it seems logical to propose that the multi-hormonal derangements in patients with HIV requires multi-hormonal replacement therapy. So it is possible that rhGH and rhIGF-1, particularly when given in combination, may improve immune function and even augment accrual of lean tissue in patients with HIV, especially when levels of other anabolic hormones are corrected to physiologic concentrations or used in combination with other hormonal therapies.

REFERENCES

1. R. Clark and P. Jardieu, In The insulin-like growth factors and their regulatory proteins, R.C. Baxter, P.D. Gluckman PD and R.G. Rosenfeld (eds.), Elsevier, Amsterdam, (1994) pp. 393-400.
2. C.J. Auernhammer and C.J. Strasburger, Eur. J. Endocrinol., 133 (1995) 635.
3. R.G. Clark, Endocr. Rev., 18 (1997) 157.
4. K. Dorshkind, In Handbook of Experimental Immunology, L.A. Herzenberg and D.M. Weir (eds.), Blackwell Science, Cambridge MA, (1996).

5. R. Clark, J. Strasser, S. McCabe, K. Robbins and P. Jardieu, J. Clin. Invest., 92 (1993) 540.
6. G. Babameto and D.P. Kotler, Gastroenterology Clinics, 26 (1997) 393.
7. R. Clark, D.L. Mortensen and L. Carlsson, Endocrine, 3 (1995) 297.
8. E. Montecino-Rodriguez, R.G. Clark, A. Johnson, L. Collins and K. Dorshkind, J. Immunol., 157 (1996) 3334.
9. M. Lambert, Ballieres Clin. Endocr. Metab., 8 (1994) 825.
10. D.M. Derry, Medical Hypotheses, 45 (1995) 121.
11. T.H. Schurmeyer, V. Muller, A. von zur Muhlen and R.E. Schmidt, Eur. J. Med. Res., 2 (1997) 220.
12. C.A. Spencer, S.M.C. Lum, J.F. Wilber, E.M. Kaptein and J.T. Nicoloff, J. Clin. Endocr. Metab., 56 (1983) 883.
13. S.A. Lieberman, G.E. Butterfield, D. Harrison and A.R. Hoffman, J. Clin. Endocrinol. Metab., 78 (1994) 404.
14. C.L. Mackall, F.T. Hakim and R.E. Gress, Immunol. Today, 18 (1997) 245.
15. D. LeRoith, J. Yanowski, E.P. Kaldjian, E.S. Jaffe, T. LeRoith, K. Purdue, B.D. Cooper, R. Pyle and W. Adler, Endocrinology, 137 (1996) 1071.
16. K.M. Sullivan, M. Mori, J. Sanders, M. Siadak, R.P. Witherspoon, C. Anansetti, F.R. Appelbaum, W. Bensinger, R. Bowden, C.D. Buckner, J. Clark, S. Crawford, H.J. Deeg, K. Doney, M. Flowers, J. Hansen, T. Loughran, P. Martin, G. McDonald, M. Pepe, F.B. Petersen, F. Schuening, P. Stewart and R. Storb, Bone Marrow Transplantation, 10 (1992) 127.
17. M.L. Hartman, P.E. Clayton, M.L. Johnson, A. Celniker, A.J. Perlman, K.G. Alberti and M.O. Thorner, J. Clin. Invest., 91 (1993) 2453.
18. S. Kupfer, L. Underwood, R. Baxter and D. Clemmons, J. Clin. Invest., 91 (1993) 391.
19. C. Grunfeld, B.M. Sherman, R.R. Cavalieri, J. Clin. Endocrinol. Metab. 67 (1988) 1111.
20. M.A. Hussain, O. Schmitz, J.O. Jorgensen, J.S. Christiansen, J. Weeke, C. Schmid and E.R. Froesch, Euro. J. Endocrinol., 134 (1996) 563.
21. J.S. LoPresti, J.C. Fried, C.A. Spencer, J.T. Nicoloff Ann. Int. Med., 110 (1989) 970.
22. K.G. Thorngren and L.I. Hansson, Acta Endocrinol., 75 (1974) 669.
23. K. Mulligan, C. Grunfeld, M.K. Hellerstein, R.A. Neese and M. Schambelan, In GHRH, GH and IGF-1: Basic and Clinical Advances, M.R. Blackman, J. Roth, S.M. Harman and J.R. Shapiro (eds.), Springer-Verlag, New York, (1995) pp. 75-91.
24. M.A. Papadakis, D. Grady, D. Black, M.J. Tierney, G.A. Gooding, M. Schambelan and C. Grunfeld, Ann. Intern. Med., 125 (1996) 873.
25. P.D. Lee, J.M. Pivarnik, J.G. Bukar, N. Muurahainen, P.S. Berry, P.R. Skolnik, J.L. Nerad, K.A. Kudsk, L. Jackson, K.J. Ellis and N. Gesundheit, J. Clin. Endocrinol. Metab., 81 (1996) 2968.
26. K. Ellis, P.D. Lee, J. Pivarnik, J.G. Bukar and N. Gesundheit, J. Clin. Endocrinol. Metab., 81 (1996) 3033.
27. W. Ricart-Engel, J.M. Fernandez-Real, F. Gonzalez-Huiz, M. del Pozo, J. Mascaro and F. Garcia-Bragado, Clin. Endocrinol., 44 (1996) 53.

Molecular Mechanisms to Regulate the
Activities of Insulin-like Growth Factors
K. Takano, N. Hizuka and S-I. Takahashi (Editors)

IGF-I Treatment of Growth Hormone Insensitivity

R. G. Rosenfeld

Department of Pediarics, Oregon Health Sciences University
3181 SW Sam Jackson Park Road, Portland, Oregon 97201, United States

1. BACKGROUND

The syndrome of growth hormone insensitivity (GHI) may be divided into primary and secondary forms (1,2). Primary GHI results from a single molecular defect of the GH-IGF axis, and includes molecular abnormalities of the GH receptor (GHR) or GH-GHR signal transduction pathway, or defects of IGF-I, the IGF-I receptor, or IGF-I signal transduction pathway. Secondary GHI may result from GH-inhibiting antibodies, or from malnutrition or a variety of chronic diseases, such as liver disease, renal failure, and diabetes. All of these conditions are characterized, by definition, with end-organ resistance to GH and are incapable, or, at best, only partially capable, of responding to exogenous GH therapy.

2. IGF-I ADMINISTRATION TO NORMAL ADULTS AND PATIENTS WITH GHD

The development of methods for production of recombinant DNA-derived human IGF-I provided a means of bypassing defects in the GH receptor or GH receptor signaling pathway, by allowing direct treatment with IGF. Initial studies involved administration of single iv injections of IGF-I, 100 ug/kg (3). Hypoglycemia was noted within 15 minutes, although it was observed that, on a molar basis, IGF-I had approximately 6% of the glucose-lowering effect of insulin. Subsequent longer-term investigations involved 6 days of iv infusion at a rate of 20 ug/kg/h, resulting in a rise in serum IGF-I concentrations from 150 to 700 ng/ml, without evidence of hypoglycemia (4). The rise in serum IGF-I was sufficient to suppress serum GH concentrations, increase creatinine clearance and decrease plasma urea. A half-life of 12-15 hours was observed for IGF-I as part of the 150 kDa ternary complex. Six normal subjects and two patients with GH deficiency (GHD) were administered 100 ug/kg IGF-I by daily sc injections for 7 days; normal subjects were found to have a 2.5-fold increase in serum IGF-I concentrations (5). Wilton et al. (6) reported that daily sc injections of 40ug/kg to normal subjects produced a steady state IGF-I level 150 ng/ml above baseline, without evidence of symptomatic hypoglycemia.

3. IGF-I ADMNISTRATION IN GHI DUE TO GHRD

3.1. Short-term studies in GHI due to GHRD

As noted above, serum IGF-I was found to have a half-life of 12-15 hours, largely due to the "protective" effect of IGFBP-3 and, particularly, the ternary complex of (IGF)-(IGFBP-3)-(acid-labile subunit, ALS). Since IGFBP-3 and ALS, like IGF-I, are markedly GH-dependent, concentrations of all three peptides are extremely low in patients with GHI due to GH receptor deficiency (GHRD) (This is not the case for patients with GHI resulting from mutations/deletions of the IGF-I gene, who have normal serum concentrations of both IGFBP-3 and ALS). The deficiency of IGFBP-3 and ALS posed practical problems for IGF-I therapy of GHRD for two reasons: 1) low serum concentrations of IGFBPs would increase the potential for rapid rises in free IGF-I, with resulting hypoglycemia, and 2) low concentrations of IGFBPs would enhance the rapid clearance of IGF-I from serum, thereby decreasing the serum half-life of the peptide.

In the first study of administration of IGF-I to patients with GHRD, administration of 75 ug/kg iv bolus to nine individuals 11-33 years of age, following a prolonged fast, resulted in acute symptomatic hypoglycemia (7). Serum glucose concentrations fell 55% within 30 minutes of injection, and hypoglycemia persisted until food was administered at 2 hours. A concomitant rise in serum GH concentrations, presumably the result of hypoglycemia, was observed, but any counter-regulatory actions of GH could not be expected in patients with GHR defects. Subsequent subcutaneous administration of IGF-I, at a dosage of 150 ug/kg/d for 7 days, did not produce hypoglycemia (8). Walker et al. (9) administered IGF-I for 11 days to a 9-year-old child with GHRD, and, although asymptomatic hypoglycemia was observed, significant anabolic effects were also noted.

Six Ecuadorian adults with GHRD were administered sc IGF-I at a dose of 40 ug/kg every 12 hours for 7 days; the frequency of administration was employed in an effort to overcome the shortened half-life of IGF-I resulting from the low serum concentrations of IGFBP-3 and ALS (10). Regular feeding was used and no hypoglycemia was observed, although 2h postprandial serum insulin concentrations were suppressed. The mean integrated 24-hour serum GH concentration was suppressed, as were the number of GH peaks, the area under the serum GH curve, and the degree of clonidine-stimulated GH release. A mean peak serum IGF-I of 253 $\pm$ 11 ng/ml was observed, typically attained 2-6 hours after injection; the mean trough for IGF-I was 137 $\pm$ 8 ng/ml. Serum IGFBP-3 concentrations did not rise with IGF-I treatment, resulting in a short half-life for IGF-I and necessitating the q12-hour administration.

3.2 Long-term studies in GHI due to GHRD

Initial reports suggested that IGF-I treatment of children with GHRD resulted in growth rates comparable to those observed with GH treatment of naive patients with GHD. Laron et al. reported that five GHRD children treated for 3-10 months with single daily sc injections of IGF-I at a dosage of 150 ug/kg/d, has an annualized growth rate of 8.8-13.6 cm/yr (11). Walker et al. (12) observed that the patient previously reported by them (9) had sustained growth on twice daily injections of 120 ug/kg IGF-I, with an acceleration of growth from 6.5 to 11.4 cm/yr during the nine months of treatment. Backeljauw and Underwood (13) described catch-up growth in five GHRD and three GHD type-IA children treated for 3-18 months with twice daily sc injections of 80-120 ug/kg.

Subsequent studies have not been as optimistic about the growth response of GHRD patients to IGF-I (Table 1). In reports from Israel, year-one growth rates have averaged 8.2 cm/yr, with a decline to 6.0 cm/yr during the second year of treatment (14). Backeljauw et al. (15) reported a first year growth rate of 9.3 cm/yr, declining to 6.2 cm/yr during the second year, in five patients with GHRD and three patients with anti-GH antibodies. In a multicenter international study, growth rates in years 1 and 2 were 8.6 and 6.4 cm/yr, respectively (16). Two separate studies conducted in a homogeneous GHRD population in Ecuador (17, 18) were consistent with investigations reported from Israel, North Carolina and Europe (international). All in all, 62 children with GHRD have been treated with IGF-I for a minimum of one year (Table 1).

Table 1
IGF-I treatment of GHI

Study Group (reference)	#	IGF-I Dose (ug/kg)	Growth Velocity (cm/yr) (SD) Pre-Rx	Year 1	Year 2
Israel (14)	9	150-200/d	4.6 (1.3)	8.2 (0.8)	-
	6*	150-200/d	5.0 (1.2)	8.3 (1.0)	6.0 (1.3)
Multicenter Study Group (16)	26	40-120 bid	NA	8.5 (2.1)	-
	18*	40-120 bid	NA	8.6 (1.7)	6.4 (2.2)
N. Carolina (15)	5	80-120 bid	4.2 (0.9)	9.3	6.2
Ecuador (17)	15	120 bid	3.4 (1.4)	8.8 (1.1)	6.4 (1.1)
(18)	7	80 bid	3.0 (1.8)	9.1 (2.2)	5.6 (2.1)

* Subjects included in prior cohort for year 1 data.
NA = not available; bid = twice per day
Adapted from A.L. Rosenbloom, R.G. Rosenfeld, and J. Guevara-Aguirre, Pediatr. Clin. N. Amer. 44 (1997) 423 (reference 19).

The studies from Ecuador are noteworthy for several reasons (17, 18). The initial investigation, using 120 ug/kg bid, employed a control group, which received twice daily injections of placebo for a period of six months. Although a clinically and statistically significant difference in growth rates was found in IGF-I- vs. placebo-treated subjects, 3 of the 9 control subjects accelerated their growth during the 6-months of placebo administration, presumably reflecting the impact of improved nutrition, a phenomenon also noted by Crosnier et al. (20). Interestingly, the incidence of hypoglycemia was not different between the IGF-I and placebo groups. The second study from Ecuador employed a lower dosage of IGF-I (80 ug/kg bid), but the growth rates achieved were similar to those previously reported with 120 ug/kg bid. To date, inadequate data are available on dose-response effects to IGF-I in this group of patients. Furthermore, although pharmacokinetic data support the use of twice daily injections of IGF-I, this has not been tested critically, and, consequently, the optimal scheduling and dosage for IGF-I in these patients have yet to be determined. The failure of serum concentrations of IGFBP-3 or ALS to rise with prolonged IGF-I administration supports, but does not prove, the need for bid administration (21, 22).

Studies from Ecuador also permitted comparison of treatment of naive GHRD patients with IGF-I to treatment of naive, age-controlled GHD patients with GH (18) (Table 2).

Table 2
Comparison of IGF-I treatment of GHRD and GH treatment of GHD

	GHRD	GHD	P
Growth velocity (cm/yr) (SD)			
Year 1	8.9 (1.5)	10.9 (1.6)	<0.0001
Year 2	6.1 (1.5)	8.1 (2.2)	<0.02
Year 3	5.7 (1.4)	8.3 (1.9)	<0.001
Increment of growth velocity above baseline (cm/yr)			
Year 1	5.5 (1.4)	8.8 (2.5)	<0.0001
Year 2	2.9 (1.6)	6.1 (3.1)	<0.01
Year 3	2.9 (1.2)	6.5 (2.4)	<0.005
Change in height SD score*	1.4 (0.6)	2.2 (1.0)	<0.01
Change in height age*	1.7 (0.8)	2.5 (0.7)	<0.01
Change in bone age*	2.0 (1.1)	2.6 (1.3)	NS

*Over 2 years of treatment
Adapted from J. Guevara-Aguirre, et al., J Clin Endocrinol Metab 81 (1997) 629 (reference 18).

These findings suggest that IGF-I, by itself, does not have the full growth-promoting actions of GH. This conclusion must be tempered by the realization that dose-response data for IGF-I treatment of GHI are inadequate, although the Ecuadorian results suggest a plateauing of the response at 80 ug/kg bid (18). The failure of IGF-I to fully duplicate the growth-promoting action of GH has several possible explanations: 1) GH treatment of GHD normalizes serum concentrations of IGFBP-3 and ALS, thereby normalizing the half-life and clearance rate of IGF-I; this effect is not duplicated by IGF-I administration; 2) GH administration may stimulate local production of IGF-I at the growth plate, an effect not reproduced by systemic administration of IGF-I; 3) GH may have important anabolic effects independent of its stimulation of the IGF system. Further studies are clearly necessary, to help tease apart the actions of GH and IGF-I.

4. IGF-I TREATMENT OF OTHER FORMS OF GHI

Experience with long-term use of IGF-I for treatment of GHI has largely been limited to patients with GHRD or with GHD-IA with anti-GH antibodies. Both of these disorders constitute forms of "IGF deficiency," and, accordingly, IGF-I is appropriate therapy. The recent identification of a patient with GHI and IGF deficiency resulting from a partial deletion of the IGF-I gene has provided an unique model for IGF-I therapy (23). The phenotype of this patient differs somewhat from that of classical GHRD, in that, in addition to severe postnatal growth failure, the patient had intrauterine growth retardation, dysmorphic features, a hearing disorder, and mental retardation. Importantly, serum IGFBP-3 and ALS concentrations were normal, as were IGF-I pharmacokinetics. Preliminary results to date indicate a satisfactory growth response to IGF-I therapy.

The use of IGF-I to treat secondary forms of GHI, resulting from chronic disease or catabolic conditions, has been limited. Short-term studies have supported the use of combined IGF-I and GH in calorically-restricted normal volunteers (24). Whether this simply represents the consequences of the higher serum IGF-I concentrations achieved with combined therapy, or indicates advantages of simultaneous elevation of serum levels of IGF-I, IGFBP-3 and ALS, or the anabolic consequences of hyperinsulinemia, remains to be determined.

5. SIDE EFFECTS OF IGF-I THERAPY

To date, side effects associated with IGF-I treatment of GHI have been mild (15, 17). Hypoglycemia, representing the insulin-like effects of IGF-I, may be observed, especially in patients with low serum concentrations of IGFBP-3 and ALS. Clinical hypoglycemia can be avoided, generally, by administering IGF-I together with meals. Other observed side effects include parotid swelling, lymphoid hyperplasia, headache, and papilledema associated with pseudotumor cerebri. Tachycardia may be noted during the early phase of treatment (25). Normalization of craniofacial abnormalities in GHRD has been observed with IGF-I treatment, suggesting the possibility that prolonged therapy with high dosages of IGF-I may result in acromegalic features (26). Theoretical concerns remain about potential tumor-promoting actions of IGF-I, although no such cases have been reported to date.

REFERENCES

1. R.G. Rosenfeld, A.L. Rosenbloom and J. Guevara-Aguirre. Endocr. Rev., 15 (1994) 369.
2. Z. Laron, W. Blum, P. Chatelain, et al. J. Pediatr., 122 (1993) 241.
3. H.P. Guler, C. Schmid, J. Zapf and R. Froesch. Proc. Natl. Acad. Sci. USA, 86 (1989) 2868.
4. H.P. Guler, J. Zapf and E.R. Froesch. N. Engl. J. Med., 317 (1987) 137.
5. K. Takano, N. Hizuka, K. Shizume, K. Asakawa, I. Fukuda and H. Demura. Growth Regul., 1(1991) 23.
6. P. Wilton, A. Sietnicks, R. Gunnarsson, L. Berger and A. Grahnen. Acta Paediatr. Scand. [Suppl.], 377 (1991) 111.
7. Z. Laron, B. Erster, B. Klinger and S. Anin. Lancet, 2 (1988) 1170.
8. Z. Laron, B. Klinger, J.T. Jensen and B. Erster. Clin. Endocrinol., 35 (1991) 145.
9. J.L. Walker, M. Ginalska-Malinowska, T.E. Romer, J.B. Pucilowska and L.E. Underwood. N. Engl. J. Med., 324 (1991) 1483.
10. M.A. Vaccarello, F.B. Diamond, Jr., J. Guevara-Aguirre, et al., J. Clin. Endocrinol. Metab., 77 (1993) 273.
11. Z. Laron, S. Anin, Y. Klipper-Auerbach and B. Klinger. Lancet, 339 (1992) 1258.
12. J. Walker, J.J. Van Wyk and L.E. Underwood. J. Pediatr., 121 (1992) 641.
13. P.F. Backeljauw and L.E. Underwood. Pediatr. Res., 33 (1993) S56 (abstract).
14. B. Klinger and Z. Laron. J. Pediatr. Endocrinol. Metab. 8 (1995) 149.
15. P.F. Backeljauw and L.E. Underwood. J. Clin. Endocrinol. Metab., 81 (1996) 3312.
16. M.B. Ranke, M.O. Savage, P.G. Chatelain, et al. Horm. Res. 44 (1995) 253.
17. J. Guevara-Aguirre, O. Vasconez, V. Martinez, et al. J. Clin. Endocrinol. Metab., 80 (1995) 1393.
18. J. Guevara-Aguirre, A.L. Rosenbloom, O. Vasconez, et al. J. Clin. Endocrinol. Metab. 82 (1997) 629.
19. A.L. Rosenbloom, R.G. Rosenfeld and J. Guevara-Aguirre. Pediatr. Clin. North Amer., 44 (1997) 423.
20. H. Crosnier, M. Gourmelen, C. Prevot, et al. J. Clin. Endocrinol. Metab. 76 (1993) 248.
21. K.F. Wilson, P.J. Fielder, J. Guevara-Aguirre, et al. Clin. Endocrinol. 42 (1995) 399.
22. S. Gargosky, K. Wilson, P. Fielder, et al. J. Clin. Endocrinol. Metab. 77 (1993) 1683.
23. K.A. Woods, C. Camacho-Hubner, M.O. Savage and A.J.L. Clark. N. Engl. J. Med., 335 (1996) 1363.
24. S.R. Kupfer, L.E. Underwood, R.C. Baxter and D.R. Clemmons. J. Clin. Invest. 91 (1993) 391.
25. O. Vasconez, V. Martinez, A. Martinez, et al. Acta Paediatr. [Suppl.], 399 (1994) 137.
26. J. Leonard, M. Samuels, A.M. Cotterill and M.O. Savage. Acta Paediatr. [Suppl.] 399 (1994)140.

Molecular Mechanisms to Regulate the
Activities of Insulin-like Growth Factors
K. Takano, N. Hizuka and S-I. Takahashi (Editors)

IGF-I therapy for patients with extreme insulin resistance syndromes in Japan

M. Kasuga[a] and Extreme Insulin Resistance Syndromes Research Group*

[a]The Second Department of Internal Medicine, Kobe University School of Medicine, Kobe

1. INTRODUCTION

Extreme insulin resistance syndromes (EIRS) are pathologically defined as primary defects in insulin actions at the receptor or postreceptor sites. The condition is commonly associated with hyperinsulinemia and ovarian masculinization. Despite a marked increase of insulin secretion, some patients develop frank diabetes mellitus that does not respond adequately to insulin therapy.

Insulin-like growth factor-I (IGF-I) exerts metabolic activities similar to those of insulin probably through both insulin and IGF-I receptors. Recently, human IGF-I produced by recombinant DNA technology has been available in large quantities and been getting more attentions on its clinical potentials for many sorts of diseases, especially those of endocrinological and metabolic abnormalities. The therapeutic potentials of IGF-I for EIRS had been evaluated in small number of patients. The results obtained after one-year treatment suggested that recombinant human IGF-I could be used clinically as a hypoglycemic agent in EIRS patients in whom an excess amount of insulin was ineffective [1, 2].

* The members of EIRS Research Group: H.Kuzuya; Diabetes Center, Kyoto National Hospital, Kyoto / M.Kobayashi; The first Department of Internal Medicine, Faculty of Medicine, Toyama Medical and Pharmaceutical University,Toyama / T. Kadowaki; The Third Department of Medicine, Faculty of Medicine,University of Tokyo, Tokyo / H. Makino; The Department of Clinical Laboratories, Ehime University School of Medicine, Ehime / N. Matsuura; The Department of Pediatrics, School of Medicine, Kitazato University, Kanagawa / Y. Sakamoto; The Third Department of Internal Medicine,Teikyo University School of Medicine,Chiba / M.Okuyama; Inazawa Municipal Hospital , Aichi / S. Nagataki; Radiation Effects Research Foundation , Hiroshima / H.Imura; Kyoto University, Kyoto.

We organized the EIRS Research Group just after the time when the first IGF-I product was launched on February, 1995 in Japan for promoting reasonable usage of IGF-I for EIRS.

We describe here the current status of EIRS therapy with IGF-I in Japan.

2. DOSAGE AND ADMINISTRATION

IGF-I was manufactured by Fujisawa Pharmaceutical Co., Ltd., Osaka Japan using a recombinant DNA technology and had an identical amino acid sequence to that of endogenous human IGF-I. The product is provided in a form of 10mg lyophilized powder in a vial for subcutaneous injection. For use, the product is reconstituted with 1 ml of isotonic sodium chloride solution.

3. DIAGNOSTIC CRITERIA

EIRS Research Group generated the diagnostic criteria for Type A and Type B insulin resistance syndromes (A-IRS and B-IRS), congenital generalized lipodystrophy (CGL) and Leprechaunism (Lep) .The criteria for A-IRS required any evidences of insulin resistance proven by hyperinsulinemia or decreased response of blood glucose to insulin, acanthosis nigricans and/or masculinization (hirsutism , polycystic ovary etc.), and abnormalities in insulin receptor verified at molecular or genetic levels. On the other hand, the diagnosis for B-IRS was based on the existence of anti-insulin receptor antibodies in the serum. CGL and Lep were based on the characteristics of physical signs and endocrinological features of these diseases.

4. CHARACTERISTICS OF PATIENTS WITH EIRS (Table 1)

Twenty patients have been treated with IGF-I so far, including 8 of CGL , 5 of A-IRS, 5 of Lep and 2 of B-IRS, in which seven are males and thirteen are females , and their ages from 1 to 44. Nine out of 18 patients (50%) had markedly elevated blood glucose and were uncontrollable by insulin in various severity levels. Insulin concentrations also significantly increased in about half patients. Acanthosis nigricans appeared in 13 out of 17 patients (76%) and other associated abnormal features were observed in some patients including hirsutism, primary amenorrhea, clitomegaly or hypergenitalism, osteo/odontodysplasia, arthrogryposis, splenohepatomegaly as well as the common characteristic feature of face in Lep and the lack of fat tissue in CGL.

Table 1

Characteristics of Patients with EIRS

Patients		Sex	Age (yr)	Clinical Features before Treatment				
				Glucose Conc.	Insulin Conc.	Insulin Resistance	Acanthosis Nigricans	Other Associated Features
A-IRS	1	M	40	High *	High **	Severe	Present	Hirsutism, Short length of fingers
	2	F	11	Normal	High	Moderate	Present	
	3	F	16	High	High	Moderate	Absent	Primary amenorrhea
	4	F	16	High	High	Moderate	Present	
	5	M	13	High	High	Severe	Present	Hirsutism(mild)
B-IRS	1	F	19	High	Normal	Severe	N.D. ***	
	2	F	44	High	High	Severe	Present	Hirsutism
Lep	1	F	1	N.D.	N.D.	N.D.	N.D.	
	2	M	2	Normal	High	Severe	Present	Hirsutism, Hypergenitalism
	3	F	7	Normal	High	Severe	Present	Hirsutism, Odontodysplasia
	4	M	1	N.D.	N.D.	N.D.	N.D.	
	5	F	44	Normal	Normal	Mild	Absent	Osteo/Odontodysplasia
CGL	1	F	14	High	High	Severe	Present	Hirsutism, Clitomegaly, Odontodysplasia
	2	M	16	High	High	Severe	Present	
	3	F	6	Marginal	High	Mild	Absent	Hirsutism, Clitomegaly, Splenohepatomegaly, Odontodysplasia
	4	F	5	Marginal	Normal	Severe	Absent	Hirsutism, Clitomegaly, Splenohepatomegaly
	5	M	36	Marginal	Marginal	Severe	Present	hypergenitalism
	6	F	12	High	Marginal	Moderate	Present	Hirsutism(detectable)
	7	M	9	Marginal	High	Moderate	Present	Delayed Intelligence(mild)
	8	F	17	Normal	Marginal	Moderate	Present (Pigmentation)	Osteodysplasea, Arthrogryposis

* These data present the status of fasting blood glucose levels in patients except those of postprandial in Leprechaunism , and represent as follows ; more than 140 and 200 mg/dl (High), 110 - 140 mg/dl and 120-200 mg/dl (Marginal), and less than 110 and 120 mg/dl (Normal) in fasting and postprandial conditions, respectively.

** These data present the status of fasting serum insulin levels in patients except those of postprandial in Leprechaunism, and represent as follows; more than 30 and 100 μU/ml (High), 13-30 and 30-100 μU/ml (Marginal), and less than 13 and 30 μU/ml (Normal) in fasting and postprandial conditions, respectively.

*** N.D.: Data are not present.

5. EFFICACY AND SAFETY OF IGF-I (Table 2)

In the therapies, IGF-I was subcutaneously administered to these patients in a dose of 0.01~0.4mg/kg once or twice a day. The maximum period of therapies reached over 7

Table 2

Efficacy and Safety of IGF-I

Patients		Dosage (mg/kg x (times))	Treatment Period (yr;mon)	Clinical Efficacy of IGF-I Treatment				Side Effect
				Glucose Conc.	Insulin Conc.	Acanthosis Nigricans	Hirsutism	
A-IRS	1	0.06-0.13 x(2)	2;6	↓↓*	↓↓*	x **	x**	Not Observed
	2	0.04-0.1 x(1-2)	5;7	→	→	○	–	Not Observed
	3	0.1-0.2 x(1)	2;5	↓↓	↓↓↓	–	–	Not Observed
	4	0.27-0.54 x(1-2)	0;3	→	→	x	–	Pain at Inj. Site
	5	0.1-0.2 x(2)	6;4	↓↓	→	◎	x	Hypoglycemia
B-IRS	1	0.02-0.16 x(1)	S***	↓↓	↓↓	N.D.	–	Headache, Nausea
	2	0.19 x(2)	S	↓↓↓	→	x	x	Not Observed
Lep	1	N.D.	N.D.	N.D.	N.D.	N.D.	N.D.	
	2	0.05-0.1 x(2)	5;9	→	→	○	x	Not Observed
	3	0.1 x(2)	6;11	↓	↓↓↓	○	○	Not Observed
	4	N.D.	N.D.	N.D.	N.D.	N.D.	N.D.	
	5	0.12 x(1)	S	↓↓	↓↓	–	–	Not Observed
CGL	1	0.1-0.4 x(1-2)	2;7	↓↓↓	↓↓↓	◎	◎	Edema, Redness and Heat Sensation at Inj.Site, Liver Cell Adenoma
	2	0.1-0.2 x(2)	0;5	↓↓	↓↓↓	○	–	Pain and Swelling at Inj. Site, Eosinophilia, Thrombocytopenia
	3	0.15 x(1)	2;4	↓↓	→	–	○	Not Observed
	4	0.17 x(1)	1;4	↓	↓	–	x	Not Observed
	5	0.01-0.05 x(1)	0;3	?	→	x	–	Swelling at Inj.Site, Palpitation, clammy Sweat
	6	0.05-0.12 x(2)	3;11	↓↓	→	○	○	Not Observed
	7	0.05-0.1 x(2)	2;2	↓	↓↓	○	–	Not Observed
	8	0.17 x(1)	2;1	→	→	◎	–	Unconsciousness

* : These symbols represent as follows , remarkable decrease(↓ ↓ ↓), signifi cant decrease(↓ ↓), mild or moderate decrease(↓), no change or deterioration(→), unable of judgement(?) and no data(N.D.), respectively.

** : These symbols represent as follows, remarkable improvement(◎), significant improvement(○), no change(X), absence of abnormalities(—) and no data(N.D.), respectively.

***: Single IGF-I was only administered.

years. Changes of diabetic control were periodically assessed by fasting and postprandial blood glucose, HbA1c, fructosamine, insulin and C-peptide as well as usual laboratory tests .

Among the 20 patients, thirteen are maintained IGF-I therapy at present, in which blood glucose and/or insulin markedly decreased, and acanthosis nigricans or hirsutism apparently came to ameliorate in some, but not all of patients.

On the other hand, eight patients have been dropped out mainly due to avert the pains or swellings accompanied by injections and unfavorable episodes such as edema, anemia, headache and nausea regardless of drug-related or not.

6. DISCUSSION

The results obtained from our two years survey supportably suggest that IGF-I could be clinically useful as a hypoglycemic agent in patients with EIRS, nevertheless global evaluation of efficacy needs more experiences with larger number of patients and longer period of treatment. In addition, IGF-I also exerted marked blood insulin lowering effect in diabetic and nondiabetic conditions (e.g. leprechaunism), which would be of great value to prevent any adverse effects of insulin derived from long-sustained hyperinsulinemia and maybe also to improve insulin sensitivity.

On the other hand, severe episodes related to IGF-I have not been observed so far. However, because of stimulating cell growth in a variety of tissues, IGF-I should be used carefully with adequate monitoring , especially during a long -term treatment of EIRS.

7. REFERENCES

1. H.Kuzuya , N. Matsuura, M. Sakamoto, H. Makino, Y. Sakamoto, T. Kadowaki, Y.Suzuki, M. Kobayashi, Y. Akazawa, M. Nomura, Y. Yoshimasa, M. Kasuga, K. Goji, S. Nagataki, H. Oyasu, and H. Imura, Diabetes, 42, 696 (1993).
2. H.Kuzuya, Clin. Pediatr. Endocrinol., 3(Suppl 5), 127 (1994).

Molecular Mechanisms to Regulate the
Activities of Insulin-like Growth Factors
K. Takano, N. Hizuka and S-I. Takahashi (Editors)

Insulin-like Growth Factor-I (IGF-I): Therapeutic Potential in Neuromuscular and Other Neurological Diseases

J.M. Farah, Jr.
Cephalon, Inc., 145 Brandywine Parkway, West Chester, PA 19380, USA

1. INTRODUCTION

The neurotrophic properties of the IGFs are among the pleiotropic effects of these polypeptides which have been discussed during this symposium by Doctors Calissano, Gluckman, Powell-Braxton and others, and reviewed recently (Dore et al, 1997; Feldman et al, 1997). Based upon experimental evidence in vitro and in vivo, a general confidence exists that the IGF systems are involved not only in the development of the nervous system but also in the response of the nervous system and its associated target organs to injury and stress (Logan et al, 1995). The purpose of my presentation is to consider with you the preclinical evidence which supported a rationale for clinical evaluation of recombinant human IGF-I (rhIGF-I, mecasermin) in amyotrophic lateral sclerosis (ALS) and the initiation of clinical studies in other neurological disorders by Cephalon and with our corporate partners at Chiron Corporation in Emeryville, California and at Kyowa Hakko Kogyo here in Tokyo. As implied in the title of my presentation, I will focus on the results of two pivotal clinical trials of IGF-I in ALS and attempt to provide you some perspective on how we and others interpret the outcomes. At the outset, it is important to note that my ability to offer this perspective on the potential neurological utility of rhIGF-I is predicated upon the input and diligence of my colleagues at Cephalon, at Chiron and at Kyowa, of our dedicated collaborators in the scientific and neurology community and of those unselfish patients who participate in clinical studies of IGF-I in the hope of offering therapeutic alternatives for others suffering from these terribly debilitating diseases.

When we consider the potential therapeutic uses of IGF-I in nervous system disease, our corporate efforts have been directed to clinical conditions in which a) scientific rationale for use of rhIGF-I was sound, b) rhIGF-I could be delivered to diseased elements of the neuraxis at what might be safe and tolerated dose levels, and c) practical trial designs could be implemented that offered meaningful clinical outcomes. Conditions which fulfill such criteria include diseases affecting motor, sensory and/or associative neuronal functions of the central and peripheral nervous systems. I will review certain examples in which rhIGF-I is under evaluation in ongoing clinical initiatives.

2. NEUROMUSCULAR DISEASES

2.1. Preclinical Rationale

In order to perform integrative and coordinating functions within the organism, neurons communicate with other neurons, with neuroglia, and with cells of target tissues such as skeletal muscle fibers and sensory end organs. IGF-I has been shown to be trophic for each of these elements of nervous system function (see Dore et al, 1997). IGF-I promotes the survival (Neff et al, 1993), axonal regeneration (Nachemson et al, 1990; Near et al, 1992), and terminal neuritic sprouting (Ang et al, 1993) of neurons, with an abundance of evidence accumulated in cholinergic motor neurons (see Lewis et al, 1993). A unifying feature of these beneficial effects is that, regardless of the experimental model, whenever a neuron was placed at risk of dying or loss of function, IGF-I has been shown to promote the survival and maintain the phenotype of the cell. For example, IGF-I has been shown to enhance collateral sprouting and consolidation of neuromuscular junctions (Caroni & Grande, 1990). These are crucial elements of a motor neuron's response to injury of the motor unit (a single motor neuron together with the skeletal muscle fibers which it innervates). The dual effects of IGF-I as promoter of both neuritic sprouting and stabilization of neuromuscular junctions offers a unique opportunity to enhance compensatory responses of the surviving motor neurons in disorders like ALS, spinal muscular atrophy and post-polio syndrome (see Jubelt & Drucker, 1993). Another characteristic of IGF-I is that the factor not only promotes differentiation of myoblasts but also increases protein synthesis and decreases protein catabolism in skeletal muscle (Tomas et al, 1993). These net anabolic effects of IGF-I in skeletal muscle indicate that exogenous IGF-I may retard the atrophy and functional loss which characteristically occurs secondary to denervation in neuromuscular disorders as seen in the Wobbler mouse (Hantai et al, 1995). Together with the neurotrophic features of this molecule, IGF-I appears uniquely capable of benefiting all elements of the motor unit as represented in Figure 1.

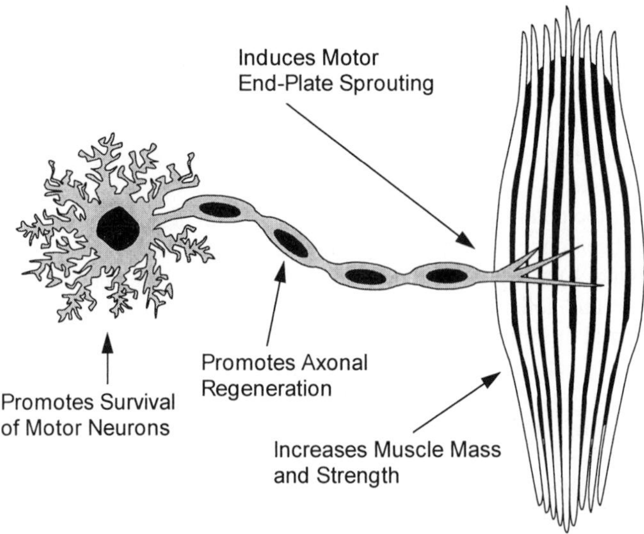

Figure 1. Schematic of lower motor neuron, ensheathing Schwann cells and innervated skeletal muscle fiber

Therefore, IGF-I seemed an appropriate candidate for treating neuromuscular disorders in which one or more elements of the motor unit were affected. In ALS patient autopsy specimens, IGF binding sites appear to be upregulated throughout the spinal cord, with significant increases in the cervical, thoracic and sacral ventral horn (Adem et al., 1994; Dore et al, 1996). Since this increase in IGF-I binding occurs without notable changes in either tissue levels of IGFs or circulating IGF-I (see Dore et al, 1996), a systemic deficiency in endogenous IGF-I is unlikely to account for the alteration in putative IGF receptor sites or to contribute to disease pathogenesis. Rather, as in many injured tissues, elements of the IGF system may be upregulated in ALS as part of an endogenous response to the disease condition. In addition to the concepts discussed below, this suggested that a critical element of the motor unit may be exquisitely sensitive to IGF-I in ALS patients. Although the cell bodies of lower motor neurons are within the blood brain barrier, their terminals in skeletal muscle emerge from both the blood-brain and blood-nerve barriers for access to agents in the periphery.

An important consideration prior to initiating clinical trials with rhIGF-I was whether blood levels of IGF-I could be achieved in vivo in association with a therapeutically beneficial dosing regimen. Rodents subjected to experimental damage of the sciatic nerve, a mixed motor and sensory nerve to the hindlimb, exhibit spontaneous recovery over a two to three week period. This recovery is associated with increased expression of endogenous IGF immunoreactivity at or near the injury site (Hansson et al, 1986). Local or systemic

IGF-I Enhances Nerve Regeneration

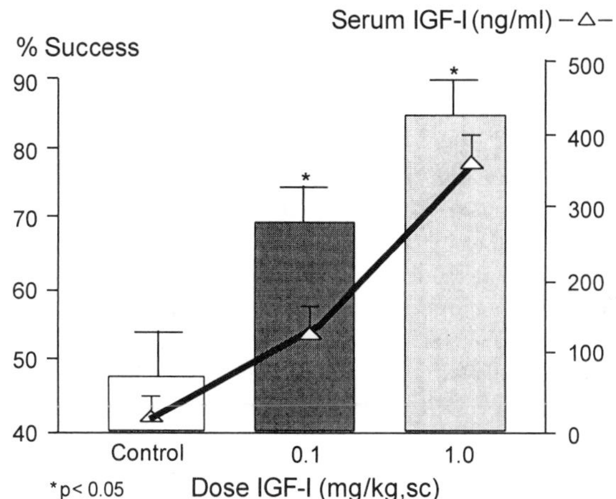

From Contreras et al., J. Pharmacol. Exp. Ther., 1995

Figure 2. Effect of rhIGF-I on recovery of motor function 10 days after sciatic nerve injury in mice. Subcutaneous injections of rhIGF-I or placebo (Control) were administered daily following sciatic crush injury; ability of the mice to maintain hindlimb grip on an inverted wire screen was scored in 10 consecutive tests. Serum levels of total IGF-I were measured by radioimmunoassay and are expressed as mean ± SE of peak concentrations after injection.

administration of exogenous IGF-I has been shown to accelerate recovery in this peripheral nerve injury model (Near et al, 1992; Contreras et al, 1995) and has proven to be a reasonable means of linking efficacious IGF-I dose levels in animals with changes in circulating IGF-I. The response to daily subcutaneous administration of rhIGF-I after nerve crush in mice was evident within 10 days after initiation of treatment as demonstrated by Contreras and colleagues and redrawn in Figure 2. Those injured mice treated daily with 0.1 and 1.0 mg/kg/d rhIGF-I exhibited a dose-related improvement in their ability to grip an inverted wire-mesh container compared with injured mice treated with placebo in which the failure rate was greater than 50%. Notable in Figure 2 was the significant and concentration-related increase in total serum IGF-I levels which occurred in association with therapeutically effective amounts of rhIGF-I. Higher doses elevated serum IGF-I levels to an even greater extent; however, there was no further efficacy in this or other injury models and acute hypoglycemia became dose-limiting in the rodents.

In separate cohorts of mice, subcutaneous administration of rhIGF-I also increased circulating IGF-I in a dose-dependent manner along with an associated increase in serum levels of IGF Binding Protein-2 (IGFBP-2) (Bhat et al, 1997). The rise in IGFBP-2 levels occured without significant alterations in either IGFBP-1 or IGFBP-3. Because the increase in serum IGFBP-2 occured at doses of systemic rhIGF-I which accelerated recovery in sciatic crush and other models of nervous system injury, and treatment of human subjects with rhIGF-I also increases serum IGFBP-2 (Baxter et al, 1993), it is conceivable that circulating IGFBP-2 could be a surrogate marker of pharmacologically active dosing with rhIGF-I in human clinical trials.

In phase 1 clinical studies of rhIGF-I with normal volunteers or ALS patients, we determined if increases in blood levels of total IGF-I could be achieved which were comparable to those observed in the rodent models of nervous system injury (Stong & Raskin, 1993). Subcutaneous doses up to 0.10 mg/kg/d were found to raise human serum IGF-I to the desired level of at least 400 ng/ml with only two of eight high dose volunteers developing signs or symptoms of acute hypoglycemia. For future clinical experiments with rhIGF-I, the risk of hypoglycemic episodes was reduced by dividing the daily dose into equal post-prandial injections following the morning and evening meals. Therefore, a regimen of rhIGF-I treatment was identified which was thought to be sufficiently benign for human testing at doses which would be tolerated and potentially useful for therapeutic purposes. Thus a second criteria was fulfilled in considering the possible use of IGF-I for treatment of neurolocial disorders.

2.2. Amyotrophic Lateral Sclerosis (ALS)

During the last five years, a number of clinical trials have evaluated factors with neurotrophic activity as potential agents in the treatment of ALS or motor neuron disease (Yuen & Mobley, 1996). ALS is a devastating neuromuscular disease which has been used as an entre to the study of neurotrophic and other factors in clinical medicine for several reasons. First, neurons die in ALS and the neurotrophic factors and certain small molecules have been characterized for their motor neuron-sparing activities. Second, systemically administered polypeptides like rhIGF-I are able to access the terminal neuritic sprouts of lower motor neurons, despite the spinal cord location of the cell bodies which are inside the blood-brain barrier. Third, the lack of medical alternatives in a disease with such severe morbidity and mortality has driven great interest among patients and their care givers and physicians in clinical trial participation. Fourth, a relatively predictable clinical course in ALS has facilitated clinical trial design and logistics. Finally, among a number of motor neuron disorders, ALS has the highest incidence in the general population which is comparable to that of multiple sclerosis [MS]. Due to the rapid lethality of ALS, however, the prevalence of this motor neuron disease (4-10/100,000) is far less than that of MS in the general population. In summary, the opportunity for testing promising drug candidates has been best in ALS of all such neurodegenerative conditions.

ALS is a relentlessly progressive, debilitating and ultimately fatal disease of unknown etiology. The disease is defined by the death of upper motor neurons (glutamatergic) arising in parietal cortex of the brain together with their targets, the lower motor neurons (cholinergic) arising in the brainstem and ventral spinal cord. These lower motor neurons extend into the periphery (where they may gain access to circulating neurotrophic polypeptides which otherwise poorly penetrate the blood-brain barrier) and innervate skeletal muscle responsible for voluntary limb, trunk, orofacial and respiratory function. The greatest morbidity of the disease is associated with lower motor neuron death which denervates the skeletal muscle, leading to muscle atrophy and loss of function subserved by the affected muscle group.

The typical ALS patient presents during the fifth decade of life with weakness, fasciculations or spasticity in the extremities. Sensory, autonomic, cognitive and ocular functions are spared. ALS patients become extremely impaired by their disease, rely on a high level of support by caregivers, and die from respiratory failure within three to five years of disease onset (Ringel et al, 1993). Many ALS patients remain undiagnosed or misdiagnosed for up to a year after their first symptoms. By the time a patient has received an accurate diagnosis, it is estimated that, for the affected muscle groups, at least 50% of the lower motor neurons have been lost (Bromberg et al, 1994). Neighboring neurites of surviving motor neurons sprout collaterals to orphaned skeletal muscle fibers as part of a compensatory response (Wohlfart, 1957). This temporarily sustains function in the affected muscle group (DeBelleroche et al, 1996) and probably explains the rather linear loss of function typically observed during much of the course of the disease (see Felice, 1997). The increased metabolic load on these surviving motor neurons is thought to place such cells at even greater risk of death (Bromberg et al, 1994). Therefore, the therapeutic rationale for evaluating rhIGF-I in ALS patients was based on the premise that intervention at a time of possible or probable diagnosis might enhance the survival and sustain the function of the remaining motor neurons that would be otherwise lost in the rapidly progressive course of this disease. Because rhIGF-I was hypothesized to maintain the integrity of surviving motor units and retard the loss of additional motor units (Lewis et al, 1993), clinical trial design required assessments that would accurately detect changes in the loss of function and health-related quality of life and the progression to death of ALS patients.

2.3. ALS Trial Design & Methods

Two double-blind, placebo-controlled, multicenter studies of subcutaneously administered rhIGF-I were conducted in 449 ALS patients between January 1993 and August of 1995. These studies involved eight clinical centers each in North America (trial CEP-1200, n=266 patients) and Europe (trial CEP-1202, n=183 patients). In addition to the differences in total number of patients, there were three arms with approximately equal randomization to each arm in the North American study (90 placebo, 89 rhIGF-I 0.05

mg/kg/d and 87 rhIGF-I 0.10 mg/kg/d) and only two arms in the European study with a 2:1 randomization in favor of rhIGF-I treatment (124 rhIGF-I 0.10 mg/kg/d and 59 placebo). Both the North American and European trials used a disease-specific scale, the Appel ALS (AALS) rating scale, as the global clinimetric assessment. This scale captures the findings of a thorough neuromuscular examination into a single number which reflects the severity of impairment and disability resulting from underlying disease pathology (Appel et al, 1987). Objective assessments were acquired for each patient's upper and lower limb function, respiratory function, overall muscle strength and bulbar (speech and swallowing) function. These elements comprise the AALS total score which is minimally 30 for individuals with completely normal neuromuscular function and maximally 164 for ALS patients who are entirely disabled by their disease. Patients with AALS scores of 40-60 were eligible for a screening period of 2-3 months. A minimum accumulation of 5 AALS points on screen was required as evidence of progressive disease and eligibility for randomization to double-blind treatment. The rate of change of the monthly AALS total score is unique for each patient, linear during the majority of active disease, and tracks progression based upon correlations with independent measures of disease morbidity and mortality (Appel et al, 1987; Haverkamp et al, 1995, Felice, 1997).

In North America, the primary outcome was specified as rate of the change in the AALS total score over time (i.e., the slope); in Europe, the primary outcome was specified as the change from baseline in the AALS total score. Secondary outcomes included time to advanced disease (AALS $\geq$ 115 points or forced vital capacity [FVC] <39%) or time to a new milestone of functional disability (i.e., an additional 20 AALS points over baseline). In addition, the Sickness Impact Profile™ (SIP), which is a disease-non-specific questionnaire focused on patient-perceived quality of life and survival, were used as measures of the ultimate handicaps resulting from the disease. Only in the North American trial was survival a prespecified secondary outcome.

In brief, subjects enrolled for the clinical trials of rhIGF-I were adults (>20 years of age) with probable or definite diagnosis of classical, non-familial ALS and no confounding health conditions such as other neuromuscular disorder, diabetes, malignancies or heart disease. The demographics of patients randomized in the studies were not significantly different across treatment arms and were comparable to those expected in the overall ALS population with similar levels of disease progression (Lai et al, 1997; Borasio et al, 1996).

2.4. Results of rhIGF-I in ALS

As reported recently for the North American Study (Lai et al, 1997), the higher dose of 0.10 mg/kg per day of subcutaneous rhIGF-I reduced the signs and symptoms of disease progression and lowered the patient-perceived deterioration in quality of life. The mean AALS slope was 26% lower in the higher dose rhIGF-I treatment group compared to the AALS slope in the placebo group; the lower dose group was intermediate between the two (3.1 vs 3.8 vs 4.2 mean AALS points per month for, respectively, 0.10 mg/kg/d rhIGF-I vs 0.05 mg/kg/d rhIGF-I vs placebo). This result in the primary outcome indicates that patients receiving higher dose rhIGF-I accumulated functional deficit at a lower rate that those receiving placebo. The change in the severity of disease impairment and disability, is best illustrated in a monthly analysis of the AALS assessments. As illustrated in Figure 3, at two months after randomization, 0.10 mg/kg/d rhIGF-I significantly decreased the change from baseline AALS in a manner consistent with a dose-related effect throughout the nine month double-blind protocol.

Because the rate of change in the AALS (the AALS slope) has previously been asserted to be a surrogate for disease progression and a predictor of survival (Havercamp et al, 1995), a reduction in the AALS slope would be expected to delay the time an individual patient takes to reach a new stage of advanced disease whether assessed by overall functional deficit, respiratory insufficiency or death. In the North American study, the relative risk of advanced

rhIGF-I in ALS: North American Trial Results

Change from Baseline in AALS Total Score

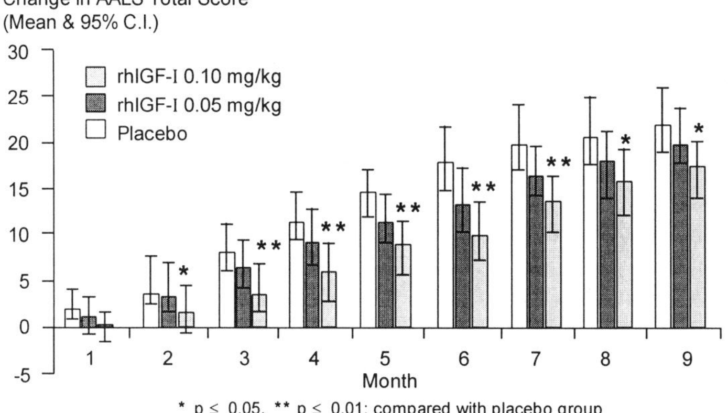

Figure 3. Monthly changes in Appel ALS Rating Scale (AALS) total score from baseline for patients in the three treatment groups (mean baseline scores were 71.5 $\pm$ 1.5 placebo, 70.5 $\pm$ 1.3 rhIGF-I 0.05 mg/kg and 68.9 $\pm$ 1.3 rhIGF-I 0.10 mg/kg). The bars depict the mean change $\pm$ 95% confidence interval at each point in time, with the last observation carried forward. Statistical significance was reached starting at month 2 in the comparison of the high-dose group (n=86) versus placebo (n=88). **reprinted with permission from American Academy of Neurology and Lippincott-Raven Publishers; Lai et al, 1997**

disease (AALS $\geq$115 or FVC <39%) or of progression through a milestone of functional disability (20 point progression on the AALS) was reduced by 44 % and 50%, respectively, by 0.10 mg/kg/d rhIGF-I compared to placebo. Patient status was followed for nearly three years after randomization in the North American study and analyses were performed to determine the effect of a 9-month delay to rhIGF-I treatment. This was possible because a majority of patients (>75%) in all treatment groups elected to use rhIGF-I on an open-label basis after completing the double-blind portion of the trial or disqualifying from further double-blind treatment due to protocol-specified criteria. As indicated in Figure 4, patients in the North American trial who received the higher dose of 0.10 mg/kg/d rhIGF-I during the double-blind phase of the study appeared to have a slightly reduced risk of death from disease at any point in time (median survival of 17 versus 20 months in placebo). In post-hoc analysis of these data in which risk factors for death (advanced age and lower respiratory capacity at baseline) were taken into account, there was a significant advantage imparted by earlier treatment with the higher dose of rhIGF-I (Lai et al, 1997). Therefore, across several study outcomes in the North American trial, higher dose rhIGF-I treatment consistently imparted a statistically signficant benefit over placebo treatment. This occured in the context of relatively limited drug-associated adverse experiences in a disease with abundant adverse experiences. These largely disease-associated adverse experiences (e.g., weakness,

rhIGF-I in ALS: North American Trial Results

Kaplan Meier Survival Analysis

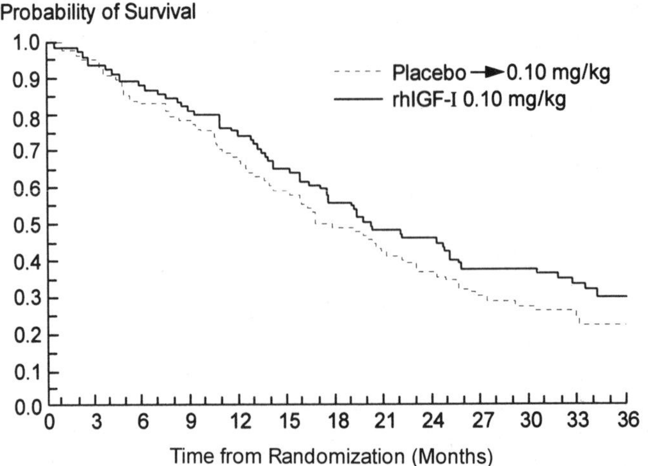

Figure 4. Extended survival analysis showing the proportion of patients surviving over time in the placebo (n=90 at baseline) and high dose recombinant human insulin-like growth factor-I (rhIGF-I) treatment group (n=87 at baseline) through July 31, 1996. This unadjusted plot shows the separation of the survival curves in favor of patients initially randomized to rhIGF-I treatment. **reprinted with permission from American Academy of Neurology and Lippincott-Raven Publishers; Lai et al, 1997**

coordination abnormalies, headache, dyspnea) were distributed comparably among the treatment arms of the study. The most frequent adverse experiences associated with rhIGF-I treatment were injection-related complaints. Patients receiving rhIGF-I treatment registered significantly more injection site inflammation than placebo-treated patients (24% vs 33% vs 7% for 0.10 mg/kg/d, 0.05 mg/kg/d and placebo, respectively). Additionally, a higher percentage of rhIGF-I treated patients registered significant hair growth changes (15% vs 7% vs 3% for 0.10 mg/kg/d, 0.05 mg/kg/d and placebo, respectively), knee pain (11% vs 10% vs 1% for 0.10 mg/kg/d, 0.05 mg/kg/d and placebo, respectively), facial edema (9% vs 1% vs 1% for 0.10 mg/kg/d, 0.05 mg/kg/d and placebo, respectively), sinusitis (8% vs 13% vs 3% for 0.10 mg/kg/d, 0.05 mg/kg/d and placebo, respectively) and lung findings (5% vs 17% vs 4% for 0.10 mg/kg/d, 0.05 mg/kg/d and placebo, respectively). These adverse experiences did not appear dose-related, were never frequent enough at each trial site to jeapordize the blinding of the study, and were tolerated by most patients throughout the trial. In the European study, all adverse experiences were comparably distributed among treatment groups with none significantly associated with rhIGF-I treatment.

As a consequence, if the efficacy results of the North American trial were the only data to rely upon, there might be widespread enthusiasm over the prospects for treating ALS with rhIGF-I. Although the numerical means in the European trial were directionally consistent with all but one of the outcomes in North America, none of the prespecified primary or secondary outcomes reached statistical significance between the 0.10 mg/kg/d rhIGF-I and placebo groups in Europe (Borasio et al, 1996). Furthermore, additional deaths among patients on rhIGF-I during the double-blind phase of the European trial raised concerns that rhIGF-I might be deleterious to ALS patients. As the database on probability of survival matured, however, it was clear that median survivals (19 months) were comparable in both treatment arms of the European Study. Indeed, it appeared that more patients at higher risk for death by disease (older with lower respiratory capacity at baseline) were randomized to the rhIGF-I treatment arm (Borasio et al, submitted) and might explain the higher frequency of deaths in the rhIGF-I versus the placebo arm during the double-blind period.

As a result of concerns that the European trial results did not replicate the North American efficacy outcomes, the neurology community, regulatory authorities and the sponsoring corporations have been attempting to determine which of the two studies embody the true signal concerning effects of rhIGF-I treatment in ALS. One way that trial investigators have approached this conundrum is by exploiting the very similar nature of trial designs in both North America and Europe. Leigh and colleagues performed a post-hoc analysis of pooled results from the key treatment arms from both continents and determined that, although the absolute magnitude of the rhIGF-I treatment outcomes were smaller, compared to placebo, 0.10 mg/kg/d rhIGF-I significantly reduced the AALS slope, the accumulation of AALS points over time and reduced the risk of advanced disease (Leigh et al, 1997). An interesting observation from these post-hoc analyses is that the functional benefits of rhIGF-I during the 9 month double-blind were nested among patients who were progressing most rapidly through disease (pre-treatment AALS slope >4 points per month). Furthermore, the effects of rhIGF-I appeared in the domains of the AALS total score assessing bulbar function, upper extremity function and overall muscular strength. This has been encouraging as the pooled analysis incorporates data from more patients overall. Nevertheless, additional clinical trial results may be needed to resolve the doubt for some neuromuscular clinicians and their ALS patients.

At this time, open-label treatment protocols are being enrolled in the United States and in Europe as a means of offering ALS patients early access to rhIGF-I and obtaining additional safety information on this potential treatment for ALS. Kyowa Hakko Kogyo is conducting a phase 3 double-blind, placebo-controlled study of rhIGF-I to determine safety and efficacy of this factor in Japanese ALS patients.

2.4. Post-Polio Syndrome (PPS)

The post-polio syndrome occurs in 24-60% of polio survivors many decades after the original loss of lower motor neurons due to infection. Typically, these patients experienced partial to total recovery from the initial neuromuscular disability (Cashman et al, 1987). PPS is included in the category of motor neuron diseases because its clinical symptoms and histological signs are related to the antecedent death of lower motor neurons. Subsequent dysfunction and fatigue in PPS patients is often associated with previously affected muscle groups (Dalakis and Illa, 1991). As already discussed, there is evidence that orphaned skeletal muscle fibers receive collateral innervation from residual motor neurons, but the stability of these connections may be undermined decades after a surviving motor neuron expanded its terminal field to support the function of additional muscle fiber groups (Dalakis & Illa, 1991; Jubelt & Drucker, 1993). If the neuromuscular fatigue associated with the syndrome relates to this presumed underlying pathology, it is possible that rhIGF-I could support the integrity of the neuromuscular junctions as implied from animal studies (Caroni & Grandes, 1990). A pilot study of rhIGF-I treatment for neuromuscular fatigue and recovery was conducted in 22 PPS patients. After three months of double-blind treatment, preliminary analysis indicates that rhIGF-I significantly increased maximal contraction upon recovery from fatiquing exercises. Nevertheless, rhIGF-I treatment failed to alter muscle fatiguability, muscle metabolism or other clinical and biopsy outcomes in the study (Miller et al, 1997). Whether the preliminary results of this study are sufficient to motivate additional studies remains open to consideration by neuromuscular clinicians, their PPS patients, and the sponsoring firms.

3. IGF-I FOR OTHER NEUROLOGICAL DISORDERS

IGF-I may be an attractive candidate for treating other neurological conditions, especially those degenerative conditions of neurons in which the neuron or its processes have access to circulating levels of exogenous rhIGF-I. Some of these disorders included the toxic neuropathies, painful neuropathies of known and unknown origin, post-polio syndrome and multiple sclerosis. For example, IGF-I may be a useful adjunct therapy in patients undergoing chemotherapy (Contreras et al, in press), where neurotoxic properties of very useful anti-neoplastic agents, such as cisplatin, vincristine and paclitaxel, compromise sensory or motor neuron function and become dose-limiting side effects (Rowinsky et al, 1993; Chaudry et al, 1994). IGF-I may also reverse terminal degeneration of sensory neurons in painful neuropathies. Finally, because of benecifical effects on oligodendroglia and new information on the ability of IGF-I to reduce perivascular lesions, IGF-I may have utility in the treatment of inflammatory demyelinating disorders such as multiple sclerosis.

3.1. Neurosensory Disorders

Idiopathic small fiber painful neuropathy (ISFPN) is a rare but highly disabling disorder in 1-2% of peripheral neuropathy cases reviewed in a major midwestern US database. As the name implies, ISFPN arises from unknown etiology and its diagnosis occurs by exclusion of other sensory neuropathies (Windebank et al, 1990; Smith et al, 1993). Typical patients are middle aged and present with burning, numbness or shooting pain in their distal extremities. These symptoms progress over months to years and are exacerbated by activity and surface contact of the affected limb. The pain is restricted largely to the feet and extending to no more than mid-calf. There is sometimes autonomic involvement evident as inappropriate sweating responses. Usually, the painful nature of this neuropathy disables to the extent that sufferers cannot apply pressure to their feet and often discontinue employment for medical disability.

It is believed that the primary pathology in ISFPN is distal degeneration of small, mostly unmyelinated, nociceptive neurons (Chalk et al., 1993; Dyck et al, 1976). This axonal atrophy and degeneration accompanied by unsuccessful regenerative neurite regrowth by the distal sensory axons (Smith et al, 1993). IGF-I may be useful in the treatment of ISFPN for the following reasons. Like neurons of the central nervous system, sensory neurons express functional IGF-I receptors (Windebank et all, 1995) and physiologic concentrations of IGF-I enhance the survival in vitro of chick sympathetic and rat sensory neurons (Zackenfels and Rohrer, 1993; Russell et al, 1997). IGF-I promotes neuritic outgrowth and myelination of adult rat dorsal root ganglia neurons in vitro (Fernyhough et al, 1993; Russell et al, 1997); locally infused IGF-I increases the rate of sensory nerve regeneration in injured sciatic nerve (Kanje et al, 1989). Because both the peripheral sensory endings as well as the cell bodies are outside the blood-nerve barrier, circulating IGF-I should be able to reach multiple aspects of the nociceptive and autonomic neurons involved in ISFPN. As in motor neuron disorders, this evidence indicates that exogenously administered IGF-I might access the diseased peripheral neuron, promote distal axonal sprouting and regeneration and normalizing nociceptive and, to a lesser extent, autonomic, function in patients with ISFPN.

A single site double-blind, placebo-controlled study of rhIGF-I is ongoing in a collaborative effort to determine if daily treatment with rhIGF-I (0.1 mg/kg, sc) reduces the patients' perception of pain over a 6-month treatment period. Additional measures include electrophysiologic, clinimetric and autonomic evaluations to determine the potential usefulness of rhIGF-I in this syndrome. If rhIGF-I is determined to be safe, tolerated, and efficacious in reducing the painful neuropathy in these patients, there may be additional indications in which rhIGF-I may be studied, particularly those in which small, unmyelinated peripheral neurons are injured.

3.2. Inflammatory & Demyelinating Nervous System Disorders

IGF-I stimulates oligodendrocyte proliferation and differentiation, and promotes survival and myelination in vitro (Mozell & McMorris, 1991). Additionally, IGF-I is a potent regulator of myelin gene activity and may act as an autocrine regulator of oligodendrocyte development in vivo (see Webster, 1997). Mice overexpressing IGF-I have greater brain mass with increased oligodendrocytes and CNS myelin content (Carson et al, 1993; Mathews et al, 1988), whereas, mice lacking expression of IGF-I or the IGF-I receptor, display decreased brain mass with reduced oligodendrocyte number and myelin content (Beck et al, 1995; Liu et al, 1993). Together, these findings and others beyond the scope of this presentation demonstrate the importance of IGF-I in oligodendrocyte survival, differentiation and myelination.

Muliple sclerosis (MS) is the best characterized of the inflammatory and demyelinating nervous system diseases. In an acute rodent model of MS (i.e., experimental autoimmune encephalomyelitis, EAE), reactive astrocytes synthesize high levels of IGF-I and oligodendrocytes express IGF-I receptors at the time of myelin sheath regeneration (Komoly et al., 1994; Liu et al., 1994; Yao et al., 1995a). At lesion sites in EAE, astrocytes synthesize high levels of IGF-I mRNA and protein in association with upregulation of IGF-I receptor indicating that the IGF system is involved in the endogenous response of the brain to inflammatory injury. As in the CNS, IGF-I is upregulated in Schwann cells after peripheral nerve injury (Hansson et al, 1986) and enhances ensheathment of cultured neurons by Schwann cells (Cheng et al, 1997).

In experimental rodent studies with rhIGF-I, Webster and colleagues have found that administration of exogenous rhIGF-I in EAE: a) reduced the weight loss and clinical disability observed in rats and mice, b) limited the appearance of leukocytes, edema and serum proteins in the perivascular space of spinal cord and brain stem lesions, c) increased the expression of myelin proteins and d) decreased the amount of demyelination in the area of CNS perivascular lesions (Liu et al, 1995;Yao et al., 1995b; Yao et al, 1996; Liu et al., 1997). The finding that exogenous IGF-I reduced the size and number of perivascular lesions in EAE in addition to reducing evidence of demyelination indicate significant effects on two pathological hallmarks of early plaque formation in MS. In no small part, the demonstration of rhIGF-I benefits in animal models of inflammatory and demyelinating disorders led to initiation of a clinical trial to determine if open-label rhIGF-I treatment will be safe in MS patients whose blood-brain barriers are known to be compromised by disease pathology and who may have active autoimmune disease underlying their inflammatory demyelination. An early effect which might signal hope for use of rhIGF-I in conditions such as MS would be the reduction of the size or numbers of lesions which can now be objectively measured in patients using magnetic resonance imaging and appropriate contrast agents.

4. SUMMARY

Trophic factors play an essential role in neuronal development and maintenance of motor neurons. While deficits in trophic factors may not exist in motor neuron diseases, recent observations suggest that binding sites for IGF-I may be overexpressed in motor neuron disease. Agents capable of interacting with the IGF receptor or its signalling pathways may exert pharmacologically beneficial effects in the palliative treatment of ALS in particular and neurodegenerative disorders in general. Recombinant human IGF-I was shown in a North American double-blind, placebo-controlled study to reduce the rate of progression of the signs and symptoms of ALS. This observation was supported by reduced risk of disease specific signs and symptoms of advancing disease on the higher of two doses of rhIGF-I. While a second clinical study in Europe failed to demonstrate a significant effect of rhIGF-I treatment on any prespecified disease assessment, the effects of rhIGF-I were statistically significant upon post-hoc analyses of multiple outcomes when the data from the North America and European studies were pooled. This post-hoc analysis was performed to assist in attributing valence to the overall effect observed by rhIGF-I treatment in ALS.

Additional clinical studies will be required to further demonstrate to the neurology community and patients suffering neurodegenerative diseases the potential for benefit from rhIGF-I treatment. Significant additional investment of resources may be needed to demonstrate the safety, tolerability and effectiveness of rhIGF-I in any neurological conditions for which a scientifically sound rationale has been articulated. A spectrum of neurological disorders may be approached using rhIGF-I treatment due to the pleiotrophic effects of this factor and the relatively benign adverse experiences noted in acute and chronic clinical studies to date.

ACKNOWLEDGEMENTS

The author is grateful to E. Alexander, F. Baldino, B. Brooks, M. Bromberg, E. Feldman, J. Frank, H. McFarland, R. Miller, H. Webster and A. Windebank for their education regarding the neurological disorders discussed and the nature of the potential effects of IGF-I in each. The careful review of and commentary on this manuscript from V. Egel, K. Ingalls, C. Joyce, and G. Mayor greatly improved its readability; their input is appreciated greatly.

REFERENCES

1. Adem A, Ekblom J, Gillber G-G, Jossan SS, Hoog A, Winblad B, Aquilonius S-M, Wang L-H, Sara V. J Neural Trans 97:73-84, 1994.
2. Ang LC, Bhaumick B, Juurlink BHJ. Dev Brain Res 74:83-88, 1993.
3. Appel V, Stewart SS, Smith G, Appel SH. Ann Neurol 22:328-333, 1987.
4. Baxter RC, Hizuka N, Takano K, Holman SR, Asakawa K. Acta Endocrinol 128:101-108, 1993.
5. Beck KD, Powell-Braxton L, Widmer HR, et al.. Neuron 14:717-730, 1995.
6. Bhat RV, Engber TM, Zhu Y, Miller MS, Contreras PC. J Pharmacol Exp Ther 281:522-30, 1997.
7. Borasio GD, DeJong JMBV, Emile J, Guiloff R, Jerusalem F, Leigh N, Murphy M, Robberecht W, Silani V, Wokke J, the European ALS/IGF-I Study Group. J Neurol 243 (Suppl 2):S26, 1996.
8. Bromberg MB, Nau KL, Foreshew DA. IN Rose FC (ed) ALS - from Charcot to the present and into the future. The Forbes H. Norris (1928-1994) Memorial Volume. Smith-Gordon-Nishimura, London, 1994 pp 267-277.
9. Caroni P, Grandes P. J Cell Biol 110:1037-1017, 1990.
10. Carson MJ, Behringer RR, Brinster RL, et al.. Neuron 10:729-740, 1993.
11. Cashman NR, Maselli R, Wollman RL, Roos R, Siman R, Antel JP. N Eng J Med 317:7-12, 1987.
12. Chaudhry V, Rowinsky EK, Sartorius E, et al.. Ann Neurol 35:304-311, 1994.
13. Chalk CH, Lennon VA, Stevens JC, Windebank AJ. Neurol 43:3309-2211, 1993.
14. Cheng H-L, Russel JW, Feldman EL. Soc Neurosci Abs 27:585, 1997.
15. Contreras PC, Steffler C, Yu E, Callison K, Stong D, Grebow P, Vaught JL. J Pharmacol Exp Ther 274:1443-1449, 1995.
16. Contreras PC, Vaught JL, Gruner JA, Brosnan C, Steffler C, Arezzo JC, Lewis ME, Kessler JA, Apfel SC. Brain Res, in press.
17. Dalakas M, Illa Iy. Adv Neurol 56:495-511, 1991.
18. DeBelleroche J, Orrell RW, Virgo L. J Neuropathol Exp Neurol 55:747-757, 1996.
19. Dore S, Krieger C, Kar S, Quirion R. Mol Brain Res 41:128-133, 1996.
20. Dore S, Kar S, Quirion R. Trends Neurosci 20:326-331, 1997.
21. Dyck PJ, Lambert EH, O'Brien PC. Neurol 41:799-807, 1976.
22. Feldman EL, Sullivan KA, Kim B, Russell JW. Neurobiol Disease 4:201-214, 1997.
23. Felice KJ. Muscle Nerve 20:179-185, 1997.
24. Fernyhough P, Willars GB, Lindsay RM, Tomlinson DR. Brain Res 607:117-124, 1993.
25. Hansson HA, Dahlin LB, Danielsen N, Fryklund L, Nachemson AK, Polleryd P, Rozell B, Skottner A, Stemme S, Lundborg G.. Acta Physiol Scand 126:609-614, 1986.
26. Hantai D, Akaaboune M, Lagord C, et al.. J Neurol Sci 129(suppl):130-139, 1995.
27. Haverkamp LJ, Appel V, Appel SH. Brain 118:77-719, 1995.
28. Jubelt B, Drucker J. Seminars Neurol 13:283-290, 1993
29. Kanje M, Skottner A, Sjoberg J, Lundborg G. Brain Res 486:396-398, 1989.
30. Komoly S, Hudson LD, Webster HdeF, et al.. PNAS 89:1894-1898, 1994.

31. Lai EC, Felice KJ, Festoff BW, Gawel MJ, Gelinas DF, Kratz R, Murphy MF, Natter HM, Noris FH, Rudnicki SA, North America ALS/IGF-I Study Group. Neurol 49:1621-1629, 1997.
32. Leigh N, The North American and European ALS/IGF-I Study Groups. Neurol 48 (Suppl 2):A217-A218, 1997.
33. Lewis, ME, Neff NT, Contreras PC, Stong DB, Oppenheim RW, Grebow PE and Vaught JL. Exp Neurol 124:73-88, 1993.
34. Liu J-P, Baker J, Perkins AS, Robertson EJ, Efstratiadis A. Cell 75:59-72, 1993.
35. Liu X, Yao D-L, Bondy C, et al.. Mol Cell Neurosci 5:418-430, 1994.
36. Liu X, Yao D-L, Webster HdeF. Multiple Sclerosis 1:2-9, 1995.
37. Logan A, Oliver JJ, Berry M. Prog Growth Factor Res 5:379-405, 1995
38. Mathews LS, Hammer RE, Behringer RR, et al.. Endocrinology 123:2827-2833, 1988.
39. Miller RG, Gelinas DF, Kent-Braun J, Dobbins T, Dao H, Dalakas M. Neurol 48 (Suppl 2):A217, 1997.
40. Mozell RL, McMorris FA. J Neurosci Res 30:382-390, 1991.
41. Nachemson AK, Lundborg G, Hansson HA. Growth Factors 3:309-314, 1990.
42. Near SL, Whalen LR, Miller JA, Ishii DN. Proc Natl Acad Sci 89:11716-11720, 1992.
43. Neff NT, Prevette D, Houenou L, Lewis ME, Glicksman MA, Yin QW, Oppenheim RW. J Neurobiol 24:1578-1588, 1993.
44. Ringel SP, Murphy JR, Alderson MK, Bryan W, England JD, Miller RG, Petajan JH, Smith SA, Roelofs RI, Citer F, Lee MY, Brinkmann JR, Almada A, Gappmaier E, Graves MJ, Herbelin L, Mednoza M, Mylar D, Smith P, Yu P. Neurol 43:1316-1322, 1993.
45. Russell JW, van Golen C, Parekh A, Windebank AJ, Feldman EL. J Periph Nerv System 2:297A, 1997.
46. Rowinsky EK, Chaudhry V, Fonsiere AA, et al.. J Oncol 11:2010-2020, 1993.
47. Smith BE, Windebank AJ, Dyck PJ. In Peripheral Neuropathy. PJ Dyck, Pk Thomas, JW Griffin, PA Low, JF Poduslo, (eds) Saunders, Philadelphia, 1993, pp 1525-1532.
48. Stong DB, Raskin R. Ann NY Acad Sci 692:317-320, 1993.
49. Tomas FM, Chandler CS, Knowles SE, Francis GL, Martin AA, Lemmy AB, Read LC, Ballard JF. IN: J.S. Bond and A.J. Barrett (eds) Proteolysis and protein turnover. Portland, Portland 1993, pp 141-148.
50. Webster HdeF. Multiple Sclerosis 3:113-120, 1997.
51. Windebank AJ, Blexrud MD, Dyck PJ, Daube JR, Karnes JL. Neurol 40:584-591, 1990.
52. Windebank AJ, Schenone A, Gross L, Feldman EL. Neurol 45:A279, 1995.
53. Wohlfart G. Neurol 7:124-134, 1957.
54. Yao D-L, West NR, Bondy CA, Brenner M, Hudson LD, Zhou J, Collins GH, Webster HdeF. J Neurosci Res 40:647-659, 1995a.
55. Yao D-L, Liu X, Hudson LD, Webster HdeF. Proc Natl Acad Sci USA 92:6190-6194, 1995b.
56. Yao D-L, Liu X, Hudson LD, Webster HdeF. Life Sci 58:1301-1306, 1996.
57. Yuen EC, Mobley C. Ann Neurol 40:346-354, 1996.
58. Zackenfels K, Rohrer H. Ann NY Acad Sci 692:302-304, 1993.

Molecular Mechanisms to Regulate the
Activities of Insulin-like Growth Factors
K. Takano, N. Hizuka and S-I. Takahashi (Editors)

Directions for research into the insulin-like growth factor system as the millennium approaches; Closing remarks to the IVth IGF Symposium

E. Martin Spencer[a] and Robert Sapolsky[b]

[a]Laboratory of Growth & Development, Davies Hospital, San Francisco CA 94114, USA[1]

[b]Department of Biological Sciences, Stanford University, Stanford CA 94305, USA

1. INTRODUCTION

Early in sulfation factor/somatomedin research it was not difficult to foresee that molecules with growth-promoting and insulin-like properties might be biologically important in a variety of areas. But what could not be envisioned was the ubiquitous nature and functions of insulin-like growth factor (IGF) gene expression, the critical regulatory roles of the seven or more IGF binding proteins (IGFBPs), the IGFBP proteases, and the unique clinical applications that could develop. Thus, the earliest IGF Symposia focused on methodology, plasma and tissue levels in a variety of conditions, nomenclature, purification (chemistry), receptor characterization and actions, mitogenic activity in different cell systems, and in vivo data testing the somatomedin hypothesis. The IInd IGF Symposium in 1991 was the last meeting where any attempt at a comprehensive review was possible. Even then the enormous breath of the IGF System was apparent. The meeting began with Zena Werb discussing the IGF System at the 2 cell embryonic stage and ended with two hot topics on clinical use of IGF-I for diabetes and wound healing (1). The Sydney meeting was more sophisticated in its presentation of basic science and ramifications. Now, at the IVth IGF Symposium, we are seeing movements toward understanding the evolutionary history, mechanisms of action of all the components of the IGF System, and application of our knowledge to clinical problems.

2. DOES THE IGF GENE IMPROVE FITNESS?

The fact that the IGF gene has been incorporated into the functioning of so many diverse systems in the animal kingdom suggests that it confers a selective advantage. We proposed to test this hypothesis by examining the plasma IGF-I levels in males of a tribe of free-ranging baboons habituated to human observation, but without interference in their ecosystem (2). Dr.

[1]Support provided by: EMS, NIH 27345-17; RS, Guggenheim Foundation. Approval by Office of the President and the Ministry of Tourism and Wildflife, Republic of Kenya. Field assistance by Richard Kones, Hudson Oyaro, and Lisa Share.

Sapolsky has studied this tribe in the Masai Mara National Reserve for over 20 years investigating the physiological factors contributing to the dominance hierarchy of the males.

Samples of blood were obtained from unsuspecting animals after being tranquilized by a dart. The plasma IGF-I levels were measured on the males by RIA. Samples were extracted by acid-ethanol and cryopreciptated before measurement. This technique was validated for baboon plasma by comparison with samples analyzed after Sephadex G-50 chromatography in 1 M acetic acid, the gold standard.

Plasma IGF-I values from males over a 7 year period, from 1980 to 1986, were plotted against their rank. Figure 1 shows the relationship of plasma IGF-I levels to rank for *all* data points. *The data is perhaps better appreciated in Figure 2 where the means of the values for each rank over the 7 years is plotted against the rank, r = 0.79.*

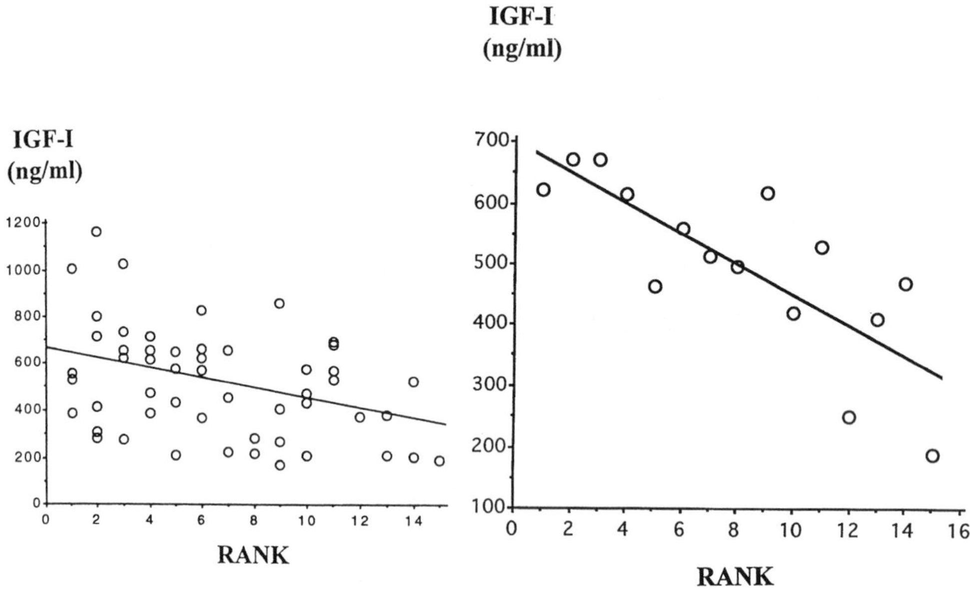

Figure 1. Relationship between IGF-I plasma levels and dominance rank. y = -21.3x + 665. F = 7.812; degrees of freedom = 1/53; P<0.007; r = .35 Reprinted with permission from the Am. J. Physiol. 273 (1997) R1346.

Figure 2. Relationship between the *mean* plasma IGF-I levels and dominance rank. Data for each rank represents the mean of 7 distinct annual sessions of study. y = -24.8x + 698. P <0.001; r = .79.

This relationship was not explained by age, since elderly males and young males within 2 years of puberty were excluded. Likewise, nutrition, weight, or levels of testosterone, glucocorticoid, and IGF binding proteins could not be correlated to the IGF-I levels. The significance of the plasma IGF-I level to dominance is best illustrated by serial measurements on baboon 261 over this 7 year span. Figure 3 shows the amazing concordance of plasma IGF-I and rank in this baboon.

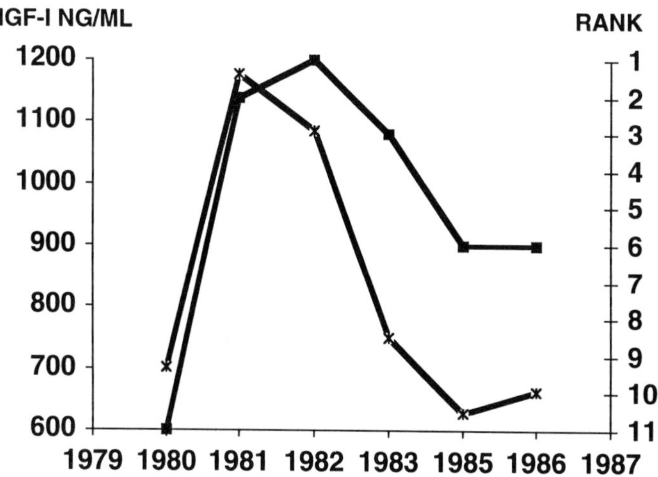

Figure 3. The relationship of plasma IGF-I levels to rank in one male baboon over 7 years. Reprinted with permission from the J. Physiology. 273, (1997) R1346.

Collectively, the data suggest, but do not establish, that differences in plasma IGF-I levels may be a major factor in dominance of male baboons by improving fitness.

3. APPLICATIONS OF THE IGF SYSTEM TO CLINICAL PROBLEMS

Our glimpses of possible unique clinical applications depend on the basic work done by many in diverse systems. I focus on the clinical applications now because they signify a maturing of IGF research and also for humanitarian reasons, the ability to prevent and cure disease. I will present several examples, but possibly none could be more impressive than IGF-I stimulating growth in children with GH receptor dysfunction (GHRD) or achieving metabolic control in certain types of insulin resistance, for without IGF-I neither could be attained.

3.1 Diabetes Mellitus

IGF-I therapy is established for type A insulin resistant diabetes. Kasuga of Kobe University Medical School has shared his experience at the Symposium with this indication. In addition, in is special lecture at the beginning of the symposium Froesch pointed out the therapeutic potential of IGF-I in other types of diabetes. IGF-I has many properties that would be favorable to type 2 diabetics. IGF-I increases insulin sensitivity, decreases insulin secretion and lowers blood sugar in diabetics. Therapeutic doses in vivo also improve the plasma lipid profile, decrease protein degradation and increase lipid oxidation.

3.2 Postmenopausal Osteoporosis

The unique role for IGF-I on longitudinal skeletal growth was the original observation of Daughaday. This was followed up by in vivo and in vitro work on bone formation in several labs including, Spencer et al., Mohan et al., Froesch et al, Canalis et al. and Sommer et al. (3). These studies have collectively defined a major role for IGF-I in bone formation, especially at the time of menopause when plasma IGF-I levels decline. Indeed, we and several other authors have found that women with established postmenopausal osteoporosis have significantly lower plasma IGF-I levels than their unfractured counterparts (4). Thus, Celtrix and Green Cross are testing IGF-I in postmenopausal osteoporosis. The ramifications of a positive result are tremendous because of the enormity of the problem world-wide. They could set the stage for a more physiological replacement therapy using IGF-I at menopause to prevent rather than treat fractures in at risk populations. In addition to positive effects on bone formation, this therapy should improve muscle strength and decrease the number and/or seriousness of falls. A unique aspect of the therapy offered by Celtrix is the use of IGFBP-3 in a 1:1 complex with IGF-I to improve pharmacokinetics and potentiate IGF-I action (5).

3.3 Nervous System

The neurotropic properties of IGFs have led to a much publicized clinical trial of IGF-I therapy in patients with **amyotrophic lateral sclerosis** (ALS). The significance of the small effect is under discussion but three points should be made: 1) ALS is a uniformly fatal disease. 2) There is no other therapy. 3) Administration as a complex with a potentiating IGFBP may eliminate doubts as to efficacy. However, there are other even more exciting applications of the neurotropic properties of IGFs. Gluckman and his collaborators and also Isgaard et al. have clearly demonstrated the **neuroprotective effect** of IGF-I in animal models of anoxic brain damage. The data presented in this Symposium indicate the importance of clinical trials with IGF-I to lessen the devastating impact of stroke.

3.4 Wound Healing

My lab has focused on the critical role of the IGF System in wound healing (6). Deletion of IGF-I from experimental wounds delays healing, whereas the addition of IGF-I to experimental wounds hastens tissue repair. However, the administration of IGF-I in complex with IGFBP-3 results in a synergistic increase in wound healing in both rat and porcine wound healing models (5). The effectiveness of IGF-I has led to the discovery of multiple sites of action and sources of wound IGF-I. The initial source of IGF-I comes from stores in the alpha granules of platelets which also releases IGFBP-3 in a 1:1 ratio (7). Wound IGF-I recruits macrophages which are regarded as the directors of the subsequent tissue repair process (8). Macrophages also release IGF-I into the wound, which induces fibroblasts to proliferate, release IGF-I and synthesize collagen. IGF-I also participates in the angiogenic response and in stimulating epithelial closure of cutaneous wounds. The wound healing response provides multiple models to study the action of IGF-I and the potentiating effect of simultaneously administered IGFBP-3. Currently Celtrix is evaluating the use of the IGF-I IGFBP-3 complex to promote **donor site healing in burned patients**. This should be successful in view of the positive results obtained with GH. However, there should be added benefits from IGF-I use. First it is a more potent anabolic agent ideal for the catabolic burn state. Second, the combination with IGFBP-3 should obtain better results than with GH alone. Third, higher concentrations of IGF-I can be delivered to the wounds.

The potent anabolic properties are also being tested clinically in **post-operative hip surgery** patients (Celtrix), but may also be useful in other catabolic states. Trials for AIDS cachexia using IGF-I alone were disappointing.

3.5 Muscle

The effect of IGF-I on myocyte proliferation and differentiation has led to suggestions that IGF-I might significantly improve muscle strength in clinical situations such as chronic obstructive pulmonary disease (COPD) where muscle fatigue may significantly limit respiratory function. Reference has already been made to IGF-I improving muscle function in elderly women and thereby preventing/minimizing falls. This work has been extended to cardiac function first reported at the III Symposium in 1991 by Florini et al and Elahi et al. (1). Clark et al. established that IGF-I improved cardiac function in rats. Now at this Symposium Froesch has reported on human applications of IGF-I in **cardiac failure** where IGF-I exerts a positive inotropic effect on the left ventricle. The inherent vasodilatory property of IGF-I may add to its inotropic property and further decrease the work of the left ventricle in congestive heart failure. It is clear that IGF-I is ready for clinical testing. Although GH may be efficacious for dilatory cardiomyopathy, IGF-I may prove to have a wider spectrum of activity. Even in dilatory cardiomyopathy, IGF-I may be superior for the reasons cited above for wound healing.

3.6 The Immune System

GH has long been thought to have immunoregulatory properties. The research on IGF-I was spearheaded by the Zurich group who showed in hypophysectomized rats that IGF-I replacement had a greater effect on spleen and thymus size than GH. With closer attention to the mechanism of action of GH it appears that IGF-I, which is produced locally in lymphoid tissue, is responsible for these effects. They include lymphopoiesis, lymphocyte maturation, and lymphocyte function (reviewed by Clark et al. in this Symposium). Our lab has added that IGF-I potentiates antigen activation of CD4 and CD8 cells and increases macrophage numbers in wounds (8,9). We strongly suspect that cytotoxic lymphocyte production is under control of IGF-I. No clinical tests have been proposed to utilize these properties.

3.7 The Gastrointestinal Tract

Leanna Read presented her group's data on the intestinal mucosal action of IGF-I in the gut at the IInd IGF Symposium. Subsequent work has extended their original observations, and suggest strongly that IGF-I may be a potent therapy to stimulate proliferation of GI mucosa in short bowel syndrome and other conditions characterized by deficient mucosa. Our lab has also been able to independently demonstrate the strong mucosa mitogenic action of IGF-I in the jejunum and ileum of the rat. The importance of this effect of IGF-I was also demonstrated when IGF-I markedly decreased the mucosal translocation of bacteria in injured rats. This can be translated into decreasing posttraumatic and postoperative sepsis resulting from the failure of the intestinal mucosa to filter out enteric bacterial pathogens. Clearly, clinical trials are in order to use the GI actions of IGF-I.

3.8 Reproductory Functions

A prominent part of the last three Symposia has been the role of the IGF System in reproductory functions, embryogenesis and fetal growth and development. There are several areas where IGF-I, an IGFBP, or both together could be therapeutic candidates, but as yet there are no clinical trials. At this Symposium, among others, Linda Giudice and GF Erikson speak to the regulation of the endometrial cycle, ovulation, implantation, and placental function by components of the IGF System. Other authors have contributed data on the critical roles of the IGF System on organogenesis and growth of the embryo and fetus. A therapeutic role for IGF-I in intrauterine growth retardation (IUGR) is worthy of consideration.

3.9 Cancer

More and more investigators are turning to the role of the IGF System in malignancy. Interest is high but work on employing components of the IGF System in therapy has gone slow. Although there are several possible examples, in this review we shall focus on two areas, the prostate and the breast.

Pinchas Cohen has reviewed the **prostate** system at this meeting. It appears that IGFs, the IGF-I receptor, IGFBPs, and IGFBP proteases are all involved in normal and abnormal prostate growth. He singles out IGFBP-3 as an agent which in cancer cells 1) stimulates apoptosis via inhibiting IGF-I action and via IGF-I independent mechanisms through the IGFBP-3 receptor and 2) mediates the action of other apototic agents such as TGFb and p53. However, proteolysis of IGFBP-3 destroys its protective effect. Reduced stimulation of IGFBP-3 in benign prostatic hypertrophy (BPH) may play a significant part in this hypertrophic process because it is associated with elevated IGF-II and IGF type 1 receptor mRNA levels. Consequentially IGFBP-3 may become a major new player in the therapy of abnormal prostatic growths.

IGF-I stimulates the growth of **breast cancer** cells, however, we should keep in mind tha it may be unfair to say that IGF-I causes malignancy. Actual malignant conversion probably involves several sequential steps. Since these cells are sensitive to IGF-I, some of which comes from stromal synthesis, blocking its action by lowering tissue levels, by interfering with the type 1 IGF receptor, or by interrupting IRS-1 activation or downstream signaling molecules may add to our therapeutic armamentarium (Yee et al.).

4.0 THE IMPORTANCE OF BASIC RESEARCH ON THE IGF SYSTEM TO CLINICAL APPLICATIONS

Therapy with any of the components of the IGF System poses a problem of specificity because of the widespread, diverse functions of the IGF System. Methods to overcome this limitation are very important. One mechanism to achieve selectivity would be to combine IGF-I with one of its binding proteins as Celtrix has done with their therapeutic IGF-I:IGFBP-3. IGFBP-5 might be an excellent vehicle for targeting bone because of its unique affinity for hydroxyapatite. The use of an IGFBP also may bring the added benefit of potentiation of the action of IGF-I. One should not be restricted to testing just IGFBP-3 for all IGFBPs except IGFBP-4 and -7 have shown the potential to potentiate under certain experimental conditions. In particular Clemmons et al. have focused on IGFBP-5 in skeletal tissue and its mechanism of augmenting the action of IGF-I in cultured cells.

Another unique way of realizing improved specificity is to target the post-receptor signaling pathways. Thus, more basic research to identify the various intracellular signaling mechanisms used by IGF-I and IGFBP-3 is needed. Once these have been ascertained, research can be focused on finding/designing molecules, preferably small orally active ones, that can interdict or stimulate specific pathways.

5.0 REFERENCES

1. E.M. Spencer (ed), Modern Concepts of Insulin-like Growth Factors, Elsevier, New York, 1991.
2. R. Sapolsky and E.M. Spencer, Am. J. Physiol., 273 (1997) R1346.
3. E.M. Spencer, C.C. Liu, E. Si, and G.A. Howard, Bone, 12 (1991) 21.
4. P. Ravn, K. Overgaard, E.M. Spencer, and C. Christiansen, Eur. J. Endocrinology, 132 (1995) 313.
5. A. Sommer, C.A. Maack, S.K. Spratt, D.Mascarenhas, T.J. Tressel, E.T. Rhodes,R. Lee, M. Roumas, G.P. Tatsuno, J.A. Flynn, N. Gerber, J. Taylor, H. Cudny, L. Nanney, T.K. Hunt, and E.M. Spencer, in: E. M. Spencer (ed) Modern Concepts of Insulin-like Growth Factors, Elsevier, New York, 1991, 715.
6. E.M. Spencer, G. Skover, and T.K. Hunt, in: A. Barbul, E. Pines, M. Caldwell, and T.K. Hunt, eds. Growth Factors and Other Aspects of Wound Healing: Biological and Clinical Implications. Alan R. Liss, Inc. New York, 1988, 103.
7. E.M. Spencer, A. Tokunaga, and T.K. Hunt, Endocrinology, 132 (1993) 996.
8. R.V. Mueller, T.K. Hunt, A. Tokunaga, and E.M. Spencer, Archives of Surgery, 129 (1994) 262.
9. S. Tang and E.M. Spencer, Ann. Meet. Endocrine Soc. 75th, (1993) A502.

Index of authors

Accili, D. 71
Adams, S. 327
Adesanya, O.O. 163
Angelloz-Nicoud, P. 99
Arany, E. 145

Babajko, S. 99
Barbato, C. 231
Baserga, R. 301
Bassal, S. 49
Baylink, D.J. 169
Binoux, M. 99
Blakesley, V.A. 285
Boisclair, Y. 49
Bondy, C.A. 163
Bourcier, T. 261
Bradshaw, S. 65
Busby Jr, W.H. 115
Butler, A.A. 285

Calissano, P. 231
Candy, J. 215
Canu, N. 231
Cerro, J. 65
Ciotti, M.T. 231
Clark, R.G. 351
Clarke, J. 115
Clemmons, D.R. 115
Cohen, P. 205
Collet, C. 215
Conover, C.A. 107
Cwyfan Hughes, S.C. 89

Daubas, C. 99
Daughaday, W.H. 1
de la Rosa, E.J. 155
de Pablo, F. 155
Delhanty, P.J.D. 135
Dey, B.R. 269, 291
Díaz, B. 155
Donath, M.Y. 11
Dube, M. 351
Dus, L. 231

Erickson, G.F. 185

Farah Jr, J.M. 371
Fernihough, J.K. 89
Firth, S.M. 79
Frick, K. 269, 291
Froesch, E.R. 11, 341
Fujitani, Y. 31
Furlanetto, R.W. 269, 291

Galli, C. 231
Gao, J. 39
García-de Lacoba, M. 155
Giudice, L.C. 195
Gluckman, P.D. 225
Gooch, J.L. 319
Grewal, A. 65
Grimberg, A. 205
Guan, J. 225

Han, V.K.M. 243
Hassid, A. 261
Hill, D.J. 145
Holly, J.M.P. 89
Hori, M. 31
Howarth, G.S. 331
Hussain, M.A. 11
Hwa, V. 125

Imai, Y. 115

Jackson, J.G. 319
Jardieu, P. 351

Kaburagi, Y. 279
Kadowaki, T. 279
Kajimoto, Y. 31
Kanno, H. 71
Karas, M. 285
Kasuga, M. 365
Kido, Y. 71
Kim, H.-S. 185, 125
Kobayashi, S. 251
Koval, A.P. 285

Kubo, T. 185

Lalou, C. 99
Lamb, J. 331
Lauro, D. 71
Lee, A.V. 319
LeRoith, D. 285
Li, D. 185
Lopaczynski, W. 269, 291
LoPresti, J. 351

Maile, L.A. 89
Matsell, D.G. 243
Mercanti, D. 231
Mohan, S. 169
Mohseni-Zadeh, S. 99
Montgomery, A.B. 351

Nicoloff, J.T. 351
Nissley, S.P. 269
Nissley, P. 291
Nogami, H. 251

Oh, Y. 125
Okazaki, R. 179
Ooi, G.T. 49

Pell, J.M. 145
Petrik, J. 145
Pintar, J. 65
Powell-Braxton, L. 57, 163
Prisco, M. 301

Rajah, R. 205
Read, L.C. 331
Rechler, M.M. 49
Reeve, A.E. 309
Reik, W. 145
Resnicoff, M. 301
Roberts, C.T. 23
Rosen, D. 327
Rosenfeld, R.G. 125, 359
Rother, K.I. 71

Saad, M. 351
Samathanam, C. 163
Sapolsky, R. 385
Sara, V. 215
Sattler, F. 351
Scheepens, A. 225
Schmid, C. 341
Schuller, A. 65
Shimasaki, S. 185
Shoubridge, C.A. 331
Sommer, A. 327
Spagnoli, A. 125
Spencer, C. 351
Spencer, E.M. 385
Stannard, B.S. 285
Steeb, C.-B. 331

Tamemoto, H. 279
Terry, C. 269, 291
Tobe, K. 279
Tseng, L. 39
Tsushima, T. 279

Ueki, K. 279
Umayahara, Y. 31

Vega, E. 155
Vitolo, O.V. 231
Vorwerk, P. 125

Wanek, D. 125
Weng, C.-N. 319
Williams, C.E. 225
Wilson, E. 125
Won, W. 57

Xu, S. 89

Yakar, S. 285
Yamamoto-Honda, R. 279
Yamanaka, Y. 31, 125
Yamauchi, T. 279
Yang, D.-H. 125
Yazaki, Y. 279
Yee, D. 319

Zapf, J.L. 11, 341
Zhao, H. 205
Zheng, B. 115
Zhou, J. 163
Zhu, H.H. 39
Zona, C. 231